NEW MATERIALS
PROCESSES, AND METHODS
TECHNOLOGY

NEW MATERIALS
PROCESSES, AND METHODS
TECHNOLOGY

MEL SCHWARTZ

CRC Press
Taylor & Francis Group
Boca Raton London New York

CRC Press is an imprint of the
Taylor & Francis Group, an **informa** business

A TAYLOR & FRANCIS BOOK

CRC Press
Taylor & Francis Group
6000 Broken Sound Parkway NW, Suite 300
Boca Raton, FL 33487-2742

First issued in paperback 2019

© 2006 by Taylor & Francis Group, LLC
CRC Press is an imprint of Taylor & Francis Group

No claim to original U.S. Government works

ISBN-13: 978-0-8493-2053-8 (hbk)
ISBN-13: 978-0-367-39181-2 (pbk)

Library of Congress Cataloging-in-Publication Data

Schwartz, Mel M.
New materials, processes, and methods technology / by Mel Schwartz.
p. cm.
ISBN 0-8493-2053-4 (alk. paper)
1. Materials. I. Title.

TA403.S397 2005
620.1'1--dc22

2005049921

**Visit the Taylor & Francis Web site at
http://www.taylorandfrancis.com**

**and the CRC Press Web site at
http://www.crcpress.com**

Dedication

To Carolyn and Anne-Marie, my wife and daughter, whose courage and

strength are overwhelming and gratefully acknowledged

and whose inspiration guides my everyday life.

Additionally, Heather Pauline and David Andrew — the future.

Preface

One of the most important requisites to the development and manufacture of satisfactory products at minimum cost is making a sound economic choice of engineering materials. This task is not an easy one. It is fraught with many uncertainties. Often there is a lack of complete facts on which to operate. Then, too, there are a vast number of materials from which to ferret out the best ones for the job.

So, finding the optimum materials for an application requires a rigorous engineering approach. It is only in the past couple of decades that material selection has been taken seriously and integrated into the methodology of engineering development and manufacturing. Prior to this time, materials selection was done rather casually. The general procedure was to rely almost entirely on past experience and fit the design of a product to the properties of the few materials with which the engineer was familiar. Or, often, the product development program was completed before any thought was given to materials. Under such circumstances, materials changes took place gradually and each change usually provided only small improvements.

In recent years, the tremendous increase in the variety of materials, coupled with the rise of new and more severe service requirements and the demand for lower costs, have brought an end to this haphazard approach. Today, materials selection is a complex process that operates throughout the entire span of a product's evolution, and usually involves a number of different technicians and departments. Analytical procedures have been developed to deal with the complex interaction of requirements and performance properties. And materials engineering departments and specialists who devote full time to materials selection problems are now a common adjunct to engineering or manufacturing departments.

Materials selection is not a simple function that is performed at one certain time and place in the history of a product. When and where it is done depends on a host of factors. Nevertheless, while each materials selection decision has its own individual character and its own sequence of events, there is a general pattern common to the materials selection process.

The materials selection process operates in two distinctly different areas of the engineering and manufacturing complex. First, it is an indispensable part of new product development projects. Second, it plays an important role in the review and reevaluation of existing products. In the former case, materials selection is hardware oriented and, therefore, has directed or assigned character about it. In the latter case, it is not necessarily tied to a specific product and, therefore, often has a less formal and, what might be termed, an unassigned character to it. Both roles of the materials selection process are important.

This book covers many of the newly developed materials as well as the processes that have come about in the past decade. The Materials Age is here and will be with us for a long time to come. It is an age in which the materials available to us are limitless in number and diversity. It is an age in which the microscopic similarities of all materials are more significant than their macroscopic differences. It is an age in which most advanced mechanisms are critically dependent on materials. Yet, it is an age in which new materials and processes still to come promise almost unbelievable breakthroughs in the devices of automation, miniaturization, communication, space exploration, and the maximum utilization of energy. "Engineering and technology lie at the heart of the matter," according to Robert M. White, former president, National Academy of Engineering, on how the U.S. can prosper in the global economy (*The Bridge — 25th Anniversary Issue*). The Materials Age is an age made by people. It is an age that calls for people capable of grasping and using the new science of materials, for people farseeing enough to break down the conventional barriers of the various engineering disciplines. It is an age that requires us to reappraise our ways of doing things. In education it has raised questions of how best to teach materials to engineers. In research and development it has raised questions of how to most efficiently plan and administer the vast government-sponsored programs. In industry it has raised questions of how best to handle materials problems. In technical societies it has raised questions of scope, stimulated reappraisals of programs, and spawned entirely new societies. And in the information field it has created formidable problems in collecting, evaluating, and disseminating data on materials.

Our progress in space flight, automation, electronics, and atomic energy is dependent on the solution of crucial materials problems. Likewise, in the less glamorous fields of industrial and consumer products, materials are now widely recognized as the critical factor that determines cost, quality, and performance. Dr. W. O. Baker, former vice president of research at Bell Labs, summed up the significance of materials in our present age this way: "Materials as never before must be the means through which man realizes his dreams of well being on earth, or, failing that, liberation into space."

Not only has the number of materials increased, but the combinations have increased as well. Clad metals, reinforced plastics, laminates, composites, honeycomb sandwich, nanostructures and powders, layered coatings (CVD, PVD, laser), FGMs, single crystals, fuel cells, and piezoelectric materials have been introduced as well as the systems concept to the materials user. Now we can construct materials in which the best properties of two or more constituents will be used to advantage.

The tremendous increase in the number of materials, coupled with the rise of new and more severe service requirements, is bringing about many changes in our ways of applying materials. Traditionally, the point of view of the user of materials was to fit the design or product to the properties of a material. This attitude has changed. The major concern now is finding and applying a material or materials with the right combination of properties to meet the design and service conditions. Thus, the new attitude is end-service-oriented. The next logical step, of course, is tailor-made materials. As we will read in this book, definite progress in this direction has been and will be made in the future.

No longer do scientists, researchers, and technologists accept the atomic arrangements nature gives us. Such, for example, is the case with high polymers — those aggregates of giant, chain-like molecules. High polymers that nature gave us, such as wood, leather, and glue, have been used as engineering materials for centuries. But only during the past couple of decades have we acquired sufficient understanding of their molecular structure to improve on nature. Now, by varying the chain length and degree of branching or cross-linking, materials can be produced with combinations of properties to meet specific application requirements.

In many of the materials and processes described there is the solid state approach. The impact of solid state science on engineering materials is reflected on three distinct levels. It has made possible:

1. Discovery of completely new and perhaps novel materials.
2. Major or minor improvements in existing materials that come only with a more complete understanding of the nature of the materials and their properties.
3. Understanding the ultimate theoretical limits to the properties of materials before they have been attained practically.

Finally, most of the common materials being used today have been developed over the centuries by pure art, and the early effort of solid state science was directed at trying to rationalize, in terms of the new theoretical concepts, facts and properties that were already well-known to the technologist and the engineer.

Many of the important developments in the fields of new metals, semiconductors, magnetic materials, dielectrics, and ceramics have been the result of trial-and-error or "statistical" methods of research with little benefit from the advancing science of the solid state.

On the other hand, it should be emphasized that not only has solid-state sciences advanced the various materials sciences but also that many important fundamental advances in solid state would have been impossible without the aid of the rapidly advancing technology of materials which has its scientific foundations, for the most part, in classical thermodynamics and classical inorganic chemistry. The preparation of ultra pure materials, the controlled addition of minute concentrations of impurities, and the growth of nearly perfect single crystals have been among the indispensable techniques of progress.

The lesson in this catalog of new materials is that for years the engineer relied on certain old standbys such as the steels in structural designs or on iron-silicon alloys in magnetic designs. Rare or hard-to-find metallic elements remained scientific curiosities. Now, suddenly, a host of new materials has appeared on the scene with too many combinations to enable all possibilities to be exhausted purely by trial-and-error, a fruitful avenue of progress in the past. Trial-and-error will always be the ultimate arbiter, but more basic principles are needed to guide us on what to try, and how to proceed from preliminary success to further improvements.

At the other end of the scale from metals are materials that the physicist would like to ignore, but which the engineer cannot afford to: the complex synthetic materials and organic materials. The pattern of progress in these fields is more difficult to forecast or generalize; however, several chapters will have a discussion on these synthetic materials. With these more complex materials, the problems are immeasurably more difficult and will take all the resources of modern techniques to solve and produce.

The solid-state science has left its mark and will leave its mark in some of the aforementioned fields but solid state is a basic science showing no signs of running out of motivation for discovery within itself. If there is one challenge for the years ahead, it is the challenge to the engineering scientist to modify the traditions in his training and practice to allow more room for the approach of the physicist in the practically infinite spectrum of materials problems.

The Author

Mel Schwartz has degrees in metallurgy and engineering management and has studied law, metallurgical engineering, and education. His professional experience extends over 51 years serving as a metallurgist in the U.S. Bureau of Mines; metallurgist and producibility engineer, U.S. Chemical Corps; technical manufacturing manager, chief R&D Lab, research manufacturing engineering, and senior staff engineer, Martin-Marietta Corporation for 16 years; program director, manager and director of manufacturing for R&D, and chief metals researcher, Rohr Corp for 8 years; staff engineer and specification specialist, chief metals and metals processes, and manager of manufacturing technology, Sikorsky Aircraft for 21 years. While retired Mel has been a consultant for many companies including Intel and Foster Wheeler, and is currently editor for *SAMPE Journal of Advanced Materials.*

Mel's professional awards and honors include: Inventor Achievement Awards and Inventor of the Year at Martin-Marietta; C. Adams Award & Lecture and R.D. Thomas Memorial Award from AWS; first recipient of the G. Lubin Award and an elected Fellow from SAMPE; an elected Fellow and Engineer of the Year in CT from ASM; and Jud Hall Award from SME.

Mel's other professional activities involve his appointment to ASM Technical Committees (Joining, Composites and Technical Books; Ceramics); manuscript board of review, *Journal of Metals Engineering* as peer reviewer; the Institute of Metals as peer reviewer as well as *Welding Journal*; U.S. Leader of IIW (International Institute of Welding) Commission I (Brazing & Related Processes) for 20 years and leader of IIW

Commission IV (Electron Beam/Laser and Other Specialized Processes) for 18 years.

Mel's considerable patent activity has resulted in the issuance of five patents especially aluminum dip brazing paste commercially sold as Alumibraze.

Mel has authored 16 books and over 100 technical papers and articles. Internationally known as a lecturer in Europe, the Far East, and Canada, Mel has taught in U.S. colleges (San Diego State, Yale University), ASM Institutes, McGraw-Hill Seminars, and in-house company courses.

Table of Contents

1

Introduction

What kinds of technologies will be discussed, developed, or matured or become a concern 5, 10, and 20 years from now?

For one, the technologies powering the new information and communication revolution will be ones pioneering low-power CMOS, single outline ICs (SOI), and Si-Ge-based chips as well as advances in making integrated microelectronic circuits even smaller, cheaper, and more efficient.

Two other high-impact technologies will be supermaterials and compact, long-lasting energy sources. Computers will be designing and manufacturing new materials at the molecular level, which will be used heavily in transportation, computers, energy, and communications industries. We will also see widespread use of fuel cells and batteries for portable electronic devices.

There have been innumerable material and process technical achievements and developments within the past decade. Included among these are

- The development of a series of advanced, intermetallic heat-resistant materials with increased corrosion resistance over conventional materials. The new materials show significant decreases in recession rates due to oxidation, have a significantly higher melting point, and maintain strength and creep characteristics when compared to conventional materials. They have application potential in aircraft engine exhaust ducts and engine components.
- An improved method for extruding high-temperature metallic and intermetallic alloys, which shows potential for lower production costs and increasing material yield by up to 40%.
- A new method of application of the hot melt process for producing sheets of preimpregnated phenolic material used to make carbon-carbon (C/C) or carbon-phenolic composite components. This hot melt process will reduce the use of hazardous chemicals from the composite production process by 75% while lowering the cost of prepreg materials and improving the material's properties.
- Intelligent processing, which has become the driver for improvements in processing. It promises processes controlled through the

use of sensors and control systems using artificial intelligence, neural networks, and fuzzy logic.

- Composites manufacturing, which is a materials transformation process, with each step affected by previous decisions made and routes chosen, and these, in turn, affecting decisions and process choices for subsequent steps.
- Concurrent engineering, which demands that decisions be made as early as possible and that they be made based on the coupled nature of the materials (resins, metals, ceramics, reinforcement, and fillers), configuration (orientation, architecture, and shape), and processing routes.

Resin transfer molding (RTM) is a genuine example of where concurrent engineering can best be used. The preform must be designed keeping in mind not only structural aspects but those of resin infusion and flow as well. The tool must be designed not only for the particular features of the finished product but also for the specifics of injection, cure rates, cooling, and demolding.

The injection of the resin must reflect the delicate balance between a quick fill and the integrity of the architecture, as well as tooling and preform-related aspects such as local exotherms through thick sections, differential flow rates through the preform, and resin reactivity. If all of these factors are considered before the product is designed, the RTM process itself can be designed to ensure that the product is of high quality, yet economical.

The emergence of functionally gradient materials (FGMs), which are composites in which the proportions of the constituent phases are varied in an article of interest made from the composite in order to tailor the properties locally to improve the performance of the article. Special interest relates to aerospace, where surface temperatures as high as 2000°C and temperature gradients of the order of hundreds of degrees per millimeter are envisioned.

Other scientists believe that in producing FGMs by chemical vapor deposition (CVD), the gradual modification of the mix of the vapor species from which the separate constituents of an FGM can be deposited make it possible to produce a gradual shift in the composition of the deposit. One such effort is the C-SiC system, where the purpose of the SiC is to protect the carbon structure against oxidation. If successful, such a coating could allow carbon to be used as a combustor, for example. Conditions leading to codeposition of C and SiC have been achieved.

The above advancements are only a part of the revolution that has occurred in materials development. It came about through the discovery of new materials, the development of new processing techniques, and the creation of new methods and instruments to analyze materials.

As a result, composites and superconductors and materials with entirely new characteristics not even dreamed about 30 years ago, such as buckyballs, oriented polymers, diamond films, and photonic compounds, can actually be manufactured.

"Improvements in materials characterization at the atomic level have made these discoveries possible and will continue to drive new development," said Bill Appleton, associate director of physical science and advanced materials at Oak Ridge (TN) National Laboratory.

Combining atomic precision characterization methods with supercomputing abilities makes it easier to predict properties. With this capability, you can then design and produce a new material with predetermined properties.

FGMs have attracted and continue to attract attention worldwide, as new materials that locally different functions can be added to a single material by continuously changing dispersion phase density, kinds of structure from one side of the material to the other, or changing them locally to alter its characteristics.

The development of materials that are resistant to superhigh temperatures will be needed for the development of various types of components, vehicles, and so on. For example, to completely reuse the space shuttle vehicle, FGMs might be the ideal material. Where one surface of the material would be exposed to a high-temperature (1727°C) oxidizing atmosphere, the other could be cooled to 727°C. Thus, there is a temperature difference of 1000° in the material. Currently, available materials cannot be used in such an extreme environment. Different materials stuck together will break because of the thermal stress generated within them.

Though the term "functionally gradient material" is indisputably a mouthful, talking about these newfangled materials has proven much easier than producing them.

By applying alternating thin layers of matrix materials to traditional reinforcement structures, researchers have developed a new type of material that they believe will be tougher than conventional fiber-reinforced composites. The "laminated-matrix" composites could replace metals in aircraft structural components, heat exchangers, particulate filters, and other applications requiring high-temperature, high-strength materials, according to material engineers. They say they hope the materials will replace costly reinforcing fibers with less-expensive platelets or particles. The process uses a fibrous preform made of stacked layers of cloth. "We infiltrate with one material until we get a layer of it around each fiber, then we infiltrate with another material, then we switch back to the first," explains research scientist W. Jack Lackey. "We put down as many as 50 layers." Lackey notes that the matrix materials must be chemically compatible and closely matched in their thermal-expansion properties.[1]

The recent discovery of a family of large, solid carbon molecules with great stability, the so-called "fullerenes," has considerably extended the scope and variety of carbon molecules known to exist and has opened an entirely new chapter on the physics and chemistry of carbon, with many potential applications. The fullerenes can be considered as another major allotrope of carbon and its first stable, finite, discrete molecular form. They are different, in that respect, from the other two allotropes, graphite and diamond, which are not molecular but infinite-network solids.

The fullerenes are generally arranged in the form of a geodesic spheroid and thus were named after the inventor of the geodesic dome, the renowned architect Buckminster Fuller.

These fullerenes are being developed as a drug-delivery system for cancer, AIDS, and other diseases.

The hope of scientists is that it will be possible to load these minuscule, spherical structures — each containing 60 carbon atoms arranged like the hexagonal pattern on a soccer ball — with drugs or radioactive atoms and then fire them like guided missiles at diseased cells.

"Think of a smart bomb," says Canadian scientist Uri Sagman. "Conventional chemotherapy is like carpet-bombing. You drop it from 60,000 feet and hope for the best. This goes precisely to the target."[2]

Once the province of science fiction, nanotechnology, the science of assembling materials one atom at a time, is now the rage among Silicon Valley and other developers, scientists, researchers, venture capitalists, and entrepreneurs.

"Nano," from the Greek prefix for "one billionth," requires an unimaginably tiny scale; a million or so fullerenes could fit into a grain of rice. The sturdy carbon molecule has emerged as one of the most versatile tools in the rapidly developing nanotechnology arsenal. An elongated chemical cousin called the buckytube has also shown great promise. Although computing breakthroughs such as molecular circuits may not be commercially available for another decade, the first devices for nanoscale drug delivery entered clinical trials in 2004.

The fullerenes are also known as "buckyballs."

Many properties of CVD diamond approach or equal those for natural diamond. The manufacture of solid diamond fibers by deposition on various wire and ceramic cores and the production of hollow diamond fibers by removal of the cores have been demonstrated, and there is a reasonable expectation of cost-competitive fiber manufacture in the future. The potential Young's modulus values of continuous and discontinuous diamond fibers are predicted to be substantially higher than for current commercial SiC_f, with corresponding increases in competitive stiffness. Hollow fibers may also allow the use of sensors in smart structures.

Ceramics offer attractive properties including good high-temperature strength and wear, and oxidation resistance. However, the major limitation to their use in structural applications is their inherent low fracture toughness. Methods currently being used to improve the toughness of ceramics include incorporation of reinforcing whiskers and fibers; inclusion of a phase that undergoes transformation within the stress field associated with a crack; and cermet technology (ceramic/metal composites), in which the tough metallic component absorbs energy. Improved toughness is attributed to various mechanisms including crack branching, transformation-induced residual stresses, crack bridging, and energy absorption by plastic flow.

Another possible approach to improve the toughness of ceramics is via laminated construction. This type of construction is used successfully to fabricate polymer matrix composites (PMCs). However, little attention has been given to the development of ceramic or cermet/metallic laminated composites, although this class of material has excellent potential for use in structural applications. Sophisticated coating techniques are used to produce laminated microstructures of ceramics and metals, but the cost to produce a bulk-laminated composite of useful size by any of them is prohibitive. Microinfiltrated macrolaminated composite (MIMLC) fabrication offers an economically feasible approach to produce a variety of cermet/metallic and ceramic/metallic bulk-laminated composites.

Liquid crystal polymers (LCPs) are distinguished from other plastics by their rod-like microstructure in the melt phase. Other resins have randomly oriented molecules in the melt phase, but when LCPs melt, their long, rigid molecules can align into a highly ordered configuration that produces a number of unique features.

These include low heat of crystallization, extremely high flow, and significant melt strength. The molecular structure has such a profound effect on properties and processing characteristics that LCPs are best treated as a polymer category separate from amorphous and semicrystalline resins. In spite of their unique properties, however, LCPs can be processed by all conventional thermoplastic forming techniques. They can, for instance, easily flow into molds for very thin or highly complex parts, reduce cycle time in molding, produce parts with little molded-in stress, eliminate deflashing and other secondary finishing steps, and reduce part breakage during assembly. These benefits can raise finished assembly or system yields 3% or more, which can more than offset the higher resin cost.

LCPs are ordered (crystalline) in both the solid and liquid phases (see Figure 1.1).

This order is nematic (i.e., longitudinal or parallel) and is developed by the flow or deformation in the liquid phase during filling of the mold.

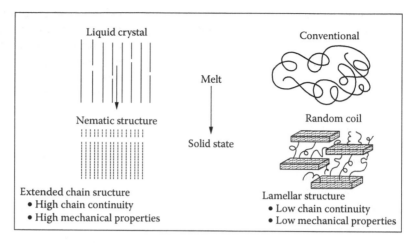

FIGURE 1.1
Liquid crystal vs. conventional polymer structure.[3]

The first commercial LCP, Kevlar fiber, was introduced in 1965. In 1985, Vectra high-performance resins were commercially available, and Xydar was introduced in 1984.[3]

Interpenetrating networks (IPNs) are one of the current "buzzwords" among polymer chemists and will continue to be in the future.

Network polymers are formed by cross-linking one linear homopolymer with a second linear homopolymer, using grafting techniques. In contrast, IPNs are generally composed of two or more cross-linked polymers with no covalent bonds between them. They are often incompatible polymers in a network form. However, at least one is polymerized or cross-linked in the immediate presence of the other, with the result that the chains are completely entangled.

IPNs are heterogeneous systems comprised of one rubbery phase and one glassy phase. This combination produces a synergetic effect on mechanical properties, yielding either high-impact strength or reinforcement, which are both dependent on phase continuity. Depending on composition, thermal properties can also be improved.

Four types of IPNs have been currently identified:

- Sequential IPNs. These start with one polymer already cross-linked. They are then swollen in another monomer, initiators and cross-linking agents are added, and the system polymerized in place.
- Simultaneous IPNs. These are formed by a homogeneous mixture of monomers, prepolymers, linear polymers, initiators, and cross-linkers. Polymerization occurs by designed independent, noninterfering reactions.

- Semi-IPNs. These are made from one linear polymer and one cross-linked polymer.

- Homo-IPNs. These are prepared from polymers that are generically identical but retain specific characteristics when interpolymerized.

Nanoengineered machines will be used in manufacturing and process-control applications.

Increasingly powerful micromachines have been developed by Sandia National Laboratories' researchers over the past 2 years — now they have added intelligence to the machines. An intelligent micromachine can signal for more power, communicate that it is operating too fast or slow, or even perform actions on an automated basis.

This type of micromachine, consisting of tiny motors fabricated with integrated circuit "brains" on individual silicon chips, has been mass-produced for general applications by researchers. The compact design enables the fabrication of entire electromechanical systems on a chip. Potential applications include air bag accelerometers, nuclear weapons' locks, and gyroscopes for civilian and military use; medical possibilities include tiny drug-delivery devices.

The manufacturing process consists of etching tiny trenches in silicon chips and fabricating the machines within these depressions. The machines are then heat-treated and submerged in silicon dioxide. The hardened silicon dioxide re-creates a level chip surface upon which circuitry is fabricated by photolithography. Removal of the silicon dioxide after completing the process frees the microengines. Working systems are manufactured with a 78% success rate, considered a reasonably high measure of production yield. Because micromachine and microcircuit performances can be optimized separately, the process can produce a wide range of micromachine systems with more powerful micromachines yet smaller transistors (see Figure 1.2).

Circuits fabricated only microns away from a machine do not experience the interference of ghost signals caused by excess electrical capacitance in long connecting wires. Thus, by applying a mechanical load, researchers can measure the capacitance change in the drive gear teeth as they move in and out and determine the machine's speed.[4]

While semiconductor researchers struggle to develop the software tools for modeling future electronic devices, their mechanical-engineering counterparts have already developed tools for designing future nanomachines.

At Oak Ridge National Laboratory, researchers are modeling nanomachines using a combination of computational chemistry and proprietary techniques. Computer-visualized nanomachine models are merely mechanical versions of their biological counterparts, such as viruses. All the

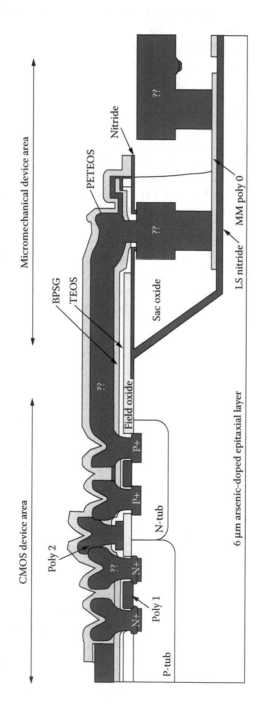

FIGURE 1.2

The trench (far right), only microns deep, makes possible fabrication of a variety of smart micromachines at Sandia National Laboratories. Microcircuitry (left), connected directly to micromachine stud, helps make automated control possible by eliminating ghost signals caused by long wiring.[4]

researchers are doing is putting together different atoms with chemical bonds that exist in nature and not breaking any basic laws of science.

The researchers have already simulated how fluids flow through nano-machines and how a simple graphite bearing and shaft composed of about 4000 atoms interact. Their goal is to show how atomic-scale devices can be made and used.

Nanotechnology will be the greatest technical innovation of the next 20 years. It will give us the ability to manipulate matter at the molecular level and will turn manufacturing on its head with its alchemy-like capabilities. It will truly transform the world into an information economy, and the manufacturing process itself will become radically automated.

What can we predict about molecular nanotechnology?

Practitioners define nanotechnology as the ability to fabricate individual devices one by one with essentially every atom in the right place. The purpose is to construct almost any structure consistent with the laws of physics and chemistry that can be specified in atomic detail — in other words, everything from spaceships to human organs is fair game.

Moreover, the cost of manufacturing such entities would not be much more than the cost of the needed raw materials and the energy required. All in all, nanotechnology could end up rewriting the entire book on manufacturing.

Researchers speculate that nanomaterials might have a structure resembling diamond fiber. Produced in exactly the shapes desired, they could sport an impressive strength-to-weight ratio. Planes, ground vehicles, and even spaceships made from them could all be orders of magnitude lighter without sacrificing strength.

Research continues for the molecular transistor, and there is no question that molecular electronics could usher in some startling developments. But consider physical structures, not just circuits, built molecule by molecule. The resulting structures would have not a single molecule out of place, other than displacement caused by thermal effects or damage by radiation. Thus they would have virtually no molecular defects to promote shearing and tearing. They would be super strong and light.

Nanotechnologists say such structures could be built through positional assembly techniques. The term refers to the idea of molecule-sized robot arms that pick up, move, and place molecules one at a time under the control of a computer. With enough molecular robot arms on the job, everyday objects could be built from the ground up.

Proponents say there are many more opportunities for materials engineered one molecule at a time and built into 3D structures. The idea is to hold molecular parts in the right position so that they join with other molecules in exactly the right way.

To position single molecules with respect to one another would require a nanoscale equivalent to computer-controlled robot arms and grippers.

Molecular robotic arms would be able to move back and forth, withdrawing atoms from "feedstock" to build any structure desired.

Nanoscale robotic arms would have to be super stiff. Such a requirement leads researchers to consider diamond as a likely arm material because its carbon-carbon bonds are especially dense and strong.

Some engineers propose using chemical vapor deposition to deposit *layers* of diamond for nanoscale applications. CVD is a process somewhat analogous to spray painting. The resulting diamond surface would be covered with hydrogen atoms, rendering it inert.

Some of the hydrogen atoms would have to be stripped off to expose reactive dangling bonds. A so-called hydrogen abstraction tool would perform this function. This tool would be molecular-sized with an affinity for hydrogen at one end and the other end serving as an inert "handle." A positioning mechanism would grab the handle and swipe the tool over hydrogen atoms and remove them.

U.S. Army researchers and scientists in academia are making progress in the study of spider dragline silk, according to recently published proceedings of the National Academy of Sciences.

The protein that lets spiders drop and helps the web to catch prey is what interests the researchers. The molecules are designed to be pulled; they are elastic and very strong. The silk can be extended 30 to 50% of its length before it breaks. It is stronger than steel and comparable in strength to Kevlar.

"The last decade has seen a significant increase in the scientific literature on spider dragline silk," according to the proceedings. "This interest is due to the impressive mechanical properties of spider dragline silk, at a time when biomaterials and biomimetics are both exciting interest in the rapidly growing field of materials research."

And why is the U.S. Army interested in this material? "The major interest is to use it as material for bulletproof vests, armor, and tethers; there are many possibilities," claims Emin Oroudjev, a researcher at the University of California at Santa Barbara.

At UC Santa Barbara, the focus is on the basic research of learning how the protein folds and how it is organized in the silk fiber. Using atomic force microscopy and a molecular puller, the researchers are getting clues from imaging and pulling the protein. These observations help the researchers to model what is happening in the silk gland when silk proteins are assembled into spider dragline silk fibers.

They found that when the protein unfolds it is modular. It has sacrificial bonds that open and then re-form when the load lifts. This follows a pattern that has been found in other load-bearing proteins.

Spider silk has crystalline parts and more rubber-like stretchy parts. The researchers found that single molecules have both, explains Helen Hansma, adjunct associate professor of physics at UC Santa Barbara.[5]

Other researchers at MIT are focusing on creating materials to make high-strength fibers that may one day see use in artificial tendons, specialty textiles, and lightweight bulletproof gear. Spider silk is a polymer with two distinct alternating regions. One region is soft and elastic; the other forms small, hard crystallites. Researchers are attempting to make a series of different synthetic polymers and study how changes in the polymer chemical structures affect its physical properties. This research is in conjunction with the study of processing techniques that will maintain the unusual properties of the material produced.[6]

Protein-like polymers that can mimic the growth of crystals that abalones and coral rely on to build their tough and beautiful shells have been developed by researchers at UC Santa Barbara. The polymers can be used to create advanced surgical implants, better computer chips, and other components that depend on microscopic construction of complex structures, according to the scientists.

Even though abalone shells are made of the same material as brittle limestone, they are extremely tough and can be used to pound nails. The shell's uniquely shaped proteins make the calcite and aragonite crystals grow in interlocking blocks, giving them their strength. Although the proteins make up only 1% of the shell's weight, the shell is 1000 times stronger than a material constructed of only crystals.

Physical properties of solids are greatly influenced by the arrangements of atoms present in the network. Nanostructured systems can offer improved and new properties compared to those of bulk materials. Consequently, in such materials, the influence of interfaces and interphases are of noticeable importance. Such interactions between the grains can lead to new physical and chemical properties.

In recent years, an increasing interest has been devoted to nanostructured composites. This attention is largely due to exciting possible applications ranging from new catalysts to the preparation of nanocomposite ceramics, with significant improvements in their properties.

Ceramic composites, obtained by dispersing nanoscale particles of one constituent into larger particles of a second constituent, have shown such improvements in mechanical properties (micro/nano composites). In particular, superplasticity has been obtained by mixing two equally fine constituents (nano/nano composites).

Different modes of preparation have been used to make nanosized particles in large enough quantities. Among them, laser ablation of solid targets and plasma- or laser-induced reactions of gaseous mixtures have been developed. Other methods are related to evaporation condensation, ball milling, self-propagating high- temperature synthesis, sol-gel, spray drying of solutions, aerosol pyrolysis, and chemical vapor decomposition at low pressure.

In the evolution of the materials field today, engineered components or devices typically in the range of 1 mm to 1 m are produced, but micromachines of the future may have dimensions as small as 10^{-5} m. Electronic circuits are already engineered as small as 10^{-7} m. At still-smaller dimensions lie the molecular and atomic configurations and defects that compromise the basic foundations of the material properties that are exploited. In the future, we will see more specific property tailoring as we learn to use these new tools, such as one that can manipulate atoms.

Advances such as these will continue over the next decade, some foreseen, others not; this book will show how, in the evolution of these new materials technologies, computational technology will play a growing role in the design of these new materials and processes; and we will control microstructure in new ways, aided by new energy sources, analytical software, and devices.

References

1. jack.lackey@gtri.edu, *Design News*, p. 9, January 22, 1996.
2. K. Breslau, Tiny Weapons with Giant Potential, *Newsweek*, pp. 72–75, June 24, 2002.
3. C. E. McChesney and J. R. Dole, Higher Performance in Liquid Crystal Polymers, *Mod. Plast.*, pp. 112–118, January 1998.
4. L. Trego, Intelligent Micromachines, *Aerospace Engrg.*, p. 35, July 1996.
5. http://www.netcomposites.com/news.asp?1259, pp. 5–6, June 24, 2002.
6. *Machine Design*, p. 41, June 19, 2003.

2

Nanotechnology

Nanoscience/Nanotechnology

Nanotechnology is the science and engineering of creating materials, functional structures, and devices on a nanometer scale. In scientific terminology "nano" means 10^{-9}: 1 nanometer (nm) is equivalent to one thousandth of a micrometer, one millionth of a millimeter, and one billionth of a meter. Nanostructured inorganic, organic, and biological materials may have existed in nature since the evolution of life on Earth. Some evident examples are micro-organisms, fine-grained minerals in rocks, nanoparticles in bacteria, and smoke. From a biological viewpoint, the DNA double helix has a diameter of about 2 nm, whereas ribosomes have a diameter of 25 nm. Atoms have a size of 1 to 4 Angstroms. Therefore, nanostructured materials (NSMs) could hold tens of thousands of atoms all together. On a micrometer scale, an excellent example is a human hair, which has a diameter of 50 to 100 μm.

Why are we so fascinated with downsizing materials to a nanoscale? The fundamental physical, chemical, and biological properties of materials are surprisingly altered as their constituent grains are decreased to a nanometer scale owing to their size, shape, surface chemistry, and topology. For example, 6 nm copper grains show 5 times more hardness than conventional copper. Cadmium selenide (CdSe) can be made to yield any color in the spectrum simply by controlling the size of its constituent grains. Therefore, nanotechnology has generated much interest in the scientific community, and it has become a very active area of research. There are several major government programs on research and development of nanostructured materials and nanotechnology in the United States, Europe, and Asia. A new vision of molecular nanotechnology will develop in the coming years, and we may see many technological breakthroughs in creating materials atom by atom. The resulting new inventions will have a widespread impact on the fields of science, engineering, and medicine.

The use of the terms *nanoscience* and *nanotechnology* has become widespread, encompassing a vast range of disciplines and technologies that are predicted to become increasingly important over the next 10 to 20 years. With applications in electronic, biomedical, chemical, materials, sensor, and information technology areas, the common factor is that the materials, techniques, devices, and, in some cases, systems have one or more critical dimensions on the *nanometer* scale.

The prefix "nano" is derived from the Greek "nanos," meaning dwarf or extremely small. Often, the prefix is used to describe particular phenomena as well as applications, such as nanoelectronics and nanomagnetics, based on the understanding that it describes the length scale that is responsible for a unique phenomenon. The nano regime is normally considered to encompass the range between 1 and 100 nm — the region intermediate between micron scale, dominated by integrated circuit (IC) technology, and the molecular scale, dominated by synthetic chemistry. To give some perspective, the lower limit, 1 nm, is 1 billionth (10^{-9}) of a meter and is roughly the length of three atoms in a row, some 50,000 times smaller than a human hair. Figure 2.1 illustrates the approximate extent of the nano-, micro-, and macroscale regimes, the size of some typical features in each, and the typical range of some useful observation techniques, from the naked eye to advanced microscopes.[1,2]

"There is plenty of room at the bottom," Richard Feynman said in 1959 at the California Institute of Technology, as he posed the challenge of placing all the world's worthwhile literature in a 5 mil cube of atomic memory, roughly the size of one poppy seed grain, and the challenge of creating tiny computers and machines that might allow humankind to move atoms around one by one. "Put the atoms down where the chemist says," Feynman enjoined, "and so you make the substance."

In the 1970s, Eric Drexler, a disciple of futuristic exploitation of the ultrasmall, began to challenge the limits of credibility with hypothetical artificially programmed atomic-sized "assemblers" for making or modifying almost anything. "Nanotechnology," a term coined by Nobuhiko Taniguchi in 1974, refers to the field of materials processing and fabrication at dimensions below 200 nm. This field has become a natural extension of the more conventional microelectronics world, where mass production now routinely churns out devices with nanoscale features. At an ESA roundtable in Noordwijk in 2001, there arose a new acronym, MNT, representing this confluence of micro- and nanotechnology.

MNT is the synthesis and integration of materials, processes, memory, and devices of submicron to millimeter sizes — a synthesis that has given rise to the new technical field of microengineering. This field integrates several distinct disciplines: traditional microelectronics infrastructure; new two- and three-dimensional (2-D and 3-D) surface processing for power, sensors, data processing, and communications; and 3-D micromachining, which forms movable microscopic machine elements for MEMS.

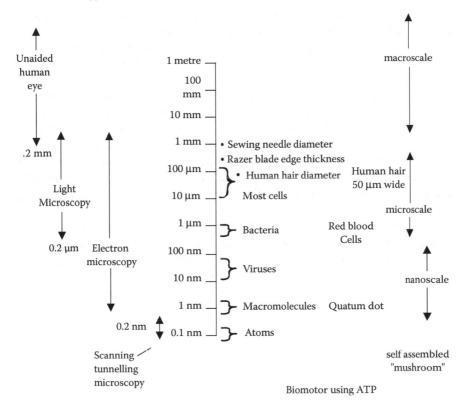

FIGURE 2.1
The scale of nano.

These technologies stem from electronic microprocessing infrastructures. When they are joined with nanomaterials processing methods, they will provide great strides in applied devices that feature low-power and applied quantum electronics in dense-pack memories. They offer ultra-high-speed computation and will be intimately connected with other circuits and MEMS in mass-producible designs.

Now many engineers are aware of microelectromechanical systems, or MEMS. These tiny machines, which are only a few steps behind the state-of-the-art microelectronic size scale, are made with materials and methods pioneered in the semiconductor industry. Feynman's added challenge to devise a controllable rotating electric motor in a half-millimeter cube was met quickly and has now been far surpassed by MEMS mass-producible gears, shafts, and tiny turbines less than half that size.

Today, nanoscience and nanotechnology promise to be a dominant force in our society in the coming decades. Initial commercial inroads have already shown how new scientific breakthroughs at this scale can change production paradigms and revolutionize businesses.

To look in more detail at the impact of nanotechnology requires an examination of what in practice nanotechnology includes, and what it excludes.

The "Nano-Universe" can be understood as being comprised of three areas: nanoscience, nanotechnologies, and nanoenabled technologies. Nanoscience is a collection of fundamental endeavors to understand the structure-property relationship and create the tools necessary to enable assembly and characterization. Nanotechnologies, such as novel electronics and new nanoscale machines, represent new economic sectors, generally pushing the envelope of feasibility, serving as inspiration and capturing the public attention.

However, it cannot be emphasized enough that the revolutionary insights provided by nanoscience will radically transform existing technology sectors, such as catalysts, coatings, consolidated materials, power generation and storage, and so on, and probably provide the greatest economic impact. This represents the last general area, nanoenabled technologies. Figure 2.2 graphically summarizes this relationship between these areas.

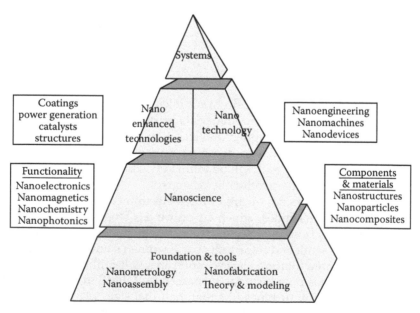

FIGURE 2.2
The relation of the nano "universe".

The National Science and Technology Council in the United States has described a vision of these novel materials created for specific purposes. They might have multiple functions and be able to sense and respond to changes in surroundings. Researchers also envision materials that might be 10 times the strength of steel or a tenth the weight of paper, paramagnetic or superconducting, perhaps optically transparent, and many with higher melting points.

Nanotechnology applied to memory storage could permit putting the entire contents of the Library of Congress in a device the size of a sugar cube.

Presidential science adviser Neal Lane says that nanoscale technology is the area of science and engineering most likely to produce the breakthroughs of tomorrow. Not just another step toward miniaturization, nanotechnology represents a dramatically new scale, where the behavior of matter is dominated by quantum mechanics.

What is understood is that nanoscience is the scale where the fundamental properties of materials are determined and can be engineered. Previous scientific breakthroughs in areas such as hard disk coating and pharmaceutical products have shown how they can revolutionize multibillion-dollar businesses. The prospects for future developments are presaged by a few recent discoveries into the organized structure of matter, such as carbon nanotubes, molecular motors, quantum dots, and molecular switches.

The main interest is with materials that have novel and significantly improved physical, chemical, and system interest at the molecular level. Some specific aspects intrigue researchers:

- Electronic and atomic interactions are influenced by variations at the nanometer scale. Therefore, structuring matter at the nanometer level should make it possible to control fundamental properties — such as magnetization, charge capacity, and catalytic activity — without changing a material's chemical composition.

- Biologic systems have a systematic organization at the nanoscale, which should permit placing artificial components and assemblies inside cells to make new structurally organized materials More biocompatible materials could result, mimicking the self-assembly methods of nature.

- Nanoscale components have a very high surface area, which should make them ideal for chemical use as catalysts and absorbents, for electrical energy delivery, and for drug delivery to human cells.

- Materials structured at the nano level could be harder yet less brittle than comparable bulk materials with the same composition.

Scientists believe nanoparticles will be too small to have surface defects, will be harder because of their high surface energy, and, as a result, will be useful for building very strong composite materials.

- With dimensions orders of magnitude smaller than microstructures, interactions will occur more rapidly, leading to more energy-efficient systems and faster performance.

After a decade-long buildup, nanotechnology is beginning to see its first commercial successes, although basic materials are grabbing the spotlight from more sensational early projections of cancer-killing nanobots and self-assembling automobiles. The development of nanoscale materials (i.e., substances with particle size between 1 and 100 nm) is a key step in the eventual production of more sophisticated nanomachines, nanoelectronics, and nanomedical devices. In fact, the move toward nanotechnology is a continuation of ongoing miniaturization efforts in many industrial sectors. Research has flowed in two directions: reducing the size of existing manufacturing materials such as metal oxides through the use of new production technologies, and developing entirely new materials, such as nanotubes and buckyballs, which are intrinsically nanosized.

The U.S. market for nanomaterials (which totaled only $125 million in 2000) is expected to surpass $1 billion in 2007 and reach $35 billion by 2020.[69] Early growth will come from numerous niche applications that span the entire U.S. manufacturing sector. These include wafer polishing abrasives and high-density data storage media for the electronics industry; improved diagnostic aids for the medical community; transparent sunscreens, stain-resistant pants, and wear-resistant flooring for consumers; cost-cutting equipment coatings for the defense industry; fuel-saving components for the auto industry; and better paper and ink for the printing industry. In the long run, however, the best opportunities are expected in health care and electronics, which together are expected to comprise nearly two thirds of the market by 2020.

Conventional products — essentially smaller versions of extant products, such as silicon dioxide, titanium dioxide, and clay — are expected to register robust growth both through the forecast period and beyond. Already, these materials have established a market presence in such applications as sunscreens and chemical mechanical polishing (CMP) slurries.

Conventional nanomaterials will account for the large majority of overall demand for the foreseeable future. However, new nanomaterials — many of which are yet to be commercialized — are expected to register even better growth, from a negligible base in 2002. Carbon nanotubes, the likely anchor product of the new materials category, are finding use now in specialty composites and in fluoropolymers.

Nanofabrication/Nanoassembly

To exploit the unique nanomaterial properties, techniques must be developed that enable manipulation of materials on the nanoscale. This is the domain of nanofabrication and nanoassembly. The general approaches can be thought of as "top-down" (nanofabrication) and "bottom-up" (nanoassembly).

The top-down approach to nanotechnology is that in which large objects are made smaller by miniaturization. Techniques that remove atoms from a structure (such as wet and dry etching) and those that add atoms to a structure (such as oxidation, diffusion, and electroplating) in controlled ways are the foundations of this approach. The top-down approach to reaching the nanoscale is best exemplified by the development of the integrated circuit, in which the lateral feature size of components such as transistors, contacts, and metal interconnects has gradually decreased from several microns in the late 1960s and early 1970s to dimensions currently approaching 100 nm. As vertical feature sizes are often already <100 nm, the components of integrated circuits today are poised to enter the nanoscale regime as true, three-dimensional nanostructures.

The alternative to the top-down approach is known as the bottom-up approach, first predicted by Feynman. This is the purview of chemistry and colloidal science and includes macromolecular synthesis, dendrimers, inclusion chemistry, surfactant mediate nanoparticle synthesis, and microemulsion approaches. The quintessential bottom-up concept was articulated by Drexler and is based on the assembly of nanoscale structures from atomic or molecular constituents; this has sometimes been referred to as "molecular engineering" or "molecular nanotechnology." Many mathematical models and computer-aided predictions of molecular components such as bearings and gears have been developed but have yet to be realized in practice. Atomic/molecular assembly is making progress by the application of techniques such as the scanning tunnelling microscope, shown in Figure 2.3, but we are a long ways off from truly modifying and fabricating nanoscale structures at the atomic or molecular level.

Atomic Bit Processing

A chronological advance in the accuracy of materials processing is shown in Figure 2.4. By extrapolating the curve of ultrahigh-precision machining in Figure 2.4, in the early years of the 21st century the attainable processing accuracy has reached the nanometer level. Moreover, the accuracy of processing is expressed by the sum of the systematic error and the random error (3 σ standard variance), as shown in the lower part of Figure 2.4.[3]

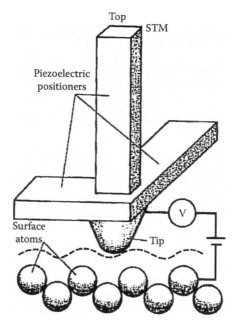

FIGURE 2.3
Schematic for scanning tunneling microscope.

It is necessary to apply a processing unit of atomic bit size to get high-precision and fine products at nanometer accuracy. Such processing includes removing, deforming, and consolidating processes of atomic bit size.[4,5]

The energy densities necessary for atomic bit processing are called the threshold of the specific processing energy.

For atomic bit processing, processing energy over the threshold of specific processing energy has to be concentrated on the target atom. In mechanical processing, energy is channeled through the sharp-edged tip of very fine and hard abrasives (such as diamond, SiC, and Al_2O_3) that have larger specific bonding energy than the workpiece. Therefore, for atomic bit processing, ordinary cutting tools cannot be applied because of the low atomic lattice bonding energy or corresponding lower indentation hardness and quick abrasion wear resulting from the high-density working stress on the tool tip. In physical processing, energy is supplied through the energetic particle beams of photons, electrons, ions, and evaporated atoms. In chemical processing, energy is applied through chemically reactive atoms, activated thermally or electrolytically. Atomic bit processing of materials via physical and chemical elementary energy particles is detailed in Table 2.1 and in the literature.[5–11]

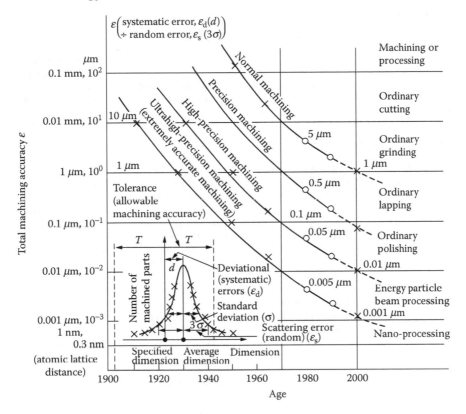

FIGURE 2.4
Achieved machining accuracy in respective ages. Processing error = difference between specified and processed dimensions. Machining accuracy ε = systematic error ε_d(d) + random error ε_s(3 σ). Accuracy in narrow sense = systematic error or deviational error (d). Precision in narrow sense = random error or scattering error (3 σ). Precision in wide sense = high accuracy. Standard variance = σ^2.[3]

Nanomaterials

There are many existing and future technologies under the nano umbrella, the majority relying on the rapid, encompassing advances in the development of materials at the nanoscale. The term *nanomaterial* is often used to encompass the wide range of nanoscale and nanostructured materials necessary to enable these technologies, including structural and functional applications. At one extreme are bulk materials that are specially processed to develop a nanostructure of some form, for example, nanocrystallinity. At the other extreme are materials for which the focus is upon individual nanoscale units, such as nanoparticles, nanoclusters, and

TABLE 2.1

Atomic Bit Processing of Materials by Energy Beam

Type	Processing Mechanism (Atomic Bit Size)	Processing Method (Energy Particle Beam)	Dimensional Accuracy Control Systems: Depth (1-D), Pattern (2-D), Form (3-D)
Atomic Cluster Removal			
Nano-cutting	Shearing slip, tensile rupture, atom cluster bit	Fine cutting with single point diamond tool (for soft materials), ultrafine lapping	Accuracy of form and surface roughness (machine tool–dependent)
Super polish	Elastic atomic failure machining, chemically reactive (solid, atomic bit, fracture) removal	Noncontact polishing EEM, magnetic fluid machining, ultrafine polishing (for hard materials)	Form accuracy (polishing tool), profiling, depth accuracy (micro)
Atomic-bit separation (removal)	Chemical decomposition (gas, liquid, solid) (thermal activation)	Chemical etching (isotropic, anisotropic), reactive plasma etching, mechanochemical machining, chemicomechanical machining (ion, electron, laser beam assisted photo resist etching)	Form accuracy (preformed profile), depth accuracy (macro), pattern accuracy (mask)
	Electrochemical decomposition (liquid, solid) (electrolytic activation)	Electrolytic polishing, electrolytic processing (etching)	Form accuracy (preformed profile), depth accuracy (macro), pattern accuracy (mask)
	Vaporizing (thermal) (gas, solid)	Electron-beam machining, laser machining, thermal-ray machining	Linear accuracy position (control), plane accuracy (macro), depth accuracy (macro), pattern accuracy (macro)
	Diffusion separation (thermal) (solid, liquid, gas)	Diffusion removal (dissolution)	Pattern accuracy (mask), form accuracy (macro), depth accuracy
	Melting separation (thermal) (liquid, gas, solid)	Melt removal	Form pattern accuracy (mask), form accuracy (macro), depth accuracy (macro)

TABLE 2.1 (continued)

Atomic Bit Processing of Materials by Energy Beam

Type	Processing Mechanism (Atomic Bit Size)	Processing Method (Energy Particle Beam)	Dimensional Accuracy Control Systems: Depth (1-D), Pattern (2-D), Form (3-D)
	Ion sputtering (solid) (dynamic)	Ion-sputter machining (reactive ion), sputter machining (reactive ion etching)	Pattern accuracy position control (mask), form accuracy (macro), depth accuracy (micro), form accuracy (preformed profile)
	Photon sputtering (solid) (dynamic)	Photo direct bond cutting (excimer, x-ray)	
Atomic-bit consolidation (accretion)	Electric-field removal of ionized surface atom	Evaporation by electric field (STM processing, AFM processing)	Atomic-scale STM control system
	Chemical deposition and bonding (gas, liquid, solid)	Chemical plating, gas phase plating, oxidation, nitridation, activated-reaction plating	Pattern accuracy (mask), thickness accuracy (macro)
	Electrochemical deposition and bonding (gas, liquid, solid)	Electroplating, anodic oxidation electroforming, electrophoresis forming	Pattern accuracy (mold mask), thickness accuracy (micro)
	Thermal deposition and bonding (gas, liquid, solid)	Vapor deposition, epitaxial growth, molecular beam epitaxy	Pattern accuracy, (mold mask), thickness accuracy (micro)
	Diffusion bonding, melting (thermal bonding)	Sintering, blistering, ion nitridation, dipping, molten plating	Pattern accuracy (mask), thickness (depth), accuracy (macro)
	Physical deposition and bonding (dynamic)	Sputtering deposition, ionized plating, cluster ion epitaxy, ion-beam deposition, ion-beam mixing	Pattern accuracy (mask), depth accuracy (micro)
	Ion implantation (dynamic)	Ion implantation (injection)	Pattern accuracy position control (mask), depth accuracy (micro)
	Electric field evaporation	STM processing	

TABLE 2.1 (continued)

Atomic Bit Processing of Materials by Energy Beam

Type	Processing Mechanism (Atomic Bit Size)	Processing Method (Energy Particle Beam)	Dimensional Accuracy Control Systems: Depth (1-D), Pattern (2-D), Form (3-D)
Atomic-bit deformation	Thermal flow	Surface tension (thermal, optical, laser, electron beam, gas high-temperature) flattening	Form accuracy (macro) (preformed profile) depth, pattern accuracy (mold)
	Viscous flow	Liquid-flow (hydro) polishing, injection molding	Form accuracy (macro), Form accuracy (micro)
	Friction flow (shearing slip) (ion rubbing)	Fine particle flow polishing (rubbing, burnishing, lapping), coining, stamping	Pattern accuracy (stamping)
	Molecular orientation		
Surface treatment	Thermal activation (electron, photon, ion, etc.)	Hardening, annealing (metal, semiconductor), glazing, solidify (resin)	Macro: open-loop control of macroscopic processing conditions or states of processing bit tools, i.e., atom, molecular of gas or liquid, ions, electrons, photons, etc.
	Mixing, deposition (electron, photo, ion, etc.)	Diffusion, mixing (ion)	
	Chemical reactivation (electron, photon, ion, etc.)	Polymerization, depolymerization	
	Energy-assisted chemical reactivation (electron-, ion-, photon-beam assisted)	Surface reactive finishing	Micro: closed-loop control of microscopic states of workpiece by means of feedback control of state of work
	Catalytic reactivation	Promotion of reaction	

EEM, elastic emission machining; STM, scanning tunneling microscope; AFM, atomic force microscope.

supramolecular structures. In between are a vast number of potential materials that combine different components in a wide variety of form and arrangement, so-called nanocomposites. In a nanostructured material, the nanoscale component may have all the functionality and be simply embedded in a supporting substrate phase, or the substrate or template may act to promote a particular form or arrangement of the nanoscale component that is essential to the development of the properties or function of interest. Most important, though, is that because of the scale, the distance between the nanoscale constituents becomes comparable to the size of the constituents. Therefore, unlike typical microscale materials and composites, the material between the nanoscale components is also nanoscopic and thus exhibits properties drastically different from its bulk counterpart. The constituent materials included in current nanomaterial research may be as conventional as semiconductors, polymers, metals, or ceramics, or may be as revolutionary as one of the emerging "giant molecular" materials such as dendrimers, biological molecules, quantum dot clusters, or fullerene-based materials.

In general, what distinguishes a nanomaterial from its chemically identical, conventional counterpart is not just having a feature on the nanoscale, but rather the breakdown of scaling laws. This means that one or more properties are (radically) different from what would be expected from a simple extrapolation of the properties of the material in the bulk and microscale region (above about 0.1 μm) down to nanometer scales.

For the emerging supramolecular nanomaterials, there are no conventional counterparts, and the target properties are radical from the outset. Here in particular there is huge potential for tailoring the material to obtain specific properties.

As if what is offered by nanomaterials is not vast enough, a wide variety of nanostructures may be conceived in which different component entities are brought together and assembled in some way into a bulk structure with some desired, very probably functional rather than structural, property. This is not to be confused with other endeavors such as nanoelectromechanical devices and systems, with which indeed there is common ground. The idea of assembling a nanostructure, using a variety of "tools," to a particular design, from components that may themselves be complex and highly designed, suggests perhaps the use of the term *functional nanostructure*. Concepts such as biomimetic design and processing principles as well as synergism with the rapidly advancing understanding and tools of biotechnology are likely to find their greatest application in such functional nanostructures and perhaps may be where the real future for the creation of truly novel materials lies.

Ultrafine microstructures having an average phase or grain size on the order of a nanometer (10^{-9} m) are classified as nanostructured materials.[12] Currently, in a wider meaning of the term, any material that contains grains or clusters below 100 nm, or layers or filaments of that dimension, can be considered to be nanostructured.[13] The interest in these materials has been stimulated by the fact that, owing to the small size of the building blocks (particle, grain, or phase) and the high surface-to-volume ratio, these materials are expected to demonstrate unique mechanical, optical, electronic, and magnetic properties.[14]

The properties of NSMs depend on the following four common micro-structural features:

1. Fine grain size and size distribution (< 100 nm)
2. The chemical composition of the constituent phases
3. The presence of interfaces, more specifically, grain boundaries, heterophase interfaces, or the free surface
4. Interactions between the constituent domains

The presence and interplay of these four features largely determine the unique properties of NSMs.

Synthesis: Metals/Intermetallics

Metals and intermetallics are made by employing either aqueous or non-aqueous methods. Fine metal powders have applications in electronic and magnetic materials, explosives, catalysts, pharmaceuticals, and powder metallurgy.[15] The two main routes of chemical synthesis for the metals and intermetallics are (1) thermal or ultrasonic decomposition of organometallic precursors to yield the respective elements or alloys, and (2) reduction of inorganic or organometallic precursors by reducing agents. Semiconductor nanoclusters are made by incorporating the particles in micelles/colloids, polymers, glasses, or zeolites or by controlled cluster fusion.

Intermetallics are defined as solid solutions of two or more metals in varying proportions. The properties of the intermetallics are unique.[16] The general chemical synthesis methods used to make intermetallics are very similar to those used to produce individual metals.

Most commonly, the intermetallics are prepared by reduction reactions. For example, Buhro et al.[17] synthesized nanocrystalline powders of TiAl, $TiAl_3$, NiAl, and Ni_3Al by the reductions of $TiCl_3$ or $NiCl_2$ with $LiAlH_4$ in a mesitylene slurry followed by heating in the solid state (< 550°C).

Various other intermetallics such as TiB_2, Ni_2B, WC-Co, Co-B, Fe-B, Ni-B, and Pd-B have been prepared by reduction reactions using $NaBH_4$ as a reducing agent on different inorganic precursors such as $TiCl_4$, $NiCl_2$, and $6H_2O$.[17]

A thermochemical processing method for preparing high-surface-area powders starting from homogeneous precursor compounds has been

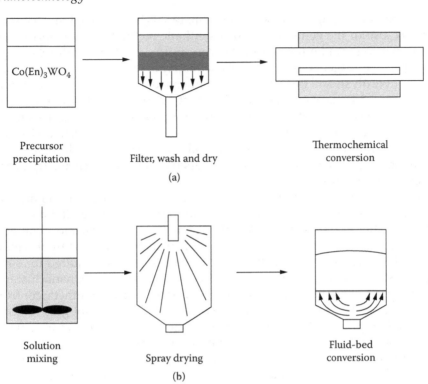

Precursor precipitation

Co(En)$_3$WO$_4$

Filter, wash and dry

Thermochemical conversion

(a)

Solution mixing

Spray drying

Fluid-bed conversion

(b)

FIGURE 2.5
Schematic diagrams of (a) laboratory scale process, (b) industrial scale process for the thermochemical processing of nanostructured powders, starting from aqueous solution mixtures.[18]

reported. The method has been applied successfully to the synthesis of nanophase WC-Co powders.[18] The method is known as "spray conversion processing" and consists of three sequential steps: (1) preparation and mixing of aqueous solutions of the precursor compounds to fix the composition of the starting solution, (2) spray drying of the starting solution to form a chemically homogeneous precursor powder, and (3) thermochemical conversion of the precursor powder to the desired nanostructured end-product powder (see Figure 2.5).

Sol-Gel Methods[19]

Sol-gel methods are commonly used to prepare nanocomposite materials, because these conversions occur readily with a wide variety of precursors and can be conducted at or near room temperature. In addition, the gel products frequently have properties ideal for desired applications.

Sol-gel processing of nanocomposite materials has attracted much interest because of the generality of this synthetic method. A wide variety of host matrices can be formed under very mild conditions with relatively inexpensive reagents using sol-gel chemistry. Conversion of a sol to a gel using hydrolysis and intermolecular condensation reactions occurs, under proper conditions, even for complex reactant mixtures affording convenient syntheses of host matrices having complex compositions. This flexibility enables the tuning of the pore size, surface area, density, dielectric constant, refractive index, and chemical composition of a host matrix. Inorganic/organic matrices are also accessible, and sol-gel chemistry even occurs within inverse micelles.

Formation of the nanoparticulate phase is achieved most commonly through reduction processes for elemental particles or precipitation reactions for chemical substances. A variety of methods are available to initiate particle formation. Nanoparticle growth is usually promoted by thermal annealing, and, for many substances, thermal energy is required for single-crystal formation. This growth process typically affords nanoclusters of the guest phase having a distribution of particle sizes. Synthetic control of nanoparticle size has been, and remains, a challenge in sol-gel processing. Although control of host matrix pore size gives some degree of control of nanoparticle size, convenient routes to monodisperse nanocomposites are still needed. Although a high degree of monodispersity might not be required for some applications, it is highly desired for most optical applications.

Compaction

Recently, there has been a significant increase in interest in fabricating ceramic materials from ultrafine powders that consist of nanosized primary particles ranging in mean diameter from 1 to 100 nm. Theoretical predictions by Frenkel[20] and Herring[21] clearly indicate that the rate of densification varies inversely as a function of particle size. Thus, based on this prediction, as particle size decreases from micrometers to nanometers, a substantial decrease in sintering time can be expected at a given temperature. Indeed, many experimental investigations support this theoretical prediction. For example, Rhodes[22] produced densely packed compacts of nanosized zirconia particles and observed sintering of the compacts to near theoretical density at much lower temperatures than are used for sintering coarse zirconia particles. Recently, Skandan et al.[23] sintered nanosized titania at 800°C, well below the sintering temperature for conventional titania powders.

These results suggest that nanosized particles as starting materials might offer considerable advantages for fabricating ceramics, especially because the reduced sintering temperatures and sintering times required may inhibit undesired grain growth, one of the most important microstructural parameters because it is directly related to materials properties.

The driving force for densification of nanosized powders is very high because of their small particle size. However, in many situations, their relatively low green-body densities require longer sintering times or higher temperatures than expected to achieve full densification. The longer sintering times and higher temperatures required promote excessive grain growth or irregular grain growth, resulting in a microstructure that does not exhibit the desired nanosized grain structure.

Silicon-Based Nanostructures

The search for suitable device and systems applications of nanostructures is an ongoing process. In this respect, the driving force to fabricate nanostructure devices can be split into three different aspects: speed, versatility, and increased integration. An important point in fabricating nanometer active areas is the opportunity to fabricate electronic devices operating at frequencies in the 100 GHz range up to the terahertz range. In optoelectronics, the realization of structures on the submicrometer scale, where quantum effects become more interesting, enables new material properties to be investigated making use of exciton confinement in a 3-D quantum box with distinct energy levels and their associated optical transitions yielding distinct emission lines. Furthermore, nanometer-sized active regions in electronic components could allow an increase of integration in circuits up to 10^{10} elements/cm^2.

The vast knowledge about silicon together with existing production lines make silicon the most favored material for many potential applications of nanostructures. The efforts in nanoelectronics are to be understood in the context of future possible needs of the silicon industry.

The need for faster and more complex microelectronic devices leads unavoidably to an increase in miniaturization of devices and, consequently, to higher packing densities in integrated circuits. Current feature sizes are about 250 nm and are expected to reach 70 nm by the year 2007. However, this trend brings with itself a design problem, because the density of aluminum wires connecting the various ports of the microchips increases dramatically, too, leading to what is called in the microelectronics industry an "electrical wire bottleneck." One strategy is to replace electrical interconnects with optical interconnects between chips and between circuit boards for ultrafast connection, with the expectation of overcoming the bottleneck problem. Many research teams worldwide are looking at hybrid gallium arsenide on silicon technology. An alternative is to investigate optical interconnects based on silicon technology. Several optical components based on silicon and silicon oxide already exist, such as photodetectors and waveguides. However, for a realistic approach, it is necessary to address two basic devices: (1) a silicon-based light emitter and (2) a silicon-based detector.

Despite their indirect band gap, silicon-based structures continue to be intensively investigated for direct integration of a light-emitting diode onto chips. There are two main geometries considered for potential silicon-based optoelectronic devices: vertical structures, such as top- or bottom-emitting devices, suitable for interplane connections, and edge-emitting devices potentially suitable for in-plane optical connections. Some device ideas are based on laterally structured quantum wires and quantum dots. However, the fabrication of one- or quasi-zero-dimensional devices for device research, as opposed to physics research, remains a significant challenge mainly because of the strong device specifications expected.[25]

In the quest to realize a variety of nanostructures using silicon and silicon-related semiconductors, the trend to miniaturization in electronics plays an overwhelming role. To some degree, this places an enormous burden on other developments that could potentially make use of nanostructures for other applications. The range of other applications comes under the umbrella of nanotechnology, as demonstrated by the work on nanotips.

The wealth of technological expertise developed collectively in silicon electronics ensures that any relevant scientific development in this area that survives the acid test of industrial suitability and device specifications could be made available to potential users because the connections to the outside world are provided by existing input and output Si-based circuitry and technology. Whereas this may well be the case in sensors and optical devices, there are significant challenges remaining in the area of quantum nanoscale electronic circuits, which is bound to keep the Si nanoelectronics community very active for years to come.

Researchers at the University of California, San Diego[26] discovered how to transfer the optical properties of silicon-crystal sensors to plastic. This could lead to flexible, implantable devices able to monitor the delivery of drugs within the body, strains on a weak joint, or the healing of sutures.

Although silicon has many benefits, it has downsides as well. It is not particularly biocompatible or flexible, and it can corrode. You need something that possesses all three traits for medical applications.

Researchers treat a silicon wafer with an electrochemical etch, producing a porous silicon chip with a precise array of nanometer-sized holes. This gives the chip optical properties similar to photonic crystals — a crystal with a periodic structure that can control the transmission of light much as a semiconductor controls the transmission of electrons.

Molten or dissolved plastic is cast into the pores of the finished silicon photonic chip. The chip mold dissolves, leaving behind a flexible biocompatible "replica" of the porous silicon chip. It is essentially a similar process to the one used in making a plastic toy from a mold; however, what is left behind is a flexible, biocompatible nanostructure with the properties of a photonic crystal. Scientists can fabricate polymers to respond to specific wavelengths that penetrate deep within the body.

Hybrid Semiconductors — Molecular Nanoelectronics

Ultradense integrated circuits with features smaller than 10 nm would provide enormous benefits for all information technologies, including computing, networking, and signal processing. However, recent results indicate that the current very large-scale integrated circuit paradigm based on complementary metal oxide semiconductor (CMOS) technology cannot be extended into this region. The main reason is that below a 10 nm gate length, the sensitivity of silicon field-effect transistor parameters — most importantly, the gate-voltage threshold — to the inevitable random variations of device size grows exponentially. This sensitivity may send fabricating costs skyrocketing and lead to the end of Moore's law some time during the next decade.[27]

The main alternative nanodevice concept — single electronics, based on controlling the motion of single electrons in solid-state structures — offers some potential advantages over CMOS technology, such as a broader choice of materials. However, the critical dimensions of single-electron transistors for room-temperature operation should be below ~1 nm, far too small for current and even forthcoming lithographic techniques, such as those based on extreme ultraviolet radiation and electron beams.

Hybrid Circuits

Many physicists and engineers believe that the impending Moore's law crisis may be resolved only by a radical paradigm shift from purely CMOS technology to hybrid semiconductor–molecular circuits, which is called CMOL, a combination of CMOS and Molecular. Such a circuit (see Figure 2.6) would combine

- A level of advanced CMOS devices fabricated by lithographic patterning
- A few layers of parallel nanowire arrays formed, for example, by nanoimprinting
- A level of molecular devices that self-assemble on nanowires from solution

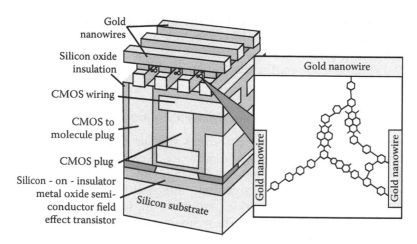

This hybrid complementary metal oxide semiconductor / molecular circuit contains a molecule that may function as a latching switch (including a single-bit memory cell) controlled by two input signals.

FIGURE 2.6

This hybrid complementary metal oxide semiconductor/molecular circuit contains a molecule that may function as a latching switch (including a single-bit memory cell) controlled by two input signals.[27,28]

The CMOL concept combines the advantages of nanoscale components, such as the reliability of CMOS circuits and the minuscule footprints of molecular devices, and the advantages of patterning techniques, which include the flexibility of traditional photolithography and the potentially low cost of nanoimprinting and chemically directed self-assembly. This combination may enable CMOL circuits of unparalleled density — up to 3×10^{12} functions/cm² — at acceptable fabrication costs.

For single-molecule components, single-electron devices are the leading candidate because, in contrast to field-effect transistors, their operation mechanism does not require high conductivity of the device-to-electrode interfaces. The CMOS layer allows CMOL circuits to circumvent one prominent drawback of single-electron transistors: their low voltage gain.

One group at the State University of New York at Stony Brook[27,28] is working to develop and implement several molecules suitable for this purpose, an example of which is shown in Figure 2.6. This arrangement essentially combines two well-known single-electron devices, the transistor and the trap.

Nanophase Materials

Nanophase materials are comprised of particles <50 nm in size, or 10,000 times smaller than the period at the end of this sentence. The basic building blocks that make up these materials are clusters containing only a few thousand atoms each Argonne scientists pioneered the development of nanophase ceramics. Together with colleagues from universities, industry, and other national laboratories, they determined the properties and characteristics of both nanophase ceramics and metals.[29] Scientists can now engineer nanophase materials to give them desired properties to meet the requirements of specific applications. Industry already uses materials of nanometer dimensions in advanced technological applications. For example, semiconductor devices are built atom layer by atom layer.

One exciting new application being developed involves fabricating miniature machines that are formed by etching single crystals of silicon or other materials. However, scientists continue to search for additional ways to improve technology. Nanophase synthesis and processing may make a variety of new technological advances possible.

Nanophase materials are assembled artificially in the laboratory. Atoms and molecules are condensed into clusters, and the clusters are then assembled into nanophase materials. The resulting materials' design flexibility and unique properties are what make nanophase materials attractive for future applications having demanding materials requirements. Nanophase materials are made quickly and in quantity — not atom layer by atom layer.

Producing Nanophase Particles

Nanophase particles are made by a gas-condensation process in which a metal or ceramic is vaporized, condensed from the vapor, and solidified into a new form. This sequence reconstitutes a ceramic or metal from a conventional coarse form into an extremely fine powder consisting of nanometer-sized atom clusters or particles. The powder is then pressed to form a bulk solid with a desired shape and size in a sealed chamber under extremely clean conditions.

The process starts with a conventional ceramic or metal source material. For example, if the goal is to produce nanophase copper, pieces of conventional copper are the source of atoms. Typically, the material is heated to slightly above its melting point so that hot atoms leave the surface of the source material, creating a vapor (see Figure 2.7).

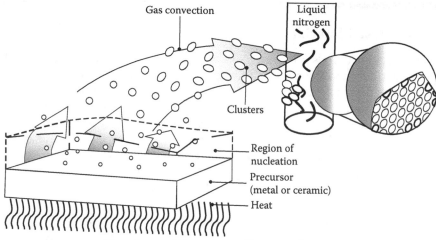

Nanophase particles are created by gas condensation.

FIGURE 2.7
Nanophase particles are created by gas condensation.[29]

When these atoms encounter colder gas atoms in the production chamber, they cool and condense to form solid material. (Helium gas is normally used because it is inert and no chemical reaction changes the material.) This condensation is similar to steam from a teapot meeting cool room air and condensing to form water droplets.

The resulting atom clusters are collected in the chamber on the surface of a hollow tube cooled by liquid nitrogen to 77°K. The chilled tube attracts the warm material naturally — cluster by cluster — by a process called thermophoresis. The ultrafine powder that collects on the tube is then removed under vacuum conditions that ensure the purity of the material. The powder is pressed into an object of the desired shape and size. This process works for ceramics as well as metals.

Nanophase ceramics also become easier to machine without cracking or breaking.

Sintering of nanophase ceramics can be done at temperatures ~600°C lower than those required for conventionally sized ceramic powders. Shrinking during sintering also is greatly reduced, so nanophase objects need less machining to achieve their final shapes.

Nanophase materials exhibit direct relationships between grain size and properties. When nanophase copper is produced by the gas-condensation process, it is up to 5 times stronger than conventional coarse-grained copper. The same is true for other metals.

Ductility of nanophase ceramics such as titanium dioxide and zinc oxide is dramatically improved. Their fine grain structure allows individual grains to be moved over one another without the ceramic breaking.

Grain Boundaries/Size

Because nanophase materials are formed by the consolidation of large numbers of small clusters or grains, the interfaces between grains — called grain boundaries — are important in determining and controlling the properties of these materials.[29]

Because nanophase grains are typically no larger than 50 nm, the mechanisms that control mechanical, magnetic, optical, and other properties in conventional materials can change significantly when grains become smaller than this. For example, pure metals are easily deformed because, at the microscopic level, atoms can move easily within the metal's crystal lattice structure.

This type of atomic motion is accomplished through the movement of lattice defects called dislocations through the structure.

When a metal is created from nanophase particles, the grain boundaries impede dislocation movement, and the small grain size makes dislocation formation difficult. Larger forces are needed to deform the material. The metal becomes harder and stronger.

On a microscopic level, ceramic nanophase grains can apparently move easily over one another because of their extremely fine grain size. Nanophase ceramics are therefore more easily deformed and less brittle than conventional ceramics. In conventional ceramics, grains do not slide easily and the materials are brittle. This inherent brittleness has been the main reason that ceramic materials have not been used more widely in applications such as engine components.

Research also has shown that the size of the particles can be manipulated in production to achieve a precisely desired property. By adapting the size and makeup of the 3-D atom cluster, a desired property can potentially be controlled for a specific application and a variety of properties can be fine-tuned within the same nanophase material.

Nanostructured Oxides[30]

Nanostructured oxides can be prepared by aerosolizing a solution of metal salts and entraining the aerosol in a carrier gas. The gas stream carries

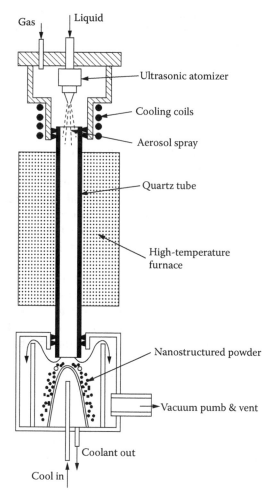

FIGURE 2.8
Schematic diagram of a thermal conversion processing system.[31]

the aerosol droplets into a hot-wall reactor, where the solvent evaporates and the precursor salts react to form the desired product. The product powder is separated from the carrier gas by filtration or impaction, or in a cyclone. Figure 2.8 shows a schematic diagram of a typical aerosol thermal conversion processing system. In such a system, the droplet temperature increases and solvent begins to evaporate as droplets flow through the hot zone of the reactor. The droplet size decreases and the dissolved salts precipitate, forming particles of dried salt.

At higher temperatures, the salts undergo pyrolysis reactions, product gases are evolved, and porous amorphous oxide particles are formed. At still higher temperatures, the amorphous oxide transforms to a porous nanocrystalline material. The transformation is driven by the decrease in free volume associated with the nanocrystalline phase. Further heating causes unwanted densification/sintering and grain growth. Thus, the particle residence time in the reactor and the temperature profile of the reactor must be carefully controlled to "freeze in" the nanostructure.

Nanostructured vanadium oxide (V_2O_5); palladium oxide (PdO); and superconducting yttrium, barium, cuprate ($YBa_2Cu_3O_{7-x}$) powders have been prepared by the aerosol method.[30]

In contrast to the above method for making nanostructured oxides, other researchers have used a solution-based method that leads, via a gel, to nanostructured Sn-doped TiO_2 powder. Tin doping stabilizes the desired rutile phase.

Nanostructured Nitrides and Carbonitrides

Silicon nitride (Si_3N_4) is an important ceramic material for potential use in high-temperature/high-stress applications such as heat engine components, cutting tool inserts, and bearing races. Ideal powder for consolidation by hot-pressing should have a particle size smaller than 1 μm. The particles should be equiaxed and have a narrow size distribution. They should also be unagglomerated and of high purity. Nanostructured Si_3N_4 powder that meets these requirements can be produced using a CO_2 laser-driven gas phase process.[30] In this method, a beam of silane (SiH_4) and ammonia (NH_3) gas intersects an intense laser beam, thereby producing a plasma flame. The atomic species that are generated in the plasma recombine to form Si_3N_4 particles having an average size of 17 nm. The product powder is collected by filtration.

Nanocomposites

Background

The general microstructural features of a nanocomposite, illustrated in Figure 2.9, consists of very fine (less than 300 nm) particles dispersed throughout larger grains (1–5 μm) of the matrix material. Nanoparticles are found both within grains and on the grain boundaries.

Several viable methods of fabricating these structures have been developed, such as using a chemical vapor deposition (CVD) route to produce

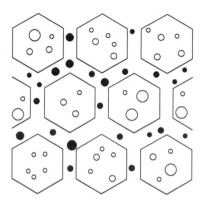

FIGURE 2.9
Schematic structure of the nanocomposite (not to scale).

an amorphous powder from which crystallization of the desired micro-structure is possible. Polymer precursors have been used in a similar way to form an amorphous body.

Both methods offer the benefits of clean processing and a homogeneous dispersion of nanoparticles throughout the matrix grains. The drawbacks are that they are expensive and are not easily applicable to the fabrication of large ceramic components. In contrast, the simple mixing of very fine powders by way of conventional milling offers a simple and cheap technique of fabricating nanocomposites. The process is also relatively straightforward to scale up for industrial applications.

Several nanocomposite systems that have been successfully fabricated are Al_2O_3/SiC, Al_2O_3/Si_3N_4, Al_2O_3/TiC, mullite/SiC, MgO/SiC, Si_3N_4/TiN, and Si_3N_4/SiC, where the first named compound is the matrix material. Volume fraction of the nanophase component has varied from 1 to 50%.

Figure 2.10 shows four kinds of nanocomposite structures that can be considered as structures that could sharply improve various mechanical characteristics and add new functions to ceramics. The in-grain nanocomposite material, the grain boundary nanocomposite, and the nano/nanocomposite can be considered three new kinds of composite materials that can be manufactured by the sintering method. Further, a structure in which an in-grain nanocomposite and grain boundary nanocomposite are achieved simultaneously also can be considered. From the viewpoint of linear fracture mechanics and metallic material design theory, the in-grain nanocomposite, as a structural material, has the most structure. By realizing this structure, it will become possible to boost strength 2 to 5 times, for example, and to raise the temperature limit by 400–700°C. But a sharp increase in fracture toughness cannot be expected. In order to achieve superstrong and supertough ceramic materials, it ultimately will be necessary to combine microcomposite materials and nanocomposite materials.

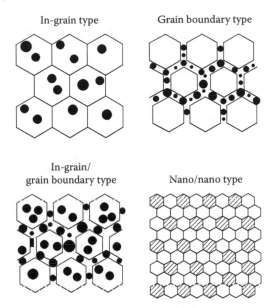

In-grain type Grain boundary type

In-grain/
grain boundary type Nano/nano type

FIGURE 2.10
Type charts of nanocomposites.

Possible micro–nano hybrid materials include whisker-reinforced, sheet grain-reinforced, and long staple-reinforced nanocomposite materials (see Figure 2.11).

Some Typical Material Types and Processing Methods

Nawa, Sekino, and Niiharat[34] reported on the fabrication of two types of Al_2O_3/Mo composites by hot-pressing a mixture of γ- or α-Al_2O_3 powder and a fine molybdenum powder. For $Al_2O_3/5$ vol% Mo composite using γ-Al_2O_3 as a starting powder, the elongated molybdenum layers were observed to surround a part of the Al_2O_3 grains, which resulted in an apparent high value of fracture toughness (7.1 MPa $m^{1/2}$). In the system using α-Al_2O_3 as a starting powder, nanometer-sized molybdenum particles were dispersed within the Al_2O_3 grains and at the grain boundaries. Thus, it was confirmed that ceramic/metal nanocomposite was successfully fabricated in the Al_2O_3/Mo composite system. With increasing molybdenum content, the elongated molybdenum particles were formed at Al_2O_3 grain boundaries. Considerable improvements of mechanical properties were observed, such as hardness of 19.2 GPa, fracture strength of 884 MPa, and toughness of 7.6 MPa $m^{1/2}$ in the composites containing 5, 7.5, and 20 vol% Mo, respectively. The fracture strength and toughness

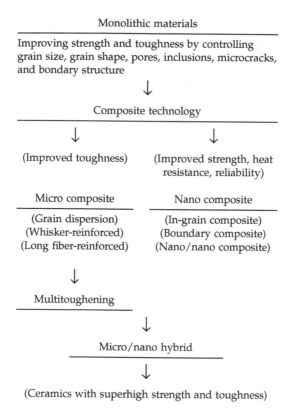

FIGURE 2.11
Development concept for high-performance structural ceramic materials.

values above were 1.5 times larger than that of the monolithic Al_2O_3 and 1.8 times larger than that of the monolithic Al_2O_3.

Metal (Fe, Cr, $Fe_{0.8}Cr_{0.2}$)/Al_2O_3 composite powders containing from 2 to 20 wt% metallic phase were prepared by selective reduction of oxide solid solutions and hot-pressing by Rousset.[35]

The microstructure of the composites are strongly dependent on the morphological and microstructural characteristics of the starting oxide. Most of the metal particles are <10 nm and are dispersed inside the matrix grains.

Massive composites were prepared by hot-pressing and were found to have significantly improved mechanical properties compared to those of alumina and metal/ceramic microcomposites, as well as a remarkable resistance to oxidation in air up to 1000°C.

The composites also contained both intragranular (≤30 nm) and intergranular (≤1 μm) metal particles. The former particles were responsible for the strong mechanical reinforcement observed at low metal content, while further enhancement was caused by the latter.

Chen, Goto, and Hirai[36] prepared SiC/Si_3N_4 nanocomposite powders by CVD using SiH_4, CH_4, WF_6, and H_2 as source gases at the reaction temperature of 1400°C.

The prepared powders consisted of SiC/W_2C composite particles and hollow β-SiC_P. The SiC/W_2C composite particles had a W_2C core and a SiC shell. The average particle diameters of the SiC/W_2C composite and hollow SiC_P increased from 18 to 30 nm and from 40 to 70 nm, respectively, with an increase of SiH_4 concentration.

The purpose of the work was to explore the dispersion of a fine W_2C or WC phase in SiC bodies, which have been shown to be effective in improving the toughness or electric properties.

Hillel et al.[37] prepared $SiC/TiC/C$ nanocomposites, chemically vapor codepositing onto a graphite substrate under atmospheric pressure in the temperature range 950–1150°C, using the $TiCl_4$-C_4H_{10}-SiH_2Cl_2-H_2 gas system.

Different characterization techniques showed that the materials codeposited at 1150°C exhibited a strongly textured (220) microstructure that decreased with the temperature of deposition. Zhaoet al.[38] did a study to assess the mechanical properties of pure Al_2O_3 and Al_2O_3 containing 5 vol% 0.15 μm SiC_P, including samples that had undergone an annealing treatment. They concluded

It is possible to obtain high strengths in Al_2O_3/SiC composites.

1. The strengthening and toughening observed in these samples are believed to derive almost exclusively from compressive surface stresses introduced by the grinding process. The addition of the SiC_P does not affect the intrinsic material toughness.

2. Annealing the composite specimens has the double effect of diminishing the compressive surface stress while also healing the surface flaws.

3. This study reinforces the need to consider all aspects of the fracture process, including toughness, strength, and intrinsic flaw distributions, before deducing any mechanism for an observed strength response.[39–44]

Nanoscale dispersions of intermetallic Ti_2Ni particles in an ordered TiNi intermetallic matrix have been produced by rapid solidification processing of near equiatomic TiNi alloys containing small amounts of Si, which utilize the principle of kinetic competition in the undercooled liquid, according to Nagarajan and Chattopadhyay.[45]

Liu et al.[46] prepared C-C/SiC nanomatrix composites by chemical vapor infiltration (CVI) methods. The aim of the researchers was to prepare C-C/SiC nanomatrix composites, but not nanocomposites, by the CVI technique through the codeposition process.

Their experimental setup used CH_3SiCl_3 liquid and C_2H_2 gas as the source, and H_2 and N_2 gases were used as the carrier and dilutant gases, respectively. The deposition temperature was selected as 1100°C, the total pressure was 1000 Pa, and CH_3SiCl_3 was carried into the furnace by bubbling H_2 carrier gas.

They concluded that CVI was an effective technique for fabricating nanomatrix composites through a codeposition process using multicomponent gas reactions. The optical texture of the C/SiC nanomatrix was composed of several sheaths of different brightness under polarized light.

Finally, the presence of SiC in the pyrocarbon matrix could significantly increase the oxidation resistance of C-C/SiC nanomatrix composites, because SiC may act as an oxidation inhibitor.

The sintering methods are the most promising for the fabrication of ceramic nanocomposites, according to Niihara and Nakahira.[40] The ceramic-based nanocomposites were prepared by the usual P/M techniques using the optimum sintering processes such as pressureless sintering, hot-pressing, and hot isostatic pressing (HIPing) in the Al_2O_3/SiC, Al_2O_3/Si_3N_4, Al_2O_3/TiC, Y_2O_3/SiC, MgO/SiC, mullite/SiC, B_4C/SiC, B_4C/TiB_2, $SiC/$ amorphous SiC, and Si_3N_4/SiC systems. The Si_3N_4/SiC nanocomposites were prepared by the liquid-phase sintering developed for the monolithic Si_3N_4 ceramics, and the solid-state sintering was applied for the other composite systems.

The mullite/SiC nanocomposites were prepared by the pressureless reaction sintering of natural kaolin-Al_2O_3-SiC powder mixtures.

In the ceramic/metal system, the nanosized metal-dispersed Al_2O_3 nanocomposites were successfully prepared by sintering of Al_2O_3 and fine metal powder mixtures and by reduction/sintering of Al_2O_3 and metal oxide mixtures. Table 2.2 reflects developed properties for the above materials.

TABLE 2.2

Improvement in Mechanical Characteristics of Ceramics by Nanocomposite

Composite Materials	Toughness $(MPam^{1/2})$	Strength (MPa)	Maximum Temperature (°C)
Al_2O_3/SiC	3.5 → 4.8	350 → 1520	800 → 1200
Al_2O_3/Si_3N_4	3.5 → 4.7	350 → 850	800 → 1300
MgO/SiC	1.2 → 4.5	340 → 700	600 → 1400
Mullite/SiC	1.2 → 3.5	150 → 700	700 → 1200
Si_3N_4/SiC	4.5 → 7.5	850 → 1550	1200 → 1400

In situ nanocomposites have been produced by the inert-gas condensation method. Typically, these nanocomposites consist of a harder particulate phase embedded in a softer metal phase. Contrary to conventional alloys, where the disperse phase is the minor phase, in these nanocomposites the disperse phase is the major phase. Using the inert-gas condensation method, the Cu-Nb nanocomposite samples produced should exhibit enhanced mechanical properties for high-temperature use.[47]

Figure 2.12 shows the room-temperature tensile strength measurements obtained, respectively, on the as-deposited and cold-pressed nanocrystalline copper, nanocrystalline copper after sintering in hydrogen for 2 h at 700°C, a Cu-4 wt% Nb produced by conventional powder processing using micron-sized particles, and Cu-40 wt% Nb nanocomposite after it had been sintered in hydrogen for 4 h at 600°C followed by HIPing at 900°C for 2 h. The value of 830 MPa measured on the nanocomposite is about 4 times the value predicted from the rule-of-mixture. On the other hand, the strength of nanocrystalline copper decreased from 500 to 100 MPa after the sintering at 700°C. Finally, the strength of Cu-Nb composite produced by the powder route was only 70 MPa.

The significant decreases in the hardness and tensile strength values of nanocrystalline copper is due to the rapid grain growth during the sintering treatments.

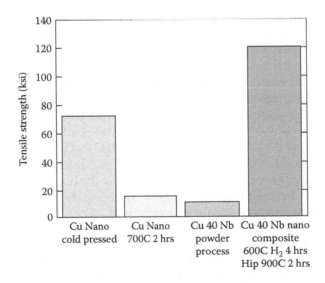

FIGURE 2.12
Bar graph showing the result of the room-temperature tensile strength measurements of a Cu-40 wt% Nb nanocomposite, compared to the corresponding powder processed Cu-Nb composite and those of nanocrystalline copper. The 828 MPa (120 ksi) for the nanocomposite is about 4 times the value predicted from the rule-of-mixture.[47]

Al_2O_3/SiC nanocomposites with a systematic variation in their SiC particle size together with monolithic alumina were produced using conventional powder processing, polymer pyrolysis, and hot-pressing by Carroll, Sternitzke, and Derby.[48]

All nanocomposites showed a clear increase in strength over similar grain size alumina but no clear dependence on the size of the SiC nano-reinforcement. However, the fracture toughness of the nanocomposites seems to increase with the SiC particle size but with values little changed from the toughness of monolithic alumina, as measured by the Vickers indentation technique.

Therefore, the increase in strength for nanocomposites can be explained by the reduction in the maximum flaw size but it is still unclear why nanocomposites show an increased surface damage resistance without any increase in toughness.

The results presented by Carroll et al.[48] show no support for Levin et al.'s model[49] of grain boundary strengthening. However, critical flaws in nanocomposites are much larger than the interparticle spacing of the nanophase. Therefore, the assumption that the strength increase can be explained by a decrease in critical flaw size is consistent with Carroll's data. A quantitative assessment of the nanocomposite effect should therefore be based on a detailed study of the processing and surface flaws present in nanocomposites and their effect on crack initiation.

Chemical vapor deposition and the precursor method are the most suitable manufacturing processes for producing a nanocomposite with very fine second phase dispersed inside grains, which have been considered the smallest construction unit of ceramics, or grain boundaries. However, it is difficult to manufacture large, complex-shaped structural ceramics by CVD at a low cost. Therefore, researchers have tried to manufacture a nanocomposite material by the conventional powder metallurgical method and found it possible to make nanocomposites with many composites, such as Al_2O_3/SiC, Al_2O_3/Si_3N_4, MgO/SiC, Al_2O_3/TiC, Y_2O_3/SiC, natural mullite/SiC, B_4C/SiC, and Si_3N_4/SiC. Initially, it was possible to make these nanocomposite materials only by the hot-press method, but now they can be made by atmospheric sintering/hot isostatic pressing combination or simple atmospheric sintering.

As a result of research on the nanocomposite process with the use of an electron microscope, it was found that conditions for the fabrication of nanocomposites by the sintering method are different from those for solid-phase and liquid-phase sintering. The following points are important for solid-phase sintering:

1. The matrix powder grain diameter should be 100–500 nm.
2. Powder with a grain diameter of 200 nm or less should be selected as nano grains to serve as the disperse phase.

3. These powders should be disperse-mixed uniformly.

4. Sintering conditions should be selected to ensure that grains will grow in the matrix phase but not in the disperse phase.

In the case of liquid-phase sintering with Si_3N_4 as the matrix, the following conditions are important:

1. Powder with a grain diameter of 100 nm or less should be selected as nano grains for the disperse phase.

2. The two powders should be mixed uniformly.

3. The disperse phase should not be dissolved in the liquid phase deriving from the sintering aid (selection of the disperse phase, sintering aid, and sintering conditions).

4. The disperse phase should become the core with β-Si_3N_4 phase deposits again from the liquid phase.

Figure 2.13 shows the high-temperature strength of Al_2O_3/SiC and MgO/SiC nanocomposite materials. Nanocomposites sharply reduce the decline in strength at high temperatures. The strength of Al_2O_3 at 1000°C was similar to that at room temperature, and the strength of MgO did not fall until the temperature reached 1400°C. Both of these represent an extraordinary improvement. It also became clear that the use of nanocomposites greatly improves high-temperature hardness and resistance to thermal shock, fatigue, and creep destruction.

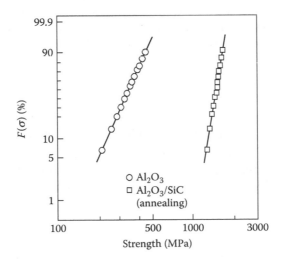

FIGURE 2.13
Weibull plot of strength of Al_2O_3/SiC nanocomposite.

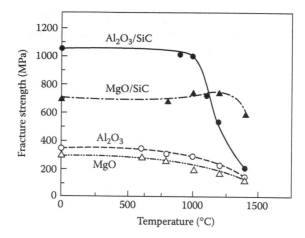

FIGURE 2.14
High-temperature strength of Al₂O₃/SiC, MgO/SiC nanocomposites.

As can be clearly seen from Table 2.2, a dramatic improvement in mechanical characteristics caused by dispersed nano grains also was observed in other nanocomposites, including Al_2O_3/TiC and mullite/SiC. This is apparently because dispersed nano grains play the following roles:

1. Local stress arising from nano grains generates subgrain boundaries, as shown in Figure 2.14, at a high temperature that allows the matrix to undergo plastic deformation during cooling from the manufacturing temperature, and achieves nanolevel miniaturization of the structure to attain high strength and high reliability. The improved strength of the Al_2O_3/SiC nanocomposite after annealing for about an hour at 1000–1300°C is because the subgrain boundaries grow farther due to annealing.

2. Nano grains cause the front end of a crack to deviate, thereby increasing fracture toughness.

3. Residual tensile stress generated around nano grains induces fractures inside grains to improve high-temperature strength.

4. Because nano grains pin the moves of dislocations, a drop in hardness at high temperatures improves resistance to creep fracture, thereby curbing deformation at high temperatures.

Modeling

Grain boundary-interface dihedral angles in Al_2O_3/SiC and Al_2O_3/TiN nanocomposites have been measured from TEM micrographs, and the ratio of the particle/matrix interfacial energy γ_i to the alumina grain boundary energy γ_{gb} has been calculated on the basis of an energy balance condition. The average ratio, γ_1/γ_{gb}, was found to be 1.21 for Al_2O_3/SiC and 0.80 for Al_2O_3/TiN. These results, together with estimates of surface energies, have been used to estimate interfacial fracture energies. Jiao et al.[50] found that the interfacial fracture energy between SiC and alumina was more than twice the alumina grain boundary fracture energy; i.e., SiC nanoparticles strengthen grain boundaries. The dominant factor in this result was the greater surface energy of SiC coated with Al_2O_3. Despite the higher surface energy of TiN and the lower γ_1/γ_{gb} ratio for Al_2O_3/TiN, it was found experimentally that TiN particles weaken grain boundaries instead of strengthening them.

Polymer Nanocomposites

The reinforcement of polymers using fillers, whether inorganic or organic, is common in the production of modern plastics.

Polymeric nanocomposites (PNCs), like other composites, are materials that are designed and processed from selected constituents. However, unlike typical advanced composites, in which a reinforcing constituent is on the order of microns, nanocomposites utilize a constituent on the order of a few nanometers, which is on the same scale as the radius of gyration of a polymer. Furthermore, the addition of a nanoscale constituent provides more than reinforcement, but utilization of the properties of the nanoscale constituent and synergism with the matrix provides additional value-added properties. These include electrical and thermal conductivity, enhanced barrier to small molecules, decreased swellability, improved flammability, optical clarity, and self-passivation response to aggressive environments.

PNCs (or polymer nanostructured materials) represent a radical alternative to conventional-filled polymers or polymer blends. In contrast to the conventional systems, where the reinforcement is on the order of microns, discrete constituents on the order of a few nanometers (~10,000 times finer than a human hair) exemplify PNCs.[51]

Uniform dispersion of these nanoscopically sized filler particles (or nanoelements) produces ultralarge interfacial area per volume between the nanoelement and host polymer. This immense internal interfacial area and the nanoscopic dimensions between nanoelements fundamentally differentiate PNCs from traditional composites and filled plastics. Thus, new combinations of properties derived from the nanoscale structure of PNCs provide opportunities to circumvent traditional performance trade-offs associated with conventional reinforced plastics, epitomizing the promise of nanoengineered materials.

A literature search provides many examples of PNCs, demonstrating substantial improvements in mechanical and physical properties. However, the nanocomposite properties discussed are generally compared to unfilled and conventional-filled polymers, but are not compared to continuous fiber-reinforced composites. Although PNCs may provide enhanced, multifunctional matrix resins, they should not be considered a potential one-for-one replacement for current state-of-the-art carbon-fiber-reinforced composites.

From both a commercial and military perspective, the value of PNC technology is not based solely on mechanical enhancements of the neat resin. Rather, it comes from providing value-added properties not present in the neat resin, without sacrificing the inherent processability and mechanical properties of the resin. Traditionally, blend or composite attempts at multifunctional materials require a trade-off among desired performance, mechanical properties, cost, and processibility.

Considering the number of potential nanoelements, polymeric resins, and applications, the field of PNCs is immense. Development of multi-component materials, whether microscale or nanoscale, must simultaneously balance four interdependent areas: constituent selection, fabrication, processing, and performance. This is still in its infancy for PNCs, but ultimately scientists will develop many perspectives dictated by the final application of the PNC. Researchers developed two main PNC fabrication methodologies: in situ routes and exfoliation. Currently, researchers in industry, government, and academia worldwide are heavily investigating exfoliation of layered silicates, carbon nanofibers/nanotube-polymer nanocomposites, and high-performance resin PNCs.

An improved melt process has been developed for the formation of PNCs with the following technology benefits:

- This method can be used to produce a broader range of nano-composite feedstocks of any molecular weight, including unfilled, glass fiber-reinforced impact-modified, pigmented, heat-stabilized, and flame-retardant versions.
- Organoclays employed exhibit low odor during compounding.
- Specific swelling/compatibilizing agents used in the invention reduce the amount of shear mixing required, resulting in less decompositon of the polymer and reduction in molecular weight.
- Process facilitates unreacted monomer removal prior to forming the nanocomposite — important because dispersed particles can interfere with the monomer removal process.
- The melt compounding method allows for the production of unfilled, glass fiber-reinforced, impact-modified, pigmented, heat-stabilized, and flame-retardant versions of polymer nanocomposites in one extrusion step.

What is unique about the process? The polymer matrix includes a melt processable polymer with a melt processing temperature equal to or greater than 220°C. Organoclays have specific swelling agents bonded to their surface. The mixture is subjected to a shear rate sufficient from evenly dispersed platelets possessing an average thickness of less than 50 Angstroms and a maximum of 100 Angstroms.

The broad range of matrix polymers are made possible by a larger selection of swelling/compatibilizing agents, each having a distinct bonding interaction with both the polymer and the platelet particles. Compounding entails no special conditions specific to polymer molecular weight distributions.

The "flowable mixture" of polymer and organoclay was selected from the group consisting of $+NH_3 R_1$, $+NH_2R_2R_3$, and $+PR_4R_5R_6R_7$, wherein:

R_1 is an organic radical having at least 12 aliphatic carbon atoms.

R_2 and R_3 are organic radicals having more than 5 carbon atoms.

R_4, R_5, R_6, and R_7 are organic radicals including at least one that has at least 8 aliphatic carbon atoms.

Compatibilizing agents are ammonium and quaternary phosphonium cation complexes containing specific numbers of aliphatic carbon atoms. These agents cover the layers of the organoclay and facilitate exfoliation. The result is reduced shear mixing and decreased decomposition of the polymer. The compatibilizing agents are selected so as to not decompose (causing chain scission, harmful vapors, or degradation of the matrix polymers).

The process developed by Honeywell[52, 53] exhibits several advantages over monomer blending and polymerizing processes of the in-reactor type. These advantages include (a) utility for a broader range of matrix polymers, (b) utility for a wider range of composites having the same matrix polymer due to a larger selection of swelling/compatibilizing agents (each having a distinct bonding interaction with both the polymer and the platelet particle), and (c) greater control over the molecular weight distribution of the matrix polymer. For example, virtually any polymer material that can be made to flow may be compounded with nanoscale particles derived from organoclays that exfoliate during mixing.

In contrast, the monomer blending and polymerizing processes of competing processes are restricted to polymers whose monomers are compatible with organoclays and can be polymerized effectively in the presence of the organoclays. With Honeywell's inventive process, the compounding entails no special conditions specific to selected polymer molecular weight distributions. On the other hand, in-reactor processes require special polymerization conditions for each selected molecular weight due to the effect of a dispersed phase on viscosity and polymerization kinetics. Virtually any loading of organoclay is possible in the process.

The most preferred organoclays are phyllosilicates possessing anegative charge on the layers (ranging from 0.2 to 0.9 charges per formula unit) and a commensurate number of exchangeable cations in the interlayer spaces. Preferred layered materials are smectite clay minerals such as montmorillonite, nontronite, beidellite, volkonskoite, hectorite, saponite, sauconite, magadiite, and kenyaite.

In order to further facilitate delamination of organoclays into platelet particles and prevent reaggregation of the particles, these layers are inter-posed with swelling/compatibilizing agents. These agents consist of a portion that bonds to the surface of the layers and another portion that bonds (or interacts favorably) with the polymer. The effective swelling/compatibilizing agent remains bonded to the surface of the layers during and after melt processing as a distinct interphase different from the bulk of the polymer matrix. Such agents interact with the surface of the layers, displacing the original metal ions and exhibiting cohesive energies sufficiently similar to that of the polymer, thereby enhancing the homogeneity of the dispersion in the polymeric matrix.[54–57]

Properties of PNCs

Nanosized clay additives boost stiffness and rigidity of olefin elastomers and nylons without compromising their elongation or toughness. The 300- to 1500-nm-long particles are less than 1 nm thick, giving them an extremely high length-to-thickness ratio — one unmatched by traditional glass, carbon, or mineral reinforcements. The high aspect ratio (the ratio of their length to their diameter) and small particle size is said to substantially improve polymer mechanical performance even at loadings of <10%. Reduced loadings let the particles disperse more easily, which helps maintain better transparency than that available with conventional reinforcements (see Table 2.3).

TABLE 2.3

Typical Properties

Properties	SEP Nylon 12	Natural Nylon 12
Hardness, Duromter, D Scale	D78	D73
Specific gravity	1.32	1.02
Notched Izod, ft-lb/in. (⅛-in. thick)	3.89	2.60
Tensile stress, kpsi @ 100% elongation	9.2	8.3
Ultimate tensile stress, kpsi	10.4	9.8
Ultimate elongation, %	135	340
Tensile modulus, kpsi	182.3	159.4
Flexural modulus, kpsi	298	180

Scientists and engineers at AFRL's Materials and Manufacturing Directorate, working with the University of Dayton Research Institute,[58] found a way to tailor the electrical conductivity (over a range of 10^{-6} S/cm to 10^2 S/cm) in polymeric materials used to build military and commercial aerospace components, with negligible impact on the polymer's mechanical properties or processability. This new technology transforms almost any commodity polymer into a multifunctional material capable of carrying or dissipating significant electrical charge — an advancement offering tremendous promise throughout the space, aerospace, automotive, and chemical industries. The researchers achieved this technological advancement by controlled dispersion of specifically designed, highly electrically conductive, yet remarkably flexible, carbon nanotubes into the supporting polymer matrix. These nanotubes have the current-carrying capacity of copper but with a comparatively much lower density.

The nanotubes used in the finished products were on the order of 50 to 150 nm in diameter with an aspect ratio of greater than 800. This high aspect ratio results in a much lower required filler content to achieve percolation (onset of conductivity) than traditional metal-filled systems. The percolation threshold for these materials is less than one half of 1% by volume. The multiwall nanotubes used in this process are available in ton quantities, which allow affordable, realistic scale-up of the resultant nanocomposites.

The electrically conductive polymer nanocomposite materials, compared to conductive metal-filled systems, offer substantial weight savings, flexibility, durability, low-temperature processability, and tailored reproducible conductivity. Researchers could use them to produce conductive paints, coatings, caulks, sealants, adhesives, fibers, thin films, thick sheets, and tubes as well as electromagnetic interference shielding for large structural components, electrostatic painting, electrostatic discharge, and optoelectronic device applications.

The materials' enhanced high-frequency absorption capability suits the cable shielding industry particularly well. Because the materials are easy to process, engineers can use them for coextrusion of coaxial cables, replacing the high-weight braided metal shielding. The technology is ready for licensing and commercialization and is easily capable of scaling to large batch production.

Some Applications of Nanotechnology

Films/Coatings

Diamondlike nanocomposites (DLNs) and nanocrystalline atomic network composites (NANCs) are a new class of potentially cost-effective coatings that are flexible, strongly adhere to a wide variety of materials (including metals, composites, ceramics, semiconductors, crystalline and glass dielectrics, and plastics), have high wear resistance and hardness (8–20 GPa), have extremely low coefficients of friction (0.03–0.1), and are stable in both oxidizing (up to 400°C) and nonoxidizing (up to 1200°C) environments. Originally developed and tested in Russia over a period of 20 years, these coatings are now being further advanced and commercialized.[59] A comparison of the properties of DLN and NANC coatings with CVD polycrystalline diamond and diamondlike carbon coatings (DLC) is shown in Table 2.4. A summary of their adhesion to various substrates, an advantage compared with DLC coatings, is shown in Table 2.5, and stability behavior is shown in Table 2.6.

TABLE 2.4

Comparison of the Properties of Coatings of CVD Polycrystalline Diamond, Diamondlike Carbon (DLC), Diamondlike Nanocomposites (DLN), and Nanocrystalline Atomic Network Composites (NANC)

Property	Diamond	DLC	DLN	Me-DLN/NANC
Min. deposition temp. (°C)	600	30	30	30
Growth rate (μm/hr)	0.1–4.0	0.1–1.0	2–10	2–10
Maximum thickness (μm)	>10	1-3	10	10
Max. substrate diameter (mm)	75	150	760	760
Hardness (GPa)	80–100	5 to >20	8–15	10–20
Adhesion[a] to quartz (N/m²)	—	5×10^7	$>10^8$	$>10^8$
Coeff. of friction	High	0.1	0.03–0.1	0.04–0.1
Roughness	Rough	Smooth	Smooth	Smooth
Flexibility/brittleness	Brittle	Brittle/ flexible	Extremely flexible	Very flexible
Electrical resistivity (Ω-cm)	10^{10}–10^{15}	10^7	1^{-7}–10^{16}	10^4–10^{16}
Maximum use temperature (°C)	>700	<300	>400	>400
			>1200 (inert)	>1200 (inert)

[a] Diamond and DLC films have good adhesions to certain substrates only, or require special surface preparation with intermediate layers. DLN and NANC have good or excellent adhesion to any substrate.

TABLE 2.5

Comparison of the Adhesion to Various Substrates of Diamondlike Carbon Coatings (DLC) and Diamondlike Nanocomposite Coatings (DLN)[59]

	Steel	Al	Ti	Cu	Au	Co	Cr	Mo	Ni	W	Si	Ge	GaAs	Glass	Plastic
DLC	++	++	++	–	–	–	–	–	–	+	+++	+	+	+++	–
DLN	+++	+++	+++	++	++	+++	+++	+++	+++	+++	+++	++	+++	++	++

Key: – no adhesions; + poor adhesion; ++ good adhesion; +++ excellent adhesion.

TABLE 2.6

Comparison of the Stability of Diamondlike Carbon Coatings (DLC), Diamondlike Nanocomposite Coatings (DLN), and Diamondlike Nanocomposite Coatings with a Glassy Metallic Phase Third Random Network (Me-DLN)[59]

	Maximum Temperature (°C) Before Destruction By:		
Coating	Oxidation	Graphitization	Thermal shock
DLC	400	350–600	N/A
DLN	500	1300	N/A
Me-DLN[a]	800	≥1300	2000

[a] Values vary with composition.

The DLN coatings are composed of two mutually stabilized, random, interpenetrating networks. One is a diamondlike carbon network stabilized by hydrogen (a-C:H, with up to 50% sp^3 bonds), and the other is a silicon network stabilized by oxygen (a-Si:O). Additional elements, such as metal atoms, can be added as glassy metallic phases in a third random network (forming Me-DLN). Carbide formation at elevated temperatures is prevented by the random network structure, even at metal concentrations of 50%. Mutual stabilization of these atomic scale networks prevents graphitization at high temperatures, enhances adhesion, and reduces internal stress in these nanocomposites. Figure 2.15 shows a schematic of a DLN composite.

The NANC type of network composite coatings are formed when 1.5–100 nm clusters or nanocrystals of transition metals or high-melting refractory compounds are incorporated into the two-network structure. The structure of both the DLN and the NANC composites can be tailored to optimize desired properties of mechanical strength, hardness, and chemical resistance.

The coatings are deposited by a proprietary process that combines plasma and ion beam technology. They can be applied on surfaces that are rough and curved as well as flat and, with equipment adaptation, even on some inner surfaces, at low temperatures between 30 and 250°C. With the present equipment, substrates up to 760 mm in diameter can be coated with coating thicknesses from 20 nm to 10 gm. The DLN and NANC coatings can also be deposited in alternate layers.

Other advantageous properties cited for the DLN coatings are their biocompatibility, high resistance to thermal shock, controllable optical properties (such as refractive index), and controllable electrical properties. The two network DLNs are dielectric materials; however, conductivity can be varied over 18–20 orders of magnitude (10^{16} to 10^{-4} Ωcm) by controlling the metal concentration in the third interpenetrating network.

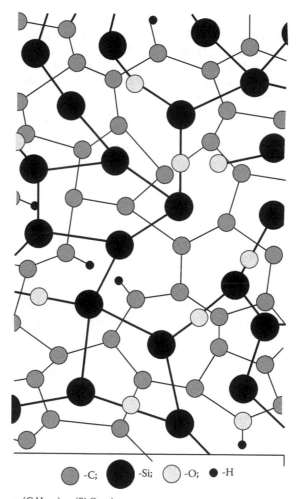

-C; -Si; -O; -H

a-$(C:H_{0.15})_{0.70}(Si:O_{0.30})_{0.30}$

FIGURE 2.15

Schematic of a typical atomic structure of a diamondlike nanocomposite (DLN) is on the left. A transmission electron micrograph showing a multilayer DLN coating on a ceramic substrate is on the right. The dark layers are a metallic tungsten-diamondlike nanocomposite (Me-DLN), and the light layers are a dielectric DLN. Defects in the microcrystalline substrate are not transmitted to the amorphous coating.[59]

Many applications are envisioned for these unusual coatings. Researchers believe these coatings will be more cost-effective than CVD polycrystalline diamond and DLC coatings.

The opportunities for the films/coatings are far reaching, covering the entire industrial base: automotive, aerospace, energy, electronics, chemical processing, personal care, business equipment, and medical segments, encompassing such items as high-energy power switches, microwave devices, plasma application components, high dielectric strength films in electronics, flat panel displays, and other areas.

One must overcome a major obstacle, which is that molecules are so fragile and unstable that if two or more touch, they fuse together and lose any special shape or properties they had in designing nanotechnology materials. The solution is a coating made of resorcinarene, an organic surfactant with a bowl-shaped head and several hydrogen-bonding tails. During the coating process, resorcinarene's curvature helps it stick to metallic nanoparticles. Researchers use chemicals to stitch them together into a polymeric cage around the nanoparticle. This porous coating lets nanoparticles interact with external substances, but not with other similarly coated particles. The process stabilizes particles up to 50 nm in diameter and those with magnetic properties. This has allowed engineers to develop materials for microelectronic devices and magnetic sensors.[2]

Nanoelectronics

To keep getting faster, electronics must eventually shrink down to the nanoscale. But making wires and switches that small is no easy feat. Biological molecules, with their ability to self-assemble, offer one promising approach. In research done at the University of Chicago, scientists have engineered proteins to form cores for gold wires only 80 nm wide.

The researchers started with yeast prions — harmless cousins of the proteins that cause mad-cow disease. Under certain conditions, these prions spontaneously form highly stable fibrils. The team at the University of Chicago genetically engineered the fibrils to bind tightly to specially prepared gold nanoparticles. The result was fibrils dotted with gold blobs. To fill gaps between the blobs, silver was added and then more gold, producing conducting wires.

The prions have advantages over other biological molecules that researchers have so far tried to use for nanoelectronics. Merging silver with DNA, for example, has allowed scientists to make nanowires, which, thanks to DNA's ability to specifically bind to complementary sequences, can then form circuitlike patterns. But the weak bonds between DNA strands tend to break easily. This is less of a problem for protein fibrils.

Other laboratories have developed proteins that can bind to about 30 different electronic, magnetic, and optical materials and then assemble the materials into structures. The goal is to integrate proteins into a self-assembly system to create a way to "grow" materials such as semiconductors — where and in whatever patterns the researchers want. Though it is impossible to predict how long it might take for the new techniques to make their way into industrial use, the research is setting a firm foundation for practical nanotechnology.

Other scientists have built what is said to be the world's first thermoelectric heat pump incorporating nanoscale materials. Thermoelectric devices convert electric power into cooling or heating, or heat into electricity. Potential applications include solid-state refrigerators and air conditioners, and more efficient and compact power sources.

The postage-stamp-sized prototype device is made of superlattices — thin-film stacks of two alternating semiconductors, bismuth telluride and antimony telluride. Each of the some 1000 film layers is about a few tens-of-a-billionth of a centimeter thick and contains up to two dozen layers of atoms.

Electric current applied to the superlattices (thermocouples) pushes heat toward one end of the circuit and cools the other end with unprecedented efficiency. Localized power densities approaching 700 W/cm^2 are possible, making them useful for everyday functions such as refrigeration and producing power. The thin-film devices react to changes in electric current in about 10 to 20 μsec or a factor of 23,000 faster than bulk thermoelectric materials.

For example, a heat pump built with the materials cooled a block of solid steel from 79 to 64°F in about 2 min, much faster than conventional refrigerators can cool. Such performance approaches cooling efficiencies of current bulk thermoelectric devices but in packages 1/40,000 the size. And ongoing improvements could boost efficiency threefold.

Eventually, superlattice heat pumps could replace most mechanical refrigerators and air-conditioning systems. Tiny heat pumps for spot-cooling microprocessors or communication lasers will likely be the first applications, followed by microscopic cooling and heating to regulate localized temperature changes on DNA microarrays.

Today, silicon microchips have features as small as 130 nm. But continuing to shrink silicon chips is getting expensive and difficult. "At some point, silicon is going to run out of steam," says John Rogers, director of nanotechnology research at Lucent Technologies' Bell Labs. "You're going to need something else." Something, Rogers says, like transistors the size of single molecules. Although still at least a decade from commercialization, chips built using these molecular transistors are the industry's best hope for building faster, cheaper computers well into this century.

"With the electronics we're talking about, we're going to make a computer that doesn't just fit in your wristwatch, not just in a button on your shirt, but in one of the fibers of your shirt," says Philip Kuekes, a computer architect at Hewlett-Packard Laboratories. Kuekes and his colleagues are designing circuits based on perpendicular arrays of tiny wires, connected at each intersection by molecular transistors. By the middle of the next decade, Kuekes says, Hewlett-Packard will demonstrate a logic circuit about as powerful as silicon-based circuits circa 1969. "We're trying to reinvent the integrated circuit — with its logic and memory and interconnects — with a consistent molecular manufacturing process," Kuekes says.

Well before the first shirt-thread computer boots up, however, companies will begin to integrate nanoelectronic components, including tiny wires and ultradense computer memory, into conventional silicon electronics. Two firms, for example, both plan to have prototype memory devices ready as early as 2004 with production models due by 2006. Devices that store a bit of data in a single molecule could provide thousands of times more storage density than the electronic memory currently used in computers.

Researchers are also working with nanoelectronics to develop new biological and chemical sensors not possible with conventional technology. One researcher is developing such sensors from silicon nanowires. Contact with even a single molecule changes the wires'electronic state. Researchers can measure that change to identify unknown molecules for purposes of diagnosis or pathogen detection.

New Bone Materials

Tissue engineers have made great strides in growing bone parts in the lab, but it is proving much more difficult to replace whole sections of leg or arm bones, which sustain constant pounding. Researchers at Rice University have developed a technique for growing bone tissue strong enough to withstand the stresses of everyday activity. Conventional bone-tissue engineering involves replacing lost bone with a biodegradable polymer scaffold seeded with cells. As the polymer degrades, new tissue develops. But in load-bearing parts of the skeleton, cells are constantly breaking down and forming new bone in response to mechanical stimuli. If the polymer scaffold placed in a patient's leg is too weak, the material falls apart under this stress. To reinforce their scaffold material, bioengineers and chemists have added nanoparticles of alumoxane (an aluminum-based compound) to a photosensitive polymer. Shining light on this blend spurs the nanoparticles to fix themselves to the polymer chains. The resulting material's compressive strength is 3 times that of the polymer alone.

Shoes

Engineers at AFRL's Materials and Manufacturing Directorate[60] use nanocomposites with nanoreinforcements 10,000 times smaller than human hair to manufacture components vital to both military and commercial systems. This technology can also benefit commercial industry through the successful design and development of dynamic new consumer products. Polymer-matrix nanocomposites could replace composite and polymer materials currently used to design and manufacture critical substructures in aircraft and space vehicles, fuel-line brackets, combustion chambers, and cryogenic storage containers, resulting in substantial weight and cost savings. Transfer of this new technology to private industry has led to the production of helium-filled pouches for athletic shoes.

The athletic shoe uses helium-filled capsules to provide greater cushioning and shock absorption over conventional shoes. This technology enabled shoe manufacturers to design a lower heel, placing the foot 25% closer to the ground than other athletic shoes. Because helium is difficult to encapsulate in plastic, nanoscale platelets are used in the capsules to prevent the gas from escaping.

Biosensors/Computational Optoelectronics

Development of a biosensor for cancer diagnostics is being pursued by Ames Research Center, which partnered with the National Cancer Institute to develop a nanoelectronic-based biopsy sampler that will circumvent conventional technologies that normally would take 1 to 3 weeks for results. The new technology would provide instantaneous results.

The focus of optoelectronics research is on modeling and simulation to understand quantum mechanical effects and the effect of radiation on device performance. It also helps in the design of quantum lasers and detectors.

Molecular Electronics

Modeling and simulation is used to generate molecular properties and create designer materials. In nanotechnology, modeling studies are designing nanoelectronic components, chemical and biosensors, and nanotube-based structures.

Inorganic Nanowires

The materials themselves — silicon, gallium arsenide, and gallium nitride — have been around for a while in a different form. Now, instead of a two-dimensional thin film, Ames Research Center is growing them as a one-dimensional wire. In one dimension, the properties seem to be much better than their two-dimensional cousins. With better properties, device integration at a much higher density will be possible.

Lubricants

People say that water and oil do not mix. They can, however, coexist as two separate molecules on the same surface. And therein lie both opportunities and challenges for government researchers who are aiding industry efforts to develop surface protecting and lubricating films that will shield super-small machines and their even tinier components from friction and wear.

Ultrathin lubricants or single-layer films will be needed for the minuscule nanotechnology gadgetry to come, from dust-sized environmental sensors to machines for repairing damaged cells. Today's lubrication systems — such as the fluorocarbon compounds and carbon overcoats used on magnetic disk drives — may not be adequate to meet the demanding performance requirements envisioned for nanotechnology applications.

With collaborators from the data-storage and lubricant industries, a government team is exploring the lubricating potential of a mixed-molecule, single-layer film. They are testing novel combinations of up to four different molecules, each one chosen to achieve desired capabilities, from wear resistance to self-repair. In one combination, for example, a particular group of molecules adheres tightly to the surface, anchoring the film and protecting against high-shear collisions. Other molecules "swim" among the anchors to prevent friction.

The next stage is the development of test methods to evaluate new materials and new combinations of materials being considered as lubricants.

Polyesters

Unsaturated polyester (UP) nanocomposites could provide greater chemical resistance, dimensional stability, and fire resistance than conventional formulations, according to scientists. Clay nanoparticles for thermoset formulations, as well as work on glass-reinforced UP nanocomposites in marine applications, are already proving promising. Chemically modified, nano-sized clay products have been used in making polymer nanocomposites.

Nanocomposites containing only very small amounts of these nanoscale particles (2–10% by weight) can result in properties that are equal to (or sometimes better than) traditional composites containing 20–35% mineral or glass. Machine wear is also reduced, and the compounds are easier to process. Because the densities of mineral and glass fillers are twice that of polymers, automotive parts and other weight-sensitive applications are where nanocomposites have great potential. Research has shown that the nanoparticles also act synergistically with other minerals and glass fiber.

Commercial applications for nanocomposites include polyamide (PA) 6 and poly-propylene for packaging and injection-molded articles, semi-crystalline PA for ultra-high barrier containers and fuel systems, epoxy electrocoat primers and high-voltage insulation, and polyolefin fire-retardant cable, electrical enclosures, and housings.

The thermoset market is where new applications are emerging in the marine, industrial, and construction markets. Unsaturated polyester nanocomposite formulations currently available are reported to provide improved chemical resistance, especially against corrosive chemicals and seawater, and they are also more dimensionally stable and fire resistant. Glass-reinforced UP nanocomposites are being used for boat accessories, where they are also said to be less prone to color fading. Better sag control is another benefit and is also seen in epoxy formulations (sag control is defined as the ability of the liquid resin to properly wet-out and adhere to glass fiber matting prior to curing). Fumed silica has traditionally been used for sag control, but nanomers are said to produce the same type of rheology as fumed silica in thermosets, providing sag control as well as property improvements in the cured part. Nanomers are also said to be easier to disperse and cost less, delivering all these benefits at little or no more cost over existing formulations.

Cellulose Materials

A one-step process for creating thermoplastic nanocomposites from cellulose fibers is the result of research at Virginia Polytechnic Institute and State University. By reacting wood pulp fibers in a solvent, then hot-pressing the partially modified pulp fibers, a semitransparent polymer sheet is formed that is made up of cellulose esters and unmodified cellulose. The addition of solvent makes the surfaces of microfibrils within each wood pulp fiber melt-flowable. The material has the virtues of thermoplastics in that it resists water, can be shaped by heat, and is expandable. But it is a nanocomposite rather than a blend, so it also retains the virtues of cellulose fibers, which nature uses as reinforcement.

"The unmodified cellulose adds strength and biodegradability, ... and it is also cheaper than the more highly processed cellulose ester materials commercially available. It comes from a renewable resource, and a low-input process has been developed both in terms of time and materials," according to researchers. It is anticipated that the thermal nanocomposites may be used where strength, resistance to moisture, and resistance to heat are important, such as for panel products in transportion industries and casings for electrical products and appliances.

Automotive

A family of thermoplastic olefin (TPO)-based nanocomposites has cut weight and improved dimensional stability, stiffness, and low temperature impact performance in prototype rear quarter and exterior door panels. The nanocomposites were a joint development between General Motors R&D and Montell North America in Troy, MI.

"Incorporating a submicron particle into a polymeric substrate, such as TPO, enhances physical properties without adding weight or loss of low-temperature properties or opacity," according to Theo Zwygers, Montell's Automotive and Industrial Business Group Technical Director.

Engineers at GM and Montell chose a family of natural clays, commonly known as smectite clays, as the nano-contributing component. The use of 5% smectite clay exfoliated in TPO provides stiffness equivalent to a traditional 25–35% talc-filled material. The new composites are produced on conventional manufacturing equipment and also improve auto recyclability potential.

Protection of Food Supply

Purdue University and U.S. Department of Agriculture engineers are using nanotechnology, magnetic beads, and other novel means to detect deadly substances in food. The goal of researchers is to keep microbial organisms such as *salmonella enteritidis, listeria, E. coli,* and others from entering the food chain at any point, whether at the farm, processing plant, or consumer's table.

Some of the projects already in the works include a detection method for fungus species in grain and one using tiny chips with electronic signaling to find listeria. Another detection method under consideration is a rapid test for predicting the toxicity of PCBs in fish.

Optics

Networking engineers are reaching the limits of what can be done with modern electronics. Communications networks have begun to switch light, or photons, instead of electrons in order to achieve higher data-transfer rates. Most engineers believe that nanosized optical switches will eventually form the foundation of an all-optical network backbone, in which mirrors or bubbles will guide streams of data-laden light without the delays associated with today's switching method of converting light to electrical signals and then back again.

Building Blocks (Carbon Tubules)

Many consider this to be the most active area of nanotech research — what many assume will be the basic building blocks of nanotech. Such tubules could be used in composite materials to add tensile strength and could be used to push breakthroughs in microelectronics by enhancing silicon circuitry. But the real hope is that carbon tubules could be used to replace silicon circuitry altogether and, in doing so, extend Moore's law, which is showing signs of exhaustion. A molecular device analogous to a transistor would be exponentially faster and more powerful than anything built from silicon.

Motors

University of California at Berkeley engineers have built a nanoscale motor. "It's the smallest synthetic motor that's ever been made," says Alex Zettl, professor of physics at UC Berkeley and faculty scientist at Lawrence Berkeley National Laboratory. "Nature is still a little bit ahead of us — there are biological motors that are equal or slightly smaller in size — but we are catching up."

The motor is about 500 nm across. The rotor is between 100 and 300 nm long. The carbon nanotube shaft to which the rotor is attached is only 5–10 nm thick.

The motor's shaft is a multiwalled nanotube, meaning it consists of nested nanotubes. Annealed both to the rotor and fixed anchors, the rigid nanotube allows the rotor to move only about 20°. However, the team was able to break the outer wall of the nested nanotubes to allow the outer tube and attached rotor to freely spin around the inner tubes as a nearly frictionless bearing.

To build the motor, Zettl and his team made a slew of multiwalled nanotubes in an electric arc and deposited them on the flat silicon oxide surface of a silicon wafer. They then identified the best from the pile with an atomic force microscope, a device capable of picking up single atoms.

A gold rotor, nanotube anchors, and opposing stators were then simultaneously patterned around the chosen nanotubes using electron beam lithography. A third stator was already buried under the silicon oxide surface. The rotor was annealed to the nanotubes and then the surface selectively etched to provide sufficient clearance for the rotor.

When the stators were charged with up to 50 V of direct current, the gold rotor deflected up to 20°. With alternating voltage, the rotor rocked back and forth, acting as a torsional oscillator. Such an oscillator, probably capable of microwave frequency oscillations from hundreds of megahertz to gigahertz, could be useful in many types of devices — in particular, communications devices such as cell phones or computers.

Because the rotor can be positioned at any angle, the motor could be used in optical circuits to redirect light (optical switching). The rotor could be rapidly flipped back and forth to create a microwave oscillator, or the spinning rotor could be used to mix liquids in microfluidic devices.[60]

Switches

Companies like Lucent Technologies, Corning, and Agilent Technologies[61] are pursuing similar research. All are seeking a way to manage the tremendous explosion of Internet traffic across thousands of miles of fiberoptic networks. "We are coming to the end of what we can do with modern electronics," says chemistry professor Larry Dalton at the University of Washington in Seattle. "We have to get to the photonics domain before we can take advantage of high bandwidth." And a key part of that, he adds, is "using nanostructure materials to greatly increase data-transfer rates."

Indeed, many engineers agree that nanosized optical switches eventually will form the foundation of an all-optical network backbone. These switches will shunt streams of light and thus billions of data packets contained within those photons, without incurring the delay of today's switches, in which the light must be converted into electrical signals. The devices also have the potential to help engineers reconfigure network traffic in a matter of nanoseconds, redirecting huge volumes of traffic across thousands of miles of networks. If successful, they will lower the management costs for carriers and ultimately the monthly bills for businesses and consumers.[62]

Agilent's nanotechnology pursuits involve the use of bubbles to refract wavelengths of light. Yet its switches are still only in the micro range: one-millionth of a meter. The company's bubble switch combines what are called photonic waveguides with the ink-jet technology used in printers. Inside the device, light travels through two tiny silica switches — the waveguides — and is refracted when it encounters a liquid bubble placed at the intersection of a waveguide and a channel, a groove running through the device. If there is no bubble at a channel intersection, the light waves pass through.

The bubble switch has several strengths: small size, reasonable switching speed, and good optical performance. Still, it has drawbacks. One is that the switches use liquid that must be heated. Network performance degrades if the device gets too hot (see Figure 2.16).

Rules and Measuring Devices

An ultra-precise method of laying chromium atoms on silicon surfaces has been developed at NIST.

Conventional lithography lenses focus light with material lenses and are limited in its precision by optical diffraction.

NIST's search has turned this paradigm around: Because both light and matter can be considered waves, it should be possible to focus matter waves with "lenses" made of light.

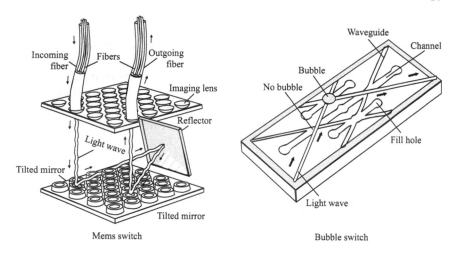

Mems switch   Bubble switch

FIGURE 2.16
The likelihood of efficient optical networks may hinge on the development of nano-size optical switches. The leading optical-switch technology suited for nanoscale is the MEMS switch, which uses movable mirrors to refract light waves in the desired direction. Similarly, engineers hope to "shrink" the bubble switch to nano-size. It uses a small bubble to refract light.[62]

Indeed, by focusing lasers over a silica surface and passing neutral chromium atoms through the field, NIST scientists have produced chromium structures with millions of parallel lines, or a series of dots with radii as small as 40 nm. The technique could be used to make nanoscale rulers and measuring devices.

Piezoceramics

A new manufacturing technology modifies the nanostructure of materials, from semiconductors to metals and dielectrics, enhancing their electro-mechanical characteristics and reliability. In particular, the process provides for improved performance of piezoceramics in a variety of applications.

This new, enhanced piezoceramic offers size and power advantages for ignition systems, electrostatics, and power-generation devices. The material's uniqueness stems from a patented technology called "nanofiber-in-nanopore," which changes the piezoceramic's microstructure to create a greatly improved piezoelectric effect and higher mechanical strength over traditional piezoceramics. The metallic nanofibers can be used to improve the characteristics of a number of materials, but the process offers the greatest advantages in piezoceramics; that is, a material that converts motion (for example, vibrations caused by an introduced frequency) into electricity.

One of the most promising uses is an ignition system in automobiles, boilers, and furnaces. Compared with traditional coil ignition systems, a piezoceramic system offers optimal combustion enabled by a sequential series of ignition pulses rather than a single spark. It also offers reduced emissions and increased power resulting from more efficient combustion, less ignition equipment (no distributor, ignition coil, high-voltage cables) for greater system reliability, and reduced electromagnetic interference.

The enhanced electrical characteristics of the piezoceramic technology enable it to be used as a high-voltage transformer charging a high-voltage capacitor to deliver a single spark; as a high-voltage igniter delivering a series of high-voltage sparks in a frequency range from 0.1 Hz to 100 kHz; and as a high-voltage discharger, providing a stable and continuous discharge across a gap.

Piezoceramic elements are resistant to vibration and the effects of fuel, solvents, antifreeze, oil, and other chemicals. The ability to provide a continuous discharge enables the use of otherwise difficult-to-ignite fuel/air mixtures (for example, water/fuel emulsions containing a high percentage of water).

Other potential applications include use as a piezotransformer in electrosurgical and electrostatic systems. Piezoceramic electrosurgical instruments are capable of delivering precisely controlled high voltage for micron-level tissue cauterization to minimize collateral damage and provide the surgeon with maximum control. Piezoceramics, when used in high-voltage electrostatic sprayers, create extremely fine droplets for uniform coverage and increased surface adhesion. Enhanced piezoceramics are ideal for any mass-produced high-voltage system, including electro-optical systems, signal equipment, capacitors, and their chargers.

Clothing

The U.S. Army opened the Institute for Soldier Nanotechnology (ISN) at the Massachusetts Institute of Technology (MIT), which will develop combat gear using materials the size of atoms. ISN scientists envision uniforms lined with a slurry of fluids that respond to magnetic fields, creating an armor system that can become stiffer during combat. The impact of nanotechnology is to have the ability to have a uniform that protects you totally against your environment.

One type of innovation being developed is for uniforms to change colors on demand to camouflage soldiers in changing environments. Other developments are for uniforms that have radio communication materials woven into the uniform fabric, and fuel cells the size of transistor radios.

It is predicted that the technologies developed could reduce the typical weight of standard soldier gear by more than half, from 120 to 60 lb.

Clusters

Atomic clusters represent new building blocks for the fabrication of nano-structured materials and films. Clusters are of particular interest because they have properties that lie between the atomic regime and the bulk (condensed matter) states. Their electronic, optical, chemical, and physical properties change with their size, due to the confinement of electrons in a small volume[63], and generally tend towards the properties of the solid state at very large sizes. For example, the transition from an electrically insulating to metallic state has been found to occur at a size of approximately 100 atoms for gold clusters.[64]

Over the last few years, atomic clusters have begun to show rich potential for exploitation in a number of areas. The fact that their properties are size dependent suggests that materials could be specifically tailored for a wide range of applications areas, from electronics and computing to sensors and catalysis.[65] Currently work illustrates the potential of gold clusters as building blocks for nanoscale circuitry; researchers are very much concerned with the scientific basis for these technological opportunities, and much work still needs to be done at this level. But it is already clear that gold has an important place in the expanding world of science and technology on the nanometer scale.

Cosmetics

The largest current commercial application for nanotechnology is not robotics, electronics, or computers. It is not what anyone would expect: It is cosmetics.

Cosmetics manufacturers are using nanoparticles to enhance their products' performance. Suntan lotion, for example, contains nanoparticles that more efficiently block the ultraviolet rays. In addition, cosmetics makers discovered that putting carbon nanofibers into nail polish makes it harder and more resistant to chipping. It becomes shinier and brighter, and it lasts longer.

Nanowire Arrays

Nearly every month, researchers develop new methods of producing smaller nanocircuit components. Many of these involve creating master patterns with electron beam lithography and then stamping out components, including fine wires, with a die pressed into temporarily molten material.[66] Such methods are limited by the resolution obtainable with electron beams — currently around 20 nm in diameter for wires.

The latest method, which achieves diameters as small as 8 nm, or about 80 atoms across, avoids this limitation by forgoing electron beam lithography. Developed jointly by researchers at the University of California campuses at Los Angeles and Santa Barbara and Caltech scientists, the method uses molecular-beam epitaxy (MBE) to form wire arrays.[67]

Alternating 8-nm-thick layers of gallium arsenide and aluminum gallium arsenide are laid down first with MBE. The finished layers are rotated 90° and the aluminum gallium arsenide layers selectively etched out. The template is then rotated another 36° and exposed to a stream of metal ions, which form a thin layer of metal on each exposed gallium arsenide layer to create the wires.

Next, the superlattice is placed face down onto an adhesive, and the gallium arsenide is etched away, leaving the wire array attached to the adhesive. If desired, the adhesive can be removed with oxygen plasma to leave the wires free. The result is an array of up to 40 wires 2 to 3 mm long and as small as 8 nm wide.

The wire arrays have many applications, including as etch masks for producing similar sized wires out of semiconductors. Two arrays can be laid down at right angles to one another to form crossbar-array circuits with junction densities as high as $10^{11}/cm^2$. In addition, the team showed that suspending the wires across a 750-nm trench formed a micromechanical oscillator with a resonant frequency of 162 MHz.

Nanobatteries

A University of Tulsa chemistry professor, Dale Teeters, and two former students, Nina Korzhova and Lane Fisher, have been awarded U.S. Patent #6,586,133 for a method of making nanobatteries for use in tiny machines.[68] The invention is a manufacturing process that can build, charge, and test nanobatteries.

The method includes filling the pores of a porous membrane with an electrolyte and capping the pores with electrodes. The manufacturing process begins with an aluminum sheet that is placed in acid solution under an electric current, resulting in an aluminum oxide membrane. When the metal is dissolved, a honeycomb structure results. The pores are then filled with an electrolyte, which in this case is a plastic-like polymer. Next, the filled pores are capped on both sides with electrodes: ceramic or carbon particles.

Each battery packs as much as 3.5 V. Key tools in the process are a scanning electron microscope and an atomic force microscope. The microscope's custom-made, electrically conducting cantilever tip is touched to the electrode so that the battery can be charged and tested.[68]

Future and Potential

Nanotechnology has been described as the manufacturing technology of the 21st century for two reasons. First, it gives thorough control of the structure of matter. This control is obviously very desirable for all sorts of manufacturing. Second, nanotechnology provides this capability at low cost. One of the main reasons for low cost is the use of self-replicating systems. Self-replicating systems are an integral and important part of nanotechnology.

Challenges

There are many opportunities awaiting us with nanotechnologies, but there are many challenges, such as those articulated by the National Nanotechnology Initiative (see Figure 2.17). Overall, these challenges may be grouped into two general categories (fundamental challenges and technological challenges) that must be first overcome before such opportunities are brought to reality.

As with any material, a fundamental understanding of the predictive processing-structure-property relationships provide the foundation for manipulation and exploitation of nanomaterials. The basic research performed against this goal will provide pervasive benefits to multiple applications of nanomaterials. To do this we need to understand the unique processes through which nanomaterials are made, which often bring in such diverse phenomena as colloidal chemistry and percolation behaviors. Characterization techniques effective and accurate at the nanoscale must be developed or refined. The size-dependent properties of the constituents must be determined. Theory and modeling tools need to be refined to adequately capture the simultaneous influence of multiple length and time scales as well as the boundary where continuum descriptions of

---

- Nanostructured materials "by design" — stronger, lighter, harder, self-repairing, and safer
- Nano-electronics, optoelectronics and magnetic
- Advanced healthcare, therapeutics and diagnostics
- Nanoscale processes for environmental improvement
- Efficient energy conversion and storage
- Microcraft space exploration and industrialization
- Bio-nanosensor devices for communicable disease and biological threat detection
- Economical and safe transportation
- National security

FIGURE 2.17
Grand challenges of the national nanotechnology initiative.

physical behavior fail to apply. Furthermore, the myriad of processing factors influencing the final structure need to be elucidated to provide reproducibility. And techniques to provide detection of assembly errors and means for correction on the nanoscale will underpin true utilization of bottom-up assembly.

On the technological level, the integration of these new concepts into current design and the identification of the impact of nanomaterials must be performed. Furthermore, design paradigms may be changed, perhaps to truly exploit methods that design the material in parallel with the application. New engineering tools that enable maximum utilization of these enhancements must be developed and validated. Issues such as life-prediction, life cycle cost, and cost-benefit analysis with regard to alternatives need to be investigated. And sufficient investment in infrastructure, manufacturability, and scalability is mandatory if the lab-scale successes are to be moved to production.

Yet perhaps the greatest challenge facing this emerging field will be the focusing of the large community involved in its research to ensure that developments build upon each other in an effective manner.

Self-Assembly

As researchers begin trying to build devices and novel materials at the nanoscale (a nanometer is a billionth of a meter, the size of a few atoms), they are facing a massive challenge. Although it is proving possible, in many cases, to push molecules around to form tiny structures and even functioning devices, efficiently mass-producing anything with nanoscale features is another matter altogether. But what if millions of these nano building blocks did the heavy lifting and assembled themselves into the desired structures, avoiding the use of expensive and elaborate manufacturing instruments?

Self-assembly has become one of the holy grails of nanotechnology, and scientists in numerous labs are working to transform it into an effective nano engineering tool. In some sense, self-assembly is nothing new: Biology does it all the time. And for decades, scientists have studied "supramolecular" chemistry, learning not only how molecules bind to one another but how large numbers of molecules can team up to form structures; in fact, the concept of self-assembly largely grew out of chemists' attempts to make molecules that aggregated spontaneously into specific configurations, in the same way biological molecules form complex cell membranes.

But now, with an expanding understanding of how molecules and small particles interact with one another, researchers can begin to predict how such elements might self-assemble into larger, useful structures like the transistors on a semiconductor chip. "Self-assembly provides a very general route to fabricating structures from components too small or too numerous to be handled robotically," says George Whitesides, a chemist at Harvard University and pioneer in the field.

To better understand how self-assembly works, Whitesides and his coworkers have recently shown that selectively coating the surfaces of microscopic gold plates with a sticky organic film can, under the proper conditions, trigger thousands of such plates to self-assemble into three-dimensional structures. So far, Whitesides's team has created a relatively large functional electronic circuit using a similar technique. The next step will involve shrinking the circuit to the micrometer scale, creating more complex three-dimensional structures out of silicon. Although micrometer-sized electronic components are nothing new — they are made all the time — Whitesides's experiments could provide valuable clues as to how to better manipulate self-assembly.

Nature itself is also providing scientists with a model of how to create self-assembling electronic devices. Materials scientist Angela Belcher at the University of Texas at Austin sorted through billions of different proteins to find ones that recognize and bind to different types of inorganic materials. For instance, one end of the protein might bind to a specific metal particle and the other end might stick to the surface of a semiconductor such as gallium arsenide. Given the right prompts, the proteins could direct nano-sized particles of inorganic materials to form various structures. These protein-mediated building blocks could have any number of technological applications, in making such things as biomedical sensors, high-density magnetic storage disks, or microprocessors.

Chemists at many laboratories are also attempting to develop self-assembled molecular computers. If they succeed, however, it will take years.

Meanwhile, less ambitiously, other researchers are making rapid strides in using self-assembly to build increasingly complex — and increasingly small — three-dimensional structures that could be compatible with existing devices. For instance, certain features of a disk drive, like the storage medium, could be created using self-assembly, while larger components needed to connect the device to the outside world would be made using conventional techniques. "We hope that self-assembly will be able to inexpensively replace certain stages in the production of materials and devices, where control is needed at the molecular level," says engineer Christopher Murray of the Nanoscale-Science Division of IBM Research in Yorktown Heights, NY.

Next Decade

New tools create new technologies, which then create the next generation of tools in the quest for greater knowledge and control of physical and chemical processes. Manipulators built on tiny flexure springs have excellent capabilities for ultraprecision positioning, having no friction, backlash, or play. Actuated by piezoelectric or electromagnetic forces, they might offer fine control to less than 1 nm. Emerging very-large-scale integrated fabrication will demand routine accuracies of 10 nm, which may be verified by laser interferometer servo control with resolution to 1.25 nm. X-ray interferometry will enhance the resolution to unheard-of accuracies, finer than 0.01 nm.

Nanofabrication researchers are now achieving success by placing selected well-knit molecular "blocks" onto supporting surfaces that provide chemical bonds strong enough to prevent disruption by thermal agitation. These blocks have been integrated into new, stable configurations, such as hexagonal rings, that do not form naturally. This approach works at room temperature and offers the flexibility of a molecular "subassembly" component, which is then handled as an individual entity in further processing by a higher level "assembler." The key is not just the properties of the resulting nanostructures, but the fabrication process and the options for achieving desired configurations in future nanomaterials.

"Nanotechnology will follow in the same path as videotape recording, compact disk technology, and personal computers. There will be a huge range of applications that have not even been thought of yet. But while the possibilities seem endless, so does the time to market for many real-world applications of nanotechnology. However, one product that could be to market soon is a 36- or 40-in. flat-panel TV using carbon nanotubes developed by Samsung. The company hopes it will be to market in 2 to 3 years," says Dr. Meyya Meyyappan. "If Samsung can get the price of a flat-panel TV down — and I think they can with carbon nanotubes — that could be the first mass-market use of nanotechnology."

For the most part, though, the expected benefits of nanotechnology in electronics and computing, composites, and sensors are going to take a while, according to Meyyappan."Every technology, from the time the scientists start shaking and baking in the lab, takes a good 10 to 15 years. It's one thing to have a technology, and another thing to develop a product. They are always miles apart. There is not a lot people can do to accelerate that time."

References

1. Vaia, R. A., Benson Tolle, T., Schmitt, G. F., Imeson, D., and Jones, R. J., Nanoscience and Nanotechnology: Materials Revolution for the 21st Century, *SAMPE J.*, 37(6), 24–31, 2001.
2. Krummenacker, M. and Lewis, J., Prospects in Nanotechnology: Toward Molecular Manufacturing, *Proceedings of the 1st General Conference on Nanotechnology: Development, Applications, and Opportunities*, November 11–14, 1992, Palo Alto, CA, New York: John Wiley & Sons, 1995, p. 297.

3. Taniguchi, N., The State of the Art of Nanotechnology for Processing of Ultraprecision and Ultrafine Products, 1993 Annual Meeting, The American Society for Precision Engineering, Seattle, WA, November 9, 1993, *Precision Engrg.*, 16(1), 5–24, 1994.
4. Taniguchi, N., Analysis of Mechanism of Various Materials Working Based on the Concept of Working Energy, *Scientific Papers of the Institute of Physics and Chemical Research*, 61, 3, 1963.
5. Taniguchi, N. Atomic Bit Machining by Energy Beam Process, *Precision Engrg.*, 7, 3, 1985.
6. Taniguchi, N., Research and Development of Energy Beam Processing of Materials in Japan, *Bull. JSPE*, 18, 2, 1984.
7. Taniguchi, N. and Miyazaki, T., Background and Development of Nanotechnology on Advanced Intelligent Industry — Current Status of Ultra Precision and Ultra Fine Materials Processing, presented at the 1st International Seminar on Nanotechnology, Tokyo Sciences University, Noda, Japan, 1989.
8. Taniguchi, N. et al., *Energy Beam Processing of Materials — Advanced Manufacturing Using Various Energy Sources*, Oxford Science Publications, 1989.
9. Taniguchi, N., Advanced Concept of Nanotechnology — Atomic Bit Processing Due to Energy Particle Beam, *Proceedings of the 16th Seiken Symposium, "Nanotechnology,"* Research Laboratory of Precision Machinery and Electronics, Tokyo Institute of Technology, Japan, 1991.
10. Ishikawa, J. (Ed.), *Proceedings of the 2nd Workshop on Beam Engineering*, Japan: Kyoto University, 1991.
11. Crandall, B. C. and Lewis, J., Nanotechnology — Research and Perspectives, *Papers from 1st Foresight Conference on Nanotechnology*, Cambridge, MA: The MIT Press, 1992, p. 381.
12. (a) Gleiter, H., *Adv. Mater.*, 4, 474, 1992, (b) Birringer, R. and Gleiter, H., in *Encyclopedia of Materials Science and Engineering*, R. W. Cahn (Ed.), Suppl. Vol. 1, Oxford, U.K.: Pergamon Press, 1988, p. 339.
13. Siegel, R. W., *Nanostruct. Mater.*, 3, 1, 1993.
14. Siegel, R. W., *Mater. Sci. Eng.*, B19, 37, 1993.
15. (a) Ichinose, K. et al., *Superfine Particle Technology*, London: Springer-Verlag, 1992, (b) Xu, Q. and Anderson, M. A., *J. Am. Ceram. Soc.*, 77, 1939, 1994, (c) Gonsalves, K. E. et al., *Adv. Mater.* 6, 291, 1994, (d) Kear, B. H. and McCandlish, L. F., *J. Adv. Mater.*, 10, 11, 1993.
16. (a) Tracy, M. J. and Groza, J. R., *Nanostruct. Mater.*, 1, 369, 1992, (b) Higashi, K., Mukai, T., Tanimura, S., Inoue, A., Masumoto, K., Kita, K., Ohtera, K., and Nagahora, J., *Nanostruct. Mater.*, 26, 191, 1992.
17. (a) Buhro, W. E., Haber, J. A., Waller, B. E., and Trentler, T. J., *Am. Chem. Soc. Symp. Ser.* 210, 20, 1995, (b) Haber, J. A., Crane, J. L., Buhro, W. E., Frey, C. A., Sastry, S. M. L., Balbach, J. L., and Conradi, M. S., *Adv. Mater.*, 8, 163, 1996.
18. (a) Kear, B. H. and McCandlish, L. E., *Nanostruct. Mater.*, 3, 19, 1993, (b) McCandlish, L. E. and Polizzotti, R. S., *Solid State Ionics*, 32/33, 795, 1989.
19. Kwiatkowski, K. C. and Lukehart, C. M., Nanocomposites Prepared by Sol-Gel Methods: Synthesis and Characterization, in *Handbook of Nanostructured Materials and Nanotechnology*, H. S. Nalwa (Ed.), Concise Edition, 2002, Academic Press, Chap. 2, pp. 57–91.
20. Frenkel, J., *J. Phys.* (USSR) 8, 386, 1945.

21. Herring, C., *J. Appl. Phys.*, 21, 301, 1950.
22. Rhodes, R. H., *J. Am. Ceram. Soc.*, 64, 19, 1981.
23. Skandan, G., Hahn, H., and Parker, J. C., *Scr. Metall.*, 25, 2389, 1991.
24. Gonzalez, E. J. and Piermarini, G. J., Low-Temperature Compaction on Nano-size Powders, in *Handbook of Nanostructured Materials and Nanotechnology*, H. S. Nalwa (Ed.), Concise Edition, 2002, Academic Press, Chap. 3, pp. 93–127.
25. Sidiki, T. P. and Sotomayor Torres, C. M., Silicon-Based Nanostructures, in *Handbook of Nanostructured Materials and Nanotechnology*, H. S. Nalwa (Ed.), Concise Edition, 2002, Academic Press, Chap 10, pp. 387–443.
26. *Machine Design*, May 22, 2003, p. 41.
27. Likharev, K. K., *Electronics below 10 nm: Nano and Giga Challenges in Micro-electronics*, Amsterdam: Elsevier; 2003.
28. Likharev, K. K. and Mayr, A. et al., CrossNets; High-Performance Neuro-morphic Architectures for CMOL Circuits, Sixth Molecular-Scale Electronics Conf., Key West, FL, December 2002, New York Academy of Sciences.
29. Siegel, R. W. and Eastman, J., Creating Materials with Nanophase Technol-ogy, *Ceramic Industry*, January 1994, pp. 31–33.
30. McCandlish, L. E., Chemical Processing of Nanostructured Materials, *Mat. Tech.*, 8(9/10), 193–197, 1993.
31. Xiao, T. D., Gonsalves, K. E., Strutt, P. R., and Klemens, P. G., *J. Mat. Sci.*, 28, 1334, 1993.
32. Walker, C., Borsa, C., and Todd, R., Nanocomposites, *Ceramic Technology International*, London: Sterling Publications Ltd., 1995, pp. 46–49.
33. Shinbara, H., Development of Nano Composite Ceramic Structural Materi-als, Univ. of Osaka, July 1992, pp. 1–10. Selected papers from *Centennial Issue of Ceramic Society of Japan — Science and Technology*, May 1992, p. 98.
34. Nawa, M., Sekino, T., and Niihara, K., Fabrication and Mechanical-Behavior of Al_2O_3/Mo Nanocomposites, *J. Mater. Sci.*, 29(12), 3185–3192, 1994.
35. Rousset, A., Alumina-Metal (Fe, Cr, Fe $_{0.8}$ Cr$_{0.2}$) Nanocomposites, *J. Solid State Chem.*, III(1), 164–171, 1994.
36. Chen, L., Goto, T., and Hirai, T., Preparation of SiC-W_2C Nanocomposite Powders by Chemical Vapour Deposition of the SiH_4-CH_4-WF_6-H_2 System, *J. Mater. Sci.*, 28(20), 5543–5547, 1993.
37. Hillel, R., Maline, M., Gourbilleau, F. et al., Microstructure of Chemically Vapour Codeposited SiC-TiC-C Nanocomposites, *Mater. Sci. Eng.*, A168(2), 183–187, 1993.
38. Zhao, J., Stearns, L. C., Harmer, M. P. et al., Mechanical Behavior of Alumina-Silicon Carbide Nanocomposites, *Am. Ceram. Soc. J.*, 76(2), 503–510, 1993.
39. Niihara, K. and Nakahira, A., Structural Ceramic Nanocomposites, in *Ce-ramics: Toward the 21st Century*, The Ceram. Soc. of Japan, 1991, pp. 404–417.
40. Niihara, K., Nakahira, A., and Sekino, T., New Nanocomposite Structural Ceramics, *MRS Symp. Proc.*, Vol. 286, *Nanophase and Nanocomposite Materials*, S. Komarneni, J. C. Parker, and G. J. Thomas (Eds.), December 1–3, 1992, pp. 405–412.
41. Hirvonen, J.-P., Lappalainen, R., Kattelus, H., et al., Structure, Mechanical Properties, and Oxidation Behavior of Nanolayered $MoSi_2$/SiC Coatings, *MRS Symp. Proc.*, Vol. 286, *Nanophase and Nanocomposite Materials*, S. Komar-neni, J. C. Parker, and G. J. Thomas (Eds.) December 1–3, 1992, pp. 373–378.

42. Roy, R., Nanocomposites: Retrospect and Prospect, *MRS Symp. Proc.*, Vol. 286, *Nanophase and Nanocomposite Materials*, S. Komarneni, J. C. Parker, and G. J. Thomas (Eds.), December 1–3, 1992, pp. 241–250.
43. Hoffman, D., Komarneni, S., and Roy, R., Preparation of a Diphasic Photosensitive Xerogel, *J. Mat. Sci. Lett.*, 3, 439–442, 1984.
44. Preparation of Nanocomposites, NERAC, Inc., Tolland, CT, 96(22), 3, 1996.
45. Nagarajan, R. and Chattopadhyay, K., Intermetallic $Ti_2Ni/TiNi$ Nanocomposite by Rapid Solidification, *Acta Metal Mater.*, 42(3), 947–958, 1994.
46. Takeyuma, M. and Liu, C. T., *J. Mater. Res.*, 5, 1189, 1990.
47. Provenzano, V. and Holtz, R., Production of Nanocomposites with Enhanced Properties for High Temperature Applications, IH, April 1994, pp. 47–53.
48. Carroll, L., Sternitzke, M., Derby, B., SiC Particle Size Effects in Alumina-Based Nanocomposites, *Acta Materialia*, 44(11), 4543–4552, 1996.
49. Levin, I., Kaplan, W. D., Brandon, D. G., and Layyous, A. A., *J. Am. Ceram. Soc.*, 78, 254–256, 1995.
50. Jiao, S., Jenkins, M. L., and Davidge, R. W., Interfacial Fracture Energy-Mechanical Behaviour Relationship in Al_2O_3/SiC and Al_2O_3/TiN Nanocomposites, *Acta Materialia*, 45(1), 149–156, 1997.
51. Vaia, R. A., Polymer Nanocomposites, AFRL Materials and Manufacturing Directorate, Nonmetallic Materials Division, AFRL Technology Horizons, September 2002, pp. 41–42.
52. U.S. Patent 5747560, Melt Process Formation of Polymer Nanocomposite of Exfoliated Layered Material, 1998.
53. U.S. Patent 5385776, Nanocomposites of Gamma Phase Polymers Containing Inorganic Particulate Material, 1995.
54. Giannelis, E., Go Lighter with Solvent-Free Silicate-Polymer Composites, High-Tech Materials Alert, May 10, 1996, Englewood, NJ: Tech Insights, Inc., p. 2.
55. Lichtenhan, J., Promising but Difficult Silsesquioxane Becomes Viable, High-Tech Materials Alert, May 10, 1996, Englewood, NJ: Tech Insights, p. 3.
56. Rice, B. P., Chen, C., and Cloos, L., Carbon Fiber Composites: Organoclay-Aerospace Epoxy Nanocomposites, Part I, *SAMPE J.*, 37(5), 7–9, 2001.
57. Chen, C. and Curliss, D., Resin Matrix Composites: Organoclay-Aerospace Epoxy Nanocomposites, Part II, *SAMPE J.*, 37(5), 11–18, 2001.
58. Alexander, Jr., M. D., Wang, C.-S., and Meltzer, Jr., P., Electrically Conductive Polymer Nanocomposite Materials, AFRL Materials and Manufacturing Directorate, Air Expeditionary Forces Technologies Division, and Anteon Corp., AFRL Technology Horizons, September 2002, pp. 44–45.
59. Goel, A., Versatile, Cost-Effective, Strongly Adherent, Highly Stable, Wear-Resistant Coatings, *Mat. Tech.*, 8(5/6), 85–93, 1993.
60. Nanotechnology — Motors, NASA Tech Briefs Insider, http://link.abpi.net/l.php?20030729A5, accessed July 29, 2003.
61. AFRL Technology Horizons, December 2001, p. 3.
62. Bruno, L., Bright Lights, Red Herring, June 15 and July 1, 2001, pp. 46–58.
63. Moskovits, M., *Annu. Rev. Phys. Chem.*, 42, 465–499, 1991.
64. Wertheim, G. K., *Phase Transitions*, 24–26, 203, 1990.
65. Khanna, S. N. and Jena, P., *Phys. Rev. Lett.*, 69, 1664, 1992.
66. *The Industrial Physicist*, December 2002–January 2003, p. 9.

67. *Science*, 300, 112, 2003.
68. U.S. Patent 6586133 at http://www.uspto.gov.
69. Roco, M. C., *Nanobriefs*, 5–6, 2003.

Bibliography

Brave New Nanoworld, http://www.ornl.gov/ORNLReview/rev32_3/brave.htm.
Freer, R., Nanoceramics, *British Ceramic Proceedings*, No. 51, 200, 1993.
Froes, F. H. and Suryanarayana, C., Nanocrystalline Metals for Structural Applications, *JOM*, 41(6), 12–17, 1989.
Froes, F. H., Suryanarayana, C., Chen, G.-H., Frefer, A., and Hyde, G. R., Nanostructure Processing for Titanium-Based Materials, *JOM*, 44(5), 26–29, 1992.
Lian, J., Baudelet, B., and Nazarov, A. A., Model for the Prediction of the Mechanical Behaviour of Nanocrystalline Materials, *Mater. Sci. Eng.*, A172(1–2), 23–29, 1993.
Lowe, T., The Revolution in Nanometals, *AM&P*, January 2002, pp. 63–65.
Manthiram, A., Bourell, D. L., and Marcus, H. L., Nanophase Materials in Solid Freeform Fabrication, *JOM*, 45(11), 66–70, 1993.
Mat. Tech., 8, 181–192, 1993.
Mayo, M. J., Hague, D. C., and Chen, D.-J., Processing Nanocrystalline Ceramics for Applications in Superplasticity, *Mater. Sci. and Eng.*, A166, 145–159, 1993.
Reihs, K., Nanostructures in Industrial Materials, *Thin Solid Films*, 264(2), 135–140, 1995.
Shull, R. D., Nanometer-Scale Materials and Technology, *JOM*, November 1993, pp. 6061.
Surinach, S., Malagelada, J., and Baro, M. D., Thermodynamic Properties of Nanocrystalline Ni$_3$Al-Based Alloys Prepared by Mechanical Attrition, *Mater. Sci. and Eng.*, A168, 161–164, 1993.
Tolles, W. M., Nanoscience and Nanotechnology, Naval Research Laboratory, NRL-MR-1003-92-6989, May 1992.
Vendange, V. and Colomban, P., Elaboration and Thermal Stability of (Alumina, Aluminosilicate/Iron, Cobalt, Nickel) Magnetic Nanocomposites Prepared through a Sol-Gel Route, *Mater. Sci. Eng.*, A168(2), 199–203, 1993.

3

Carbon-Carbon Composites

Introduction

The element carbon is well recognized as having some anomalous properties, including its ability to embrace the three major divisions of materials science: polymers, ceramics, and metals. It forms strong interatomic bonds with itself, leading to organic chemistry and polymers; the element in either of its allotropic forms, diamond or graphite, is highly refractory and remains solid above the melting point of most ceramics, while the graphite crystal form has electrical and thermal transport properties comparable with typical metals.

Although the tetrahedral crystal structure known as diamond has technological and industrial applications, this book is concerned with carbon in its hexagonal or graphite crystalline form. Apart from diamond, all other forms of carbon such as charcoal, coke, soot, carbon black, carbon fibers, and even the recently discovered fullerenes[1] are distorted versions of the graphite form. The term *carbon as a solid* has been suggested to embrace the numerous different types,[2] but has not gained wide acceptance.

The distinction in nomenclature between carbon and graphite can be confusing, especially to newcomers to the field. Clearly, carbon is the element and all so-called graphite is composed of carbon, but the term *graphite* should be applied strictly only to those carbons with a perfect hexagonal structure. Such perfection is rarely achieved in manufactured graphite, which is a heterogeneous agglomeration of near-perfect crystallites intermingled with less well-ordered areas. Nevertheless, the word *graphite* has come to be accepted in common usage as applying to carbons that approach the perfect crystal structure. Similarly, the term *graphitization* is used to denote high-temperature heat treatment of carbons with the aim of obtaining a more graphite-like structure. The extent to which this can be achieved varies widely depending on the structure of the starting carbon and has led to the classification of carbons into *graphitizing* or *nongraphitizing*.[3]* Such a distinction is oversimplistic, and there is a

---

* The terms *soft* and *hard* carbon are sometimes used with the same intended meaning.[4]

whole spectrum of potential graphitizability among the array of possible carbon precursors.[4-5] The term *heat treatment* is to be preferred to *graphitization* unless it is known that a near-perfect graphitic structure will be achieved.

Graphite and Carbon-Carbon Composites

Carbon in the near-perfect hexagonal crystal form, commonly referred to as polycrystalline graphite, has found a wide diversity of industrial uses, and a bewildering number of different grades are available from various manufacturers worldwide. The uses are based on the unusual combination of attractive properties that graphite displays. Its good electrical conductivity finds uses as electrodes in electrorefining and contact brushes in electric motors, its high thermal conductivity in heat exchangers, its chemical inertness and high-temperature resistance in metallurgical foundry processes, and its inherent lubricity in seals and wire drawing dies.

Despite these manifold uses and attractive properties, graphite as a material suffers from one serious drawback, which is its mechanical weakness. Typically, the tensile strength of a high-grade polycrystalline graphite at room temperature is of the order of 35 MPa, compared with 135 MPa for a brittle metal (cast iron). At 2500°C, the tensile strength of graphite increases to 55 MPa, whereas most metals have melted. Consequently, any attempts to take advantage of its thermal capability in structural situations necessitates massively thick sections to cope with the loads. Moreover, the compressive strength of graphite is typically 90 MPa at room temperature, rising to 180 MPa at 2500°C, so structural designs must be unconventional to take advantage of the higher compressive strength. Alternatively, graphite is used in conjunction with a strong but heavy structural material, such as steel, to give mechanical support. These mechanical disadvantages are especially evident in aerospace uses of graphite where its low density, high heat of ablation, and excellent thermal resistance are otherwise invaluable. Because of the weight limitation imposed by the environment on aerospace components, massive design concepts cannot be used, nor can graphite always be stressed in the compressive mode and resort must be made to other methods to deal with its low tensile strength.

The concept of using fibrous reinforcement to improve the strength of graphite dates from the early 1960s and coincides with the availability of carbon fibers and also with an upsurge in interest in the use of other fiber-reinforced composite materials, particularly glass fiber-reinforced thermosetting resins. Thus, a new family of materials called carbon fiber-reinforced carbon composites, carbon-carbon composites, or, simply, carbon-carbon (C/C) was created. It was unique at the time in having the same material (albeit in different forms) for both fiber and matrix constituents of the composite. Since that time a few other homocomposites such as silicon carbide-silicon carbide and alumina-alumina have emerged but C/C is by far the best known and the most mature example. All these composites address the recognized need to toughen what are fundamentally brittle but refractory materials by fibrous reinforcement.

It cannot be too strongly emphasized that C/C is not a single material but a concept that gives rise to a wide range of materials tailored to suit various applications, just as there are many different grades of graphite for different purposes. The common objective of C/C composites is to capitalize on the many attractive properties of graphite and combine them with the mechanical properties associated with fiber-reinforced composites. Tensile strengths up to 900 MPa have been achieved with C/C,[6] which represents an increase of nearly 30-fold on a high-grade graphite.

C/C composites, so called because they combine carbon-fiber reinforcement in an all-carbon matrix, can best be viewed as part of the broader category of carbon-fiber-based composites, all of which seek to utilize the light weight and exceptional strength and stiffness of carbon fibers. However, in C/C, the structural benefits of carbon-fiber reinforcement are combined with the refractoriness of an all-carbon materials system, making C/C composites the material of choice for severe-environment applications, such as atmospheric reentry, solid rocket motor exhausts, and disk brakes in high-performance military and commercial aircraft. Their dimensional stability, laser hardness, and low outgassing also make them ideal candidates for various space structural applications.

Such mechanical and refractory properties are not met by the various bulk graphites for two reasons: 1) Graphites are very flaw sensitive and, therefore, brittle, and (2) graphites are difficult to fabricate into large sizes and complex shapes. These difficulties are largely overcome by taking advantage of the "two-phase principle of material structure and strength."[7]

In the classical two-phase, or composite, materials system, a high-strength, high-modulus, discontinuous-reinforcement phase is carried in a low-modulus, continuous-matrix phase; e.g., graphite fibers in a thermoplastic-resin matrix. The stress in a composite structure having fiber reinforcement that is continuous in length is carried in proportion to the moduli of the constituent phases, weighted by their respective volume fractions. Therefore, the much stiffer (higher-modulus) fibers will be the principal load bearers, and the matrix, in addition to having the task of binding together the composite, will deform under load and distribute the majority of stress to the fibers. At the same time, because the brittle carbon fibers are isolated, the possibility that an individual fiber failure will lead to propagation and catastrophic failure is practically eliminated.

Another major benefit of composites is that they permit the construction of complex geometries, and in such a way that different amounts of the load-carrying fibers can be oriented in specific directions to accommodate the design loads of the final structure. Closely associated with this "tailoring" feature of composites is that carbon-fiber technology enables exploitation of the exceptional basal-plane stiffness (and strength, in principle, although this is still much farther from realization) of sp^2 bonded carbon atoms; that is, the fibers are not isotropic, but rather have their graphite basal planes oriented preferentially in the fiber axial direction.

For very-high-temperature carbon-fiber-composite applications, say, above 2000°C, even for brief periods of time, it is necessary to employ a carbon matrix; however, like the fiber, the carbon matrix is also brittle. When fiber-matrix bonding is very strong in C/C, brittle fracture is frequently observed. The explanation is that strong bonding permits the development of high crack tip stresses at the fiber-matrix interface; cracks that initiate in either fiber or matrix can then propagate through the composite. However, if the matrix or the fiber-matrix interface is very weak, or microcracked, then the primary advancing crack can be deflected at such weakened interfaces or cracks. This is the Cook-Gordon theory[8] for strengthening of brittle solids, which states, more specifically, that if the ratio of the adhesive strength of the interface to the general cohesive strength of the solid is in the right range, large increases in the strength and toughness of otherwise brittle solids may result. Therefore, good fiber strength utilization in a brittle-matrix composite like C/C depends on control of the matrix and interfacial structures.

Carbon fiber-reinforced carbon composites form a very specialized group of materials. They may be considered as a development of the family of carbon fiber-reinforced polymer composites, which have become ever-more prevalent in modern engineering. Since the early 1960s, a large number of so-called "advanced materials" have appeared on the scene. C/C is arguably the most successful of all these products, finding many and varied applications. In the field of Formula 1 motor racing, for example, the present levels of performance simply could not be achieved without the use of C/C brakes and clutches. Despite the materials' obvious assets, they have not, and will not, reach their full potential until their inherent problems of excessive production costs and oxidation resistance have been addressed properly.

C/C composites were first introduced in the aerospace industry as a replacement for polycrystalline graphite in the nose tip of rockets. This was due to the failure of graphite by intensive heating and eroding during the reentry of rockets. The thermal and mechanical properties of carbonaceous materials have significantly improved due to the addition of reinforcing fibers to the bulk carbon. This has allowed C/C composites to be successfully used in different engineering applications. These include military, aerospace, industrial, commercial, and medical applications.

Carbon fiber-reinforced carbon composites are made of synthetic pure elemental carbon. Carbon is a unique solid that can be made to possess the widest variety of structures and properties. This makes C/C composites intermediate materials. They exhibit properties taken from metals, polymers, and ceramic or glass.[9] C/C composites as engineering materials can be tailored to have properties designed to fit the user's needs.

In general, a C/C composite material consists of a carbonaceous matrix reinforced with carbon fibers in the form of continuous filament yarn, cloth, macerated or chopped fibers, or three-dimensional woven reinforcements.[10] Many different architectures have been used for the reinforcement of C/C composites. These were outlined[11] as random fibers, unidirectional fibers, braided yarns, stacked two-dimensional (2-D) fabrics, pierced fabrics to provide increased interlaminar shear properties, orthogonal three-dimensional (3-D) geometries in either Cartesian or cylindrical coordinates, or multi-directional weaves designed to improve the off axis properties and to maximize the empty spaces that occur at filament cross-over locations. The combination of matrix resins with carbon fibers as well as the different architectures produce extraordinary properties that characterize this class of materials. This includes superior stiffness, better fatigue strength, high-heat resistance, low shrinkage, low thermal expansion coefficient, high heat storage capacity, and good chemical resistance.

The coarse orthotropic texture of the relatively large yarn bundles coupled with the extreme anisotropy of the graphitic matrix play a major role in determining the microstructural character and behavior of the composite. An important consequence of such anisotropy is the microcracking network induced by thermal stresses during manufacturing.[11] Figure 3.1 illustrates a minimechanical feature of 3-D C/C composites showing these cracks.[12] These microstructural complexities violate the fundamental assumptions of the continuum upon which fracture mechanics is founded. The few reported studies that attempted to characterize the fracture resistance of C/C composites attest to the inappropriateness of conventional fracture mechanics methods.

In spite of the many favorable properties that C/C composites possess, they remain a very difficult class of material to conduct fracture testing on.[13,14] The fracture behavior of C/C composites, manufactured from carbon matrix with carbon fiber reinforcement, is complex. Severe heterogeneities in the strength fields of many composites cause a wide scatter in their long-term strength and fracture toughness. Accordingly, an unconventional approach for fracture toughness and lifetime evaluation in such composites ought to be found.

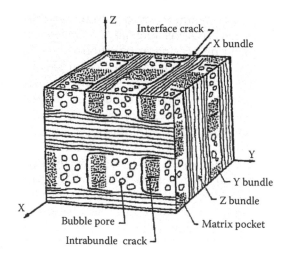

FIGURE 3.1
Schematic of a 3-D composite showing some mini-mechanical features.[12]

Carbon Fiber and Matrix Materials

It is axiomatic that C/C composites cannot exist without carbon fibers. They are arguably the most important constituents as they play a dominant role in many of the properties of C/Cs.

C/C composites are strongly influenced by the materials of both their matrices and fibers. Fabrication of C/C composites requires a better understanding of the interrelationship between the fibers and the matrix that binds them. The selection of the appropriate materials to produce these composites affects their processing techniques and also their final structure.

Carbon Fibers

Although carbon fibers are nowadays synonymous with high-strength materials, the earliest fibers were valued for their electrical properties. The popular attribution of the discovery of carbon filaments to Thomas Edison in 1880[15] appears to be predated by Joseph Swan, who exhibited a carbon filament incandescent lamp in 1878.[16]

Various improvements were obtained over the next 20 to 30 years, but by 1910 carbon filaments in lamps were superseded by tungsten. Carbon fibers received little attention until the mid-1950s, when rocket technology brought a need for reinforcing fibers lighter and more refractory than glass or textiles.

The properties of a fiber to be suitable for the formation of carbon fiber have been summarized as[17]

1. Sufficient integrity to hold the fiber together throughout all processing stages
2. Nonmelting
3. Capable of giving a high yield of carbon product
4. Capable of giving rise to good crystal alignment
5. Available at a reasonable cost

Some of these factors are almost mutually exclusive (e.g., it is necessary to have some fluidity during carbonization to permit good crystal alignment, but this contradicts the requirement for a nonfusible fiber). No single raw material meets all the requirements, and modifications or treatments have to be carried out in order to achieve them.

Carbon fibers are the backbone of any C/C composite. The latest advancements in C/C composites are largely attributed to the improvement in carbon fiber properties such as stiffness, modulus of elasticity, fatigue strength, and lower thermal expansion coefficient. High-strength, high-modulus carbon fibers are about 7 to 8 μm in diameter and consist of small crystallites of graphite, an allotropic form of carbon.[18] In graphite, single-crystal carbon atoms are arranged in hexagonal arrays. Very strong covalent bonds are effective in the layer planes, while weak Van der Waals forces exist between the layers. High-modulus and high-strength fibers are achieved when these later planes are aligned parallel to the fiber's axis. The arrangement of the layer planes in the direction perpendicular to the axis of fibers influences their transverse and shear properties.

Carbon and graphite fibers are produced either from fibrous organic precursors or pitch. Fibrous organic precursors include rayon, polyacrylonitrile (PAN), and acrylic. PAN-type carbon fibers have a thin skin of circumferential layer planes and a core with random crystallites. Mesophase pitch-based fibers have radially oriented layer structures. Three basic fiber morphologies are shown in Figure 3.2.[19] These different structures greatly influence the properties of the fibers.

The mechanical properties of the various types of carbon fibers are summarized in Figure 3.3.[20]

Two major carbon fiber types emerged from PAN precursors in the early stages of the development. They arose from recognition of the fact that the modulus of the fiber increased progressively up to the highest heat treatment temperatures (HTT) of 2700°C (high modulus [HM] or Type I fibers), whereas the strength reached a maximum at temperatures in the

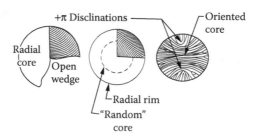

FIGURE 3.2
Three types of high-modulus fibers spun from mesophase pitch.

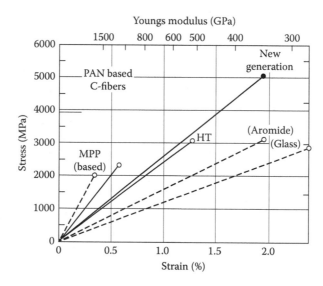

FIGURE 3.3
Mechanical properties of various types of carbon fibers.

range 1400–1500°C (HT or Type II fibers), as illustrated in Figure 3.4. The increase in modulus is accounted for by improvement in the crystalline perfection and alignment, although, with a density of 1.9 g/cm³, HM fiber is not highly graphitic. Maximization of the tensile strength at lower temperatures is explained by the fact that the strength is limited by flaws. As well as flaws arising from adventitious inclusions and voids, thermal contraction from the processing temperature leads to the formation of longitudinal cracks, all of which limit the strength.

Both HM and HT fibers have found uses in C/C composites[21] but, for the reinforcement of thermosetting resins, their low elongation of less than 1% at the point of tensile failure was less acceptable. A third type of PAN-based carbon fiber (A or Type III fiber) with a lower HTT than HT but

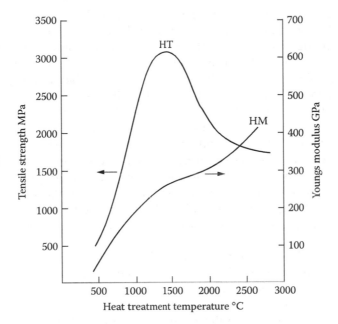

FIGURE 3.4
Strength and modulus of PAN-based carbon fibers.

with a higher failure strain was later developed to meet this requirement, but it does not find wide application in C/C. During the last 15 years, a number of other PAN-based carbon fibers have been developed in attempts to optimize strength, modulus, elongation, and cost. They are termed intermediate modulus (IM) fibers and have arisen from research into fiber structure/property relationships along with improvements to production processes. More recently, there has been interest in the preparation of thinner, PAN-based fibers with the aim of improving mechanical properties. Hitherto, almost all PAN precursors have been based on a textile fiber with a diameter of 13 μm, which, after carbonization shrinkage, yields carbon fiber with a diameter of 7–8 μm. The new fibers, with diameters of 4–5 μm, seek to make use of the fact that strength-limiting flaws are volume-related phenomena; hence, a narrower fiber should be expected to contain fewer flaws and show better strength and strain properties. There is thus a variety of carbon fibers based on PAN from which to make a choice for C/C preparation.

Apart from rayon and PAN, no other textile fiber has found acceptance as a source of high-performance carbon fiber. There is, however, a third carbon fiber precursor that has gained importance because its potential has provided high-quality carbon fibers at lower cost than those from PAN or rayon. Pitch is found as a residue from the carbonization of coal and the refining of petroleum. There is an abundance of it available at low cost. It is well known in the carbon industry as the principal binder material in the manufacture of artificial graphite, and many different grades are produced. It can be spun into fibers from the melt but, like PAN, is fusible and needs to be oxidatively treated before being suitable for carbonization in fiber form. Chemically, pitch is a complex mixture of low molecular weight aliphatics, condensed aromatic ring molecules, heterocyclic structures, and inorganic compounds: Condensed, planar aromatic systems with four and five rings have the latent graphite layer configuration that, coupled with the fluidity of pitch, give a good expectation of a graphitic structure in its carbonized and graphitized form. Because of this favorable, inherent structure, it is not essential to stretch pitch fibers during carbonization in order to obtain molecular alignment. This is an attractive advantage over PAN and offers further scope for reduced cost.

Unlike PAN-based fibers, the strength of pitch-based fibers does not pass through a maximum during heat treatment, but the strength and the modulus continually increase as the HTT is raised, as shown in Figure 3.5. Currently, pitch-based carbon fibers are available with a tensile strength approaching 2000 MPa (similar to the PAN-based HT fibers), a modulus

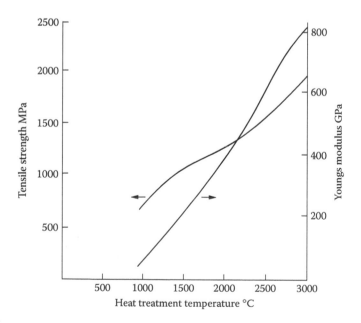

FIGURE 3.5
Strength and modulus of pitch-based carbon fibers.[22]

exceeding 800 GPa (twice that of HM fiber), and a density of 2.15 g cm^{-3}, approaching the theoretical value of 2.26 gcm^{-3} for the perfect graphite crystal. Such highly crystalline fibers suffer from low elongation but show improvements in electrical and thermal conductivity and better oxidation resistance than PAN-based fibers, which can be of considerable benefit in certain applications.

Matrix Materials

The selection of the matrix materials is of great importance for the tailored design properties of C/C composites. The microstructure of the matrix can influence the performance of the fibers and consequently influence the properties of the composites.

In a broad sense, any organic material that can be pyrolyzed in an inert atmosphere to produce a tenacious carbonaceous residue, or char, could serve as a matrix material (see Table 3.1). However, the char yield and properties of the chars obtained from different organic polymers vary widely, depending on the nature of the precursor. In practice, three classes of matrix precursor materials are used for C/C composites: thermosetting resins, petroleum and coal tar pitches, and high-carbon-content hydrocarbon gases for chemical vapor infiltration (CVI).

TABLE 3.1

List of Ingredients for C/C Composite Materials

Carbon Fibers	Matrix Types	Infiltrated and Densified Materials
PAN fibers	Thermosetting resins:	Thermosetting and
High-elasticity type	Phenol resin	thermoplasticizing
High-tensile-strength type	Furan resin	resins:
Pitch fibers	Epoxy resin	Phenol resin
High elasticity type	Polyimide resin	Furan resin
(petroleum, carbon type)	COPNA resin	Pitch
Rayon fibers	Thermoplasticizing resins:	CVD ingredients:
	Pitch:	Methane
	Isotropic	Propane
	Anisotropic	Benzene
	Pitch + phenol resins	Dichloroethylene
	Aromatic polymers	
	Reinforcement materials:	
	Cokes	
	Carbon black	
	Natural graphite	
	Meso carbon	

There are two related, but somewhat different, functions for the matrix material in C/C composites. One is to stiffen, or rigidize, the fabric reinforcement prior to densification processing, and the second is to densify the then-rigidized structure. Although each of the three basic matrix material types can be used for both purposes, initial rigidization is commonly accomplished with thermosetting resins.

Two types will be considered next.[23]

Thermosetting Resins

Thermosetting resins are suitable materials for the matrix of C/C composites due to their ease of impregnation. Most thermosetting resins polymerize at low temperatures (<250°C) to form a highly cross-linked thermosetting, amorphous type. Further heating up to temperatures approaching 3000°C causes these forms to graphitize. Resin systems that have these properties include phenolics, furyls, furan, and some selected epoxies.[24–26]

Phenolic resins are widely used in the production of C/C composites due to their high-heat resistance, dimensional stability and tolerances, and creep resistance. These resins are classified as condensation polymerization polymers and are derived from the reaction of phenols with aldehydes. Phenolics are chosen for processing C/C composites due to their high char strength. This allows them to be used for specific applications that would degrade most organic resins.[24,27] Furyl esters also have a high char yield and are often used to reimpregnate the composites after the phenolic resin has been pyrolized. Furan is another thermosetting resin that forms a dense, homogeneous crystalline structure when pyrolized.[27]

Tar Pitch Resins

Petroleum and coal tar pitches are used as a matrix precursor for C/C composites. Coal tar pitches, a byproduct of the coke oven, have a low softening point, low viscosity (in the melt phase), and high graphitic yield. The graphitic yield of pitch can be greatly increased by pyrolysis under high gas pressure.[28]

Pitch goes through several stages during carbonization. These stages are volatilization, polymerization, cleavage, and rearrangement of the molecular structure. In carbonization, spheres are formed within the resin that exhibit a highly oriented structure similar to liquid crystals. The formation of these spheres has been designated as the mesophase and is important in the processing of C/C composites. Pressure cycling and oxidation processes are used to control the mesophase behavior during processing. Further heating solidifies these spheres, and at temperatures approaching 2500°C, the graphite structure is reached.[25]

Carbon Fiber Structure

The external appearance of carbon fibers from the three principal precursors shows some characteristic features, which are readily observed by optical microscopy and which are useful in identifying fiber origins. Rayon-based carbon fibers have an irregular, crinkled cross-section, as shown in Figure 3.6a. PAN-based fibers come in two distinct cross-sectional shapes, according to the method used for the initial PAN preparation. Those that have been wet-spun from solution into a coagulating bath have a circular cross-section that is retained in the carbon form, to the extent that minute indentations from the spinning jets can usually be seen. PAN fibers spun from solution into warm air (dry-spun fibers) have a characteristic dumbbell or dogbone cross-section. Carbon fibers from dry-spun PAN retain this shape (as in Figure 3.6b), which is sometimes distorted into a kidney shape, as in Figure 3.6c. Pitch-based carbon fibers are basically circular in cross-section, although early fibers were characterized by longitudinal flaws, which led to a "missing segment" appearance, as in Figure 3.6d.

Pitch is not a high molecular weight polymer; consequently, the long, ribbon-like structure has not been suggested for carbon fibers from pitch. Later developments, particularly the spinning of mesophase, led to carbon fibers with a mixed structure of radial, circumferential, and randomly oriented crystallites retaining the original circular cross-section. The high aromaticity of pitch leads to carbon fibers with a more highly graphitic structure and larger crystallite dimensions than those from rayon or PAN.

It will be evident from the foregoing that there are a substantial number of different carbon fibers with properties dependent upon their origin and conditions of manufacture. In addition to the major classifications based on precursor and heat treatment, manufacturers have developed a wide range of specialized fibers for various end applications, so much so that

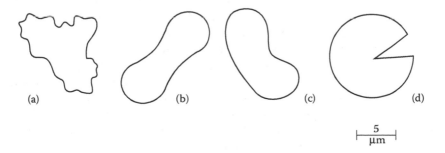

5 μm

FIGURE 3.6
Cross-sectional shapes of some carbon fibers. Key: (a) Rayon-based; (b) Dogbone PAN-based; (c) Distorted dogbone PAN-based; (d) Early pitch-based.

there is a publication devoted exclusively to the description of the available types.[29] A list of the fibers most commonly used in C/C with some of their relevant properties is given in Table 3.2.

Rayon-Based Carbon Fibers

The majority of fiber types are manufactured from polyacrylonitrile precursors, but a growing number are available from pitch. Pitches are relatively low-cost byproducts, and fibers derived from them have a high potential for technological improvement. Thus, their share of the industry will probably increase. Rayon-based carbon fibers were the first to be produced for structural applications, and a few are still available. These are of the low-strength, low-modulus variety and remain of interest owing to their relatively low thermal conductivity and good flexibility. They played an important part in the 1960s and 1970s in U.S. defense and space programs as the first C/C composites to be developed and incorporated in rayon-based carbon fibers. Many of these are still in service. With respect to their use in C/C composites, a summary of the essential properties of available fibers is given in Table 3.3.

Figure 3.7 shows the basic elements required for producing carbon filaments from rayon.[30] The first low-temperature treatment takes place typically at around 300°C and converts the structure to a form that is stable to higher processing temperatures. The process involves polymerization and the formation of cross-links. The rayon may be subjected to a chemical treatment before the first-stage oxidation exposure. The chemical bath can be an aqueous ammonium chloride solution or a dilute solution of phosphoric acid in denatured ethanol. The chemical treatment serves to reduce the time for the low-temperature step from several hours to around 5 min. Of the fiber mass, 50–60% is lost to decomposition products such as H_2O, CO, and CO_2 during oxidation. The carbonization step, resulting in further weight loss, is usually carried out at ~1500°C. The yield after carbonization is typically 20–25% of the original polymer weight. At this stage the fibers have an essentially isotropic structure. The mechanical properties of carbonized rayon are poor as a direct result of the poor alignment of the graphene layers. Stretching of the fibers during heat treatment to graphitization temperatures significantly increases both strength and modulus, but is an expensive process. The combination of poor mechanical properties, low carbon yield, and expense of graphitization has meant that ex-rayon carbon fibers have generally not proved competitive in the marketplace, although they are used extensively in ablative technology. This is due to their poor through-thickness thermal conductivity and because their composites yield high inter laminar shear

TABLE 3.2

Properties of Some Carbon Fibers Used in C/C Composites

| Manufacturer | Hercules (USA) PAN | | | Toray (Japan) PAN | | | | BASF (FRG) PAN | | Amoco (USA) Pitch | | | Graphite Single Crystal |
| Precursor | | | | | | | | | | | | | |
Grade	AS4	IM6	HMS4	T300	T800	M40	M50	G40	GY70	P55	P75	P100	
Diameter, μm	8	5	8	7	7	7	7	7	8	10	10	10	
Density, g cm^{-3}	1.79	1.76	1.80	1.76	1.81	1.81	1.91	1.77	1.96	2.0	2.0	2.15	2.26
Tensile strength, GPa[a]	3.7	5.1	2.3	3.5	5.6	2.7	2.4	5.0	1.9	1.9	2.1	2.2	100[b]
Young's modulus, GPa[a]	230	280	360	230	300	400	500	300	520	380	520	700	1000[b]
Elongation, %[a]	1.6	1.7	0.8	1.5	1.9	0.6	0.5	1.6	0.3	0.5	0.4	0.3	10[b]
Electrical resistivity, μΩ m[a]	—	—	—	18	14	11	9.5	13	6.5	8.5	7	2.5	0.4[b]
Linear coefficient of thermal expansion, ×10^{-6} K^{-1}[a]	—	—	—	-0.6	-0.75	-0.75	-0.7	—	—	-1.3	-1.4	-1.45	-1.8[b]
Thermal conductivity, Wm^{-1} K^{-1}[a]	—	—	—	10	15	45	100	—	175	100	150	300	2000

[a] Properties in the axial fiber direction.
[b] Properties in the ab plane.

TABLE 3.3

A Compilation of Carbon Fiber Property Data

Precursor		Axial Tensile Strength (GPa)	Axial Tensile Modulus (GPa)	Axial Comp. Strength (GPa)	Density (g cm⁻³)	Axial Thermal Conductivity (W m⁻¹ K⁻¹)	Axial Electrical Resistivity (μΩ m)
Rayon		0.25–0.7	25–40		1.5	3.5–4.0	35–60
Isotropic pitch		0.8–1.0	40		1.6	15–40	
Mesophase pitch		1.4–3.9	160–965	0.45–1.15	1.9–2.2	120–1100	2–13
PAN	HT	3.0–5.0	210–250	2.7–2.9	1.7	10–25	18
7 μm	HM	2.3–3.5	360–490	1.6	1.9	70	10
	UHM	1.9	570	1.0	1.96		7
PAN	HT	5.1–5.8	280–310	2.75	1.8	15	14
5 μm	HM	3.9–4.5	435–590		1.85		

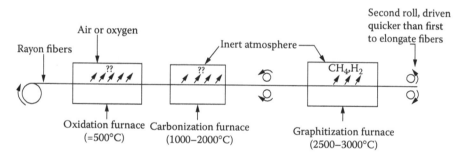

FIGURE 3.7
Basic elements required to produce carbon fibers from rayon.[31]

TABLE 3.4

Typical Properties of Rayon-Based Carbon Fibers

Axial	Tensile strength	1.0 GPa
	Tensile modulus	41.0 GPa
	Elongation to break	2.5%
	Electrical resistivity	20 Ωm
Bulk	Density	1.6 g cm⁻³
	Fiber diameter	8.5 μm
	Carbon assay	99%

strengths. Table 3.4 lists typical rayon-based carbon fiber properties. A full review of the conversion of rayon to carbon fibers is given in the book by Gill.[31]

Pan-Based Carbon Fibers

In the mid-1960s it was discovered that the PAN structure could be stabilized by an oxidation process that assisted controlled thermal decomposition during a second carbonization stage, enabling the production of carbon fibers with superior mechanical properties to those made from rayon. The late 1960s and early 1970s saw a rapid transformation from batch methods to continuous processing. The tow size (filament count) varies considerably between manufacturers, especially in the case where PAN is the chosen precursor. Tow sizes are becoming increasingly standardized at 1, 3, 6, 12, and 15k for high-tech aerospace and sporting goods markets, 6 and 12k being by far the most popular. Larger tows (24k and upward to 320k) are produced primarily for automotive applications and chopped fiber molding compounds.

The fibers derived from PAN precursors can generally be subdivided into three categories:

1. Low modulus (LM) (190–120 GPa) commercial-quality fibers.
2. Intermediate modulus (IM) (220–250 GPa) fibers. These fibers are of high quality, possess the highest tensile strength and strain to failure (1.2–1.4% increasing to 1.5–1.8% for specialized high-strain grades), and are the favored grade in aircraft and racing car manufacturing.
3. High modulus (HM) (360–400 GPa) fibers. These fibers offer improved stiffness at the expense of strength and strain to failure (0.5–0.8%).

The reason for the variety of properties obtained from PAN-based carbon fibers is to be found in the processing parameters used in the conversion of polymer to carbon, and this variety leads to the ability of tailoring those properties to the desired requirements. The PAN precursors to carbon fibers are selected on the basis of a fairly high degree of C/C orientation within the polymer chains. The conversion to carbon involves an initial oxidation stage, carbonization, and a high-temperature graphitization treatment. During each of those processes the C/C orientation is maintained along the fiber axis while competing degradative chemical reactions proceed, involving the loss of all noncarbon heteroatoms. The orientation is significantly improved during the higher temperature treatment as a result of crystallite growth and crystallite axial alignment and may be further enhanced by applying tension or stretching. The Young's modulus of the fibers is directly related to the preferred orientation of the graphene layers, hence control of that orientation will allow a tailoring of the modulus of the fiber.

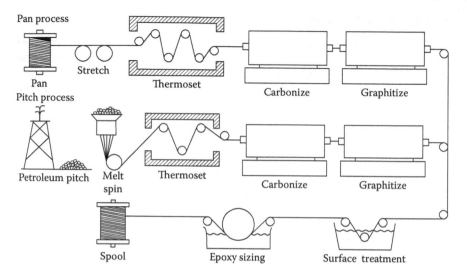

FIGURE 3.8

The processing sequence for PAN and mesophase pitch-based precursor carbon fibers shows the similarity of the two processes. The PAN process obtains highly oriented carbon chains by hot stretching of the polymer chains prior to carbonization, while the high degree of orientation in pitch is a natural consequence of the mesophase (liquid crystal).[33,34]

Polyacrylonitrile can be spun into well-oriented polymer fibers, and the chemistry of the process of conversion of PAN to carbon fibers is extremely complex. Figure 3.8 shows the major steps in the continuous production process.[33,34]

Pitch-Derived Carbon Fibers

Pitches are isotropic mixtures of polyaromatic molecules obtained as by-products of coal tar and petroleum processing. They are relatively easy to melt or spin into fibers. Unfortunately, the fibers so formed are of low modulus and strength as a direct result of their isotropic structure. Their mechanical properties are not improved even when they are carbonized to high temperatures, unless there is hot stretching at very high temperatures of between 2700 and 3000°C — a very costly and impractical process. Although pitches are isotropic in character, a number of processes have been developed to convert the pitch into a mesophase, or liquid cry system.[25,35,36]

Commercial pitches are complex mixtures of aromatic compounds that, when heated to temperatures of around 400°C, undergo dehydrogenated condensation reactions to form planar aromatic molecules that aggregate into a liquid crystalline phase known as the mesophase. When the mesophase content reaches approximately 40%, a phase inversion occurs so that this highly anisotropic material becomes the continuous phase. The discovery that mesophase pitch could be used as a precursor for carbon fibers led to an expectation that the process would create a low-high-performance fiber.

The carbon yield of mesophase pitch is much higher than that from either rayon or PAN, so the pitch process ought to be more efficient.[37] Despite the obvious cost advantages, high-performance pitch-based carbon fibers are unable to compete with PAN-derived fibers except in the very specialized ultrahigh-modulus ($E > 400$ GPa) market. Hughes[38] suggests that the unexpectedly high cost results from the extensive purification necessary to remove flaw-inducing particles from the precursor. The most significant cost in the production of pitch-based fibers appears to arise, however, from the difficulties incurred during processing. The problem arises because the melt spinning of mesophase pitch is far removed from "conventional" melt spinning processes.[39]

The term *melt spinning* is, in fact, something of a misnomer due in the main to the terminology of the textile industry, whence carbon fiber production has evolved. It would be far more accurate to describe the process as melt extrusion. A typical melt spinning process and the process variables that require to be controlled involve the precursor, which is generally melted in an extruder that pumps the melt into a *spin pack*. The spin pack contains a filter to remove solid particles from the melt. After filtration the melt leaves the bottom of the spin pack via a spinnerette. The spinnerette is a plate containing a large number of individual capillaries. Melt leaving the capillaries cools and forms a filament. The solidified fiber, whose cooling is often aided by an *air-quench*, is finally wound on a spinning spool.

If a precursor pitch is to be melt-spinnable it must be capable of forming an unbroken filament between the capillary exit of the spinnerette and the winding-up device. Should the melt temperature be too high, the extruding jet of material will disintegrate into droplets, destroying the filament. On the other hand, if the tensile stress within the filament exceeds the tensile strength of the material at any point along the threadline, the fiber will break as a result of cohesive fracture. For any given material there will be a processing window of spinning conditions that permit the formation of an unbroken filament. The extreme temperature dependence of the viscosity of mesophase pitch makes melt spinning a very difficult process to control. The processing window for the manufacture of pitch-based carbon fibers is extremely small.[39] The melt spinning of mesophase pitch thus requires very accurate and expensive process control.

After extrusion, the fibers are thermoset by oxidation in dry air by being slowly heated to temperatures in the range 200–300°C, followed by rapid cooling in an argon (inert) atmosphere. The fibers are subsequently carbonized and graphitized to between 2500 and 2700°C. Typical properties of mesophase pitch-based fibers are given in Table 3.5.

TABLE 3.5

Properties of Mesophase Pitch-Based Carbon Fibers

Property	Low Modulus	High Modulus	Ultra-High Modulus
Axial			
Tensile strength (GN m^{-2})	1.4	1.7	2.2
Tensile modulus (GN m^{-2})	160	380	725
Elongation to break (%)	0.9	0.4	0.3
Thermal conductivity (W m^{-1} K^{-1})	—	100	520
Electrical resistivity (Ωm)	13	7.5	2.5
CTE at 21°C (10^{-6} K^{-1})	—	−0.9	−1.6
Transverse			
Tensile modulus (GN m^{-2})	—	21	—
CTE at 50°C (10^{-6} K^{-1})		7.8	
Bulk			
Density (g cm^{-3})	1.9	2.0	2.15
Fibre diameter (μm)	11	10	10
Carbon assay (%)	>97	>99	>99

Fiber Structure and Architecture

The modulus of a carbon fiber is determined by the degree of preferred orientation of the graphene layers along the fiber axis. The tensile strength of the fiber is governed by both axial and radial textures along with any flaws present in the structure. Wetting of the fibers by the composite matrix and the strength of the interfacial bond are strongly influenced by the orientation of the graphene layers at the fiber surface.

The composites designer, in addition to being able to choose from a wide variety of fiber types, also has a large number of fiber architectures available. For high-performance C/C applications, continuous (in length) fiber reinforcement is integrated to produce either a 2-D or 3-D fabric preform. According to Ko,[40] a fabric preform is defined as "an integrated fibrous structure produced by fiber entanglement or yarn interlacing, interlooping, intertwining, or [nonwoven] multiaxial placement."

The preform may be dry (i.e., unimpregnated), as in 3-D orthogonal block structures, in which the x, y, and z yarns are laid in straight to produce a structure having about 60% void volume (see Figure 3.9a). The yarns may also be pultruded (i.e., impregnated with a resin binder and formed into rigid rods).

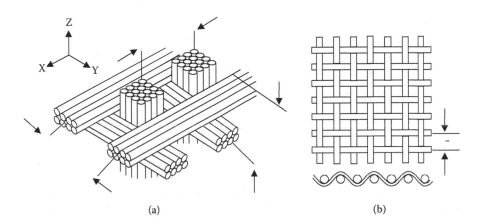

FIGURE 3.9
Schematic of (a) 3-D block construction and (b) 2-D plain-weave fabric (McAllister and Lachman).[11]

Alternatively, the fabrics may be impregnated with a thermosetting-resin binder and then the fabric plies laid up to produce the desired component (Figure 3.9b). Such a structure is still termed 2-D because of the lack of through-thickness reinforcement.

To produce a C/C, the carbon-resin composite is baked, or fired, to pyrolyze the organic matrix. If the fabric is initially impregnated with a state-of-the-art phenol formaldehyde resin system, we can expect to obtain a C/C part with approximately 25% residual porosity after baking. However, experience has shown that such porosity is excessive and that significant improvement in properties will follow if the porosity is reduced to values in the 5–15% range, depending on the particular type of structure. Therefore, not only in the dry preform, but also in the pyrolyzed "prepreg" fabric, additional volume increments of carbon matrix must be introduced into the C/C structure. The introduction can be achieved by one or a combination of three densification processes: CVI, use of coal tar and petroleum pitches, and use of thermosetting resins.

Weaving

Carbon fibers, with diameters in the range of 5–10 μm, are rarely, if ever, handled singly. Manufacturers normally supply fibers in aligned bundles called tows, which typically contain 1,000, 3,000, 6,000, or 12,000 individual

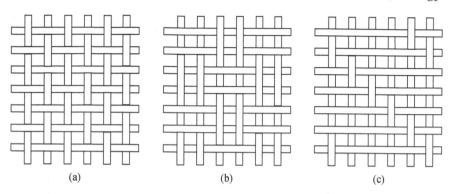

FIGURE 3.10
Weave patterns for fabrics. Key: (a) Plain weave; (b) 2X2 twill; (c) 5 harness satin.

fibers, described as 1K, 3K, 6K, or 12K tows. Untwisted tows can be spread to give a unidirectional tape, which is a useful form for certain composite fabrication processes such as tape winding, but in the interest of maintaining the coherence of the tow, it is more usual for it to be twisted. The amount of twist can vary from less than one to several hundred turns per meter of length. Tows can be used directly to make composites or may be woven into conventional fabrics in a number of different weave patterns such as those shown in Figure 3.10, which are common in the textile industry. Fabrics offer a valuable method for providing bidirectional fiber reinforcement in composites, including C/C, but because of the brittleness of carbon fibers compared to textiles, the method of weaving has to be modified. Instead of the usual shuttle carrying the weft thread between the warps, a system of rapiers is used to give single passes of the weft tow, sometimes accompanied by water lubrication. In these fabrics the mechanical properties of the fibers are slightly degraded because of the convolutions involved in passing tows of fiber over and under each other. Satin weave has an advantage over plain or twill weaves because the length of nonconvoluted fiber is greater and the straight domains allow a closer approximation to the full mechanical properties of the fiber.

To aid the weaving process and reduce friction and fiber damage, it is usual for tows to be coated with a small quantity (typically 0.5 to 2 w/w) of a sizing material, typically uncured epoxy resin applied from an organic solvent solution. The presence of the epoxy size is not detrimental in resin matrix composites, but it can be troublesome in C/C and is usually removed by solvent washing or by heat. Sizing must not be confused with surface treatment, which is another process applied to carbon fibers. This is an oxidative treatment of the fiber surface carried out by wet chemical, electrolytic, or gas phase reactions. Its purpose is to provide, on the chemically inert carbon surface, active polar sites that improve bonding of the fiber to resinous matrices. This step is very important in carbon resin/resin composites, because their performance depends upon efficient stress transfer between matrix and fiber, which is assisted by strong bonding. The mechanics of C/C do not depend upon the same mechanism, and strong fiber-matrix bonding can be positively detrimental in certain stress modes. It is, therefore, necessary to consider the end application carefully before deciding whether surface-treated fibers should be used in C/C. It has been shown that surface-treated fibers can be effectively detreated by heating in an inert atmosphere at temperatures in the region of 1300–1600°C.[11,22]

As well as formal two-dimensional weaving, carbon fibers can be converted into nonwoven fabrics in which the fibers may be aligned, to give a unidirectional sheet, or randomly distributed as in paper or felt. In these constructions, the straightness of the fibers gives the potential to develop properties close to the theoretical values for true unidirectional fiber. Certain other textile processes, such as knitting, where the fibers are very highly convoluted, cannot be applied directly to carbon fibers because of their brittleness, but PAN-derived fibers can be knitted in the Oriented Preform Fiber (OPF) form, where the fiber has textile behavior, then converted by heat to carbon. OPF can also be converted into staple yarns by stretch-breaking and spinning, then woven or knitted into fabrics and heat-treated. The mechanical properties of carbon fabrics made by this method fall short of those of continuous, aligned fibers, but much finer fabrics can be woven from these yarns than those made by weaving carbon fiber tows.

Thus, as well as many different grades of carbon fiber, there are also many different forms in which they may be obtained. This situation poses a problem of choice of the most appropriate type and form of fiber for each particular composite and its end use.

Multidirectional Preforms

The major advantage of multidirectional C/C composites is the freedom to orient selected fiber types and amounts to accommodate the design loads of the final structural component. The disadvantages of multidirectional fabrication technology are the cost of producing the fiber preform, the size limitation on components as dictated by available equipment size, and the difficulty of matrix impregnation between the three-dimensional fiber arrays.[42]

The simplest type of multidirectional preform is based on a three-directional orthogonal construction and is normally used to weave rectangular, block-type preforms (see Figure 3.11). This type of preform consists of multiple yarn bundles located on Cartesian coordinates. Each of the bundles is kept straight so that the maximum structural capability of the fiber is maintained. The preforms are described by yarn type, number of yarns

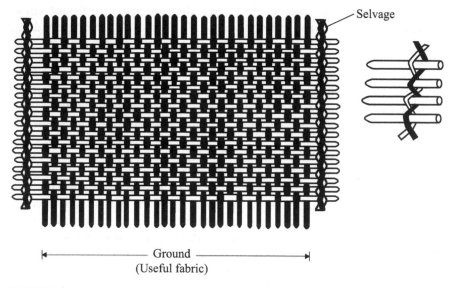

FIGURE 3.11
Full-width plain weave fabric showing selvage.

per site, spacing between adjacent sites, volume fraction of yarn in each direction, and preform density. A number of modifications of the basic three-directional orthogonal construction are available in order to achieve more isotropic preforms. This is accomplished, as one would imagine, by introducing yarns in additional directions. Based on geometric principles, a variety of fiber orientations ranging from orthogonal, 3-directional to 11-directional reinforcements may be produced.

Other types of preforms include the following:

- Polar weave preforms, which are used to form cylinders and other shapes of revolution. They are 3-D constructions with yarns oriented on polar coordinates in the radial, axial, and circumferential directions. Preforms of this geometry normally contain 50 vol.% of fibers that can be introduced equally in the three directions. Although originally developed as thick-walled cylinders, polar weave preforms may now be fabricated in a number of "body of revolution" shapes such as cylinders, cylinders/cones, and convergent/divergent sections. A two-step process may be used to form nonaxisymmetric shapes such as leading edges and conical/rectangular transitions.

- Angle (or warp) interlocks are multilayered fabrics in which the warp yarns travel from one surface of the fabric to the other. Up to eight layers of fabric may be held together, creating a thick 2-D fabric.

Should higher in-plane strength be required, additional *stuffer* yarns may be added to create a quasi-3-D fabric.

Both stitched fabric and needled felt may be considered 3-D preforms although, despite having reinforcement in all three dimensions, the amount of fiber in the interply direction is more often than not negligible. The interply properties of both these fabrics are, thus, seldom significantly better than the matrix-dominated properties of 2-D composites.

The majority of multidirectional preforms used in C/C manufacture are represented by the orthogonal or polar constructions or by some modification of these constructions. The techniques used to manufacture the preforms include weaving dry yarns,[43] piercing fabrics,[44] assembling resin-rigidized yarns,[45] and modified filament winding technology.[46] Infiltration or impregnation of the carbon matrix becomes increasingly difficult with increasing thickness of preform and complexity of reinforcement. Chemical vapor deposition (CVD) can be used only on relatively thin sections (up to a few centimeters). Vacuum impregnation with polymeric resins followed by pyrolysis is a possible method, but by far the most efficient is the hot isostatic pressure impregnation carbonization (HIPIC) process using petroleum or coal tar pitch as a matrix precursor.

Braiding

Braiding is a textile process known for its simplicity, in which two or more systems of yarns are intertwined in the bias direction to form an integrated structure. Braided structures differ from woven fabrics in the method of yarn introduction into the fabric and in the manner by which the yarns are interlaced (see Figure 3.12). Braiding is similar in many ways to filament winding. Dry fiber tows or prepreg tapes may be braided over a rotating and removable mandrel in a controlled manner to form a variety of shapes, fiber orientations, and fiber volume fractions. Braiding cannot achieve as high a fiber volume fraction as filament winding, but braids can assume more complex shapes. The interlaced nature of braids provides a higher level of structural integrity, essential for ease of handling, joining, and damage resistance. The low interlaminar properties of composites can be lessened by use of a three-dimensional braiding process.[47] The current trend in braiding technology is to expand to large-diameter braiding, develop more sophisticated techniques for braiding over complex-shaped mandrels, multidirectional braiding, production of near-net-shape preforms (see Figure 3.13), and the extensive use of computer-aided design and manufacturing (CAD/CAM).

A typical braiding machine consists of a track plate, spool carrier, former, and take-up device. In some cases, a reversing ring is used to ensure uniform tension on the braiding yarns. The resulting braid geometry is defined by the braiding angle, θ, which is half the angle of the

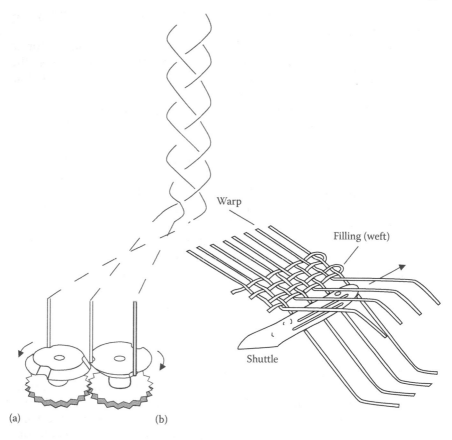

FIGURE 3.12

Fabric techniques: (a) braided; (b) woven.

interlacing between yarn systems, with respect to the braiding (or machine) direction. The tightness of the braided structure is reflected in the frequency of interlacings. The distance between interlacing points is known as the pick spacing. The width or diameter of the braid (flat or tubular) is represented as d. Should longitudinal reinforcement be required, a third system of yarns may be inserted between the braiding yarns to produce a triaxial braid with $0° \pm \theta$ fiber orientation. If there is a need for structures having more than three yarn thicknesses, several layers (plies) of fabric can be braided over each other to produce the required thickness. For a higher level of through-thickness reinforcement, multiple track braiding, or pin braiding, or three-dimensional braiding, can be used to fabricate structures in an integrated manner. The various criteria and braiding classifications are shown in Table 3.6. A braided structure having two braiding yarn systems with or without a third laid-in

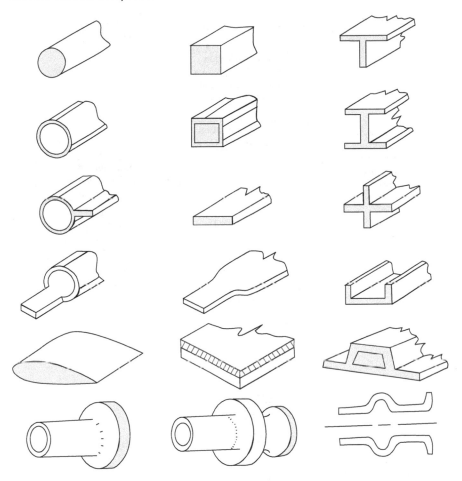

FIGURE 3.13
Net shape structures produced by three-dimensional braiding.

TABLE 3.6

Braiding Classifications

Parameter	Levels		
Yarn axes	Biaxial	Triaxial	Multiaxial
Dimension of braid	Two-dimensional	Three-dimensional	Three-dimensional
Shaping	Formed shape	—	Net shape
Direction of braiding	Horizontal	Vertical	Inverted vertical
Construction of braid	1/1	2/2	3/3

yarn is defined as two-dimensional braiding. When three or more systems of braiding yarns are involved in forming an integrally braided structure, it is known as three-dimensional braiding. The three-dimensional braiding

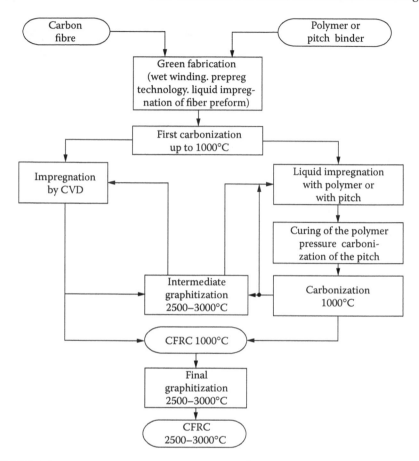

FIGURE 3.14
Production scheme of C/C composites.[22]

system can produce structures in a wide variety of complex shapes, as shown in Figure 3.14. The structures may be produced as thick as desired by proper selection of the sizes of the yarn bundles. Fiber orientation can be chosen and 0° longitudinal reinforcements can be added as desired.

Matrix Precursors

Fibers, however strong and stiff, have virtually no structural capability, and in order to use their excellent mechanical properties in practical materials, it is necessary to combine them with a matrix to obtain a fiber-reinforced composite. The most widely used carbon fiber-reinforced composites have a resin matrix, and a strong technology exists for fabricating these composites by liquid phase methods using solution or molten processing. Because carbon is neither soluble nor fusible, these techniques cannot be used directly for the preparation of carbon matrices but existing resin technologies can be used for C/C fabrication. There are two distinctly different methods for forming the matrix of a C/C composite, both of which have been in use since the mid-1960s. The first method is based on liquid carbonaceous materials that are coated onto or impregnated into fiber tows and subsequently converted into a residue of carbon by heating in an inert atmosphere. In the other method, the carbon fibers are first heated in an inert atmosphere, then exposed to gaseous organic compounds, usually hydrocarbons, which decompose to deposit a coating of pyrolytic carbon on the fiber surfaces. Figure 3.15 is a flowchart showing the two methods and their inter-relationship. These methods will now be considered in a little more detail.

The first method for C/C densification, chemical vapor infiltration[48,49] or chemical vapor deposition, involves the passage of a hydrocarbon gas, typically methane, through the porous preform at temperatures in the 1000–1200°C range, with resulting deposition of carbon in the open porosity.

Densification of C/C composites by the CVD technique can be achieved by three methods. The first method involves isothermal heating of the porous preform structure, in an induction furnace susceptor, to a temperature between 950 and 1100°C. The atmosphere in the furnace is kept at pressures ranging from 100 to 20,000 Pa. Under these conditions the reactant hydrocarbon gas diffuses into the open pores of the preform, depositing its carbon content on the surfaces of the preform. The isothermal heating method requires from 60 to 120 h for each densification cycle (see Figure 3.15a). The carbon deposits produced by the isothermal heating method are of high density, high modulus, and highly graphitizable. Another advantage of this method is that many preforms can be densified at the same time in the same furnace. A major disadvantage is the long cycle time.

In the second method, a pressure differential is created along the thickness of an isothermally heated preform. The hydrocarbon gas is forced to infiltrate through the open pores of the preform. The pressure differential reduces the infiltration time of the hydrocarbon and produces a uniform carbon deposit on the substrate of the preform.

The third method of densifying C/C composite preforms by CVD is based on maintaining a thermal gradient across the thickness of the preform.[50] The pressure inside the furnace is kept atmospheric. The hydrocarbon gas flows through the preform surface. The preform is kept at a temperature below the threshold pyrolysis temperature of the gas. In this method, the induction coil and susceptor are tailored (see Figure 3.15b) to the geometry of the substrate. Carbon depositions occur first on the

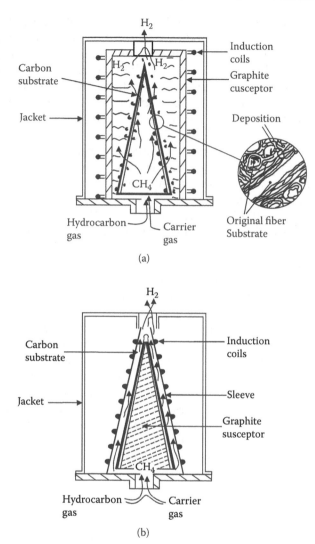

FIGURE 3.15
(a) Schematic of the induction furnace susceptor (isothermal CVD technique). (b) Schematic of the induction furnace susceptor (thermal gradient CVD technique).

outer surface of the substrate. The densification of C/C composite by the thermal gradient method is generally faster than the above two methods. The major disadvantages of this method are the low reproducibility, the nonuniform deposition of carbons within the same preform, the need for a separate coil for each preform, and the inability of producing more than one item at a time.[24, 50, 51]

The properties of C/C composites manufactured by CVD methods are largely influenced by their morphological features and the processing conditions, particularly the reaction temperature.[51]

The second method for C/C densification is the use of coal tar and petroleum pitches. Because they are thermoplastic, pitches are used mostly for redensification; that is, further densifying of a C/C structure that has been "rigidized" by an earlier impregnation/densification step (e.g., a resin-impregnated fabric preform) or that has sufficient rigidity from the friction between the elements of the woven structure (e.g., 3-D braided preform).[50, 51]

Pitches are unique in passing through a liquid-crystalline transformation at temperatures between about 350 and 550°C.[52] In this transformation, large lamellar molecules formed by the reactions of thermal cracking and aromatic polymerization are aligned parallel to form an optically anisotropic liquid crystal known as the carbonaceous mesophase.[53] One of the features of a mesophase-based matrix is high bulk density, which is achievable because the matrix density can approach the value for single-crystal graphite, 2.26 g/cm^3.

For pitches carbonized at atmospheric pressure, coke yields are of the order of 50–60%, impregnant densities are ~1.35 g/cm^3, and, as we have noted, densities for pitch-derived matrices are ~2.2 g/cm^3. From these values, we calculate volumetric densification efficiencies of only 30–40% at atmospheric pressure.[54] By resorting to so-called hot isostatic pressure impregnation carbonization, to pressures of about 100.3 MPa, carbon yields of pitches can be increased to almost 90%.[41] But even with HIPIC, volumetric filling is only 55%. Therefore, given a preform with initial porosity of 45%, typical for many 3-D woven structures, three cycles at maximum densification efficiency would be required to reduce the porosity to 4%. With current HIPIC procedures, however, it is found that at least five cycles at 100.3 MPa are required to achieve this same level of porosity.

Some researchers consider the above fabrication of C/C composites as the liquid impregnation process (LIP), which involves the impregnation of the preform structure with pitch or other organic materials (precursors) and recarbonization. In order to achieve the desired properties of the composite (high density, high strength and modulus, and low thermal conductivity), the impregnation-recarbonization step must be repeated. Optimization of the number of cycles with respect to the process conditions as well as the fiber and precursors is necessary.[55] An important property of the precursors for LIP is the high carbon yield/low weight loss during carbonization (see Table 3.7).

Matrices of C/C composites produced by LIP can be derived either by thermosetting resin precursors or pitches. Typical thermosetting resins are polybenzimidazole, polyphenylene, biphenol formaldehyde, furfuryl alcohol, phenol formaldehyde (phenolic), evoxv novalac, and polyimide.

TABLE 3.7

Comparison of Characteristic Features and Properties of Carbon Matrices from Vapor Phase, Pitch, and Resin Precursors

Characteristic Feature or Property	Type of Carbon Matrix		
	CVD	Pitch	Resin
Density	High approx. 2000 kg m^{-3}, except for isotropic form.	High increases with HTT up to value for graphite.	Low 1300–1600 kg m^{-3}.
Carbon yield	C directly deposited — no further thermal degradation. Varies according to pitch composition 50–80% w/w.	About 50% for phenolics increasing to 85% for polyphenylenes.	
Porosity	Low, except for isotropic form. Laminar fissures.	Macro-sized gas entrapment pores plus shrinkage and thermal stress fissures.	High microporosity (pore diameter <1.0 nm) becomes closed above 1000°C. Macrovoids may be evident, due to vapour evolution during curing of resin.
Microstructure (in the bulk state)	Varies from isotropic to highly orientated laminar forms.	Macro-domains (1–100 µm) showing preferred orientation developed from the mesophase state.	Isotropic — except on the nanoscale, i.e., BSUs are randomly orientated.
Orientational effects within fiber preforms	Strong orientation of laminar matrix with fiber surfaces. Orientation also on crack surfaces within other matrix types if subsequently treated to CVI.	Preferred orientation of lamellae with fiber surfaces — increases as HTT develops the graphitic structure. Modified by pressure pyrolysis.	Preferred orientation at fiber surfaces — but to a much lower extent than other precursors — increases with HTT.
Graphitizability/ crystallinity	Linear forms highly graphitizable, can have L_c and L_a >200 nm after 3000°C HTT.	Highly graphitizable L_a, L_c usually less than 100 nm.	Normally non-graphitizing.
Purity/ composition control	Controlled by gas phase composition — enables other elements to be incorporated into deposit (e.g., Si, B)	Controlled by source — can be high purity. Not easy to incorporate other elements except as powders.	As for pitch.

TABLE 3.7 (continued)

Comparison of Characteristic Features and Properties of Carbon Matrices from Vapor Phase, Pitch, and Resin Precursors

Characteristic Feature or Property	Type of Carbon Matrix		
	CVD	Pitch	Resin
Reactivity to oxidizing gases	Very low reactivity for highly oriented pyrographite.	Low reactivity decreases with HTT.	Usually high reactivity due to micropore network — decreases with HTT but still relatively high.
Thermal expansion	Depends on preferred orientation — can be highly anisotropic, approaching values for the crystal in the two major directions	Depends on domain orientation — expansion partly accommodated by lamellar cracks. $1-5 \times 10^{-6}$ K^{-1}	Isotropic in bulk. $\sim 3 \times 10^{-6}$ K^{-1}
Thermal and electrical conductivity	Determined by preferred orientation — approaching single crystal graphite values.	Depends on domain orientation, HTT and internal porosity, increasing with HTT.	Isotropic in bulk.
Young's modulus	7–40 GPa depending upon structure.	5–10 GPa depending on grain size, porosity, and degree of graphitization but up to 14 GPa for very fine (1 μm) grain size materials.	10–30 GPa
Strength	Depends on microstructure and degree of graphitization. 10–500 mPa.	Depends on porosity and pore geometry. 10–50 mPa for most polycrystalline carbons/graphites rising to 120 mPa for very fine grain (1 μm) graphites.	Approximately 80–150 mPa for glassy carbons. Lower for resin carbons, depending on porosity.
Failure strain	0.3–2.0% depending on structure. Higher values at highest deposition/HT temperatures.	Up to about 0.3% depending on grain size and degree of graphitization.	Up to about 0.4% (largest values for glassy carbon at HTT).

Thermosetting resins can be carbonized easily at low pressures. In general, they exhibit low viscosity, high wetting to the preform surfaces, and curability before carbonization. This enables them to inhibit the loss of the liquid during further heating up.[9] In this respect, thermosetting resins are superior to coal tar pitches.

One last group of materials that can be used for C/C composites are the thermosetting resins (see Table 3.7). Thermosetting resin matrices can be processed either by pyrolyzing an impregnated dry preform or by manufacturing composite articles from resin impregnated tapes, strands, woven fabrics, or short fibers. The technique by which matrices derived from thermosetting resins are processed involves the impregnation of the preform under vacuum followed by curing and postcuring processes. The composite is then pyrolyzed at temperatures between 650 and 1000°C. The structure is then transferred to an induction furnace and heated to temperatures in the range of 2600 to 2750°C. Heating rates depend on the size and the shape of the composite. The impregnation-heating cycle is repeated in order to densify the structure. The density of the composite is largely influenced by the number of impregnation-heating cycles.[24] After several cycles (about seven), the strength of the matrix approaches that of the already pyrolyzed preform, resulting in a high-strength composite.

Resin-impregnated tapes, strands, woven fabrics, and short fibers are also commonly used in fabricating C/C composites. A classical technique for fabricating rocket nozzles is from 3-D woven fabrics. The fabrication of the rocket nozzles[26] involves molding and curing the woven fabric after being impregnated with a phenolic resin. This is followed by trimming and pyrolyzing to drive off gases and moisture from the fabric. The composite is then impregnated with furfuryl alcohol and pyrolyzed. The density and strength increase with each cycle. A ceramic coating consisting of silica and alumina in powder form is applied to the composite. The composite is then heated to a temperature of about 1650°C. As a result of this, silicon carbide is formed on the surface. A further treatment with tetraethyl-orthosilicate and curing forms silicon dioxide residue on the surface to protect the carbon. A final product of a rocket nozzle fabricated, from 3-D woven fabric, with thermosetting resin impregnation is shown in Figure 3.16.[9]

Thermosetting resin precursor matrices are generally amorphous and isotropic in their properties and microstructure when they are heat-treated to temperatures below 600°C.[11]

Several new techniques have been developed and are in production.[56] A newly patented preformed yarn method involves producing yarn consisting of a bundle of carbon fibers in a matrix precursor of coke and pitch binder powders. The yarn is encased in a flexible thermoplastic sleeve to contain the powders during handling and subsequent processing. This simplified manufacturing process ensures better penetration of the binder into the carbon fiber bundle, thereby resulting in higher strength than conventional C/C composites. Furthermore, the matrix powders are often

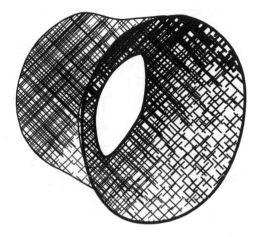

FIGURE 3.16
A final product of a rocket fabricated from 3-D woven fabric, with a thermosetting resin precursor.

derived from the residue of petroleum processing, which can have beneficial environmental effects.

Research has shown that unidirectional C/C composites manufactured by this simplified method achieve strengths of between 414 and 670 MPa, compared with 70 MPa for conventional C/C composites. They also show significantly higher flexural moduli.

The preformed yarn may be woven into a sheet or chopped to fill a mold, and is then hot-pressed to form the composite product. Because the reinforcing fibers are homogeneously dispersed in the matrix, properties of the resulting composite are quite uniform.

The preformed yarns have excellent workability and processability. For example, pipe-, rod-, and crucible-shaped preforms may be readily fabricated, as well as unidirectional sheets, cloth sheets, thick textiles, tapes, and chopped yarns. Workability of the preformed yarns allows the material to be tailored to meet various design loads. This method is also well suited for the production of complex shapes and components having small radii of curvature.

The composites are available in unmachined form, as well as finished products such as coil springs, rods, shelves, bolts and nuts, link conveyor belts, roller conveyor parts, and furnace parts. Figure 3.17 reflects the comparison of the preformed yarn process and previously described conventional methods.

The two remaining processes show promise to overcome overcoating problems and slash processing times from weeks/months to hours.

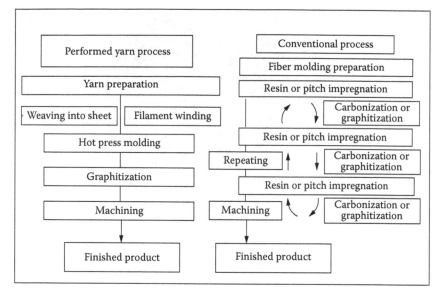

FIGURE 3.17
The preformed yarn process reduces the number of steps required to produce composites, compared with conventional methods.

The first technique (RD) uses liquid immersion rather than vapor or resin impregnation. To start, a carbon fiber preform is put in a reactor that contains a carbon-containing liquid precursor. The reactor is relatively small, maybe 5 cm on a side; the preform is immersed in liquid and resistively heated by electricity. This boils the surrounding liquid at temperatures high enough to crack off its hydrogen atoms and deposit carbon on the preform.

Using a liquid provides several benefits; most important, it eliminates overcoating. The boiling vapor around the part draws heat away from the edges. This cools the outer surface, so deposition occurs in the center, where the preform remains hottest.

Because the process automatically densifies the center first, it shoves open the processing window. As a result, one does not have to limit deposition speed to keep from overcoating the preform. In fact, the turbulence of the bubbling precursor liquid actually speeds deposition.

Fully dense, 32.5-cm-diameter brake disks have been produced in only 8 h. The result allows for fast turnaround times and will free aircraft maintenance organizations from carrying C/C disk inventories to ensure a replacement.

The RD process should also lower capital costs. Because the preform is resistively heated instead of the entire reactor, it eliminates the need for costly furnaces. The RD process also seems amenable to automation and simple closed-loop control.

The other process has supercharged a conventional CVI system, called flow-thermal gradient CVI. The process can produce 1-cm-thick parts in about 2.75 h, compared to several hundred hours for conventional CVI. Instead of placing a preform in a furnace and letting things happen with slow diffusion, gas is forced through it.

The process is based on a system developed about 15 years ago to produce ceramic composites. The process uses pressure to force carbon-containing propylene, propane, or methane vapor through preforms heated to 1200°C.

The key to faster processing lies in creation of a temperature gradient within the preform. This allows for the flow of vapor over the part without worrying about overcoating the outer edges. It also gives a wider processing window and the ability to tailor C/C microstructure. It is possible to pick the temperature, pressure, and reagent conditions that give the characteristics required. Therefore, researchers can tailor the structure and improve properties like thermal conductivity or stiffness. It may also allow one to add special graphitization catalysts and oxidation inhibitors to the fiber-reinforced composite as it forms.

So far, researchers have produced more than a dozen C/C parts, some as thick as 2 cm, comparable in quality to those produced by conventional processes. Fixturing has been the greatest barrier to commercialization. Holding the preform so that gas flows over it properly requires complex equipment.

Both the above processes promise faster processing times, and that should translate into lower prices. Although no one has truly solved C/Cs oxidation problems, a lower price tag should open more applications in structures, turbines, and avionics (thermal management).

C/C Significant Properties

Mechanical Properties

The structural properties of continuous fiber advanced composites are generally controlled by the properties, volume fraction, and geometry of the fibers. C/C composites, however, are by nature very complex as a result of the physical and chemical changes and interactions that occur during processing. Pyrolysis of an organic precursor, for example, to form a carbon matrix involves a 50% reduction in volume. Shrinkage of this magnitude can create severe damage to the composite by the buildup of large process-related stresses.[59] The transformation from a carbon fiber/organic interface to a C/C fiber interfacial bond will vary considerably, dependent on materials and process variables. Extended heat treatments, process-induced stresses, and fiber-matrix interactions are very likely to affect adversely the primary properties of the fiber reinforcement.[60, 61] Severe thermal stresses, due to thermal expansion mismatches between fibers and matrix, and between different matrix structures, will result in a time-dependent deterioration of composite mechanical properties during heat treatment and service cycles.

In keeping with all fiber-reinforced materials, the mechanical properties of a C/C composite show marked anisotropy as a result of the anisotropy in the fibers. For the case of unidirectional (UD) fibers, the maximum stress theory states that the tensile strength of the fiber will be manifest when the angle (θ) between the applied load and fiber axis is no more than $4°$. If θ exceeds $4°$, the strength (σ_{11}) will be governed by the shear strength (τ) and may be expressed as

$$\sigma_{11} = \tau/\sin \theta \cos \theta. \qquad (3.1)$$

Should θ exceed $24°$, the strength approaches that of the matrix.

The composite's tensile strength (σ_c) in the fiber axis of a 1-D material can be expressed according to a simple rule of mixture:

$$\sigma_c = \sigma_f V_f + \sigma_m (1 - V_f) \qquad (3.2)$$

where σ_f is the fiber strength, σ_m the matrix strength, and V_f the volume fraction of fibers. The strength of the matrix can generally be neglected when compared to the strength of the fibers such that the strength of a composite is proportional to the fiber content.

It is very difficult, in practical applications, to realize the theoretical strength of a composite due to twisting and distortion of the fibers, variations in fiber orientation, and stress concentrations associated with the method of test. Furthermore, should the fiber content exceed 70%, uniform impregnation by the matrix becomes impossible, resulting in a fall in strength. The large-scale, irregular porosity, poor fiber/matrix interface, and inherent brittleness of the system have the result that the C/C never attains its theoretical strength, values of 50–60% being typical. As a general rule of thumb, it is usually assumed that a balanced woven fabric-reinforced material (or 0/90° UD lay-up) will have roughly half the strength of a UD material, and a quasi-isotropic laminate one third its strength.

The strength of C/C composite materials does not generally follow the simple law of mixtures relationship. The weak interfacial bonding results in an inefficient transfer of applied loads onto the fiber reinforcement, such that theoretical strengths are seldom, if ever, achieved. Mechanical properties of C/C tend to vary between 10 and 60% lower than the values calculated from the rule of mixtures. It is possible to increase the strength of the fiber/matrix interface but, as previously discussed, this manifests itself in extreme brittleness of the composite.

The mechanical properties of C/C are very much dominated by the properties of the fibers, which depend on precursor type and processing conditions. Microstructural variations in shape, porosity, and cross-section can occur within filaments. Variations in fiber architecture on the thermo-mechanical properties of C/C are not too well understood, but it is postulated that they may result in a heterogeneous distribution within the composite. Such a distribution may result in understresses in some fibers while overstressing others at adjacent locations.[62] The key factor in optimum fiber utilization is good alignment with respect to the axis of the load.[63] Stress-strain curves in 2- and 3-D composites show reductions in fiber properties and behave nonlinearly as a consequence of misalignment[64] due to the "crimping" that occurs as a result of the weaving process. The fibers are not aligned with the principal stress axis, and their effective strengths are thus reduced.

The highest mechanical properties are achieved by using HM PAN or mesophase pitch-based fibers, preferably without surface treatment. If not surface-treated, HM fibers show very low adhesion to the resin. The bulk cross-sectional shrinkage on pyrolysis is therefore very low. In the case of IM and isotropic fibers, especially those that have been surface-treated, the first carbonization causes a bulk shrinkage as a result of good fiber/matrix adhesion. The shrinkage of the resin is resisted by the low compressibility of the fibers. As a result, a high degree of compressive prestress is built up in the fibers, which reduces the strength of the composite.

The major influence on the mechanical properties of C/C produced from thermosetting resin precursors is the heat-treatment temperature (HTT).[65] At an HTT of below 2000°C, the strength is low and brittle fracture the predominant failure mode. The matrix so formed is a brittle, isotropic glassy carbon in which propagating cracks readily sever the fibers without changing direction. At HTTs above approximately 2400°C, a graphite component begins to develop at the fiber/matrix interface. An apparent pseudo-plastic, "ductile" fracture behavior is the consequence. The strength and modulus increase up to an HTT of around 2750°C.

Processing at 2800°C and above tends to reduce the strength due to thermal damage to the fibers.[66] A graphical representation of the effect of HTT on the flexural strength of furan-based composites is shown in Figure 3.18.[65] The ductile behavior of the graphitized composites may be illustrated by considering the energy of fracture.[67] At an HTT of 1000°C, this is measured at 0.3 KJ m[2], whereas at 2700°C it is around 4 KJ m[2], more than an order of magnitude higher.

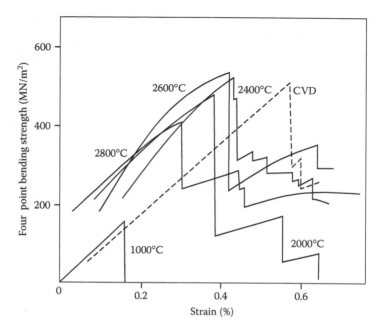

FIGURE 3.18
Fracture behavior of a C/C composite made from a furan precursor, with respect to HTT.

The properties of pitch-based C/C are controlled not only by the carbon yield of the matrix precursor, but also by the microstructure. Pitch carbonized under relatively low pressures results in a well graphitizable matrix carbon with a sheath structure parallel to the fiber surface and thus in anisotropic matrix properties.

High-Temperature Mechanical Properties

One of the major assets of a C/C composite is that of retention of mechanical properties at high temperatures, as shown in Figure 3.19.[69, 70] It can be seen from Figure 3.19 that up to 1500°C, there is a 10–20% increase in strength and modulus compared to ambient temperatures.

Fatigue Properties

Very little work is presented in the open literature concerning the fatigue performance of C/C. Fitzer and Heym[71] showed a lifetime of 10^7 cycles at a stress of 40% of the static bending strength. Other studies (Woeler diagrams) by Schunk[72] show that the fatigue endurance properties of C/C appear, therefore, to be outstanding.

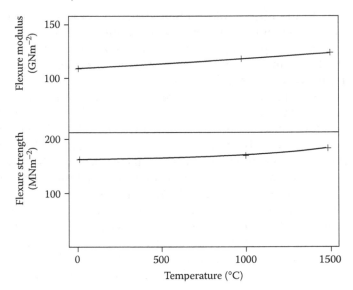

FIGURE 3.19
Change in flexure strength and modulus of a C/C composite made from a thermoset precursor with respect to operating temperature.[69]

A question arises, however, as to whether the pore opening and closing mechanism, which must occur under fatigue loading, will continue indefinitely. The pore volume is increased during longtime cycling. A probability exists that in local regions, the fracture strength of the matrix will be exceeded. The result will be the fracturing of the matrix in those areas and its loss as dust from the composite. This phenomenon, for obvious reasons, is known as *dusting-out*. Dusting-out has been observed at high temperatures (~1400°C) in centrifugal loading operations (fan blades). There is therefore a severe limitation on the longtime application of C/C under high alternating stresses.

Fracture Toughness

Materials such as C/C that have internal flaws or voids as a consequence of fabrication processes must be evaluated for fracture toughness prior to consideration for a specific application. The toughness is interpreted as the ability of a material to resist crack propagation from existing voids. Very few data have been reported to date on the fracture toughness of C/C, due in the main to the difficulty in making accurate and meaningful measurements.

The various fracture toughness parameters are found to depend strongly on the type of carbon fiber used and the orientation of the initial crack with respect to the fiber reinforcement structure. In a two-dimensional, fabric-reinforced composite, severe crack blunting and delamination are observed when crack propagation is perpendicular to the fibers. K_{1c} values of around 7.6 MN m$^{3/2}$ have been calculated in this orientation.[58, 74] Such results are, however, invalid because linear elastic fracture mechanics is not applicable to pseudoplastic failure. When crack propagation occurs collinear with the initial crack, such as between the layers of a 2-D material, fracture toughness parameters can be measured; 1- and 2-D laminates are particularly susceptible to this mode of failure. The interlaminar fracture toughness of a phenolic-derived material was found to be of the order of 0.11 KJ m^2. This value is very low, around half that measured for even the most brittle epoxy composites.[58] The extremely brittle nature of the matrix and the presence of voids and shrinkage cracks may result in failure of components during processing. Cracks and other defects formed during the first pyrolysis cycle may grow as a result of thermal loading of subsequent cycles such that the materials are extremely unpredictable. A 2-D composite may thus fail by delamination without warning during the processing.

Thermal Properties

Next to mechanical properties, the most important characteristics of a C/C composite are thermal conduction and thermal expansion. The number of papers relating to these is only exceeded by the number of papers on mechanical properties.

C/C composites present the opportunity to tailor thermo-physical properties into carbon materials. The complexity of possible options can be gauged by brief consideration of the different fiber/matrix/processing permutations possible:

1. Fibers: rayon, PAN, isotropic, and mesophase pitch (plus variations in heat treatment)
2. Matrix: thermoset pyrolysis, CVD, and thermoplastic/pitch pyrolysis
3. Fiber architecture: felt, UD, fabrics, and 3-D structures

With such a range of options, a very wide range of thermal conductivities is possible.

A range of different carbon microstructures is possible from the CVD process, each of which has considerable influence over the thermo-mechanical properties of the composite. Liebermann and Pierson at the Sandia Laboratories have made several investigations into the CVD process/property relationships for different C/C composite materials.[75]

TABLE 3.8

CVD Matrix Microstructure/Thermal Conductivity Relationships[58]

Matrix Microstructure	Crystallite Size (Å)	Thermal Conductivity at 350°C (W m⁻¹ K⁻¹)
Smooth laminar	125	25
Rough laminar	385	96
Isotropic	90	25

The correlation of matrix microstructure with the thermophysical properties is relatively straightforward, as shown in Table 3.8. The rough laminar microstructure is clearly a good deal more graphitizable than the smooth laminar and isotropic microstructures, and thus exhibits the highest thermal conductivity. A whole range of other studies validates Liebermann's theories on CVD matrix microstructure/property relationships.

Pitch-derived carbon matrix-carbon fiber composites are generally processed via the hot isostatic pressure impregnation carbonization technique (see Figure 3.20), of pitch into a dry carbon fiber preform. Repeated HIPIC cycles are required to build the composite matrix up to an acceptably high

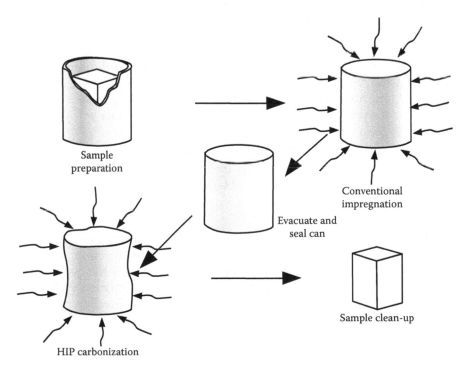

FIGURE 3.20
Schematic diagram of HIPIC process.

TABLE 3.9

Thermal Conductivity of HIPIC-Processed 3-D
C/C Materials

Thermal Conductivity (W m^{-1} K^{-1})	Battelle 3-D Carbon–Carbon		Morton Thiokol 3-D Carbon–Carbon	
	X–Y	Z	X	Y
At 30°C	149	246	202	171
At 800°C	—	—	91	80
At 1600°C	44	60	—	—

density/low porosity for deployment in severe ablative environments. Two sets of thermal conductivity data have been published on 3-D HIPIC-processed composites.

Workers at the Battelle Institute[76] (the originators of the HIP process) fabricated a 3-D weave from T-50 fiber, with a 2-2-6 fiber distribution (X-Y-Z), and then used four HIPIC cycles with a coal tar pitch impregnant to produce a composite of density 1.9 g cm^{-3}. The other documented 3-D composite is a body optimized for ring strength: a 9-2-9 (X-Y-Z) preform based on T-50 PAN fiber, HIPIC processed to a density of 1.9 g cm^{-3}. The thermal conductivity data are presented in Table 3.9.

The values are higher than those developed for phenolic-derived materials of similar fiber architecture. They reflect the more graphitic, thermally conducting nature of the HIPIC-derived matrix. It is worth noting, however, that the thermal conductivity of these materials is, at best, only a marginal improvement on those displayed by conventional synthetic graphites, despite their far superior mechanical properties.

Thermal Conductivity at High Temperatures

There are a number of reports in the literature concerning the high-temperature thermal conductivity of C/C.[78, 79]

Thermal Expansion

Thermal expansion is an important property when using materials at high temperatures. The thermal expansion in the fiber axis direction is chiefly governed by the thermal expansion of the fiber. That is to say, thermal expansion of a composite material in which a highly oriented PAN fiber is used resembles the expansion behavior in the planar direction of graphite (see Figure 3.21).[78, 79]

Processing Effects on Mechanical Properties

Regarding processing effects, Perry and Adams reported tensile properties of a 1:1:2 construction fine weave, 3-D C/C woven with high-modulus

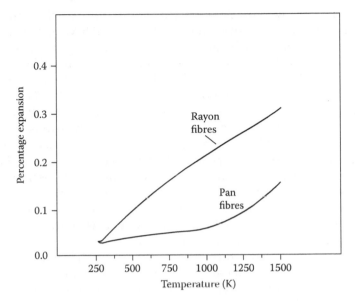

FIGURE 3.21
Thermal expansion of a CVD C/C.[58,78]

rayon-based fibers and densified by multiple resin impregnation and atmospheric pressure carbonization cycles.[80] The tensile strength and modulus in the Z direction (192 MPa and 88 GPa) after 13 densification cycles were only slightly larger than those after 7 cycles (177 MPa and 84 GPa). After graphitization, these values fell to 105 MPa and 57 GPa, whereas the same properties in the X and Y directions increased slightly. This behavior is attributed to shrinkage-induced gaps around the Z fiber tows.

A study of processing effects in 3-D C/C by Girard[81] showed that in a relatively coarse weave block of PAN-based fibers, densification by CVD gave a density of 1.6 to 1.7 g cm⁻³, tensile strength of 70 MPa, and moduli of 50 to 70 GPa. Densification by a combination of CVD and resin impregnation/carbonization raised the density from 1.8 to 1.9 g cm⁻³, tensile strength to 110 MPa, and modulus to 120 GPa.

The effects of matrix precursors in a pierced fabric 3-D block reported by McAllister and Taverna[82] showed that densification using phenolic resin at atmospheric pressure gave a Z direction flexural strength of 90 MPa. A combination of phenolic resin and CVD densification increased this value to 109 MPa, while densification with a synthetic pitch composed of anthraquinone and acenaphthylene raised it to 154 MPa.

Mullen and Roy[83] investigated a number of matrix-forming methods in 3-D C/C cylinders made by the rigidized tow method. They found that the highest density was obtained by a combination of isothermal CVD and phenolic resin, whereas differential pressure CVD gave the highest hoop tensile and axial compressive strengths (136 and 115 MPa, respectively).

McAllister and Taverna[82] showed an effect of fiber type in pierced fabric 3-D C/C, where a change from low-modulus rayon-based fibers to high-modulus rayon-based fibers in the X-Y fabric increased the tensile strength from 35 to 105 MPa and the modulus from 11 to 58 GPa. These authors also reported an increase in tensile strength from 80 to 98 MPa when the Z component was changed from dry tow to preformed rods containing the same number of fibers. This effect is attributable to fiber damage in piercing with dry tow compared with the protective effect of the resin matrix when using pultruded rods. Another effect was observed in this study when the size of the Z tow was increased to such an extent that damage occurred when forcing the larger tows through the fabric layers. Although the modulus in the Z direction was increased from 35 to 40 GPa, the strength was reduced from 80 to 75 MPa.

The same study compared the mechanical properties in the X-Y direction of 3-D composites prepared by the pierced fabric and orthogonal weaving routes using the same high-modulus rayon-based fibers. The results indicate that although the fiber content in the pierced fabric composite is higher, tensile strengths and Young's moduli are similar. The straightness of the fibers in the orthogonal process compared to the convoluted X-Y fibers in pierced fabrics makes for more efficient utilization of fiber properties.

The rather modest strengths are ascribable to the densification process, which used low-pressure impregnation and carbonization with resin and pitch. Comparison with the values for 3-D orthogonally woven C/Cs made with the same high modulus rayon-based fibers but densified by the HIPIC process to a slightly higher density of 1.95 g cm^{-3} shows the benefits of high-pressure processing. This also shows that fiber volume fraction has a dominant effect on composite properties. The decreasing strength and modulus values follow the decrease in fiber volume fraction for two types of 3-D orthogonal and a 7-D composite. It is interesting to note that at 45° to the X-Y directions, where there is no fiber, the strength is essentially that of the matrix and is in line with that for a medium-grade graphite. This deficiency is overcome in 5- and 7-D constructions but at the expense of reduced property values in X and Y directions.

Oxidation Protection

The physical and mechanical properties of C/C composites make them attractive in applications where strength, light weight, toughness, and low thermal conductivity at high temperatures are required. In an inert atmosphere or in a vacuum, C/C composites are superior to other materials in retaining their properties. A serious problem that faces the design engineers is the oxidation behavior of C/C composites at elevated temperatures.

In general, carbon in any form reacts with oxygen and burns rapidly at temperatures above 500°C, which becomes progressively more severe as the temperature rises until, at about 800°C, the rate of oxidation is limited only by the diffusion rate of oxygen through the surrounding gas to the carbon surface.[84] In certain one-off applications like rocket motors, this is not very important; however, in space vehicles and gas turbine engine development, extended lifetimes of around 100 H are required.

The solution to this problem has been the goal of many researchers, and the wide diversity of protection systems investigated and patented is evidence that no definitive system has yet been found. One high-profile success is the high-temperature thermal protection components of the space shuttle, where the nose cone and wing leading-edge components can experience temperatures up to 1650°C.

The following points all need due consideration when evaluating prospective protection systems:

1. The C/C will cycle from ambient to its working temperature.
2. It may spend minutes or hours at that temperature.
3. It will have to perform in various gas compositions (hot or cold) containing water vapor, oxygen, carbon monoxide, carbon dioxide, and other impurities present in rocket or engine exhaust, as well as at varying gas pressures and high gas velocities.
4. At temperatures above 1700°C, many materials normally accepted as stable can start to decompose or undergo solid/solid reactions, breaking down the coating integrity.
5. Is the strength of the C/C affected (weakened or strengthened) by the system?

In developing a successful oxidation protection system for carbon-based materials, there are a number of factors that must be considered, as depicted schematically in Figure 3.22. The primary aim is to apply a coating system that isolates the composite from the oxidizing environment. This coating system must have at least one major component that acts as an efficient barrier to oxygen. The primary oxygen barrier must ideally exhibit low oxygen permeability and ought to encapsulate the carbon with few or no defects through which the oxidizing species can ingress. Additionally, it must possess low volatility to prevent excessive ablation in high-velocity gas streams. A good level of adherence to the substrate must be achieved without excessive substrate penetration. The internal layers

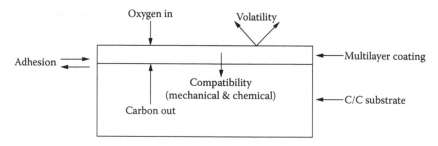

FIGURE 3.22
Considerations when designing a C/C oxidation protection system.

must also prevent outward diffusion of carbon to avoid carbothermic reduction of the oxides in the outermost layers. Finally, all of the various interfaces must exhibit chemical and mechanical compatibility.

In designing protection systems for high-performance, structural C/C for long periods of operation, mechanical compatibility (i.e., the avoidance of coating spallation) becomes the overriding issue. The composite has a considerably lower coefficient of thermal expansion (CTE) within fiber-reinforced lamina than any ceramic exhibiting a symmetric crystal structure. Any applied coatings therefore contain microcracks because the coating process is carried out at elevated temperatures.

The oxidation threshold for C/C is around 370°C, which may be improved to approximately 600°C by the incorporation of refractory particulate inhibitors. The intrinsic protection range of the coating is defined by the microcracking temperature and the limiting use temperature of the coating. In this range, the cracks are mechanically closed and sealed by the oxidation products. Flaws always exist in the primary oxygen barrier due to fabrication imperfections, due to thermal expansion mismatch with the substrate, or as a result of service stresses. To date, the most successful solution to the cracking problem has been the employment of a sealant glass to fill any cracks in the primary coating. For a successful full range of oxidation protection, the glass sealant must be capable of operating from ~600°C to the microcracking temperature of the primary oxygen barrier.

Low-Temperature Protection (up to 1000°C)

In many of the aerospace applications for C/C, the main operating temperature is in the regime above 1000°C. However, any component will have to pass through the lower temperature regime on start-up to, or shutdown from, its operating temperature. Protection in this temperature zone is most effectively achieved with lower melting glazes but, in selecting an appropriate glaze for the task, the interaction between substrate and coating needs careful consideration. A large number of oxides that form the basis of glasses have been found to have positive catalytic effects on the oxidation rates of carbon;[84] materials found to have inhibiting effects are very few in number. Two such candidates are additives based upon boron and phosphorus.[85] The volatility and reactivity of phosphorus oxide require that it be mixed with another substance (e.g., silica or alumina) to form a stable glass, while boron forms a stable oxide glass (mp 450°C, bp 2040°C) on its own in dry air.

$$B_2O_3 + H_2O \rightarrow 2HBO_2 \qquad (3.3)$$

The performance of boric oxide glazes has been tested by McKee by applying glazes 200–500 μm thick to the external surfaces and also vacuum impregnating the porosities of C/C with liquid organo-borates.

A significant factor in the success of the boric oxide glaze is the mobility of the glass, any breaching of the protective layer being rapidly sealed by flow of the bulk glass.

A variant of this method of oxidation protection has been patented by Marin.[86] In this process, boron and silicon are dispersed as fine powders (0.5–2.0 μm) of silica and boron, or boron nitride, in a thermosetting resin that is painted onto the surfaces to be protected and cured. Under oxidizing conditions, the additives are exposed and react to form oxides, which combine to form a glaze.

A system that has been used in the past for monolithic carbon and graphite has been to incorporate certain compounds of silicon, boron, zirconium, and hafnium during the manufacturing stages.[87–89] Under oxidizing conditions, these are gradually exposed and accumulate on the surface, where they slow down the rate of oxidation. They are not actually protective in their action, but, more strictly, inhibiting to the oxidation process because part of the carbon has to be consumed before the system begins to work.

More recently, this system has been applied to C/C composites by adding the inhibitors to the matrix portion of the structure.[90–93] It has also been used within a more comprehensive system including multiple external coatings.[94]

A pure borate glass is only able to give primary protection for limited times above around 1100°C due to the volatility of B_2O_3. Although limited in application, the work clearly demonstrated the effectiveness of boron additions (preferably in unoxidized form to ease processing) to the body of a carbon in laying the foundations for further development.

Protection Up to 1800°C

The higher temperatures mean deeper consideration has to be given to the interaction, physically, chemically, and mechanically, of the substrate/coating combination. The coating must provide an effective barrier to the inward diffusion of oxygen and the outward diffusion of carbon, it must have low volatility to prevent erosion in high-velocity gas streams, and the mechanical bond to the substrate must be strong.

The first major problem to be aware of is the very low coefficient of thermal expansion of C/C composites, especially parallel to the fibers. Figures in the range $0-1 \times 10^{-6 \circ} K^1$ are accepted values for composites made with high-modulus fibers.[95] If the composite has an unreinforced axis (e.g., bidirectional C/C), the CTE will be around $10 \times 10^{-6 \circ} K^1$ along this axis.

Based upon many considerations, the most promising materials for protection are silicon nitride and silicon carbide. Both are chemically compatible with carbon, and in an oxidizing atmosphere they generate a thin silica film that has a low permeability to oxygen.

A great many coatings, other than glasses and glass formers, have been added to C/C for high-temperature operation in oxidizing environments. Chown and his team investigated the suitability of coatings of SiC, produced both by CVD and direct reaction with molten silicon.[96] A number of other refractory carbide and boride coatings have also been investigated by Chown[96] and Fitzer et al.[97] The results obtained indicated that silicon carbide was capable of providing reliable protection for long time periods, provided the temperature remained below 1700°C and flaws in the coatings could be eliminated. SiC coatings are found to decompose rapidly above 1700°C, whereas a sintered coating of zirconium carbide and boride (ZrC and ZrB_2) will allow short-term protection up to temperatures of roughly 2200°C.

The temperature limit observed for SiC appears to apply to all silicon-based materials used in oxidation protection. Molybdenum disilicide ($MoSi_2$) and silicon nitride (Si_3N_4) also show this limitation, which is probably endemic to the silicon family of materials. Higher-temperature refractories based on zirconium and hafnium are able to operate at temperatures well above silicon, but only for limited periods of time as a result of the rapid rates of diffusion of oxygen through these oxides.

There are many methods for applying coatings to C/C, such as chemical vapor deposition (CVD), chemical vapor infiltration (CVI), chemical vapor reaction (CVR), impregnation and pyrolysis, dipping, and painting. The bulk of the systems investigated fall into the first three categories of chemical vapor methods and generally form the basis of the systems that are operated by commercial companies.

Chemical Vapor Deposition

Both silicon carbide and silicon nitride can be deposited very easily by CVD. Silicon carbide is very commonly deposited from methyltrichlorosilane (MTS) and hydrogen at temperatures between 1100° and 1400°C and at reduced pressures around 1–1.5 KPa. Silicon carbide can also be deposited from silicon tetrachloride or trichlorosilane and methane.

Chemical Vapor Infiltration

This process is a form of CVD where the conditions are modified to encourage the formation of internal coating within the body of a porous substrate before the open pores are sealed by external deposits. This is usually achieved by using lower temperatures, pressures, and gas concentrations than for purely overlay coating. Some original work on protecting C/C by this method deposited oxidation-resistant materials such as boron nitride, titanium carbide, and silicon carbide into C/C composites.[98-100] Although these papers are not concerned with C/C in its accepted form (i.e., major constituent carbon), the technique has possible applications in consolidating the outer layers of a composite with silicon carbide followed by overlay coating as a single-step operation, thereby providing a strong bond between the coating and substrate.

A different application of this method uses the infiltration of boron carbide followed by silicon carbide as thin, oxidation-resistant coatings (1–5 μm) on the fibers of a preform, followed by conventional consolidation to complete the composite fabrication.[101]

Chemical Vapor Reaction

This third classification of chemical vapor reactions is characterized by the deposited material reacting with the substrate to form a compound. In this instance, a silicon-containing vapor reacts with the carbon substrate to form a silicon carbide layer.

In its simplest form, this process is carried out by exposing a carbon particle to silicon vapor in a vacuum chamber at a temperature in the range 1700–2000°C; by adjusting the process time, it is possible to control the coating thickness to form coatings 50–100 μm thick.

An alternative method is that used for the first-stage treatment of the nose cone and wing leading-edge components of the space shuttle.[102,103] Here the C/C component is encapsulated in a powder pack comprising silicon carbide, silicon, and alumina and heated to 1750–1850°C under an inert atmosphere for a period of 4–7 h. This produces a coating thickness of 125–750 μm with no apparent dimensional change to the coated component. The mechanism for this coating formation is a combination of reactions that initially generate high-temperature species through one equilibrium within the pack, then these interact at the carbon surface to form silicon carbide.

The manufacturers of the shuttle C/C protection system have, more recently, made modifications to the diffusion processing packs. In one variation, the alumina has been omitted from the pack and a small proportion of boron added, while in another some of the silicon carbide has been replaced with extra silicon. The boron has been included to benefit the low-temperature protection characteristics of the coating, while the higher silicon pack produces a coating containing excess silicon, which is claimed to lower the CTE of the layer and reduce the degree of cracking on cooling.

Protection Beyond 1800°C

At temperatures above 1800°C, the choice of materials is becoming limited. Silicon-based ceramics are no longer stable and degrade, and the refractory metals, along with their borides and carbides, are all converted to their respective oxides. Apart from these oxides, the only other materials resistant to oxidation at these high temperatures are iridium and certain platinum group alloys. Therefore, long-term protection in this region will have to rely on refractory oxide layers of Si, Al, Hf, Zr, Y, Th, Be; metallic coatings of iridium or platinum group alloys; or multiplex combinations of these.

After considering the various oxides, refractory borides, and silicides (see Table 3.10), one can take all this information and combine it into a general oxide-based system of protection, represented by Figure 3.23. The system comprises four layers: an external, erosion-resistant layer; an oxygen barrier layer; an interaction barrier; and, finally, a carbon diffusion barrier. Candidate materials for each of these layers are listed in Table 3.10. These materials have been selected based on chemical compatibility and thermal stability criteria only; no consideration was given to expansion mismatch problems.

Other protection means are based on metal-based systems. Examination of the periodic table reveals 14 metals with melting points above 1800°C. However, only two are feasible to use (iridium and ruthenium), and measurements on recession rates in air have shown iridium to be better than ruthenium by a factor of ~10.[106]

The melting point of pure iridium is 2443°C but it also forms a eutectic with carbon, which lowers its mp to 2250–2280°C. Small quantities of

TABLE 3.10

Ceramic Materials Selection for Protection Above 1800°C

Erosion Barrier		Oxygen Barrier	Reaction Barrier	Carbon Barrier
Al_2O_3	SiO_2	ZrO_2	ZrC	HfB_2
ZrO_2	Al_2O_3 (Sapphire)[a]	HfO_2	HfC	TaB_2[a]
		TaO_2[a]	TaC	ZrN[a]
			TiC	HfN[a]
			ZrB_2	TaN[a]

[a] Speculative.

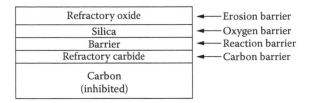

FIGURE 3.23
Projected system for oxidation protection of C/C above 1800°C.

impurities such as 1% Fe and traces of Si, Al, Ni, or Cr will lower its mp even further to around 2110°C. Its upper limit for protection is limited by the volatility of its oxides (IrO_3, IrO_2, and IrO), which becomes significant before the mp. This unique collection of properties makes it the only one-component system capable of protecting C/C up to ~2100°C. Beyond this temperature, it can be considered in a multilayered system where a top coating of zirconia or hafnia would suppress the evaporation of volatile oxides.

Three examples of such combinations whose oxidation performances have been tested are Ir-Hf, Ir-Al, and $ZrPt_3$.[107–109] The iridium-based systems were found to be effective against oxidation only at compositions having more than 50 atomic % of the protective oxide-forming component; below this ratio, the oxide scale was porous and not an effective barrier.

Both iridium and hafnium can be deposited by CVD. Iridium has been more widely investigated, and coatings have been deposited by hydrogen reduction of a range of precursors such as IrF_6, $IrCl_3$, and Ir (acetylacetonate)$_3$.

For long-term oxidation protection, researchers have proposed and are developing a four-layer protection system that is a little more detailed though quite similar to the system in Figure 3.24. The system would consist of refractory oxide/modified SiO_2 glass/inner refractory oxide/ refractory carbide (see Figure 3.24). A multilayer coating such as this, while possessing the required chemical stability, would present serious problems in deposition of consistent quality. In addition, the high CTE of the protection system compared to the substrate will create severe problems of mechanical compatibility in thermal shock situations, and these must be overcome.

One of the major obstacles to long-term protection and a truly reusable composite materials system is the spallation of coatings due to thermal mismatch with the substrate under cyclic conditions. The most feasible way to avoid this occurrence would be to replace the discrete coating/ composite interface with a functionally graded system. In such an artifact, there would be a hybrid region between protection and composite, thus minimizing thermal expansion differences.

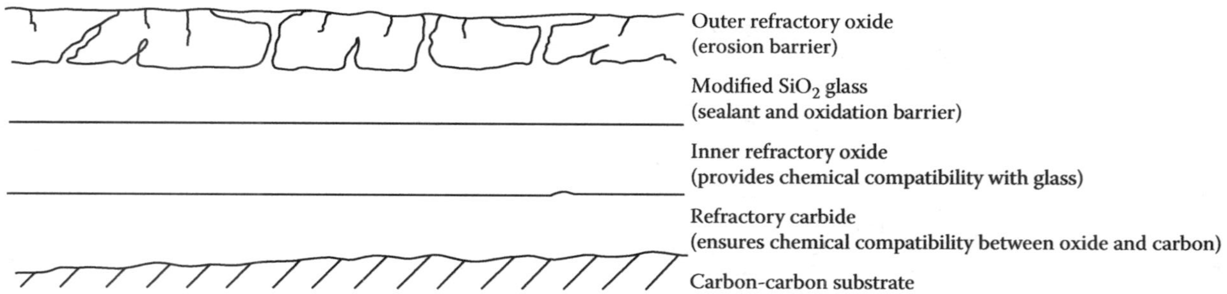

Outer refractory oxide
(erosion barrier)

Modified SiO_2 glass
(sealant and oxidation barrier)

Inner refractory oxide
(provides chemical compatibility with glass)

Refractory carbide
(ensures chemical compatibility between oxide and carbon)

Carbon–carbon substrate

FIGURE 3.24
Idealized coating system for long-term oxidation protection of C/C at temperatures of 1800°C and above.

NDE of Coatings

Nondestructive examination (NDE) techniques have a key role in the future of protected C/C composites. The fabrication of these composites is an expensive undertaking per se, and adding oxidation inhibitors to the matrix during manufacture followed by a multiplex protection system applied to the outside adds further to the cost and makes it essential to obtain the highest degree of reproducibility from the fabrication processes. It is, therefore, essential to have a reliable quality assurance technique to examine and reject suspect parts without wasting valuable good parts. Further along the line, a system will be required to evaluate the condition of in-service components as part of the routine maintenance schedule.

A large number of NDE methods are available at the present time, with the more advanced techniques already being used for composite examination. For the purpose of coating examination, only those that are able to detect internal delaminations, debonding of the protective coating, and oxidation of the matrix will be considered. These include X-radiography, infrared thermography, ultrasonics, and eddy currents.

The oldest technique is X-ray inspection. A high-energy X-ray beam is directed through the component onto an imaging medium (film, camera, or counter), and an image is formed according to the variations in transmittance of the component. Areas of matrix oxidation or delamination would have a higher transmittance and produce a contrasting zone within the general image. A variation on the X-ray method is to focus a narrower beam onto the substrate and observe the back-scattered radiation. The intensity of this is related to the material density at the point of scattering and can be used to detect delamination and oxidation cavities. The advantage of this second method is that access to the component is only required from one side, making it ideal for routine examinations of in-service parts.[110]

A more recent development of the X-ray technique is X-ray tomography. In this system, the component is scanned from a number of different angles with an X-ray camera and the images submitted to a computer analysis that can generate a real-time, three-dimensional representation of the object. The same image variations as X-radiography apply, with the result that delaminations and voids can be mapped very accurately. This analysis is not one-sided and is also very expensive.

Infrared thermography uses a high-power heat source, such as a laser or IR lamp, to irradiate the surface of the component and an IR-sensitive detector that records an image created by variations in the thermal conductivity and diffusivity of the sample. A delamination or void forms a discontinuity in the conducting path and a consequent more-rapid heating of material in front of it. Analysis of the heat flux propagation through the material produces information on spatial distribution, size, depth, and thickness of the defects.[110-111] However, the resolution of this technique decreases with increasing material thickness as a consequence of general thermal conduction of the material and results in an upper limit of ~10 mm thickness for satisfactory resolution.

Ultrasound NDE has the drawback that a coupling medium is usually necessary to conduct the source waves to the material and any return signals back to the detector. In most cases, the medium is water, either in a tank or as a jet, and because the effect of moisture on C/C protection systems is still a problem area, deliberately immersing a protected component in water is to be avoided.

Of the remaining techniques, only eddy current measurement is of significant use for protected C/C.[110] This procedure involves measuring the changes in eddy currents induced in the material by an alternating magnetic field produced by an exciter coil. The material must, therefore, be an electrical conductor.

In practice, a coil is placed against the material and a sinusoidal current of adjustable frequency is passed through it. The opposing magnetic field generated by the induced eddy currents causes changes in the impedance of the coil, which are recorded and analyzed. All properties that affect the eddy current are detected (e.g., cracks in coatings and coating thickness variations). The depth of analysis obtainable depends on the electrical characteristics of the material and the frequency of the current used; the higher the frequency, the shallower the depth of penetration. In general, the analysis is sensitive to a depth of approximately 500 μm. Examination by this technique is very localized, but can be adapted for larger areas by constructing a scanning system. The technique benefits from one-side access and so assists in-service inspection and maintenance.

A minor application can be anticipated for the technique of neutronography in the examination of inhibited C/C composites. In this procedure, the sample is irradiated with a beam of neutrons and an image constructed from the transmitted fraction. Boron happens to be one of the most efficient absorbers of neutrons (hence, its use in control rods in nuclear reactors), and this property could be used as a quality control check to assess its uniformity of distribution in inhibited C/Cs that incorporate boron additives in the matrix phase.

Fabrication Processes/Techniques

One of the key technologies in designing C/C composite parts is how they are joined or attached to one another and to adjoining structures.

TABLE 3.11

Characteristics of Candidate High-Temperature
Fastener Materials

Fastener Material	Advantages	Disadvantages
Refractory metals	Strong	Needs coatings
	Large database	Heavy
		Physical/chemical incompatibilities
Carbon/Carbon	Physical/chemical compatibilities	Limited database
	Light weight	Needs coating
		Low shear strength
Ceramics	No coating required	Low fracture toughness
	Light weight	Limited database
	Low CTE	

It is assumed that C/C composites are being used at temperatures greater than those that practical engineering metallic structures can survive (>982°C).

Mechanical Fastening

C/C composite mechanical fastening efforts have, in many ways, duplicated the manner in which high-temperature metallic structures are fastened together. The main materials for fasteners are listed in Table 3.11, along with major advantages and disadvantages.

C/C fasteners have typically been of 2-D, 3-D, and 4-D construction. Weaving and braiding have been used to obtain 3-D and 4-D constructions, with 3-D providing the best strength. However, the highest interlaminar shear strength of 3-D has been only about 41 MPa at 1371°C.

Bushings

These can be either compliant or noncompliant. One type of compliant bushing is an alumina-borosilicate fabric that is impregnated with liquid phenolic-containing inhibitors. This type of bushing provides some protection to the hole walls and more importantly provides a "seating" mechanism for the fastener against the relatively rough surface of the C/C coating.

Noncompliant bushings have been made out of Si_3N_4. They are pressfit and provide for greater bearing area and strength.

Tolerances

It is very desirable to have a tight fit in a C/C joint. The accumulated tolerances on the following items make this either impossible or very difficult: coating of the hole, bushing (if required), fastener diameter, and cumulative effects of these for two mating parts.

Locking Fasteners

To withstand the severe dynamic and acoustic environments requires some type of locking feature. The normal assistance of high initial torque-up for standard metal joints is difficult in C/C joints. Metal fasteners tend to loosen due to adverse CTE effects, and nonmetal fasteners are inherently unable to sustain high initial torque-ups. For coated refractory metal fasteners, other locking methods that are used for standard metal fasteners (such as swaging effects) are not practical because the protective coatings will be damaged. Lock wiring is also not practical because it must be coated and would probably fracture during any severe twisting. For C/C and ceramic fasteners, high-temperature cements, applied to the threads, appear to be a feasible locking mechanism.

Brazing

Depending on the specific application, joints accomplished by brazing methods can have a number of advantages over mechanical fastener joining schemes. These advantages may include the following: (a) The size and weight of brazed assemblies can be significantly less; (b) brazing permits the use of designs involving small, compact, multipart components; and (c) the use of high-temperature braze materials allows components with brazed joints to function at higher operating temperatures.

The various applications briefly discussed above represent a wide range of operating temperatures. Therefore, a number of different braze materials and techniques have been developed in order to achieve system design requirements and to enhance the capabilities of the joined components. Some of the materials investigated for brazing C/C and graphite include silver-based filler metals, gold-based filler metals, zirconium metal, hafnium metal, and hafnium-diboride/hafnium-carbide powder mixtures. Other researchers have investigated brazing and diffusion bonding of C/C composites using titanium metal, silver-based filler metals, copper-based filler metals, various silicides ($MoSi_2$, $BiSi_2$, and $TiSi_2$), and various proprietary materials.[112–116]

Because C/C composite materials are new, very little research on high- to ultrahigh-temperature brazing applications has been published. However, a wealth of information exists for related materials applications. Graphites, carbons, nonoxide ceramics, and refractory metals are often brazed for moderate- to ultrahigh-temperature applications. Braze filler materials used in these applications could be suitable for brazing C/C composite materials.[112–115]

Liquid-solid-phase joining (brazing) has been much more successful than solid-phase joining (diffusion welding).

Brazing materials for high-temperature applications can be selected from ceramic, refractory-compound, or metallic systems. Ceramic brazes are typically glassy oxides and are not suitable for carbon structures at elevated temperatures because of the reaction of oxygen and carbon to form carbon monoxide gas. Refractory-compound brazes for ultrahigh-temperature use are typically refractory silicides, borides, or carbides. Metallic brazes can be composed of noble metals, active metals, refractory metals, or combinations of these. Because of the "unlimited" available carbon in C/C composite materials, graphites, and carbons, a eutectic braze of a metal carbide plus graphite is also possible for use at ultrahigh temperatures.[113, 116]

Graphite is not readily wet by most conventional filler metals. Filler metals used to join graphite should contain strong carbide formers because the bonding mechanism depends on carbide formation.

Noble metal filler metals can be made from combinations of gold, silver, platinum, palladium, cobalt, nickel, and copper. They generally do not wet graphite, but if the graphite is pretreated with an active or refractory metal to form carbides, such as those listed in Table 3.12, then these filler metal may wet the graphite (carbide) surface and form a satisfactory joint.[117] Table 3.13 lists several commercially available braze filler metals

TABLE 3.12

Thermal Data for Active Metals, Refractory Metals, and Their Carbides [°C = 5/9(°F − 32)].[117–119,123,124]

		Melting Point (°F)				Carbide
Metal	Carbide	Metal	Eut. 1 M + C1	Carbide C2	Eut. 2 C2 + G	CTE/F
Ti	TiC	3035	3000	5550	5030	5.5
Zr	ZrC	3366	3335	6188	5270	4.4
Hf	HfC	4032	3990	7106	5760	5.5
V	VC	3450	3000	4890	4760	3.7
Nb	NbC	4474	4270	6330	5980	4.1
Ta	TaC	5425	5150	7015	6235	3.7
Cr	Cr_3C_2	3407	2790	3440	P	5.5
Mo	MoC	4730	3990	4700	4685	3.7
W	WC	6170	4910	5030	P	2.9

Abbreviations: C1 — carbide in equilibrium with metal; C2 — carbide in equilibrium with graphite; CTE — coefficient of thermal expansion; Eut — eutectic; G — graphite; M — metal; P — peritectic.

TABLE 3.13

Commercially Available Noble Filler Metal Alloys That
May Wet Pretreated Graphite [°C = 5/9(°F − 32)][118,123]

Braze Filler Metal	Temperature °F		AWS Designation
	Solidus	Liquidus	
Re	5756	5756	
Ru	4532	4532	
Rh	3574	3574	
Pt	3216	3216	
Pt-40Ir	3542	3614	
Pt-40Rh	3515	3542	
Pt-20Pd-5Au	2993	3083	
Pt-60Cu	2192	2282	
Pd	2826	2826	
Pd-70Ni	2354	2408	
Pd-36Ni-10Cr	2250	2300	
Pd-40Ni	2260	2260	
Pd-35Co	2246	2255	
Pd-82Cu	1976	1994	
Au	1945	1945	
Au-35Pd	2601	2624	
Au-25Pd	2516	2570	
Au-13Pd	2300	2381	
Au-34Pd-36Ni	2075	2136	
Au-25Pd-25Ni	2016	2050	
Au-65Cu	1814	1850	BAu-3
Au-62.5Cu	1814	1814	BAu-1
Au-18Ni	1742	1742	BAu-4
Ag-33Pd-3Mn	2100	2250	
Ag-20Pd	1958	2120	
Ag-27Pt	1823	2120	
Ag-7.5Cu	1435	1635	BAg-19
Ni-4.5Si-3.1B	1800	1900	BNi-3
Ni-4Si-2B-1Fe	1800	1950	BNi-4
Ni-23Mn-7Si-4Cu	1800	1850	BNi-8
Cu	1980	1980	Bcu-1

that may be suitable for brazing a surface pretreated with an active or refractory metal to form carbides.[112,117,118]

Active filler metals can be formulated by combining varying amounts of the carbide-forming metals in Table 3.12 with the noble filler metals in Table 3.13.

Braze filler metals for ultrahigh-temperature use could be produced from graphite and metal carbides by placing a carbide-forming metal foil between the graphite pieces and heating above the metal carbide and graphite eutectic temperature. In general, carbides are brittle and have a CTE mismatch with C/C composites and graphite materials, but if the carbides are alloyed with graphite, the resulting eutectic microstructure

- Join C-C components, coated with a HfB_2-HfC reaction-sintered material, to metal or graphitic sleeves using a moderate temperature braze alloy (ex. Au-Pd)

Potential application:
  Rocket propulsion assemblies

Example:

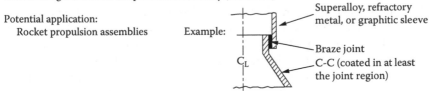

Superalloy, refractory metal, or graphitic sleeve

Braze joint

C-C (coated in at least the joint region)

- For making attachments to C-C components:
  (1) Coat C-C components with HfB_2-HfC reaction-sintered material,
  (2) Plasma-spray a metal layer to allow next step,
  (3) Attach metal foils, wires, tubes, etc. by spot welding or electron-beam micro-welding

Potential application:
  Satellite radiator panels

Example:

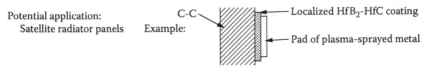

Localized HfB_2-HfC coating

Pad of plasma-sprayed metal

FIGURE 3.25
Two joining/attachment concepts for continuing technology development efforts.

may have improved toughness and improved CTE match. The improved toughness of the eutectic structure was proven by Rossi[118–120] on bulk metal carbide and carbon eutectic samples, but loss of flexural strength was also observed for high-carbon contents in the titanium and zirconium systems evaluated; see Figure 3.25.

Components assembled by braze joining methods have been designed primarily for various rocket propulsion applications, including hot gas valve component joints and injector/thrust-chamber joints. Joining schemes for assembly of C/C parts for the National Aerospace Plane program have also been developed, as have joining schemes for the assembly of C/C and graphite armor tiles for use in the large vacuum-vessel/plasma chamber for the Doublet III fusion research device.[112–125]

Another unique approach to brazing C/C composites to refractory metals for actively cooled hypersonic vehicle structures was conducted by Khatri et al.,[127] where they developed successful brazing methods and materials to address the thermal expansion mismatch problems associated with the joining of 2-D C/C composites to refractory metals for high-temperature heat exchanger assemblies. Upon cooling from brazing temperature, thermal expansion mismatch between C/C composite and refractory metal adherents results in the formation of deleterious residual stress portions throughout the joint. Interlaminar shear stresses often result in delamination of the C/C composites. Residual tensile stress in the metal can also trigger transverse crack formation. Even barring material failure, the residual stress pattern would compromise the ability of the joint to withstand the harsh thermal/ acoustic loads associated with hypersonic service vehicle. Using finite element modeling and iterative process trials, they developed a unique multilayered joint design. A compliant interlayer accommodates the CTE mismatch via plastic flow, while a high stiffness interlayer reacts shear stresses to "off-load" the 2-D C/C. Local 2-D reinforcement of the C/C composite further increases the interlaminar strength and z-direction thermal conductivity of 2-D C/C.

To join C/C to Mo-Re, Khatri et al. developed a multiple interlayer technology to minimize the residual stresses in the C/C composite. Additionally, they used a patented Z-fiber process to reinforce the 2-D C/C composite in a direction normal to the fiber plane. The effect of Z-fiber reinforcement was twofold. First, the Z-fibers improve the interlaminar strength of the C/C composite, thereby improving the braze joint strength. Second, if high thermal conductivity Z-fibers are utilized, the through-thickness thermal conductivity of the C/C composite is also improved.

The approach is to use interlayers. Interlayers provide multiple benefits. They permit a buildup of braze thickness, reducing peak stresses. More importantly, proper selection of the interlayer material and thickness will enable very high shear stresses to be absorbed by the interlayer and isolated from the C/C. The interlayer geometry used in this program is shown in Figure 3.26. This concept incorporates into the braze joints a compatible pure copper interlayer to accommodate the CTE mismatch and a high-strength tungsten interlayer to react the shear stresses and off-load the C/C composite.

The best results of C/C to C/C composites brazed to themselves were found by using Khatri et al.[127] VSi braze material at 1700°C and <5 sec to braze. The C/C to Mo-Re joints were fabricated using silver-ABA and gold-ABA brazes and the aforementioned copper and tungsten interlayers. The test results showed that the strengths of the specimen brazed with silver-ABA were higher than those brazed with gold-ABA. In fact, many of the specimens failed due to crushing of the specimen ends and not near the braze interface The highest strengths were obtained by using the Mo-Re z-pins in the adherent C/C composites. The lower strengths of the

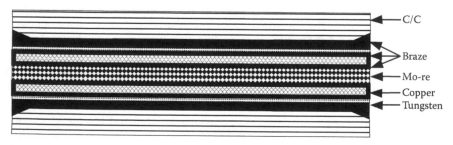

FIGURE 3.26
Braze joint geometry to join C/C composites to Mo-Re.

gold-ABA-brazed specimen are due to the lower wetting characteristics of the C/C composite by gold braze as compared to silver braze. Any presence of porosity between interlayers can be eliminated by using higher amounts of filler metal.

High-Energy Electron Beam (HEEB)

Goodman, Birx, and Dave[128] conducted a series of experiments to demonstrate the capabilities of induction accelerators for industrial material processing applications (see Figure 3.27). Joining of C/C composites utilizes the deep penetrating power of high electrons to melt a braze interlayer. Their results showed that strong joints at high throughput power (HEEB) can join C/C to itself and to high-temperature metal substrates.

Advantages include:

- A localized heating process with an accurately controllable heat source that preferentially deposits power in a thin high-density braze interlayer. The heat can be applied where needed to produce strong carbide-based joints, without heating the entire structure.
- Heat-affected-zone in a HEEB-joined C/C composite is sufficiently small that edge thermal stresses can be relieved by the ductile braze joint.

This fast, high-throughput process, consistent with the processing of a variety of C/C braze filler metals and geometries to near-net shape, utilizes the large electron beam power (50 kW–1 MW), which allows the braze region to be heated before heat is conducted to the interior of the structure. By defocussing or sweeping the beam, a wide, uniform area can be irradiated. The heating occurs over several seconds to a minute, allowing the C/C joint edges to heat, the braze filler metal to flow and wet the joint, and a strong carbide-based bond to form.

Brazing of several forms of C/C has been demonstrated by Goodman et al.,[128] including C/C made using pitch resin precursor and C/C produced by chemical vapor deposition. These materials have been joined in C/C "sandwiches" and also to high-temperature metallic substrates including nickel-based Haynes 188 superalloy and moly-50% rhenium. Wetting of the carbon is accomplished either using a carbiding element (e.g., titanium) or by premetallizing the C/C surface. Sublimation of a thin layer (~50 μm) of titanium is sufficient to ensure uniform wetting.

The C/C to moly-rhenium braze of single lap joints showed that the underside of the Mo-Re was heated to a constant temperature and the top surface of the C/C was cooled by directed air. The use of infrared photography imaging method allowed for the prediction of braze shape,

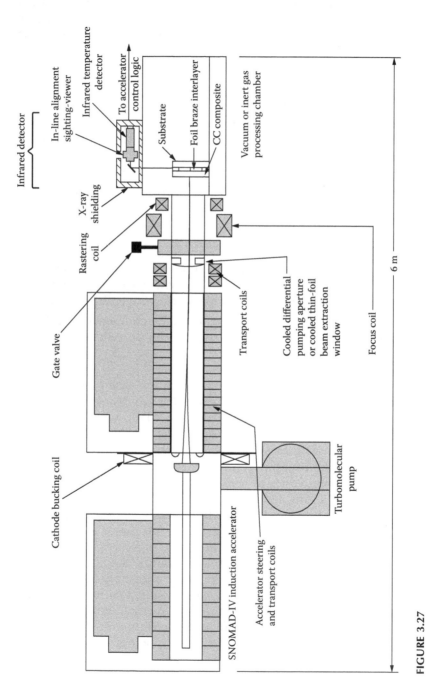

FIGURE 3.27
Design of HEEB material processing system.[128]

percentage area bonding, and thermal conductivity. When the braze joint was pulled apart to measure shear strength, the braze joint shape closely approximated those of the central, hottest contours.

The results of single lap joint shear tests showed that the HEEB joints are stronger than the C/C material, and the shear failure mode was always between layers of C/C rather than at the braze joint. The shear test results were close to estimates of the C/C interlaminar strength and slightly higher than recorded using conventional brazing and similar C/C materials.[129]

Other work for space applications that used brazing incorporated refractory metal tubes for cooling into a C/C composite heat exchanger.[130]

Applications

The microtexture, that is to say the microstructure and morphology, of the various types of C/C composite will differ according to the type of raw materials and the processing conditions. Further complication will arise from the use of subtle treatments such as surface modification of the fibers and inclusion of oxidation protection. All such differences in microstructure exert a considerable influence on the properties of the materials.

The architecture of the carbon fiber reinforcement can take many forms: Random short fibers, unidirectional (UD) continuous fibers, braided structures, laminated fabrics, orthogonal 3-D weaves (in either Cartesian or cylindrical coordinates), or multidirectional structures may be used. Commercial carbon fibers are manufactured from rayon, acrylic, and pitch precursors. The mechanical and other properties of these fibers vary considerably, again depending on raw materials and processing conditions. High-strength fibers are usually obtained from acrylic precursors, whereas pitch is preferred for the fibers of highest modulus. The carbon fiber is generally post-treated to improve weaving characteristics or to modify the bonding of fiber to matrix.

The matrix may vary from an isotropic, glassy carbon produced from the pyrolysis of a thermoset resin, through to a highly oriented anisotropic graphitic carbon arising from mesophase, developed during the carbonization of pitch. A matrix material may also be formed by chemical vapor deposition (CVD) of carbon from the cracking of hydrocarbons. The structure of the CVD matrix depends very much on the processing conditions. All of the production routes used to manufacture C/C are slow, inefficient, multistep batch processes. As a result, many finished products tend to be made using a variety of methods. Composites fabricated by the thermoset resin pyrolysis route, for example, are often densified using CVD or a mixture of pitch and a different resin from that used in the initial cycle.

The number of possible combinations of parameters is almost limitless. The theme developed throughout this chapter on C/Cs as a family of materials whose properties may be tailored to suit a specific application (see Figure 3.28) will thus become apparent.

Brakes and Clutches

Approximately 63% by volume of the C/C produced in the world is used in aircraft braking systems.

It is now commercially advantageous to employ C/C brakes on civil subsonic aircraft. Furthermore, the use of C/C has been exploited or postulated for a number of land vehicles such as racing cars, high-speed trains, and even main battle tanks (MBTs).

Aircraft brakes today are not significantly different in design from the modern brakes first introduced in the 1950s. The biggest difference is in the type of material used for the friction material in the brakes. The original brakes used steel against metal ceramic linings. The aircraft brake assemblies shown in Figure 3.29 have several rotating disks called rotors and stationary disks called stators that are located between two additional stationary disks known as end plates. The disks are collectively called the brake stack. The rotors are keyed to a wheel that is generally made from aluminum. The stators and pressure plates are keyed to a steel torque tube. The end plate is attached to the torque tube by a spline or a floating back plate assembly. The torque tube is bolted to the aluminum brake housing. Braking action occurs when fluid pressure is metered simultaneously to several hydraulic pistons carried in the brake housing, which clamps the brake stack together. The resulting brake torque is transmitted by the torque tube to the brake housing, and from the housing to the landing gear via the torque take-out.

A number of performance parameters are important in aircraft brake design. These include peak torque, oxidation, and stability. Such variables are controlled by engineering design, composition, and processing conditions of the friction material. The two primary advantages of C/C are heat capacity (2.5 times greater than that of steel) and high strength at elevated temperatures (twice that of steel). The result is a 40% saving in weight compared to metal brakes and a doubling of their service life when measured as landings per overhaul (LPO). A wide-bodied airliner such as Airbus A320 would expect to complete 2500 LPO when fitted with C/C brakes as opposed to around 1500 using a metal equivalent (see Tables 3.14 and 3.15).

Three distinct types of C/C composites are currently used in braking systems; see Figure 3.30. These are carbon fabric laminates, semirandom chopped carbon fibers, and laminated carbon fiber felts with crossply

Applications → Requirements ↓	Non-structural — Major existing products		Non-structural — Other/commercial				Structural — Advanced applications/R&D		
	Aircraft brakes	Rocket nozzles/ reentry	Automotive friction	Biomedical	Energy/ thermal/ nuclear	Expendable engines, missile	Turbine/ multi cycle engines, man-rated	Hyperve-locity vehicles, TPS/ structures	Spacecraft thermal shields
Light weight	✓	✓	✓	✓		✓	✓	✓	✓
Low cost △1	✓		✓	✓	✓	✓			
Large size △2		✓			✓			✓	✓
Friction properties	✓		✓						
Controlled ablation		✓		✓					
High strength and, or stiffness				✓	✓	✓	✓	✓	✓
Tailorable thermal properties (e.g., high conductivity)	✓	✓	✓		✓	✓	✓	✓	
Oxidation resistance, multiple cycle							✓	✓	

Drivers: △1 Low cost is <$1,000/LB △2 Large is >30 FT2 planform area or length >3.0 FT

FIGURE 3.28
Key requirements for C/C applications.

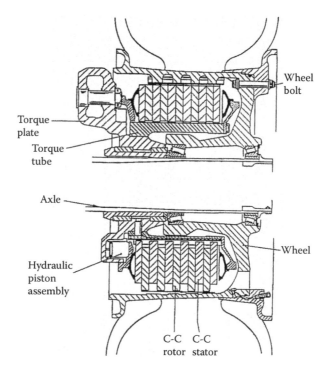

FIGURE 3.29
Schematic diagram of a C/C aircraft disc brake system.[58]

TABLE 3.14

Carbon Brake Advantages Over Steel Brakes[131]

Major Benefits

- Weight savings of 33%
- Superior RTO performance
  - Flatter torque curves (less brake fade)
  - No friction material melting
  - No welding
- Longer brake life
- Carbon strength retention (RTO conditions)
- Dimensional stability (no warping)

Increased Design Considerations

- Temperature control
  - More components
  - Increased complexities (heat shields, insulator materials)
- Environmental factors (runway deicers)
- Corrosion control (wheel protective coatings)

TABLE 3.15

Comparison of Thermal Control Design Features Typical Steel vs. Carbon Brake Installations[131]

	Steel	Carbon
Heat Shields		
Wheel	X	X
Torque tube	n/a	X
Axle	n/a	X
Structural Insulators		
Piston insulators	X	X
Piston actuator pad	n/a	X
Torque tube pedestal	n/a	X
Backing plate pad	n/a	X
Piston housing to torque tube interface	X	X
Wheel drive key to wheel drive interfaces	n/a	X
Convective cooling enhancement		
Cooling vents in torque tube	n/a	X
Wheel and piston housing windows	X	X
	(baseline)	(increased)

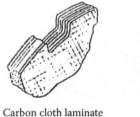

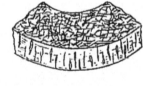

Carbon cloth laminate   Chopped fiber   Laminated matte
semi-random orientation

FIGURE 3.30
The three types of carbon fiber reinforcement used in C/C composite brakes.

reinforcement. The matrix consists exclusively of pyrolytic carbon or a combination of pyrolytic and glassy carbons. The pyrolytic carbon matrix is obtained via a CVD process. The glassy carbon is produced from the carbonization of a high char yield resin, generally a phenolic. The resin is used in the consolidation of carbon fibers in the shape of a disc or in the densification of the porous disc by resin impregnation in the final stages of processing.

C/C composite brake discs must be protected from oxidation, which occurs from repeated excursions to high temperatures during braking. Oxidation protection is achieved by the application of a "trowelable" glass-forming penetrant to block the active sites on nonrubbing surfaces or by adding an inhibitor during fabrication.

For protection against oxidation, coatings are used on the C/C such as Si_3N_4 and SiC. In one method, when the carbon part is ready to be treated for oxidation protection, it is placed in a restraint and loaded in a graphite-coated retort, encapsulated in a mixture of powders, and heated to a high temperature. The resulting reaction converts the powders to a vapor state that reacts on the outer layer to form a coating.

For the foreseeable future, aircraft braking systems as they are known today will remain. Refinements are being made to the C/C material and to the thermal management of the braking system.[132]

C/C brakes were introduced into Formula 1 motor racing in the early 1980s. Today, brakes of this type are universally used in the sport. In contrast to aircraft brakes, those employed in Formula 1 are of the disc and pad type, having evolved from the metal brakes they replaced. The brakes are operated by hydraulic callipers pressing the C/C pads against the ventilated discs (Figure 3.31).[133] Disc systems are used, however, in the clutches of Formula 1 vehicles. The operation of the C/C brakes depends on their environment and on cooling. It is important that the

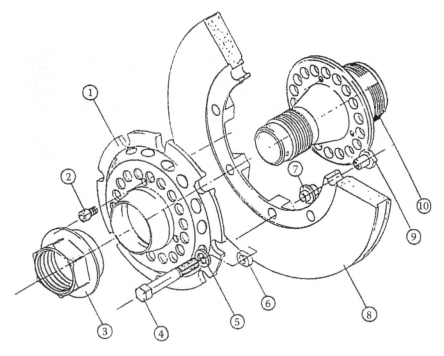

FIGURE 3.31
Exploded diagram of Formula 1 C/C brake system: 1, disc bell; 2, bell-retaining screw; 3, extended wheel nut; 4, disc/disc bell bolt; 5, retaining washer; 6, drive lug; 7, mounting bush; 8, C/C disc; 9, mounting nut; 10, axle.

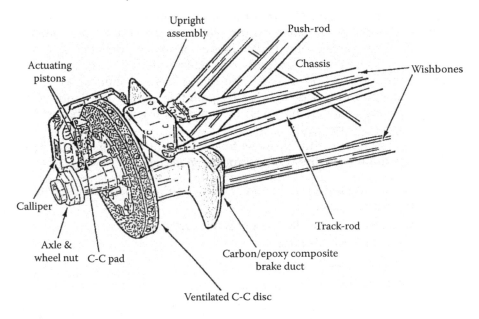

FIGURE 3.32
Configuration of brakes and ventilation ducts on a Formula 1 racing car.

geometry of the upright and cooling assembly complies with specifications concerning the disc assembly (Figure 3.32). The essential requirement is to provide adequate passages for cooling air within the uprights, airflow being directed into the vent holes in the disc. A cross-section of at least 70 cm^2 of airflow is generally required for each disc.

The major advantages of C/C braking materials are their light weight, excellent thermomechanical performance, and inertness. Disadvantages are those inherent to all C/C applications: cost and oxidation. Aircraft and racing cars use C/C friction materials to obtain improved performance at lower weight than a metal alternative. The low density of C/C coupled with the tendency towards lower costs suggests possible applications in other forms of transport or machinery. The French have already tested such systems on their TGV Atlantique high-speed passenger trains. The thermomechanical stability of C/C allows it to absorb much more energy over a short period of time than a metal or organic brake. The use of C/C in brakes in emergency situations is a very important consideration. One could imagine the development of emergency brakes in high-momentum situations such as stopping trains and large road vehicles over short distances or preventing injury and damage due to the failure of lifting gear in mine shafts, for example. C/C brakes and clutches employed in military vehicles (mainly in Japan) improve airportability or allow a more efficient distribution of the vehicle's weight into the area where it is most required, that of armored protection. The cost of C/C would generally be considered prohibitive for use in domestic vehicles. Work has been carried out, with a degree of success, developing motor car brakes operating carbon pads on cast-iron discs.[134] It is difficult to foresee the introduction of C/C into the brakes and clutches of urban motor transport unless demanded by pollution control legislation, C/C particles being relatively harmless by comparison with the asbestos and organic materials presently used.

For the future, it is thus possible to identify a number of development objectives to improve the C/C composite materials.

1. An increase in thermal conductivity especially perpendicular to the friction surface
2. An increase in the static coefficient of friction
3. A greater strain to failure
4. Improved specific strength
5. A reduction in costs

One of the major reasons for adopting carbon brakes is a saving in weight. Designers are therefore reluctant to improve strength at the cost of increased weight. A critical design consideration is the material's strength in the "fully worn" condition. An increase in specific strength is thus sought. Simultaneously, the volume of the heat sink is critical and hence a high-density composite is required that would possess the additional benefit of a higher thermal conductivity. Similarly, the use of continuous, three-dimensionally woven fibers would also increase thermal conductivity as well as specific strength and toughness. The improved toughness of a continuous fiber reinforcement network would be less sensitive to the non-uniform distribution of thermomechanical loading. An increase in the static coefficient of friction of the C/C would eliminate the unwanted high torques developed in aircraft brakes during taxiing or racing cars when "cold." It is difficult to imagine, however, how this could be achieved.

Evidence suggests, then, that the optimum material configuration for brakes would be a 3-D polar weave with a high-density pitch-derived matrix. In reality, however, it is unlikely that such a material would be introduced because it would not be economically viable, costing 7 times that for the essentially 2-D discontinuous material that, although not ideal, is adequate for the task. It is not inconceivable that such a product form will be introduced into the relatively cost-insensitive Formula 1 application, where the intrinsically more thermodynamically stable material would also alleviate oxidation problems. The brakes presently used in Formula 1 are somewhat inefficient in both materials and operation. High-density 3-D material might perhaps allow later braking into corners.

Furthermore, the development of a three-disc (two stators, one rotor) system along similar lines to an aircraft brake would reduce the risk of oxidation to the friction surface, ensure a more uniformly distributed braking pressure, and allow a greater friction contact surface area, provided the thermal conductivity was sufficiently high to prevent overheating. The brake unit could then be packaged in a smaller configuration with a greatly reduced aerodynamic cross-section and lower inertia, which would be manifest in improved acceleration out of a bend. Both discs and pads would benefit from improved oxidation protection to the nonrubbing surfaces, especially within ventilation holes.

Most of today's modern transportation runs on C/C braking systems. With the advent of huge jets, aviation manufacturers had to find better ways to quickly stop such large loads. As a result, carbon composite materials have been developed for disc brake systems that have a high capacity for heat and high-temperature strength.

Development of inventive processes for infiltrating ceramic additives into C/C disc brake preforms are effective in achieving homogeneous dispersion of the additives throughout preform cross-sections. The refractory brake material technology is therefore suited for manufacturing aerospace, automotive, and heavy truck C/C disc brakes, as well as for any application where high friction/low wear is required.

A C/C racing car clutch typically weighs less than half as much as a conventional sintered metal unit of comparable performance. Its light weight reduces the moment of inertia, the centrifugal force, and the kinetic energy developed by the clutch, enabling a high-revving engine to respond more quickly to a driver's demands. The clutch's quicker, cleaner engagement and disengagement also make more rapid shifting possible.

Now engineers have designed a clutch using C/C material that seems bound to become as popular on the racing circuits as C/C brakes. It offers significantly longer life than conventional clutches, and basically eliminates clutch failure as a problem during a race. These improvements in life and reliability can be a significant advantage in competition. And its special capabilities suggest that there may be other, nonracing applications for the C/C clutch, as well.

Perhaps the single most important characteristic of the C/C clutch is its resistance to wear. In top professional racing applications, a conventional clutch operates for a couple of warm-up runs and one race; after that, it's replaced. By contrast, C/C clutches commonly last five or more races. One C/C unit has accumulated 2000 miles, the equivalent of perhaps 7 to 10 races (remarkable in a business where engines are usually rebuilt after a few hundred miles!).

Today's C/C clutches are much more expensive than metallic clutches; however, they're worth it if you demand performance. One pro stock dragracing car uses C/C clutch surfaces to transmit the torque of its

2500 hp engine. It also uses C/C brakes to rapidly bring the car to a stop from speeds approaching 300 mph. In so doing, the brakes absorb energy at the rate of 3,000,000 ft-lb/sec, and the C/C material briefly reaches temperatures that exceed 1371°C.

We can expect C/C clutches to become commonplace where the ultimate in performance is required: in auto racing (including stock cars) and in drag racing. They're also likely to appear in big-investment, high-usage vehicles such as heavy-duty highway trucks and road-building equipment. Here, the extra cost of the clutch may be very little compared to the savings in downtime resulting from the low wear rate. Other possible applications are in equipment that is difficult to service because of inaccessibility (such as mining machinery) or that operates in an adverse environment (such as nuclear reactor control rod drives or chemical processing gear).

Pistons

Pistons for internal combustion engines have been made of such materials as cast iron, steel, and aluminum. Most pistons manufactured today are made of aluminum. However, aluminum has low strength and stiffness at elevated temperatures, and aluminum has a large coefficient of thermal expansion. A new piston concept, made of C/C refractory-composite material, has been developed that overcomes a number of the shortcomings of aluminum pistons. C/C material, developed in the early 1960s, is lighter weight than aluminum, has higher strength and stiffness than aluminum, and maintains these properties at temperatures over 1371°C. In addition, a low coefficient of thermal expansion and a high thermal conductivity give C/C material excellent resistance to thermal shock. An effort called the Advanced C/C Piston Program was started in 1986 to develop and test C/C pistons for use in two-stroke-cycle and four-stroke-cycle engines. The C/C pistons developed under this program were designed to be replacements for existing aluminum pistons, use standard piston pin assemblies, and use standard ring sets. The purpose of the engine tests was to show that pistons made of C/C material could be successfully operated in a two-stroke-cycle engine and a four-stroke-cycle engine.[135]

C/C pistons can potentially enable engines to be more reliable, to be more efficient (lower hydrocarbon emissions, greater fuel efficiency), and to have greater power output. By utilizing the unique characteristics of C/C material — very low expansion rate, low weight, high strength and stiffness at elevated temperatures, and high thermal conductivity — a C/C piston can (1) have greater resistance to structural damage caused by overheating, lean air-fuel mixture conditions, and detonation, (2) be designed to be lighter-weight than an aluminum piston, thus reducing the reciprocating mass of an engine, and (3) be operated in a higher combustion temperature environment without failure.

The state-of-the-art arrangement shown in Figure 3.33 requires relatively large clearances to account for thermal expansion and distortions due to thermal gradients in the engine. The large clearance creates a crevice that could entrap the fuel mixture that escapes combustion and is expelled as unburned hydrocarbon. The large clearance also allows the piston to rock as it reciprocates in the cylinder, particularly during cold start-up. Poor ring performance results when the piston is in the cocked position. Piston rocking during cold start-ups can also create unwanted noise. If a C/C piston replaces the aluminum piston in the cast-iron sleeve, clearance varies with engine temperature in a manner opposite that which occurs with the aluminum piston because the C/C piston does not grow while the cast-iron sleeve does. Minimum clearance must therefore be at cold start-up instead of at operating temperature. The clearance at operating temperature is therefore dictated by the minimum temperature that the engine is expected to encounter. This is an undesirable restriction. The arrangement shown on the right in Figure 3.33 alleviates this problem. Piston-to-cylinder clearance can then be set to the minimum allowed by the influence of thermal distortions of the block on the C/C sleeve. The crevice volume and piston rocking are therefore reduced over the entire

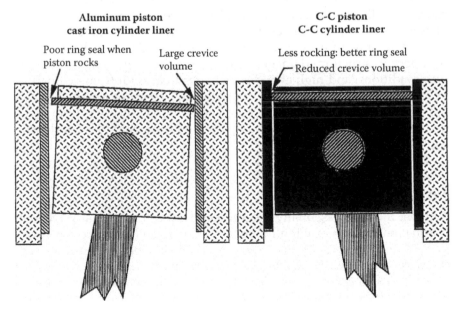

FIGURE 3.33
C/C piston/sleeve concept.

operating temperature range. Note also that the ring on the C/C piston is moved closer to the top of the piston, which further reduces the crevice volume. The higher ring position is allowed because the C/C piston retains its strength and stiffness with increasing temperature (see Figure 3.33).

The development and testing that has taken place under the Advanced C/C Piston Program[135] has shown that pistons can be manufactured from C/C refractory-composite materials and that a C/C piston can be successfully operated in an internal combustion engine. A total of eight pistons have been successfully operated in two types of engines: five pistons in a two-stroke-cycle engine and three pistons in a four-stroke-cycle engine.

C/C pistons can be designed to replace existing aluminum pistons and use standard sealing ring and piston pin assemblies. By utilizing all of the physical properties of C/C refractory-composite material, pistons can be made lighter-weight to reduce the reciprocating mass of an engine; pistons can be made to have improved structural reliability when used under the same operating conditions as aluminum pistons; and pistons can be made to have a lower piston-to-cylinder wall clearance.

C/C pistons can potentially enable high-performance engines to be more efficient, be more reliable, and have greater power output. An engine equipped with C/C pistons can operate using leaner air-fuel mixtures because these pistons can function in higher combustion temperature environments without failure. An engine using leaner air-fuel mixtures can potentially produce more power, have fewer hydrocarbon emissions, or have greater fuel efficiency. In addition, because C/C material retains its strength and stiffness at high temperatures, C/C pistons have greater resistance to structural damage caused by overheating, lean air-fuel mixture conditions, and high cylinder pressures that result from detonation.

The C/C pistons being made commercially consist of a carbon matrix reinforced with carbon fibers, and the manufacturer processes the material by CVI, which results in improved properties compared to other densification methods.

Rocket Motors

A rocket motor is essentially a venturi system, comprising a convergent portion that is embedded in the motor, a throat, and an exit cone (divergent). The exhaust gases from the propellant chamber pass out through the throat and then finally out of the exit nozzle. On average a rocket motor burns for around 30 sec. The demands on the construction materials are relatively short-lived, but extremely intense. The size of components is dependent on the rocket's payload, and a sketch of a rocket motor's construction is shown in Figure 3.34.

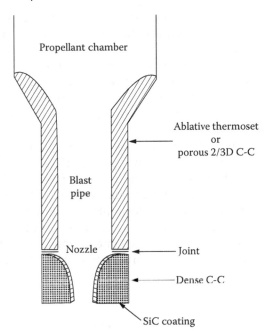

FIGURE 3.34
Schematic diagram of a typical rocket motor construction.[58]

Orthogonally woven 3-D C/C has been successfully employed in rocket throat manufacture of small radius (up to a few tens of millimeters) for a number of years. Larger components require to be made from polar woven material, either machined from cylinders or contoured shapes. Components of this type have been test fired both in the United States and France. Ongoing research and development have resulted in a successive reduction in the number of parts required such that the whole of the entry and throat can now be made as a single component, whereas previously they were separate or a number of pieces. They are now integral so that gaps, joints, or other weak points, which are sites for stress corrosion, are eliminated. Furthermore, there is no longer a difference in thermal expansion.

The criteria for selecting an exit nozzle material are even more stringent than are those for the throat. The nozzle is generally constructed out of a dense C/C composite that may be coated with a ceramic, generally SiC, to ensure good oxidation and wear resistance. It is important to note that although the burn time of the motor is relatively short, the nozzle must possess excellent dimensional stability, or else the direction of thrust cannot be properly controlled. Temperatures often exceed 2000°C, gas velocities are supersonic, and the exit gases contain uncombusted fuel and water. Such an environment would obviously severely erode unprotected C/C composites, hence the requirement for a modicum of ceramic protection. Because rocket nozzles are not typically reused they do not necessarily require oxidation protection and may be allowed to burn partially. This burning must, of course, be taken into account at the design stage.

Dense C/C is preferred because of its superior ablation resistance. A number of companies operate the HIPIC process to manufacture the rocket motor components.

The triaxial braiding of graphite fibers for the manufacture of C/C rocket nozzle component preforms has proven to be a variable alternate to other conventional labor-intensive involute preforms by automated hands-off techniques. Indications are the braided C/C structure offers unique fiber architectural design flexibility by allowing a 4-D triaxial structure in the entrance and throat regions and permitting tailoring to a triaxial membrane structure in the exit cone regions. Preliminary thermal structural characterization indicates this type material to be ideally suited for rocket motor componentry.[136]

Heat Shields (Thermal Protection Systems)

When a rocket blasts into space with velocities exceeding 27,000 km h⁻¹, the heat generated at the leading edges can lead to temperatures as high as 1400°C. Re-entry temperatures can be even higher, approaching 1700°C, and well beyond the operation temperatures of metals.[26,137] On a weight-for-weight basis, C/C can endure higher temperatures for longer periods of time than any other ablative material. Thermal shock resistance permits rapid transition from -160°C in the cold of space to close to 1700°C during re-entry without fracture. Although the space shuttle is perhaps the most well-known example of a C/C re-entry heat shield, the greatest number of parts produced are used in the nose cones of ballistic missiles. All the British, French, and American strategic nuclear missiles employ C/C heat shields that provide ablation and heat resistance within a structural member. The primary purpose of the shield is to protect the crew (in the case of an operated vehicle) or instrumentation from the searing heat of re-entry. The loading on a nose cone during re-entry is such that 3-D C/C is the most convenient material.

The surface temperature rise occurs almost immediately when the vehicle enters the earth's atmosphere. The thermal conductivity of the graphitic 3-D C/C is high enough to eliminate thermomechanical surface overload and avoid surface cracking. Additionally, the specific heat is so high that the component operates as a heat sink, absorbing the heat flux without any problems. The rate of ablation/erosion, driven by the air speed, depends critically on the grain size of the material. HIPIC-densified products are characterized by their fine grain morphology and low levels of porosity and are thus exceptionally ablation resistant. A problem arises, however, in constructing large-scale components such as those used in the leading edges and nose cone of the shuttle. It is not possible to make large thick pieces by CVD, so the only route available is using thermosetting resins. The nose cone and leading edge of the space shuttle are formed from 2-D carbon fabric impregnated with phenolic resins carbonized to around 1000°C. Reimpregnation is carried out using furan or furan/pitch blends, and usually around four cycles are required. C/C re-entry vehicles are generally protected against oxidation, especially if, as in the case of the shuttle, the parts are designed to be reusable.

Other applications for C/C composites include the heat shield material for the Russian Buran space shuttle;[138] thermal protection systems (TPS) for numerous spacecraft, scientific probers, and instrument packages;[139] and the U.S. space shuttle.[140] Due to the disaster and disintegration encountered by the Shuttle Columbia in 2003 and to give shuttle crews a better chance to survive the sort of TPS damage that brought it down, NASA is developing ways to find and fix TPS breaches in orbit, something the Columbia crew was not equipped to do.

NASA engineers have already developed a sort of high-tech caulk gun attached to a strap-on spacesuit backpack, where repair materials are mixed and ejected onto damaged ceramic tiles. Ground tests at Johnson Space Center in vacuum and an arc-jet facility designed to mimic re-entry have been positive, and astronauts have started working out repair procedures.

Repairing the reinforced C/C panels is a challenge but a technique has been developed to repair C/C rocket motor nozzles. It combines a patch of C/C, backed by adhesive and held in place with "something like a moly bolt" used to hang pictures in drywall. Astronauts would apply the patch over damaged C/C, inject a glob of plugging material behind it, and leave the bolt head outside to burn off on re-entry.

For inspecting an orbiter in flight, NASA engineers have proposed a robot arm extension patterned on one built to test plume effects from the shuttle reaction control system in proximity operations. Computer modeling has demonstrated that a camera on the end of the extension would be able to examine an orbiter's belly and leading edges, and it likely will become standard equipment on future flights.

Aero-Engine Components

The thermodynamic efficiency of heat engines such as gas turbines is greatly improved with an increase in operating temperature. There is therefore a driving force to develop C/C jet engine components. For the same power output the engines could then be made smaller, lighter, and more fuel efficient. The toughness, most especially in impact, of C/C is two orders of magnitude greater than conventional ceramics. Jet engine motors can be made in single-piece format from C/C, thus saving fabrication and assembly costs. C/C might well, therefore, be chosen as a structural material in an aircraft's basic airframe. In terms of properties, C/C is the ideal choice for hot engine components; its exploitation, though, is extremely limited due to the problems of oxidation. Until protective coatings, capable of withstanding extreme thermal cycling in aggressive oxidative atmospheres, are developed it is unlikely that the promise of C/C will ever be achieved in the theater of jet engines. A number of materials producers are involved in joint venture research programs with aero-engine manufacturers, and a selection of prototypes has been made. To date, however, none have succeeded in fabricating a suitable flying component. In view of the high toughness requirements, it is more likely that the 3-D configuration will be preferred, although production difficulties and costs may dictate a simple 2-D structure will be the first to be exploited.

Biomedical Devices

Elemental carbon is known to have the best biocompatibility of all known materials.[141] It is compatible with bones, blood, and soft tissue. The high cost of C/C compared with its competitors, be they metallic or polymeric, in the biomedical field may at first glance preclude its use. Consider, however, a rigid metal plate as frequently used in the repair of fractures. Such a device, when firmly attached to a long bone such as the femur, will inevitably result in an altered stress distribution. The modulus of the bone is generally an order of magnitude lower than, say, a stainless steel plate of roughly the same cross-sectional area.[142] A plate made from this alloy will transmit around 90% of the load, causing the underlying bone to become porotic. It is thus paramount that such plates be removed as soon as the fracture has healed. Failure to do so would most likely result in a spontaneous refracture at some later date.

Thus, although metals possess the required strength and toughness for use during the healing phase, they are far from satisfactory. C/C can be engineered in such a way as to possess mechanical properties identical to those of bone (Figure 3.35), thus eliminating the need for removal once the healing is complete. There is a considerable interest in materials such as C/C, which are considered to be "bio-active," in many areas of implant surgery. The objective is to use a material that is able to encourage the formation of bone rather than soft tissue at the interface. C/C has been shown experimentally to be biodegradable in such applications: the body gradually replacing it with bone.[137]

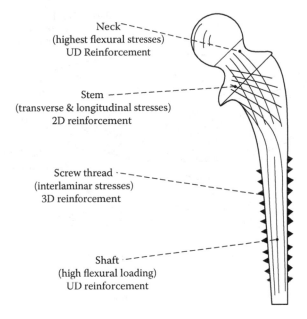

Neck~------ ~
(highest flexural stresses)
UD Reinforcement

Stem -------------
(transverse & longitudinal stresses)
2D reinforcement

Screw thread ·------
(interlaminar stresses)
3D reinforcement

Shaft ·-----------
(high flexural loading)
UD reinforcement

FIGURE 3.35
Schematic diagram of carbon–carbon hip joint replacement designed to imitate the femur structure.[58]

For obvious reasons, licenses to allow the use of materials take a long time to obtain. As a result the biomedical uses of C/C are very limited. There are reports in the literature, however, of its use in artificial hearts for animals,[141] fixations for carbon fiber artificial ligaments[143] and bone plates in osteosynthesis, and as endoprosthesis.[144]

Industrial and Miscellaneous Applications

Reflective Panels

Developers have continued work on reflective, lightweight, low-outgassing radio-antenna reflector panels that include C/C surface laminates supported by C/C core structures. Essential to the fabrication of such a panel is a technique for the densification of the surface laminate in preparation for polishing to the final mirror surface. Densification is needed to prevent "print-through" of carbon fibers on the surface. Print-through occurs when the resin-impregnated carbon fabric destined to become the C/C composite is heat-treated at 2200°C and the surface resin is decomposed,

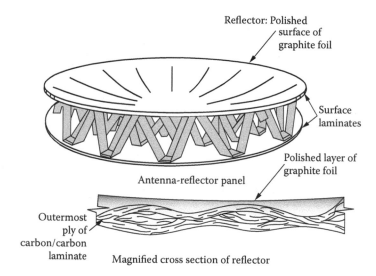

Reflector: Polished surface of graphite foil

Surface laminates

Polished layer of graphite foil

Antenna-reflector panel

Outermost ply of carbon/carbon laminate

Magnified cross section of reflector

FIGURE 3.36
The surface layer of graphite foil is polished to a smooth finish.

leaving exposed woven fabric. When the composite is properly densified, the surface can be polished to a smooth finish (see Figure 3.36).

Regardless of which surface-finishing technique is selected, the graphite foil is thermally stable enough to withstand high temperatures in processing and is thermally stable in use. It stops the propagation of surface microcracks in the underlying C/C composite. Moreover, it seals the composite, preventing any loose carbon dust from leaking out.

Ion-Accelerator Grids

Ion-accelerator grids have been fabricated from C/C composite material by using unidirectional-carbon-fiber tape as the starting material. The reason for using the tape (instead of the woven carbon-fiber cloth that was used previously) is that the tape offers the potential for making smoother, flatter grids. C/C grids are being developed as alternatives to molybdenum grids, which have been found to wear out quickly and to become so distorted by differential thermal expansion during operation that operating efficiency is degraded.

Early tests have shown that C/C composite grids last about 10 times as long in operation as molybdenum grids do. Moreover, because of the low and even negative thermal expansion of C/C composite grids, thermal distortion is no longer large enough to cause appreciable loss of efficiency. However, C/C grids made from woven carbon-fiber cloth (the conventional preform material) have been found not to exhibit the flatness and stiffness needed to maintain the precision required for ion optics.

Unidirectional-carbon-fiber tape offers several advantages over woven carbon fiber cloth: It provides greater flexibility in orientation of fibers, is thin enough that it can be stacked in multiple thin plies to fabricate thin parts with properly balanced layups, and is amenable to formation of a much smoother surface than is achievable with woven preforms. The use of unidirectional-carbon-fiber tape as the preform material was introduced in the fabrication of prototype C/C composite grids that were required to have thickness of 0.41 mm and flatness within ±0.025 mm. The proto-type grids were also required to exhibit maximum flexural stiffness along rows of holes in a hexagonal pattern.

The fabrication process devised to satisfy these requirements began with balanced layup of the unidirectional fiber tapes, with fibers in successive layers oriented in a sequence of [0°, 60°, -60°] so that the fibers in each layer were parallel to one of the rows of holes. Following layup, the carbon blanks were held in fixtures that allowed gases to pass through. The fixtured blanks were subjected to cycles of CVI and heat treatment to a temperature of 3000°C to obtain C/C composite material of maximum stiffness. There was also a long heat treatment at 2200°C to relieve stresses in the blanks while they were held flat.

Ultrasonic impact grinding has been found to be a suitable technique for machining holes in C/C sheets to make grids for ion accelerators. The holes made by ultrasonic grinding have smooth, straight walls with well-defined, sharp edges.

Heretofore, ion-accelerator grids have been made of metal. The use of C/C grids is a subject of continuing research in an effort to increase the lifetimes and improve the performances of future ion accelerators. Fabrication of such a grid typically involves drilling several thousand closely spaced holes between 1 and 2 mm in diameter.

Because C/C is relatively chemically inert, etching is not a suitable technique for making the holes.

Conventional mechanical drilling may lead to breaking away of the webbing material between the holes, in particular for open-area fractions of 70% or higher. This is caused by fibers getting caught in the drill and being pulled out of the surrounding matrix material. Laser drilling can be used, but it yields holes with crater-like shapes. Worse yet, waterjet blasting causes large chunks of material to break away, yielding holes of poor quality.

In ultrasonic impact grinding, an abrasive slurry is introduced into the region of contact between a cutting tool and a workpiece, and vibrations with amplitudes of the order of hundredths of a millimeter and frequency of the order of 20 kHz are induced in the tool by use of a piezoelectric or electromagnetic transducer excited by a signal at that frequency. The vibrations are transmitted from the tool to the abrasive particles, causing the particles to grind a hole in the workpiece. The outline of the hole makes a close match with the outline of the cutting tool.

Glass Making

The machinery used to manufacture glass bottles dispenses a small amount of molten glass known as a *gob*. The gob rolls down a channel into a mold where it is shaped by air pressure into a bottle. The equipment is designed in such a way that a damaged mold can be identified. A gob interceptor rejects the gob intended for that mold, diverting it so as not to clog up the machinery. C/C gob interceptors, although approximately 100 times more expensive than the asbestos they replace, result in fewer replacements and less frequent shutdowns and so are exceptionally cost-effective. The two-dimensional C/C is inherently less biohazardous than the asbestos and also acts as a heat shield protecting the underlying metallic components.

High-Temperature Mechanical Fasteners

The tensile strength of most ceramics and refractory metals and alloys above 750°C, is extremely poor. Hence, mechanical fixing at high temperatures is often a cause of severe problems in engineering design. Screws, nuts and bolts, and so on made from C/C experience no loss in strength at high temperatures. The load-bearing ability of C/C is less than that of metals at low to moderate temperatures but superior at high temperatures.

A new processing method involving preformed yarns has been developed in Japan that is said to considerably reduce both cost and manufacturing time.[145]

This patented preformed yarn method involves producing yarn consisting of a bundle of carbon fibers in a matrix precursor of coke and pitch binder powders. The yarn is encased in a flexible thermoplastic sleeve to contain the powders during handling and subsequent processing. This simplified manufacturing process ensures better penetration of the binder into the carbon fiber bundle, thereby resulting in higher strength than conventional C/C composites. Furthermore, the matrix powders are often derived from the residue of petroleum processing, which can have beneficial environmental effects.

Research has shown that unidirectional C/C composites manufactured by this simplified method achieve strengths of between 414 and 670 MPa, compared with 70 MPa for conventional C/C composites. They also show significantly higher flexural moduli.

The preformed yarn may be woven into sheets or chopped to fill a mold, and is then hot-pressed to form the composite product. Because the reinforcing fibers are homogeneously dispersed in the matrix, properties of the resulting composite are quite uniform.

The preformed yarns have excellent workability and processability. For example, pipe, rod, and crucible-shaped preforms may be readily fabricated, as well as unidirectional sheets, cloth sheets, thick textiles, tapes, and chopped yarns. Workability of the preformed yarns allows the material to be tailored to meet various design loads. This method is also well suited for the production of complex shapes and components having small radii of curvature.

The composites are available in unmachined form, as well as finished products such as coil springs, rods, shelves, nuts and bolts, link conveyor belts, roller conveyor parts, and furnace parts.

Molds for Forming Superplastic Metals

Superplastic forming of metals is a relatively new and extremely versatile route to the fabrication of complex shaped components, especially in the aerospace industry. Titanium alloys in particular require temperatures in the region of 1000°C and pressures of several atmospheres. C/C molds offer significant improvements over their "opposition," mild steel. Being two orders of magnitude lighter, their reduced thermal mass affords shorter processing cycles and much easier handling. Further, the negligible thermal expansion of C/C results in greater tolerances and eliminates folding of the titanium insert.

Hot-Press Dies

High-quality ceramics and metals may be produced by sintering under mechanical pressure in a hot-press operation.[146] The traditional material used in die manufacture is polycrystalline graphite. Such dies are required to be of considerable wall thickness due to the poor mechanical properties of the graphite. C/C dies are specifically designed with high hoop strengths so as to reduce considerably the size of component required. A C/C die of, for example, 130 mm ID need only have a wall thickness of 15 mm. One may thus achieve shorter heating cycles and a more uniform temperature distribution in the hot-pressed product.[147]

Hot Gas Ducts

High-temperature strength and resistance to thermal shock and temperature gradients make C/C tubes an ideal choice as the liners in the hot gas ducting in high-temperature nuclear reactors.[148]

Furnace Heating Elements and Charging Stages

Furnaces operating between 1000 and 3000°C in a nonoxidizing atmosphere tend to use graphite heating elements. The brittleness and moderate strength of graphite make such elements extremely difficult to handle and susceptible to damage, especially considering the complex geometries often required. C/C elements are far less fragile and are a considerable advantage in hot isostatic pressing equipment as only a small volume is needed, thus maximizing the working zone. Aside from its mechanical property advantages, the higher electrical resistance of C/C permits higher power operation and less voluminous electrical connections.

Charging stages allow a more effective use of furnace heating volume. Such stages have historically been produced from refractory alloys, ceramics, or graphite, depending upon operating temperature. The use of light-weight impact-resistant C/C serves to save volume and increase the lifetime of these stages. The lower thermal mass of the C/C component increases process efficiency and greatly reduces heating and cooling cycle times.

Substituting racks made of C/C (Figure 3.37) for conventional ones eliminates or minimizes previous problems and presents some added benefits. These include

- Decrease of rack weight, in some cases up to 80%. This decrease implies an increase of the net load weight, along with a higher production rate and reduced manual handling.
- Absence of rack distortion, which is the first step towards process automation.
- Minimal heat absorption by the rack, due to the small heat absorption coefficient of these materials from its reduced weight. This leads to related benefits. Heat distribution is homogeneous because little heat is absorbed by the rack, resulting in more uniform heating and better-quality workpieces.
- Shorter process times, because the thermal inertia of the whole is lower.
- Lower costs for heat production and those proportional to heat treating time.

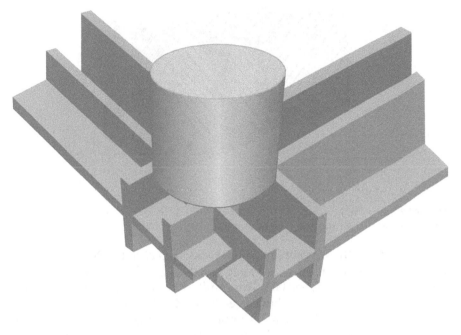

FIGURE 3.37
Typical modular construction of a C/C rack consisting of intersecting webs.[149]

- Better behavior regarding thermal stress and better heat resistance (chemically inert). This leads to a longer product life.
- Modular, flexible, and open construction optimizes heat transfer to the workpieces and expands the range of products that can be treated.[149]

Replacement of the original heat-resistant steel rack was conceived using composite materials. The resulting design was a modular rack (Figure 3.38) consisting of C/C fixtures over which the tools are located. The devices are piled up and set over a C/C structure with a defined slope, so that the tools lay at an inclination of 45° to the horizontal. The benefits achieved were

- Quality improvement by meeting required quality parameters
- A 70% increase in the net load
- Reduced manual handling, loading, and unloading of the tools
- High economic profitability and a short payback

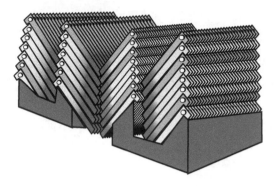

FIGURE 3.38
Rack made of C/C for the heat treatment of tools.[149]

For example, the investment for 10 racks was about $120,000. Savings from the net load increase were estimated as $40,000; those for shorter load and unload times, $50,000. This resulted in a payback period of about 1.3 years. To date, the racks have been in use for 3 years without showing damage or defects. A subsequent improvement project now provides for the automatic loading and unloading of the tools in the C/C devices and automatic stacking of the devices in the rack.

In another recent application, C/C grids made it possible to boost the net load and optimize the rack for a brazing process. The components, heat exchangers for commercial vehicles, are brazed at 1120°C in a vacuum atmosphere. The objectives of this project were to increase the net load of the existing rack and the key advantages included

- Weight reduction by combining different materials and optimally utilizing C/C
- Modular and flexible construction, allowing a damaged web to be replaced without changing the whole grid
- Easier loading and unloading of the workpieces in the grids and less demanding mounting and dismounting of the rack, which can be done easily by workers
- Net load increase of 100% up to 3000 kg with a rack weight of 100 kg
- Absence of distortion
- Reduced energy absorption leading to shorter process time
- Better brazing quality[149]

Finally, a Japanese steel company found that racks made of the new C/C composite allowed the construction of lighter-weight platforms for annealing of steel powder than conventional graphite racks. Use of the material resulted in a 22.6% saving of heating energy, as well as an 80% improvement in heating speed. In addition, the material produced more consistent heat distribution in the packed metal powder.

Nippon Denso, Japan's larger manufacturer of automotive electrical components, used the material for shelving and coil springs in its continuous heat-treating equipment, and also significantly reduced their energy costs.

The major applications for this new C/C material were parts for furnaces, ovens, and other industrial heating equipment.[145]

Other miscellaneous applications include

- Structural parts for hypersonic vehicles
- Conversion flaps, combustion liners
- Heating elements, with 2000°C working temperature
- Hardware for low-pressure and plasma CVD such as cage boats, ladders, trays, liners, cage supports, baffles
- Electrode connecting parts in electric-arc furnaces

Summary

C/C composites are a unique family of composite materials, retaining their mechanical properties to extremely high temperatures. This, along with a high degree of toughness and inertness, makes them ideal candidates for application in a large theater of advanced engineering. There are, however, two major drawbacks limiting the widespread development of C/C; primarily, the fabrication processes used are extremely inefficient, and secondly, they tend to oxidize quickly in air at temperatures as low as 400°C.

As a result, markets are limited mainly to military applications, where high premiums are paid for improved performance. Surveys of the C/C business prospects have generally predicted an increasing usage of the materials well into 2005 and beyond.

It is possible to be extremely simplistic in describing the future market prospects of C/C, splitting the business into two distinct sectors. In the commercial sector, usage such as brakes, heat exchangers, and furnace elements is constrained simply by cost. There are a whole host of applications ideally suited to the properties of C/C, provided the price is lowered as a result of more efficient fabrication. In the military sector it is performance rather than price that is of concern. Efforts to develop oxidation-resistant C/C parts for heat engine components and the heat shields of hypersonic aircraft are intensive in the drive toward a truly reusable system. The long-term reality, however, is that C/C composites will remain highly specialized performance materials. It is doubtful that there will be sufficient reduction in price or advances in technology in the short to medium term to bring C/C into widespread use.

References

1. R. Stevenson, *Chem. Br.*, 27, 685, 1991.
2. E. Fitzer, *Proceedings of the 5th London International Carbon and Graphite Conference*, London: Society of Chemical Industry, 3, 99, 1979.
3. R. E. Franklin, *Proc. R. Soc.*, Ser. A, 209, 196, 1951.
4. S. Mrozowski, *Proceedings of the 1st and 2nd Carbon Conferences*, American Carbon Society, Buffalo, NY, 1956, p. 31.
5. S. J. Mitchell and C. R. Thomas, *Carbon*, 9, 253, 1971.
6. C. R. Thomas and E. J. Walker, Materials in Aerospace, *Proceedings of the Royal Aeronautical Society International Conference*, London, 1986.
7. G. Slayter, *Sci. Am.*, 124, January 1962.
8. J. Cook and J. E. Gordon, *Proc. R. Soc.*, London, A282, 508, 1964.
9. M. M. Schwartz, *Emerging Engineering Materials*, Boca Raton, FL: Technomic Publishing Company Inc./CRC Press, 1996, pp. 206-237.
10. A. A. Watts (Ed.), *Commercial Opportunities for Advanced Composites*, Philadelphia, PA: ASTM, 1980.
11. L. E. McAllister and W. Lachman, *Handbook of Composites*, A. Kelly and S. T. Mileiko (Eds.), Vol. 4, Elsevier, 1983, p. 109.
12. J. Jortner, *Carbon*, 24, 5, 1986.
13. D. E. Walrath and D. F. Adams, Report UWME-DR-904-101-1, Mechanical Engineering Department, University of Wyoming, September 1979.
14. V. Singer, JANNAF Rocket Nozzle Materials Report 5, No. 2, Laurel, MD: Joint Army-Navy-NASA-Air Force Interagency Propulsion Committee, Chemical Propulsion Information Agency, May 1983.
15. T. Edison, U.S. Patent 223,898,1880.
16. N. P. Freestone, *Chem. Br.*, 26, 1184, 1990.
17. H. M. Ezekiel and R. G. Spain, *J. Polym. Sci., Polym. Symp.*, 19, 249, 1967.
18. D. Hull, *An Introduction to Composite Materials*, Cambridge, U.K.: Cambridge University Press, 1981.
19. J. White and M. Buechler, *Am. Chem. Soc. Symp. Ser.*, 303, 62, 1986.
20. E. Fitzer, *J. Chim. Phys.*, 81, 678, 1984.
21. J. Hill, C. R. Thomas, and E. J. Walker, Their Place in Modern Technology, *Proceedings of the International Conference on Carbon Fibres*, London: The Plastics Institute, 1974, Paper 19.
22. C. R. Thomas, *Essentials of Carbon-Carbon Composites*, London: Royal Soc. of Chemistry, 1993.
23. R. A. Meyer and S. R. Gyetway, *Am. Chem. Soc. Symp. Ser.*, 303, 380, 1986.

24. L. E. McAllister, in *Fabrication of Composites*, A. Kelly and S. T. Mileiko (Eds.), New York: Elsevier, 1983, Chap. 3.
25. J. Delmonte, in *Technology of Carbon and Graphite Fiber Composites*, New York: Van Nostrand Reinhold, 1981.
26. A. J. Klein, *Adv. Mater. Proc.*, 130(5), 64, 1986.
27. R. Prescott, *Modern Plastics Encyclopedia*, New York: McGraw-Hill, 114, 1988.
28. E. Fitzer, E. Muller, and W. Schafer, *Chem. Phys. Carbon*, 7, 237, 1971.
29. D. A. Lovell, *Worldwide Carbon Fibre Directory*, 5th ed., London: Pammac Publications, 1992.
30. J. V. Milanski and H. Katz (Eds.), *Handbook of Reinforcements for Plastics*, Amsterdam: Van Nostrand Reinhold, 1987.
31. R. M. Gill, *Carbon Fibres in Composite Materials*, London: ILIFFE Books, 1972.
32. W. Watt and B. V. Perov (Eds.), *Strong Fibres*, Amsterdam: North-Holland Publishing Co., 1985.
33. M. S. Dresselhaus, G. Dresselhaus, K. Sugihara, I. L. Spain, and H. A. Goldberg, *Graphite Fibres and Filaments*, Heidelberg: Springer-Verlag, 1988.
34. G. Henrici-Olive and S. Olive, Thermosetting resins, in *Industrial Developments, Advances in Polymer Science*, 51, Heidelberg: Springer-Verlag, 1983, p. 1.
35. J. D. Buckley and D. D. Edie, *Carbon-Carbon Materials and Composites*, Park Ridge, NJ: Noyes Publications, 1993.
36. H. F. Volk, *Proc. Symp. on Carbon Fibre Reinforced Plastics*, Bamberg, FRG, May 11, 1977.
37. J. P. Riggs, in *Encyclopedia of Polymer Science and Engineering*, 2, New York: Wiley, 1985, p. 640.
38. J. D. H. Hughes, *J. Phys. D: Appl. Phys.*, 20, 276, 1987.
39. D. D. Edie and M. G. Dunham, *Carbon*, 27 (5), 647, 1989.
40. F. Ko, *Ceram. Bull.*, 68, 401, 1989.
41. A. Kelly and S. T. Mileiko (Eds.), Fabrication of Composites, in *Handbook of Composites*, 4, Amsterdam: North-Holland Publishing Co., 1986.
42. L. E. McAllister and A. R. Taverna, *Proc. 73rd Ann. Mtg. Am. Ceram. Soc.*, Chicago, 1971.
43. R. S. Barton, *SPE J.*, 4, 31, May 1968.
44. L. E. McAllister and A. R. Taverna, *Proc. 17th Nat. SAMPE Symp.*, Paper IIIA-3, 1972.
45. P. Lamicq, *Proc. AIAA/SAE 13th Prop. Conf.*, Paper 77-882, Orlando, FL, 1977.
46. C. K. Mullen and P. J. Roy, *Proc. 17th Nat. SAMPE Symp.*, Paper IIIA-2, 1972.
47. L. R. Sanders, *SAMPE Quart.*, 38, 1977.
48. J. C. Bokros, in *Chemistry and Physics of Carbon*, Vol. 5, P. L. Walker, Jr. (Ed.), New York: Dekker, 1969, p. 1.
49. W. V. Kotlensky, in *Chemistry and Physics of Carbon*, Vol. 9, P. L. Walker, Jr. (Ed.), New York: Dekker, 1973, p. 173.
50. H. M. Stoller and E. R. Frye, 73rd Annual Meeting of the Am. Ceram. Soc., Chicago, 1971.
51. D. L. Schmidt, *SAMPE J.*, May/June 1972.
52. J. D. Brooks and J. H. Taylor, *Nature*, 206, 697, 1965.
53. J. L. White, in *Prog. Solid State Chem.*, Vol. 9, J. O. McCauldin and G. Somarjai (Eds.), Oxford, U.K.: Pergamon Press, 1975, p. 59.
54. G. S. Rellick, *Carbon*, 28, 589, 1990.

55. E. Fitzer and A. Gkogkidis, *Am. Chem. Soc. Symp. Ser.*, 303, 346, 1976.
56. Improved Process for Manufacturing Carbon/Carbon Composites, *AM&P,* 145(3), 35-36, 1994.
57. W. P. Hoffman, In Situ Rapid Densification, *AFRL Technology Horizons,* pp. 33-34, March 2003.
58. G. Savage, *Carbon-Carbon Composites,* London: Chapman & Hall, 1993.
59. E. Fitzer and A. Barger, *Proc. Int. Conf. on Carbon Fibres, Their Composites and Applications,* London, 36, 1971.
60. L. E. McAllister and A. R. Taverna, *Proc. Am. Ceram. Soc. 73rd Ann. Mtg.,* Chicago, 1971.
61. D. F. Adams, *Mat. Sci. Eng.,* 23, 55, 1976.
62. R. A. Meyer, *Proc. Carbon 1986 Conf.,* Baden-Baden, Germany, 1986.
63. J. S. Evangelides, *Proc. Army. Symp. on Solid Mech.,* U.S. Army, September 1976, p. 98.
64. J. Jortner, *J. R. & E.,* Report no. 8514, p.39.
65. D. F. Adams, *J. Comp. Mats.,* 8, 320, 1974.
66. S. Kimura, E. Yasuda, H. Tanaka, and R. Yamada, *J. Ceram. Soc. Jap.,* 83, 122, 1975.
67. J. X. Zhao, R. C. Bradt, and P. L. Walker, Jr., *Ext. Abst. 15th Biennial Conf. on Carbon,* Washington, D.C.: Am. Chem. Soc., 1981, p. 274.
68. B. Terwiesch., Ph.D. thesis, University of Karlsruhe, Germany, 1972.
69. S. Kimura, E. Yasuda, and Y. Tanabe, *Proc. Int. Conf. Comp. Mats.,* IV, 1601, 1982.
70. E. Fitzer and M. Heym, *Ext Abst. 13th Biennial Conf on Carbon,* 1977, p. 128.
71. E. Fitzer and M. Heym, *Kunstofftechnik,* 1980, 85.
72. Carbon Fibre Reinforced Carbon, Schunk promotional brochure.
73. H. C. Kim, K. J. Yoon, R. Pickering, and R. J. Sherwood, *J. Mat. Sci.,* 20, 3967, 1985.
74. E. Yasuda, H. Tanaka, and S. Kimura, *Tanso (Carbon),* 100, 3, 1980.
75. M. L. Liebermann and H. O. Pierson, *Proc. 11th Biennial Carbon Conf.,* 1973, p. 314.
76. G. M. Savage, Ph.D. thesis, University of London, 1985.
77. P. Laramee, G. Lamere, B. Prescott, R. Mitchell, P. Sottosanti, and D. Dahle, *Proc. 12th Biennial Conf. on Carbon,* Washington, D.C.: Am. Chem. Soc., 1975, p. 74.
78. H. O. Pierson, D. A. Northrop, and J. F. Smatana, *Ext. Abst. 11th Biennial Conf. on Carbon,* Washington, D.C.: Am. Chem. Soc., 1973, p. 275.
79. B. Granoff and P. Apodaca, *Ext. Abst. 11th Biennial Conf. on Carbon,* Washington, D.C.: Am. Chem. Soc., 1973, p. 273.
80. J. L. Perry and W. F. Adams, *Carbon,* 14, 61, 1976.
81. H. Girard, *Proceedings of the 5th London International Carbon and Graphite Conference,* London: Society of Chemical Industry, Vol. I, 1978, p. 483.
82. L. E. McAllister and A. R. Taverna, *Proceedings of the International Conference on Composite Materials,* Vol. 1, E. Scala, E. Anderson, I. Toth, and B. Noton (Eds.), New York: Metallurgical Society of AIME, 1976, p. 307.
83. C. K. Mullen and P. J. Roy, *Proceedings of the 17th National Symposium,* Los Angeles: Society for the Advancement of Material and Process Engineering (SAMPE), 1972, p. 111-A-2.

84. D. W. McKee, in *Chemistry and Physics of Carbon*, Vol. 16, P. L. Walker Jr. and P. A. Thrower (Eds.), New York: Marcel Dekker, 1981, p. 1.
85. D. W. McKee, *Carbon*, 24, 737, 1986.
86. G. R. Marin, U.S. Patent No. 3,936,574, 1976.
87. K. J. Zeitsch, in *Modern Ceramics*, J. E. Hove and W. C. Riley (Eds.), New York: John Wiley & Sons, 1967, p. 314.
88. S. A. Bortz, in *Ceramics in Severe Environments*, W. W. Kriegel and H. Palmour III (Eds.), New York: Plenum, 1971, p. 49.
89. E. M. Goldstein et al., *Carbon*, 4, 273, 1966.
90. L. C. Ehrenreich, U.S. Patent No. 4,119,189, 1978.
91. T. Vasilos, GB Patent No. 2,130,567, 1983.
92. D. L. Dicus et al., NASA Technical Note NASA-TN-D-8358, 1976.
93. D. W. McKee, *Carbon*, 26, 659, 1988.
94. P. B. Gray, U.S. Patents No. 4,795,677 (1989), 4,894,286 (1990), 4,937,101 (1990).
95. J. Hill et al., 11th Biennial Conference on Carbon, Extended Abstracts, American Carbon Society, Gatlinburg, TX, 1973, p. 328.
96. J. Chown, R. F. Dencon, N. Singer, and A. E. S. White, Refractory Coatings on Graphite, in *Special Ceramics*, P. Popper (Ed.), New York: Academic Press, 1963, p. 81.
97. E. Fitzer, *Carbon*, 16, 3, 1978.
98. R. Naslain et al., *Proceedings of the 4th European Conference on CVD*, Philips Centre for Manufacturing Technology, Eindhoven, Netherlands, 1983, p. 293.
99. H. Hannache et al., Proceedings of the 4th European Conference on CVD, Philips Centre for Manufacturing Technology, Eindhoven, Netherlands, 1983, p. 305.
100. H. Hannache et al., *J. Mater. Sci.*, 22, 202, 1984.
101. T. E. Strangman and R. J. Keiser, U.S. Patent No. 4,668,579, 1987.
102. D. C. Rogers et al., *Proceedings 7th National SAMPE Technical Conference*, Society for the Advancement of Material and Process Engineering, Albuquerque, NM, 1975, p. 319.
103. D. C. Rogers et al., *Proceedings 8th National SAMPE Technical Conference*, Society for the Advancement of Material and Process Engineering, Seattle, WA, 1976, p. 308.
104. D. M. Shuford, U.S. Patent No. 4,465,777, 1984.
105. D. M. Shuford, U.S. Patent No. 4,585,675, 1986.
106. H. Jehn, *J. Less-Common Met.*, 100, 321, 1984.
107. G. St. Pierre, Explanatory Research on the Protection of Carbon-Carbon Composites Against Oxidation, Department of Metallurgical Engineering and Materials Science, Ohio State University, Columbus, OH, 1988.
108. K. N. Lee and W. L. Worrell, *Oxid. Met.*, 32, 357, 1989.
109. M. D. Alvey and P. M. George, *Carbon*, 29, 523, 1991.
110. P. Plotard and C. Le Floc'h, *Proceedings ESA Symposium: Space Applications of Advanced Structural Materials*, Noordwijk, Netherlands: ESTEC, 1990, p. 171.
111. H. Tretout et al., *Proceedings ESA Symposium: Space Applications of Advanced Structural Materials*, Noordwijk, Netherlands: ESTEC, 1990, p. 181.

112. M. M. Opeka, Carbon-Carbon Composites, Fabrication, Properties, Joining, and Applications, *Proceedings, the 19th Annual Conference on Composites, Materials, and Structures,* PT2; U.S. Defense Department, 1996.

113. S. Yalof, Materials Innovation Labs Presentation from the Interagency Planning Group Meeting on Joining of Carbon-Carbon and Ceramic Matrix Composites, T. F. Kearns (Ed.), April 1987.

114. P. Dadras, Joining of Carbon-Carbon Composites by Using $MoSi_2$ and Titanium Interlayers, *14th Conf on Metal Matrix, Carbon, and Ceramic Matrix Composites Proc.,* J. D. Buckley (Ed.), NASA, 1990.

115. P. Dadras and T. Ngai, Joining of C-C Composites by Boron and Titanium Disilicide, Paper no. 5, 15th Conf. on Metal Matrix, Carbon, and Ceramic Matrix Composites, Cocoa Beach, FL, January 16, 1991.

116. H. Mizuhara, E. Huebel, and T. Oyama, High-Reliability Joining of Ceramic to Metal, *Am. Ceram. Soc. Bull.,* 68(9), 1591–1599, 1989.

117. H. E. Pattee, R. M. Evans, and R. E. Monroe, Joining Ceramics and Graphite to Other Materials, NASA, 1968.

118. M. M. Schwartz, *Ceramic Joining,* Metals Park, OH: ASM Int., 1990.

119. M. M. Schwartz, *Composites Materials Handbook,* 2nd ed., New York: McGraw-Hill, 1992.

120. K. H. Holko, Joining Methods for Carbon-Carbon Composite Structures, *Adv. Mater Manufac. Proc.,* 3(2), 247–260, 1988.

121. R. W. Rice, Joining of Ceramics, *Proc. of the 4th Army Materials Technology Conf., Advances in Joining Technology,* Chestnut Hill, MA: Brookhill, 1975, pp. 69–111.

122. J. P. Hammond and G. M. Slaughter, Bonding Graphite to Metals with Transition Pieces, *Weld. J.,* 33–40, 1971.

123. M. M. Schwartz, *Brazing,* Metals Park, OH: ASM International, 1989.

124. E. Rudy, Compendium of Phase Diagram Data, Part V, AFML-TR-65-2, 1969.

125. P. Dadras and G. M. Mehrotra, Joining of Carbon-Carbon Composites by Graphite Formation, *J. Am. Ceram. Soc.,* 77(6), 1419–1424, June 1994.

126. M. M. Opeka (Ed.), Development of Brazed Joints in Carbon-Carbon, Graphite, and Refractory Metal Components for Rocket Propulsion and Spacecraft Applications, Part 2, The 19th Conference on Composites, Materials and Structures, Carbon-Carbon Composites: Fabrication, Properties, Joining, and Applications, January 9–10, 1995, Cocoa Beach, FL, pp. 39–48.

127. S. Khatri, W. Altergott, T. Campbell, G. Freitas, and H. Croop, Brazing of Carbon-Carbon Composites to Refractory Metals for Actively Cooled Hypersonic Vehicle Structures, The 19th Conference on Composites, Materials and Structures, Carbon-Carbon Composites: Fabrication, Properties, Joining, and Applications, M. M. Opeka (Ed.), January 9–10, 1995, Cocoa Beach, FL, pp. 611–624.

128. L. Goodman, D. L. Birx, and V. R. Dave, Joining of Carbon Carbon Composites with High Energy Electron Beams, The 19th Conference on Composites, Materials and Structures, Carbon-Carbon Composites: Fabrication, Properties, Joining, and Applications, M. M. Opeka (Ed.), January 9–10, 1995, Cocoa Beach, FL, pp. 585–596.

129. A. C. Johnson, Carbon-Carbon Composites for Spacecraft Applications, Tech. Rept. TR.89-340, Naval Surface Warfare Center, 1989.

130. J. K. Weeks and J. L. Sommer, Carbon/Carbon Composites Incorporating Brazed Cooling Tubes, The 19th Conference on Composites, Materials and Structures, Carbon-Carbon Composites: Fabrication, Properties, Joining, and Applications, M. M. Opeka (Ed.), January 9–10, 1995, Cocoa Beach, FL, pp. 597–610.
131. D. J. Holt, Aircraft Braking Systems, *Aerospace Engineering*, 7–11, June 1993.
132. S. Birch, A Light Touch on the Brakes, *Aerospace Engineering*, 24–25, December 1996.
133. Carbonne Industry, *SEPCARB Brakes Assistance Manual*, France.
134. E. Fitzer, W. Fritz, A. Gkogkidis, and K. D. Morgenthaler, *Proc. Carbon 86 Conf.*, Baden-Baden, Germany, 1986.
135. M. P. Gorton, Carbon-Carbon Piston Development, Lockheed Engineering & Sciences Co., May 1994.
136. A. R. Canfield, Braided Carbon/Carbon Nozzle Development, 21st AIAA/SAE/ASME/ASEE Joint Propulsion Conference, July 8–11, 1985.
137. H. O. Pierson, *Handbook of Carbon, Graphite, Diamond and Fullerenes*, Park Ridge, NJ: Noyes Publications, 1993.
138. R. P. Dickenson and M. R. Hill, A Review of Recent Research in the Former Soviet Union on Fibrous Composites, *Composites*, 24(5), 379–382, 1993.
139. R. B. Dirling, Jr. and D. A. Eitman, Lightweight Carbon-Carbon Thermal Protection System for Starprobe, AIAA 19th Thermophysics Conference, Snowmass, CO, June 25–28, 1984, Irvine, CA: Science Applications.
140. F. Morring, Jr., Culture Shock, *Aviation Wk. Sp. Technol.*, 22–24, 2003.
141. P. J. Lamicqe, SEP International Carbon Conference, Bordeaux, France, 1984.
142. D. F. Williams, *Metals and Materials*, 7(1), 24, 1981.
143. C. Burri and R. Neugedauer, *Replacement of Ligaments by Carbon Fibres*, Munich: Springer-Verlag, 1985.
144. L. Claes, E. Fitzer, W. Huttner, and K. Kinzl, *Carbon*, 18, 383, 1980.
145. Inteque Resources Corp., *AM&P*, March 1994, pp. 35–36.
146. D. W. Richerson, *Modern Ceramic Engineering*, New York: Marcel Dekker, 1982.
147. W. Hutner, *Andernach*, 29, October 1985.
148. E. Von Gellhorn and H. Gruber, *Proc. Carbon 86 Conf.*, Baden-Baden, Germany, 1986.
149. J. Demmel, H. Lallinger, G. Kopp, and P. G. Rechea, High–Temperature Racks, *Heat Treating Progress*, 1(4), 26-32, 2001.

Bibliography

Application of C/C Composites to the Combustion Chamber of Rocket Engines, Part 1. Heating Tests of C/C Composites with High Temperature Combustion Gases, National Aerospace Lab., Amsterdam, Netherlands, *NTIS Alert*, 96(4), 12, 1995.

E. C. Botelho, N. Scherbakoff, and M. C. Rezende, Rheological Studies Applied in the Processing and Characterization of Carbon/Carbon Composites, *J. Adv. Mater.*, 4, 12, October 2001.

R. W. Burns and N. Murdie, Thermal Cycle Exposure Testing of Coated Inhibited Carbon-Carbon Composites, The 15th Conference on Metal Matrix, Carbon, and Ceramic Matrix Composites, J. D. Buckley (Ed.), NASA Langley Res. Center, Hampton, VA, NASA Conf. Publication 3133, Part 1, *Proceedings of Conf. in Cocoa Beach, FL,* January 16–18, 1991, pp. 799–810.

C/C Composites for Rocket Chamber Applications, Part 2, Fabrication and Evaluation Tests of Rocket Chamber, National Aerospace Lab., Tokyo, May 95, 21p, Feb. 15, 1996.

M. Camden, L. Simmons, and G. Maddux, Room Temperature and Elevated Temperature High Cycle Fatigue Testing of Carbon/Carbon Coupons, Structures Div., Flight Dynamics Directorate, Wright Laboratory, WPAFB, OH, The 19th Annual Conference on Composites, Materials, and Structures, Cocoa Beach, FL, January 9–10, 1995, pp. 519–530.

K. Christ and K. J. Huttinger, Carbon-Fiber-Reinforced Carbon Composites Fabricated with Mesophase Pitch, *Carbon,* 31(5), 731–750, 1993.

J. Chtopek, S. Blazewicz, and A. Powroznik, Mechanical Properties of Carbon-Carbon Composites, *Ceramics International,* 19, 251–257, 1993.

L. L. Chuk, Preventing Saltwater Corrosion of Carbon Fiber Composites, *High-Tech Materials Alert,* 3–4, March 15, 1996.

P. Dadras and T. T. Ngai, Joining of Carbon-Carbon Composites with Boron and Titanium Disilicide, The 15th Conference on Metal Matrix, Carbon, and Ceramic Matrix Composites, J. D. Buckley (Ed.), NASA Langley Res. Center, Hampton, VA, NASA Conf. Publication 3133, Part 1, *Proceedings of Conf. in Cocoa Beach, FL,* January 16–18, 1991, pp. 25–38.

T. L. Dhami, O. P. Bahl, and L. M. Manocha, Influence of Matrix Precursor on the Oxidation Behavior of Carbon-Carbon Composites, *Carbon,* 31(5), 751–756, 1993.

S. Drawin, M. Bacos, J. Dorvaux, and O. Lavigne, Oxidation Model for Carbon-Carbon Composites, ONERA, Chatillon, France, AIAA Fourth International Aerospace Planes Conf., December 1–4, 1992, Orlando, FL, pp. 1–9.

D. D. Edie, High Thermal Conductivity Carbon/Carbon Composites, Final Rept., October 1992–August 1995, Clemson University, Clemson, SC, September 1995.

D. E. Glass and C. J. Camarda, High-Temperature Testing of Components for a Carbon/Carbon Heat-Pipe-Cooled Leading Edge, NASA Contractor Rept. 1168, April 1994.

G. E. Griesheim, P. B. Pollock, and S.-C. Yen, Notch Strength and Fracture Behavior of 2-D Carbon-Carbon Composites, *J. Am. Ceram. Soc.,* 76(4), 944–956, 1993.

S.-E. Hsu, H.-D. Wu, T.-M. Wu, S.-T. Chou, K.-L. Wang, and C.-I. Chen, Oxidation Protection for 3-D Carbon-Carbon Composites, *Acta Astronautica,* 35(1), 35–41, 1995.

A. Joshi and J. S. Lee, Coatings with Particulate Dispersions for High-Temperature Oxidation Protection of Carbon and C/C Composites, *Composites Part A — Applied Science and Manufacturing,* 28(2), 181–189, 1997.

W. Kowbel and C. H. Shan, Mechanical Behavior of Carbon-Carbon Composites Made With Cold-Plasma Treated Carbon-Fibers, *Composites,* 26(11), 791–797, 1995.

Maine Studying Carbon-Carbon Composites, http://www.netcomposites.com/ news.asp?1652, accessed July 5, 2003.

K. Mumtaz, J. Echigoya, and M. Taya, Preliminary Study of Iridium Coating on Carbon/Carbon Composites, *Journal of Materials Science,* 28, 5521–5527, 1993.

H. S. Park, W. C. Choi, and K. S. Kim, Process-Microstructure Relationships of Carbon-Carbon Composites Fabricated by Isothermal Chemical-Vapor Infiltration, *J. Adv. Mater.,* 26(4), 34–40, July 1995.

J. P. Pemsler, J.K. Litchfield, R. Cooke, and M. Smith, Oxidation Resistant Coating for Carbon-Carbon Composites at Ultra-High Temperatures, Final Rept. Cont. N00014-92-C-0049, June 1, 1994.

J. Randolph, Carbon/Carbon Shield/Antenna Structure, *NASA Tech Briefs,* June 1998, p. 78.

P. O. Ransone, H. K. Rivers, and G. B. Northam, Development of Carbon-Carbon Components for Internal Combustion Engines, NASA Langley Res. Center, Hampton, VA, The 19th Annual Conference on Composites, Materials, and Structures, Cocoa Beach, FL, January 9–10, 1995, pp. 871–887.

P. Savage, Market Forecast: Carbon Fiber Composites, *High-Tech Materials Alert,* Tech Insights, 1–3, July 19, 1996.

D. L. Schmidt, K. E. Davidson, and L. S. Theibert, Evolution of Carbon-Carbon Composites (CCC), *SAMPE Journal,* 32(4), 44–50, July/August 1996.

S. T. Schwab and R. C. Graef, Repair of Oxidation Protection Coatings on Carbon-Carbon Using Preceramic Polymers, The 15th Conference on Metal Matrix, Carbon, and Ceramic Matrix Composites, J. D. Buckley (Ed.), NASA Langley Res. Center, Hampton, VA, NASA Conf. Publication 3133, *Part 1, Proceedings of Conf. in Cocoa Beach, FL,* January 16–18, 1991, pp. 781–798.

J. Takahashi, K.. Kemmochi, J. Watanabe, H. Fukuda, and R. Hayashi, Development of Ultra-High Temperature Testing Equipment and Some Mechanical and Thermal-Properties of Advanced Carbon-Carbon Composites, *Advanced Composite Materials,* 5(1), 73–86, 1995.

S. Takano, T. Uruno, T. Kinjo, P. Tlomak, and C.-P. Ju, Structure and Properties of Unidirectionally Reinforced PAN-Resin Based Carbon-Carbon Composites, *J. Mater. Sci.,* 28, 5610–5619, 1983.

H. T. Tsou and W. Kowbel, A Multilayer Plasma-Assisted CVD Coating for Oxidation Protection of Carbon-Carbon Composites, *J. Adv. Mater.,* 27(3), 9–13, April 1996.

P. G. Valentine and P. W. Trester, Reaction Sintering: A Method for Achieving Adherent High-Temperature Coatings on Carbon-Carbon Composites, The 15th Conference on Metal Matrix, Carbon, and Ceramic Matrix Composites, J. D. Buckley (Ed.), NASA Langley Res. Center, Hampton, VA., NASA Conf. Publication 3133, Part 1, *Proceedings of Conf. in Cocoa Beach, FL,* January 16–18, 1991, pp. 811–822.

W. L. Vaughan, P. O. Ransone, and C. W. Ohlhorst, Performance of Polymeric Precursor Oxidation Protection Coatings on Carbon-Carbon Composites, The 19th Annual Conference on Composites, Materials, and Structures, Cocoa Beach, FL, January 9–10, 1995, pp. 903–928.

H. Weisshaus, S. Kenig, and A. Siegmann, Effect of Materials and Processing on the Mechanical Properties of C/C Composites, *Carbon,* 29(8), 1203–1220, 1991.

J. C. Withers, W. Kowbel, C. T. Lee, and R. O. Loutfy, Economical Processing to Produce C-C Composites for Pistons, Merck Corp., P.O. Ransone, NASA Langley, Hampton, VA, The 19th Annual Conference on Composites, Materials, and Structures, Cocoa Beach, FL, January 9–10, 1995, pp. 855–865.

T.-M. Wu, K.-K. Won, W.-C. J. Wei, Processing and Measurement of Basic Properties of SiC and Cordierite Coatings of Carbon/Carbon Composite, *Surface & Coatings Technology*, 78(1–3), 64–71, 1996.

4

Shape Memory Alloys/Effect

Introduction

Have you heard or read about the term "shape memory alloys" (SMAs) or "shape memory effect" (SME) and often wondered what the term and effect meant?

SMAs are a relatively new class of metallic materials with the unique property of SME. These alloys are becoming popular because of their shape memory characteristic, by which an article deformed at low temperature will revert to its original shape when healed to a high temperature.

Applications of SMAs include aircraft hydraulic couplings and electrical connectors, antenna release actuators for satellite systems, thermo-mechanical (T/M) actuators in robotic applications, valves in fire-safety devices, implant materials in orthopedics and orthodontics, and as guidewires for steering catheters during noninvasive surgery.[1-4]

Some alloy systems exhibiting the shape memory effect are Ni-Ti, Cu-Zn-Al, Au-Cd, and Ni-Al.[1,2] Other interesting properties of NiTi-based SMA are superelasticity (SE) and biocompatibility, which have led to medical applications.

The unique ability of SMAs to remember and recover their original shape, combined with their energetic recovery power and precise shape-temperature-stress response, has stirred much excitement in this material's potential for defense and industrial applications. However, most currently used SMAs, including NiTi, Cu-Zn-Al, and Cu-Ni-Al, have a phase transformation temperature below 100°C. This unfortunately hinders the application of the SMAs in high-temperature environments, which are precisely where their application would be most useful. Many defense and industrial fields that employ shape controlling and vibration depression in high-temperature environments would greatly benefit by using high-temperature SMAs. Obviously, it is an imperative to develop high-temperature SMAs.

Since the early 1970s scientists have worked steadily to harness the potential of SMAs. At that time a few scientists began to study the effect of various metallurgical factors, such as composition and heat-treatment, on the martensite transformation temperature of the existing SMAs. They made attempts to elevate the transformation temperature of the SMAs to more than 100°C, which proved to be a stubborn barrier. This was as far as the research went for the next 20 years, and no further significant progress was made.

Recently, scientists have made inroads into research on newer, higher temperature SMAs that have a phase transformation temperature much greater than that of the typically employed SMAs. Most of the research has focused on the $(Ni_{50-x}X_x)$ Ti (X = Pd, Pt, Au) and NiMn-Ti alloy systems because it is believed that these are the candidates with the highest potential for practical applications. Of the best candidate materials, which include the $(Ni_{50-x}Pd_x)$ Ti, $(Ni_{50-x}Pt_x)$ Ti, and $(Ni_{50-x}Au_x)$ Ti systems, $(Ni_{50-x}Pd_x)$ Ti is believed to have the best potential. Disadvantages inherent in several systems are that $(Ni_{50-x}Pt_x)$ Ti is considered too expensive to be used in practice and $(Ni_{50-x}Au_x)$ Ti is not stable for thermal cycling. The transformation temperature of Ni-Mn-Ti-based alloys can reach as high as 500°C; however, because these alloys are extremely brittle and unstable, it is recognized that this alloy is also not suitable for practical usage.

One-way shape memory in and of itself was a phenomenon worth celebrating, and the realization of two-way SME has led to even more advanced, innovative concepts for applications.

One-Way Shape Memory Effect

The SME or "one-way shape memory effect" happens when the material, previously deformed in martensite, recovers its original shape while heating up to the austenite. Figure 4.1 gives an illustration of the SME. The graph on the left side represents the stability zone for two phases on a stress/temperature curve.[5] The oblique lines denote the Clausius-Clapeyron relationship, indicating the boundary between phases and transition period. Consider a strip having a flat shape as parent shape (i.e., the "memorized shape"). If a stress higher than a critical stress (σ_r) is applied, the strip can be deformed in a plastic-like way (step 1). That means that the deformation applied on the material is stable in martensite. Upon heating (step 2) up to a critical temperature, the material transforms to austenite and recovers its original shape: the parent shape. On cooling down (step 3), no more shape change is observed. Once again, a previous deformation in martensite is required to observe a macroscopic shape change. This is the reason why the term "one-way" is

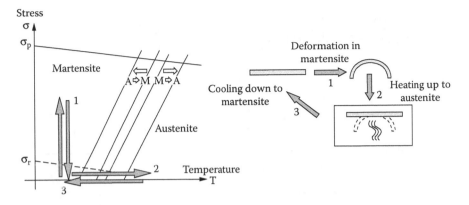

FIGURE 4.1
The "one-way" shape memory effect.

used to qualify this effect. Considering the macroscopic shape of the material, this effect is not reversible; upon cooling, the material keeps its austenitic shape. The reason is that nonoriented martensite is created upon cooling, and therefore martensite and austenite crystal structures are associated with the same specific macroscopic shape. A previous deformation in martensite is required to create some oriented martensite and thus to observe a shape change while heating up to the austenite.

A statistical explanation of this phenomenon is that randomly distributed martensite is created on cooling, and thus the resulting macroscopic deformation in regard to the austenite shape is zero. The effect of deforming the material in martensite is to modify the variants' distribution. Figure 4.2 shows a simplified phenomenological description in one dimension with two martensitic variants symbolically represented by a vertical and horizontal rectangle (M1, M2). In this model, the material is supposed to be free of any internal stress. With no stress applied on the model (set 1), the martensite variants are randomly distributed: Each variant has the same probability. When applying a force (set 2), the variants are reoriented in order to minimize the stress within the structure. The distribution between the two variants is therefore modified. Upon heating, as the austenite — symbolically represented by a square — has a higher symmetry than martensite, the two martensitic variants are transformed into the same "crystallographic" structure. It does not matter in which configuration the material is in martensite (set 1 or set 2 for instance), the parent shape will be the same. Upon cooling (set 4), if no stress is applied, the two martensitic variants have the same probability, and therefore, the macroscopic shape will remain the same as the austenite shape.

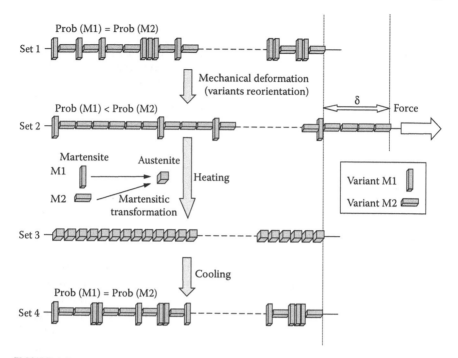

FIGURE 4.2
A phenomenologic description with two martensitic variants.

Two-Way Shape Memory Effect

Two-way shape memory alloys (TWSMAs) could memorize two configurations, as opposed to one. Although the two-way memory effect became known in the mid-1970s, a complete understanding of the whole mechanism of the phenomenon has yet not been completely explained even today. TWSMAs have shown significant potential for use in defense purposes and private industry due to these compelling features. Typical potential applications include connectors for missile guidance systems, jet fighter hydraulic couplings, tank actuators, satellite components, computer and electronic components, as well as medical and robotics usage.

Recent literature reports that a few of the current low-temperature SMAs, such as NiTi, CuAlNi, and CuZnAl, can demonstrate two-way memory effect after proper thermomechanical training. It is also reported that the two-way memory effect of those alloys strongly depends on the training procedures. However, there is little similar work that has been done on either one-way or two-way memory effects of high-temperature shape memory alloys (HTSMAs).[6]

How It Works

The SME has been attributed to martensitic transformation,[1,2,7] the same transformation that may be used to harden the surface of steel. The logical question here is, "Can we expect the SME in steel?" The answer is, "No." Even though the martensitic transformation is involved in both cases, the characteristics of the transformation in steel and SMA are different.

In both steel and SMA, martensite forms on cooling of the alloy. Upon heating, martensite tempers in steel, changing its crystal structure, whereas in SMA, martensite reverses back to its original austenitic phase. This type of transformation in SMA is known as a thermoelastic transformation. The high-temperature parent phase in nonferrous SMA is body-centered cubic (BCC, known as austenite), and on cooling, 24 variants of martensite form, with the final transformed martensite in the SMA being in the stress-free condition, unlike steel, because the variants form and grow in a self-accommodating manner.

Schematic representation of the processes involved in the SME is shown in Figure 4.3.[2] The SMA is deformed in the martensitic state below M_f.

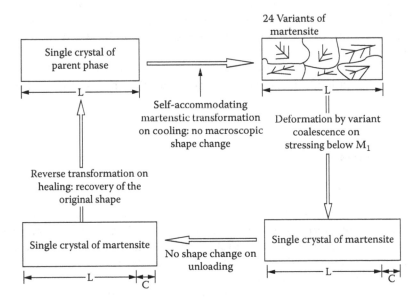

FIGURE 4.3
A schematic representation of the processes involved in the SME.[1]

Martensite deformation takes place by the growth and shrinkage of individual martensite plates, unlike in conventional materials where deformation takes place by slip. Upon heating above A_s, the martensite begins to transform to a parent austenitic phase and also recovery of prior deformation begins. Above A_f the SMA will completely transform into austenitic phase and the prior deformation will be completely recovered. The suitability of a particular SMA for an application depends on the ability of the SMA to recover substantial amounts of strain or generate significant force during the shape memory cycle. Strain recovery or force generation occurs within a temperature range that is characteristic of the alloy system, that is, austenite start temperature (A_s) and austenite finish (A_f) temperature. In other words, the transformation temperatures M_s, M_f, A_s, and A_f dictate the application range of a particular SMA.

Superelastic Effect

Known also as pseudoelastic, this effect describes material strains that are recovered isothermally to yield a mechanical shape memory behavior. The phenomenon is essentially the same as the thermal SME, although the phase transformation to austenite (A_f) occurs at temperatures below the expected operating temperature. If the austenite phase is strained with an applied load, a martensite phase is stress-induced and the twinning process occurs as if the material has been cooled to its martensitic temperature. When the applied load is removed, the material inherently prefers the austenite phase at the operating temperature and its strain is instantly recovered.

The stress-strain curve indicates a difference in stress levels during loading and unloading, known as a superelastic stress-strain hysteresis (see Figure 4.4).[5]

SMA Systems

Many systems exhibit a martensitic transformation. Generally they are subdivided into ferrous and nonferrous martensites. A classification of the nonferrous martensites was first given by Delsey et al.[8] (see Table 4.1), while ferrous alloys exhibiting an SME were first reviewed by T. Maki and T. Tamura[9] (see Table 4.2).

From all systems mentioned in both tables, in fact only one major system has become industrially successful: NiTi(X,Y), in which X, Y, are elements replacing Ni or Ti. The alloy based on NiTi, popularly known as Nitinol, is the most widely used of the available SMAs.[2]

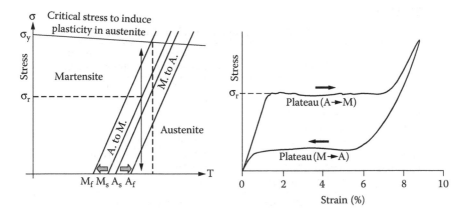

FIGURE 4.4
An illustration of the superelasticity. The right curve is the superelastic curve observed on a wire.[5]

TABLE 4.1

Classification of the Nonferrous Martensites[8]

Group	Alloy System
1. Terminal solid solutions based on an element having allotropic phases	1. Cobalt and its alloys 2. Rare earth metals and their alloys 3. Titanium, zirconium, and their alloys 4. Alkali metals and their alloys and thallium 5. Others as Pu, Ur, Hg, and alloys
2. Intermetallic solid solutions with a bcc-parent phase	1. β-Hume-Rothery phases of the Cu-, Ag-, Au-based alloys 2. β-Ni-Al alloys 3. Ni-Ti-X alloys
3. Alloys showing cubic to tetragonal trans. (incl. quasi-martensite)	1. Indium-based alloys 2. Manganese-based alloys (paramagn. ↔ antiferromagn.) 3. A15-compounds 4. Others: Ru-Ta, Ru-Nb, Y-Cu, LaCd, $LaAg_x-In_{1-x}$

Below is mainly an introduction to industrial applicable SMAs and alloy systems.[10,11]

- Fe-based alloys
- Cu-based alloys
- Ni-Ti-based alloys
- HTSMA
- Other systems

TABLE 4.2

Ferrous Alloys That Exhibit a Complete or Near-Complete SME[9]

Alloy	Composition	Crystal Structure of Martensite	Nature of Transformation[a]
Fe-Pt	≈25at%Pt	bct (α')	T.E.
	≈25at%Pt	fct	T.E.
Fe-Pd	≈30at%Pd	fct	T.E.
Fe-Ni-Co-Ti	23%Ni-10%Co-4%Ti	bct (α')	—
	33%Ni-10%Co-4%Ti	bct (α')	T.E.
Fe-Ni-C	31%Ni-0,4%C	bct (α')	Non-T.E.
Fe-Mn-Si	30%Mn-1%Si	hcp (ε)	Non-T.E.
	28~33%Mn-4~6%Si	hcp (ε)	Non-T.E.

[a] T.E.: Thermoelastic martensite, Non-T.E.: Nonthermoelastic martensite.

Fe-based Alloys[9,12,13]

In ferrous alloys, the austenite (face-centered cubic, or FCC-γ phase) can be transformed to three kinds of martensites, depending on composition or stress: γ-α' (BCC), γ→ ε (HCP), and γ→ FCT martensite.

Although an SME has been observed for all three types of transformation, most attention to develop a commercial alloy has been given to the alloys with a γ→ ε transformation. These alloys have a low stacking fault energy of austenite (Fe-Cr-Ni, Fe-high Mn alloys). The austenite to ε-martensite transformation proceeds by the a/6 {112} Schockley partial dislocations trailing a stacking fault ribbon on every [111] austenite plane and changing the crystal structure to martensite. The SME, which is of the one-way type, is mainly resulting from a reverse motion of the Schockley partial dislocations during heating.

A complete SME has been reached in both single crystals[14,15] and polycrystalline Fe-Mn-Si alloys[16,17] with a suitable amount of Mn and Si contents. A 9% shape memory strain in single crystals[15] and 5% in polycrystals[16] has been reported.

Any factors impeding the reversibility of the motion of partial dislocations will lead to an incomplete recovery and in turn a poor SME.

The internal factors hampering the recovery are alloy composition, Ne'el temperature, transformation temperature, lattice defects, and so on. External factors are applied stress and strain, deformation, recovery annealing temperature, and thermomechanical treatment.

For example, Murakami et al.[18] showed that Fe-Mn-Si alloys with 28–33% Mn and 4–6% Si exhibit a nearly perfect SME. But alloys with Mn content less than 20% have also been successfully developed.

In order to improve the corrosion resistance of commercial Fe-based alloys, Cr (less than 20%) and Ni are added.[19]

Coccia Lecis, Lenardi, and Sabatini[20] evaluated the effect of Mn-depleted surface layer on the corrosion resistance of shape memory Fe-Mn-Si-Cr alloys. In this work, they examined Fe-Mn-Si-Cr alloys possessing a good SME due to a high Mn content (28 wt pct). The addition of Cr (5 wt pct) was made in order to give fairly good corrosion resistance to the alloy. But they found that even in moderately corrosive environments, the presence of Cr does not bear any passivation.

On the other hand, they found that the alloy can acquire corrosion resistance by means of the formation of an Mn-depleted surface layer obtained by heating the alloy at high temperature (~1050°C) in air. This modified layer forms because MN is selectively oxidized with respect to the other components. The adhesion of this layer is maintained even under severe stress if the thickness of the modified region does not exceed 20 μm. Under this limit, the shape memory characteristics of the alloy are not affected, and at the same time, the specimen acquires properties of passivity comparable with one of the most common austenitic stainless steels at the presence of the same environmental conditions.

This kind of SMA modified by the oxidation process is a good candidate for SMA systems requiring corrosion resistance in moderately aggressive environments.

In a series of tests by Chung et al.,[21] the effects of thermomechanical training on martensitic transformation and SME were studied in an Fe-33.7Mn-5.22Si alloy. Experimental results showed that the volume of thermal martensite is small. The volume of reversible martensite (VRM) — that is, the martensite formed by stress-inducing, which can reverse-transform into austenite upon heating — increased more than 10 times the amount of thermal martensite by stress-inducing upon heating. The recovery temperature was found to have a crucial effect on the volume of martensite. For thermomechanical training with low recovery temperatures, VRM decreases after the third cycle. Thermomechanical training at recovery temperatures of 500 and 600°C increased the VRM by more than 1.8 times the volume of stress-inducing martensite before training, with almost complete shape recovery. The shape recovery ratio of the alloy was found to be proportional to the VRM. This was determined by integrating the internal friction peaks with respect to temperature in the reverse martensitic transformation.

Hamers and Wayman[22] studied the various Fe-based alloys showing the FCC-HCP martensitic transformation (e.g., Fe-Mn alloys). Their main effort concentrated on the addition of Co and its effect on the transformation temperatures and SME. The investigation examined electrical resistance and DSC measurements, optical microscopy, and compression and bend tests.

The alloys studied included Alloy A (70 Fe, 25 Mn, 5 Co), Alloy B (67 Fe, 25 Mn, 8 Co), Alloy C (65 Fe, 30 Mn, 5 Co), and Alloy D (62 Fe, 30 Mn, 8 Co).[23] Their results showed:

a) The transformation temperatures determined by DSC and electrical resistance measurements are very similar. The measurements indicated that the transformation temperatures depend on the Mn content, but that the Co additions have hardly any influence. On the other hand, the T/M treatment of the samples shows a significant effect on the transformation temperatures and hysteresis.

b) The microstructures comparing alloys A and B and alloys C and D are very much the same, indicating that the Ne'el temperature should be below room temperature.

c) In the present work a high recovery upon unloading in the bend test was measured and instead of decreasing SE with an increase in SME, the present alloys show an increase in SE when the SME increases.

d) The SME was measured by compression and bend tests and showed that for Fe-Mn-Co alloys the SME is rather low compared with Fe-Mn-Si alloys.

e) Finally, the results of bend samples for all four compositions clearly shows that the shape memory depends on the composition, and the difference between the hot-rolled and quenched samples indicates that the shape memory also depends on the T/M treatment.

Fe-based alloys are so far not successful SMAs. They exhibit only a (limited) one-way SME after a labor-intensive thermomechanical treatment. No significant two-way effect has been reported nor have pseudoelastic properties been seen, while only moderate damping capacity could have some interest. Therefore, the only reported successful type of applications of these Fe-based alloys are couplings. This type of application is based on the one-way effect. The recovery stresses are moderate but sufficient.[23]

Inagaki[2,19] studied the effect of Ni on the SME in Fe-Mn-Si-Cr-Ni alloys.

Cu-based Alloys[24–28]

Cu-based SMAs are derived from Cu-Zn, Cu-Al, and Cu-Sn systems. The composition range of these alloys corresponds to that of the well-known β Hume-Rothery phase. In most SMAs, this phase has a disordered BCC structure at high temperatures but orders to a B2, DO_3, or $L2_1$ form at lower temperatures. The shear elastic constant of the β phase exhibits an anomalous behavior with decreasing temperature; that is, it is lowered till the lattice instability with respect to (110) $\langle 1\,1\,0 \rangle$ shears at some temperature transforms β to martensite. The temperature of the transformation to martensite, M_s, varies with the alloy composition. The elastic anisotropy of the β phase is much higher compared to normal metals and alloys and increases further as the martensitic transformation is approached.

Cu-Zn and Cu-Al martensites are of three types, α', β', or γ', with subscript 1, 2, or 3 added to indicate the ordering schemes in β, viz. B2 (2) or DO_3 (1) or $L2_1$ (3). Some conversion from one martensite structure to another (e.g., $\beta' \rightarrow \gamma'$), may also take place. The net result is a coalescence of plates within a self-accommodating group and even coalescence of groups. Heating this deformed martensite microstructure transforms it to the β phase with the SME accompanying the structural change.

Cu-based SMAs presently used in applications are essentially derived from Cu-Zn and Cu-Al systems with elements added for various metallurgical reasons. The working martensite in these alloys is only or predominantly $\beta_{1,\,2,}$ or $_3$ type, with γ' martensite being the minor constituent in the latter case.

In selecting an alloy composition to obtain a complete β microstructure that transforms to martensite, two criteria should be taken into account: (1) The β phase must be stable over as wide a temperature range as possible. The less wide this temperature range the faster is the cooling rate required to retain the β phase without diffusional decomposition. (2) The transformation temperatures must fall within a range that satisfies the requirement for the shape memory application (–150 to 200°C). The three alloy systems mentioned in Table 4.3 satisfy these criteria. They are in use now, but within limited amounts.

Transformation Temperatures/Quenching

The transformation temperatures are, apart from the composition, also strongly influenced by other factors.

The Influence of the Composition

Several authors have attempted to quantify the M_s–composition relationship for several Cu-based alloys. An overview is given in Reference 25. Different authors give different weightage for the same element. The main reason for this discrepancy might be that the measurements have been made on samples with different thermomechanical histories (i.e., one has probably not measured on "identical samples"). Indeed, composition is not the only chemical factor affecting the M_s temperature. The type and degree

TABLE 4.3

Industrial Cu-Based Alloys[5]

Base Alloy	Composition Wt %	Ms (°C)	Hyst. (°C)	Other Alloying Elements in Solution (%)	Current Grain-Refining Elements Producing Precipitates	Remarks on the Base Alloy
Cu-Zn-Al	5–30 Zn	-190- + 100	10 (β')	Ni (~5%)	Co (CoAl); B (AlB$_2$); Zr (?); B, Cr (Cr$_x$B$_y$)	Good ductibility and reproducibility
	4–8 Al			Mn (~12%)		Prone to martensite stabilization Poor β-stability (T > 200 °C)
Cu-Al-Ni	11–14.5 Al	-140- + 200	10 (β')	Mn (~5%)	Ti (Cu, Ni)$_2$ TiAl]; B (AlB$_{12}$); Zr (?)	Low ductility
	3–5 Ni		40 (γ')			Low mar. stabilization Good β-stability
Cu-Al-Be	9–12 Al	-80- + 80	6 (β')	Ni (~5%)	B (AlB$_2$ or AlB$_{12}$) Ti (Cu$_2$TiAl)	Poor reproducibility Excellent β-stability (T > 200°C)
	0.4–1 Be					

of order of the beta and the martensite lattice also effect the M_s. Thermal treatments can, therefore, influence the transformation temperature.

The transformation temperatures of Cu-based alloys are very sensitive to minute changes of the degree of order in the β phase. Such changes are easily brought about by quenching from intermediate and high temperatures in the form of dilute disorder in an otherwise well-ordered material. The effect is noticeable in both Cu-Zn-Al-[29] and Cu-Al-Ni-[30-32] based alloys and manifests as a suppression of the transformation temperatures, thereby stabilizing the β phase relative to the martensite.

Studies by Kainuma, Takahashi, and Ishida[33] showed that ductile SMAs of the Cu-Al-Mn system have been developed by controlling the degree of order in the β phase. Additions of Mn to the binary Cu-Al alloy stabilize the β phase and widen the single-phase region to lower temperature and lower Al contents. It was shown that Cu-Al-Mn alloys with low Al contents have either the disordered A2 structure or the ordered $L2_1$ structure with a lower degree of order and that they exhibit excellent ductility.

The parent β phase of the as-quenched and aged Cu-14 at. pct Al-13 at. pct Mn alloys showed the β (A2) disordered structure and that of Cu-17 at. pct Al-10 at. pct Mn alloys an ordered β_1 ($L2_1$) structure. In the case of Cu-16 at. pct Al-10 at. pct Mn alloys, the β phase obtained by quenching changes to β_1 phase during aging at 150°C.

These alloys exhibited excellent ductility with tensile elongations up to 15% and cold workability greater than 65%, and this was mainly due to the decrease in the degree of order in the β phase.[33]

Other investigations include Xiao and Johari's[34] evaluations of the internal friction in a Cu-Zn-Al SMA in various microstructural states in a wide range of stress amplitude. Simultaneous measurement of the corresponding shape change indicated correlations between internal friction and the deformation mechanisms. They found that the observed stress-amplitude dependence of the internal friction cannot be represented by a single formalism. Rather, in different ranges of stress amplitude, where the internal friction mechanisms are different, separate formalisms are applicable.

The isothermal aging effects in an as-quenched Cu-11.88Al-5.06Ni-1.65Mn-0.96Ti (wt pct) SMA at temperatures in the range 250° to 400°C were investigated by Wei, Peng, Zou, and Yang.[35]

They found on parent aging at intermediate temperatures, the pentatomic Cu-Al-Ni-Mn-Ti alloy is more susceptible to aging effects than ternary Cu-Al-Ni alloys. The as-quenched SMA shows at least three temperature-dependent stages of microstructural evolution that correspond to different changes in the properties and performance of the alloy.

Romeo et al.[36] studied the concentration and nature of the defects introduced by quenching from different T_q temperatures and their influence on the martensitic transformation undergone by a Cu-Al-Be SMA.

Influence of Grain-Refining Elements Forming Precipitates

Cu-based SMAs exhibit a rapid grain growth at higher solutionizing temperatures. With grain sizes of the order of millimeter and the high elastic anisotropy of the β phase, they suffer intergranular cracking during quenching and plastic deformation. The problem has been solved by the addition of grain-refining elements to two SMAs. Thus Zr (0.4 to 1.2 wt%), Co (0.4 to 0.8 wt%), Ti (0.5 to 1.0 wt%), and B (0.4 to 0.2 wt%) have been added to Cu-Zn-Al alloys to reduce the grain size to the 100 μm level. Titanium is also most effective in refining the grain size in Cu-Al-Ni alloys to reduce the grain size to the 100 μm level. Titanium is also most effective in refining the grain size in Cu0Al-Ni alloys to the 50–100 μm range.[37,38]

Refining is brought about by the formation of insoluble particles that aid nucleation of the grains or retard their growth. These grain-refining elements have four direct or indirect effects on the transformation temperatures: (1) By forming intermetallics they deplete the original beta-lattice from alloying elements, which changes the transformation temperatures according to the formulas given in Table 4.4. (2) Part of these elements remain in solution within the beta matrix. Depending on the atom size, this can give rise to solid solution hardening, which decreases the M_s and eventually the other transformation temperatures.[40] (3) They can also have a chemical contribution. (4) The precipitates basically limit the grain growth during the annealing, which has an influence on the transformation temperatures as discussed in the following section.

Influence of Grain Size

Several authors have shown that a small grain size results in the stabilization of the parent phase with depression of the transformation temperatures up to 40°C.[41,42] This effect is observed in alloys with and without the special addition of grain-refining elements, which indicates the influence of the restraining effect of the grain size itself on the transformation. The lowering of the transformation temperatures is attributed to the increasing grain restraint with decreasing grain size. This is the conclusion of most authors.[43,44] Elements contributing to a high stacking fault energy will thus have an effect similar to a small grain size.

Influence of Defects

Often it is not only the effect of the grain size or the grain size thickness ratio that will account for the changes in the transformation temperatures. Annealing a sample at higher temperatures can give rise to grain growth but will also reduce the amount of defects and thus the nucleation sites. In Cu-based alloys, the situation is again complicated by the quenched-in vacancies and the size of the antiphase domains, which can also be

TABLE 4.4

High-Temperature SMAs[39]

Base System	Type of Martensite	Other Alloying Elements		As $\leq$
		Name	Reason	
Fe-Mn-Si	Non-thermoelastic $\gamma \Leftrightarrow \varepsilon$	Co, Ni, Cr	To improve the corrosion resistance.	150°C–200°C
Cu-Al-Ni	Thermoelastic $\beta 1 \Leftrightarrow 18\,R\,(2H)$	Mn, Ti, B, Zn	To improve machinability, control of transformation temperatures, Grain refinement to improve the ductility.	100°C–200°C
(Ni-X)Ti	Thermoelastic $\beta_2 \Leftrightarrow B19', B19$	X-Pt, Pd, Au, Rh	Based on B2-TiX intermetallic compounds showing martensitic transformation at very high temperatures.	150°C–500°C
		B	To reduce the grain size and improve the strength.	
Ni(Ti-X)	Thermoelastic $B_2 \Leftrightarrow B19', B19$	X = Hf, Zr	Based on NiX intermetallic compounds, forming a pseudobinary with NiTi.	120°C–350°C
Ni-Al	Thermoelastic $B_2 \Leftrightarrow 3R\,(7R)$ (L10-structure)	Cu, Co, Ag	To increase transformation temperatures.	100°C–600°C
		Fe, Co, Mn, B	To improve the ductility.	
Ni-Mn	Thermoelastic (?) $B2 \Leftrightarrow \theta$ (L10-structure)	Al, Ti, Cu for Ni	To decrease M_s and to improve the shape memory characteristics.	500°C–750°C
		Mg, Al, Si, Ti, V, Sn, Cr, Co, Fe, Mo for Mn	To increase M_2 and to improve, the shape memory characteristics.	
Zr-based intermetallics	B19' monoclinic	Ti, Ni	To improve the ductility.	200°C–900°C
CuZr	Non-thermoelastic			
Zr$_2$CuNi	Non-thermoelastic			
Zr$_2$CuCo	Thermoelastic			

regarded as strengthening the matrix. An increase in the energy of the β-phase due to a higher defect concentration such as foreign elements in solid solution, precipitates, and internal strain fields (e.g., coherency strains) causes a lowering of M_s.[44] Moreover, if the defect density is proportional to nucleation sites, a higher defect density will give rise to much smaller martensite plates.

Specific defect configurations can also be introduced by thermal cycling, and also by two-way memory training.[45] The changing character of the same dislocation in β and martensitic phases has been suggested to alter the relative phase stability of the two phases.

Ni-Ti Alloys

$Ni_{50} Ti_{50}$ is the best explored system of all SMAs and occupies almost the whole market of SMA. $Ni_{50} Ti_{50}$ is an intermetallic phase that has some solubility at higher temperatures.

The science and technology of Ni-Ti is overwhelmingly documented. The influence of composition and thermomechanical processing on the functional properties is well understood and described in literature. Therefore, only some very interesting and relevant publications[46–50] are cited.

The basic concept of processing Ni-Ti alloys is that in order to avoid plastic deformation during shape memory or pseudoelastic loading, the martensitic and the β-phase have to be strengthened. This occurs by classic methods: strain hardening during cold deformation, solution hardening, and precipitation hardening. Ni-Ti alloys have the significant advantage that these techniques can be easily applied due to an excellent ductility and a very interesting but complicated precipitation process.[51]

The compositions of the Ni-Ti SMAs are approximately between 48 and 52 at % Ni and the transformation temperatures of the B2 structure to the martensitic phase with a monoclinic B19′ structure are very sensitive to the Ni content (a decrease of about 150° for an increase of 1 at % Ni). The transformation temperatures can be chosen between -40 and +100°C.

Ni-Ti alloys show the best shape memory behavior of all SMAs. Even in the polycrystalline state, 8% shape recovery is possible, and 8% pseudoelastic strain is completely reversible above A_f, while the recovery stress is of the order of 800 MPa.

The most specific characteristics of R-phase transition which affects some Ni-Ti alloys (see page 191) is that it shows a clear one- and two-way memory effect in the order of 1% recoverable strain and that the hysteresis of the transformation is very small, only a few degrees, which creates possibilities for accurately regulating devices.

It should be noted that further cooling transforms the R-phase into B19′ martensite. During heating only the reverse martensitic transformation will be observed. To observe the reverse R-phase transition, cooling should be stopped above M_s.

It has been shown that the appearance of the R-phase depends on composition, alloying elements, and thermomechanical processing.[50] In fact, the major common point is that all effects depressing the martensitic forward transformation below room temperature will favor the appearance of the R-phase transition, which is quite stable near 30°C.

To appreciate the mechanical properties of Ni-Ti alloys, Table 4.5 compares Ni-Ti alloys to stainless steels. Ni-Ti alloys have other useful properties such as high wear resistance[56] and good corrosion resistance. Moreover, some Ni-Ti alloys have an additional martensite-like transition: the so-called R-phase transition.[52,53,57] The R-transition is a B2 ↔ rhombohedral transformation that also has second-order characteristics.[52] The R-phase transition appears upon cooling before the martensitic transformation under certain conditions of material fabrication and processing. This phase has a narrow hysteresis (nearly 1.5°C) and very good thermal cycling stability. For these reasons, the R-phase has been used in many actuator applications. However, the strain is fairly small (about 1%).

TABLE 4.5

Mechanical Properties of Ni-Ti Compared to Stainless Steel[a]

	Ni–Ti	Typical Data for a Stainless Steel
Max. reversible elastic deformation	Typ. 8%	0.8%
Mass density	6450 kg.m^{-3}	7850 kg.m^{-3}
Young's modulus (E)	M: 28–41 GPa	190–210 GPa
	A: 83 GPa	
Shear modulus (G)	M: 10–15.5 GPa	75–80 GPa
	A: 31 GPa	
Poisson's ratio (υ)	0.33	0.27–0.30
Yield stress[b]	A: 195–690 MPa	400–1600[c] MPa
	M: 70140 MPa	
Ultimate stress	895–1900 MPa	700–1900 MPa
Coefficient of thermal expansion	A: 11 × 10^{-6}/°C	8–10 10^{-6}/°C
	M: 6.6 × 10^{-6}/°C	

Note: "A" refers to austenite and "M" refers to martensite.

[a] The data for Ni-Ti, are taken from References 53 and 54. The data for stainless steel are taken from Reference 55.

[b] The yield stress for the SMA is not really a yield stress but rather a critical stress to induce martensite when in the austenitic state and a critical stress to reorient martensitic variants when in the martensitic state.

[c] These values apply typically to a spring steel.

Influence of Strain Rate on Deformation

Lin et al.[58] performed tensile tests with loading and unloading under various strain rates and for Ni-Ti SMA wires at various constant temperatures and investigated the influence of strain rate on the stress-strain-temperature relationship. A summary of the main results follows:

(1) For $\varepsilon' < 10\%/\text{min}$, the martensitic transformation stress and its reverse transformation stress did not depend on ε'. These stresses were almost constant in the transformation regions. (2) For $\varepsilon' \geq 10\%/\text{min}$, the martensitic transformation stress increased and its reverse transformation stress decreased with an increase in ε'. These stresses fluctuated in the transformation regions. (3) The influence of ε' on the transformation stress due to R-phase transformation was slight. (4) The recoverable strain energy density depended slightly on ε' but increased significantly in proportion to temperature. The dissipated strain energy density depended slightly on temperature but increased in proportion to ε' for $\varepsilon' \geq 1\%/\text{min}$. (5) The influence of ε' on deformation properties is important in the evaluation of mechanical properties of SMA elements.

Ternary Ni-Ti Alloy Systems

The addition of a third element opens even more possibilities for adapting binary Ni-Ti alloys toward more specific needs of applications. Adding a third element implies a relative replacement of Ni or Ti. Therefore, it must always be very well indicated which atom (Ni or Ti or both) is replaced by the third element.

Alloying third elements will not only influence the transformation temperatures but will also have an effect on hysteresis, strength, ductility, shape memory characteristics, and also on the B2 $\rightarrow$ (R) $\rightarrow$ B19' sequence. The influence of several elements has been already described.[59–63]

More application oriented, one can distinguish four purposes to adding third elements:

1. To decrease (Cu) or increase (Nb) the hysteresis
2. To lower the transformation temperatures (Fe, Cr, Co, Al)
3. To increase the transformation temperatures (Hf, Zr, Pd, Pt, Au)
4. To strengthen the matrix (Mo, W, O, C)

Some of the ternary alloys have been developed for large-scale applications.

Ni-Ti-Cu

Ternary Ti-Ni-Cu alloys in which mainly Ni is substituted by Cu are certainly as important as binary Ti-Ni. Increasing Cu content decreases the deformation stress for the martensite state and decreases also the pseudoelastic hysteresis without significantly affecting the M_s temperature.[64] However, more than 10% Cu addition embrittles the alloys, hampering the formability.

It should also be remembered that while Ti-Ni transforms from a B2 into a monoclinic phase, Ti-Ni-Cu exceeding 15 at% Cu transforms from a B2 into an orthorhombic phase. Ti-Ni-Cu with less than 15 at% Cu transforms in two stages.[49]

A disadvantage of most Ti-Ni-Cu alloys is that the transformation temperatures do not decrease below room temperature. In order to obtain pseudoelastic alloys at room temperature but with small hysteresis, Cr or Fe can be alloyed. This way, an Ni 39.8 Ti 49.8 Cu 10 Cr 0.4 alloy has been developed with a small hysteresis (130 MPa), one fourth compared with $Ni_{50}Ti_{50}$ and M_s below room temperature

Ti-Ni-Nb[66,67]

The inherent transformation hysteresis of Ni-Ti-Nb is larger than for binary Ni-Ti alloys. By the presence of a large dispersed volume fraction of deformable β-Nb particles, the hysteresis can be further widened by an overdeformation of stress-induced martensite, generally between M_s and M_d. Originally, Ni-Ti-Nb (more specifically Ni 47 Ti 94 Nb 9) was developed by Raychem Corp. for clamping devices. The large shift of the reverse transformation temperatures from below to above room temperature by deformation allows room storage for open couplings.

Recently, also pseudoelastic Ni-Ti-Nb alloys have been developed with three significant differences relative to binary alloys:[67]

- Stress rate is much lower
- The σ^{P-M} stresses are much higher
- The superelastic window is much larger

Ni-Ti-Zr

Studies by Pu, Tseng, and Wu[68] found that

1. Phase transformation temperatures are elevated when Zr is greater than 10 at%; however, the initial addition of Zr depresses the transformation temperatures.
2. Ni-Ti-Zr alloys display relatively poor stability during the thermal cycling process compared to Ni-Ti-Pd and Ni-Ti-Hf alloys.
3. The fully reversible strain, ε_{cr} gradually decreases as the Zr content increases. In comparison with Ni-Ti-Pd and Ni-Ti-Hf high-temperature SMAs, the Ni-Ti-Zr alloys have relatively poor SME. Further modification of SME, ductility, and stability is needed.

4. The mechanism that affects stability of Ni-Ti-Zr alloys during thermal cycling deserves further study so that this factor does not limit the potential use of the high-temperature alloy in industrial applications.

5. The alloys based on Ni-Ti-Zr demonstrate complete SME even with a Zr content of 20 at%; the fully reversible strain decreases gradually as the Zr content increases. However, the fully reversible strain of Ni-Ti-Zr alloys is much smaller than Ni-Ti-Hf alloys at the same transformation temperature. The Ni-Ti-Zr alloys also demonstrated two-way memory effect after proper thermomechanical training.

Ni-Ti-Pd

Although Ni-Ti-Pd alloys are promising candidates as high-temperature SMAs, there are also many problems that have to be addressed before this alloy achieves practical usage.

Detailed studies by Wu[69] were conducted to investigate the thermomechanical stability of Ni-Ti-Pd high-temperature SMAs. To compare with Ni-Ti binary low-temperature SMAs, a series of Ni-Ti binary alloys with different compositions were also used to study their thermomechanical stability. It was found that the transformation temperature change of Ni-Ti-Pd alloys after 1000 times cycling decreases as the Pd addition increases. In comparison with the Ni-Ti binary alloys, Ni-Ti-Pd alloy has much better stability. For example, the transformation temperature change of Ni-Ti-30 at% Pd after 1000 times cycling is only 2°C compared to 30°C for Ni-50 at% Ti alloy.

Out of this research a new theory was developed, described as *a dislocation thermodynamic* model, to describe enplanement stability of thermomechanical cycling of SMAs. This new theory extends the existing theory to successfully enplane the thermal cycling effects of SMAs. The dislocation thermodynamic model not only offers insight into the thermal cycling effect of SMAs, one of the most important phenomena of SMA, but also provides various practical approaches to improve the stability of different high-temperature SMAs.

The result of this work provides an extensive, essential, and systematic database that makes new designs and applications of Ni-Ti-Pd high-temperature SMAs possible.

Ni-Ti-Hf

Although Ni-Ti-Pd alloys possess qualities that allow high transformation temperature and good SME, they are very expensive due to the alloying element, Pd. Thus, it was necessary to continue the search for more economical materials that possess similar mechanical properties and SME as Ni-Ti-Pd.

Preliminary studies[69] found that the transformation temperature of Ni-Ti alloys increase when Ti is substituted by Hf or Zr. Because the price of the raw Hf element is only one sixth that of Pd, there is reasonable expectation that Ni-Ti-Hf alloys will prove more economical to use than Ni-Ti-Pd alloys. Thus, the development of Ni-Ti-Hf high-temperature SMAs have warranted special attention.

After 3 years of hard work, a new economical Ni50-(50-x)Ti-xHf high-temperature SMA was successfully developed. The newly developed Ni-Ti-Hf alloy has a much higher transformation temperature than the Ni-Ti-Pd and Ni-Ti-Pt alloys with the same atomic percentage of the third alloying elements. The Ni-Ti-Hf high-temperature SMAs cost 1/6 to 1/10 of the cost of Ni-Ti-Pd to produce due to the lower price of the Hf element and the smaller amount of Hf required to achieve the same transformation temperatures. Meanwhile, the SME of the new alloys is compatible with the Ni-Ti-Pd alloys. Considering the SME and the costs, the Ni-Ti-Hf alloys have been recognized as the alloys with great potential for high-temperature applications.

The experimental results show:

1. An addition of more than 5% Hf can significantly elevate the phase transformation temperature. When the concentration of Hf is 30 at%, the austenite transformation temperature can reach 450°C, but the initial addition of Hf depresses the transformation temperature.

2. The chemical driving force decreases with the Hf content in the monoclinic B19′ region and increases in the orthorhombic B19 region. The S decreases in the monoclinic martensite region and remains almost constant in the orthorhombic martensite.

3. The addition of Hf does not change the hysteresis behavior of Ni-Ti alloys. The martensite transformation in the Ni-Ti-Hf alloys is thermoelastic.

4. The Ni-Ti-Hf alloys have better stability in the thermal cycling process than Ni-Ti binary alloys. The change in transformation temperature during the thermal cycling process decreases as the Hf content increases. For the Ni-Ti-25 at% alloys, the change in transformation temperature after 100 times cycling is only 7°C compared with 30°C for Ni-Ti binary alloy.

5. The Ni-Ti-Hf alloys demonstrated an obvious two-way memory effect after proper thermomechanical training. The two-way memory effect increased as the prestrain increased, which is associated with the increasing of residual strain with increasing prestrain.

6. The high transformation temperature is harmful to the two-way memory effect, which is confirmed by the fact that the two-way memory effect deteriorates as the heating temperature increases.

7. The Ni-Ti-Hf alloy is an excellent candidate to substitute for expensive Ni-Ti-Pd alloys.

High-Temperature SMAs[39]

Actual SMAs are limited to maximal A_f temperatures of 120°C, M_s generally being below 100°C. However, because market demands on SMA have expanded greatly, the need for SMA transforming at higher temperatures than presently available is increasing. The main interested application areas are in actuators in the automobile and oil industries and in safety devices.

There is also an interest in robotics because SMAs with high transformation temperatures allow faster cooling, which would significantly increase the bandwidth in which the robot could operate.

In spite of the fact that many alloy systems show high transformation temperatures, no real large-scale applications are developed. A major breakthrough has not been reported yet mainly due to the following problems: (much) lower performance than the successful Ni-Ti alloys, stabilization of martensite, decomposition of the martensite or parent phase, and brittleness due to high elastic anisotropy or due to the presence of brittle phases or precipitates.

Another condition for a good SME is that the stress to induce martensite or the stress to reorient martensite is (much) lower than the critical stress for normal slip. Because this critical stress for slip is generally decreasing with increasing temperature, the above condition is quite difficult to fulfill, especially at high temperatures. A potential HTSMA should thus be designed at such a composition or thermomechanical treatment that strengthening mechanisms are incorporated to increase the critical stress for slip.

Table 4.4 summarizes some of the systems under investigation.[39]

Several new high-temperature two-way SMAs have been developed.[69]

Alloys of Ti-V, Ti-V-Cr, Ti-V-Fe, and Ti-Zr-Hf-Ni with various compositions have been fabricated. Others have included Ni-Mn-Ti high-temperature SMAs, which are a possible candidate as useful high-temperature SMAs. Characterized by a very wide hysteresis width (70°C), the martensite transformation is not fully reversible. The investigators[69] discovered that the Ti alloying element can successfully reduce the thermal hysteresis width and adjust transformation temperature. This makes it possible for Ni-Mn-Ti alloys to fully recover their predeformation shape. A preliminary study shows that Ni-Mn alloys possess SME that surpasses other alloys investigated (Ni-Ti-Pd and Ni-Ti-Hf). Again, the main problem of this material is its extreme brittleness.

Composite Materials

Ni-Ti-Al

In promising work by Wu,[69] an Ni-Ti-Al metal-metal composite (MMC) material was developed. The differential gradient interface layers of TaO and Ni-Ti were successfully built on an Ni-Ti wire, which was embedded in the Al matrix. Both resistance measurement and vibration measurements indicated that the layer can provide insulation of the Ni-Ti wire and can be used for vibration suppression for smart structures.[70]

Ni-Ti-TiC

Fukami-Ushiro and Dunand[71] investigated the shape-memory recovery behavior of Ni-Ti as a function of compressive mechanical prestrain and compared it to that of Ni-Ti composites containing 10 vol% and 20 vol% particulates. The following points summarize the main findings of this study:

1. Undeformed Ni-Ti-TiC composites exhibit linear allotropic expansion and linear thermal expansion coefficients smaller than those of undeformed, unreinforced Ni-Ti. The thermal expansion values are in agreement with predictions from a continuum mechanics equation assuming no relaxation of the thermal mismatch between the two phases.

2. In deformed Ni-Ti-TiC composites, the mismatch between the elastic TiC particles and the Ni-Ti matrix must be relaxed by additional matrix plastic strain: unrecoverable slip deformation or recoverable twinning. Both the SME and the two-way SME are little affected by the presence of up to 20 vol% of ceramic particles, indicating that a major part of the mismatch is relaxed by matrix twinning. The thermal recovery behavior of the composites indicates that plastic strain by slip increases as the TiC content increases: The magnitude of recoverable strain decreases, the transformation is spread over a wider temperature range, and the two-way SME is enhanced. The last effect and the decrease of all transformation temperatures can also be explained by the enhanced residual elastic stresses in the composites.

Other Types of SMAs[72]

In spite of the good biocompatibility of Ni-Ti-alloys, doubts remain on the long-term stability or on the danger of bad surface treatment leading to Ni leaching. Because Ni is known for its high allergic reaction, Ni-less SMAs could be attractive. Such alloys might be developed based on the allotropic transformation in Ti, a highly biocompatible material. Pure Ti shows an allotropic transformation from β (BCC) to (hexagonal) phase at 1155°K. Transition elements (TM) stabilize the β-phase. Thus, the temperature of the (α + β)/β transition decreases with increasing concentration of the alloying element.

β-phase Ti alloys can be martensitically transformed if they are quenched from the stable β phase. Two types of martensite, respectively α′ and α″, can be formed, depending on the composition and the solution treatment conditions.[73]

The α′-martensite is hexagonal, while a″ has an orthorhombic structure.[74] It is the a″-martensite that shows the SME. The SME was first studied in detail by Baker in a Ti-35 wt % Nb alloy.[75] Since then, several observations of SME especially in Ti-Mo-based alloys have been reported.[60,76–79,81] A systematic work on the influence of different alloying elements on the SME can be found in Reference 82. Therefore, pseudoelastic β-Ti alloys could offer an interesting alternative to Ni-Ti alloys, for example, for orthodontic wires. Such an alloy (Ti-11Mo-3Al-2V-4Nb) has been recently developed by Lei et al.[83] Good pseudoelasticity in the order of 3% has been obtained after proper cold working and heat treatment.

Fe-Mn-Si-Cr-Ni Alloys[2,19]

Chemical compositions of recently developed Fe-based SMAs are generally complex, containing relatively large amounts of Si, Mn, Cr, and Ni. Studies by the Shonan Institute of Technology[70] found:

a) In Fe-Si-Mn-Cr-Ni alloys, the Ni content giving the maximum SME was 6%.

b) The SME is significantly influenced by the cooling rate after the solution treatment. To obtain the maximum SME, rapid quenching is essential.

c) These two conditions provide an optimum initial distribution of overlapped stacking faults and martensite plates. On straining the specimens, the stress-induced γ to ε transformation can be induced with the minimum stress.

d) Training is not equally effective on all alloys. This is because the distributions of overlapped stacking faults and martensite plates introduced by training are strongly dependent on alloy contents.

Hybrid Composites[84]

SMA Fiber/Metal Matrix Composites (MMCs)[85]

The basic design approaches for the SMA fiber/MMC can be summarized in five steps: (1) The SMA fiber/MMCs are prepared and fabricated by using conventional fabrication techniques; (2) the as-fabricated composites will be heated to high temperatures to shape-memorize the fibers or to undergo some specific heat treatment for the matrix material, if necessary; (3) because SMAs have much lower stiffness at martensite stage or readily yield at the austenitic stage just above the martensitic transformation start temperature (Ms), the composites are then cooled to lower temperatures, preferably in martensite state; (4) the composites are further subjected to proper deformations at the lower temperature to enable the martensite twinning or the stress-induced martensitic transformation to occur; and (5) the prestrained composites are then heated to higher temperatures, preferably above the austenite finish temperature A_f wherein martensite detwinning or the reverse transformation from martensite to austenite takes place, and the Ti-Ni fibers will try to recover their original shapes and hence tend to shrink, introducing compressive internal residual stresses in the composites. This design concept can also be applied to polymer matrix composites (PMCs) containing SMA fibers and to the MMCs containing SMA particles.

The SMA fiber-reinforced MMCs also exhibit other improved properties. For instance, the damping capacity of the Ti-Ni fiber/Al matrix composite was measured and the results indicated that the damping capacity of the composite in the temperature range 270 to 450°K was substantially improved over the unreinforced Al. The composite was also expected to show high wear resistance.

SMA Fiber/Polymer Matrix Composites

Depending on the SMA fiber pretreatment, distribution configuration, and host matrix material, a variety of hybrid PMCs can be designed that may actively or passively control the static or dynamic properties of composite materials. Passively, as in the SMA fiber/Al matrix composites, the SMA fibers are used to strengthen the PMCs, absorb strain energy, and alleviate the residual stress and thereby improve the creep or crack resistance by stress-induced martensitic transformations. The embedded SMA fibers are usually activated by electric current heating, and hence they undergo the reverse martensitic transformation, giving rise to a change of stiffness, vibration frequency and amplitude, acoustic transmission, or shape of the composite.

Advanced composites such as graphite/epoxy and glass/epoxy composites offer high strength and stiffness at a low weight and moderate cost. However, they have poor resistance to impact damage because they lack an effective mechanism for dissipating impact strain energy such as plastic yielding in ductile metals. As a result, the composite materials dissipate relatively little energy during severe impact loading and fail in a catastrophic manner once stress exceeds the composite's ultimate strength. Typically, damage progresses from matrix cracking and delamination to fiber breakage and eventual material puncture. Various approaches to increase the impact damage resistance, and specifically the perforation resistance, of the brittle composite materials have been attempted. The popular design concept is to form a hybrid that utilizes the tougher fibers to increase the impact resistance and also the stiffer and stronger graphite fibers to carry the majority of the load. The hybrids composed of the graphite/epoxy with Kevlar, Spectrag, and S-glass fibers have demonstrated modest improvements in impact resistance.

Among various engineering materials, high-strain SMAs have a relatively high ultimate strength. They can absorb and dissipate a large amount of strain energy first through the stress-induced martensitic transformation and then through plastic yielding. Accordingly, the impact resistance of the graphite/epoxy composites may be improved by hybridizing them with SMA fibers. For example, the concept above has been developed and demonstrated that under certain load conditions the impact energy-absorbing ability of graphite and glass composites can be effectively improved by hybridizing the composites with Ti-Ni SMA fibers. Hybrid composites with improved impact and puncture resistance are very attractive because of their great potential in military and commercial civil applications.

Generally, the shape memory hybrid composite materials can be manufactured with conventional PMC fabrication methods, by laying the SMA fibers into the host composite prepreg between or merging with the reinforcing wires and then using either hot press or autoclave and several different types of cure cycles. Previously, the few attempts to incorporate embedded Ti-Ni wires directly into a PMC proved unsuccessful due to manufacturing difficulties and problems associated with interfacial bonding. To avoid the interface bonding issue, SMA wires were alternatively incorporated into polymer matrix by using coupling sleeves. Both thermoset and thermoplastic composites have been addressed.

Comparatively, fiber-reinforced thermoplastics offer some substantial advantages over fiber-reinforced thermosets because of their excellent specific stiffness, high fracture toughness, low moisture absorption, and possible fast and cost-effective manufacturing processes. However, the high process temperatures can be problematic for the embedding of SMA elements. The thermoplastic processing must be performed at higher temperatures, typically between 423 and 673°K, whereas the thermoset processing cycle of the composites is in the relatively low temperature range of room temperature to 443°K.

The thermoplastic processing cycle has some effect on the microstructure of the SMA fibers as manifested in the change in transformation temperatures and peak recovery stress: The transformation temperatures of the SMA shift upward while the peak recovery stress drops as a result of the thermoplastic processing. The thermoset processing only mildly affects the transformation characteristics of SMA fibers.

However, some dynamical properties of SMA fibers could be significantly affected. Much care should be taken to prevent shape recovery of the prestrained SMA fibers or wires during the composite cure cycle. The complexity of manufacturing the SMA composites can be greatly simplified by using the two-way SME. That is, the SMA wires will be trained to exhibit the two-way SME prior to embedding in the matrix.

SMA Particulate/Al Matrix Composites

Particulate-reinforced MMCs have attracted considerable attention because of their feasibility for mass production, promising mechanical properties, and potential high-damping capacity. In applications not requiring extreme loading or thermal conditions, such as automotive components, the discontinuously reinforced MMCs have been shown to offer substantial improvements in mechanical properties. In particular, discontinuously reinforced Al alloy MMCs provide high damping and low density and allow undesirable mechanical vibration and wave propagation to be suppressed. As in the fiber-reinforced composites, the strengthening of the composites is achieved through the introduction of compressive stresses by the reinforcing phases, due to the mismatch of the thermal expansion coefficient between the matrix and reinforcement.

The most frequently used reinforcement materials are SiC, Al_2O_3, and graphite particles. Although adding SiC and Al_2O_3 to the Al matrix can provide substantial gains in specific stiffness and strength, the resulting changes in damping capacity may be either positive or negative. Graphite particles may produce a remarkable increase in damping capacity, but at the expense of elastic modulus. More recently,[84] the concept of strengthening the Al MMCs has been proposed by the SME of dispersed Ti-Ni SMA particles. The strengthening mechanism is similar to that in the SMA fiber-reinforced composites: The prestrained SMA particles will try to recover the original shape upon the reverse transformation from martensite to parent (austenite) state by heating and hence will generate compressive stresses in the matrix along the prestrain direction, which in turn enhances the tensile properties of the composite at the austenitic stage. In the light of the well-known transformation toughening concept, some adaptive properties such as self-relaxation of internal stresses can also be approached by incorporating SMA particles in some matrix materials.

Because SMAs have a comparatively high loss factor value in the martensite phase state, an improvement in the damping capacity of the SMA particulates-reinforced composites is expected at the martensite stage. Accordingly, SMA particles may be used as stress or vibration wave absorbers in paints, joints, adhesives, polymer composites, and building materials.

Shape-memory particulate-reinforced composites can be fabricated by consolidating Al and SMA particulates or prealloyed powders via the powder metallurgical route. SMA particulates may be prepared with conventional processes, such as the atomization method and spray or rapid solidification process, which can produce powders with sizes ranging from nanometers to micrometers. However, few reports on the production of SMA particles are recorded in the open literature. Recently, a procedure was developed to prepare Ti-Ni-Cu SMA particulates through hydrogenating-ball milling-dehydrogenating. The ternary Ti-Ni-Cu alloys, where there is a substitution of Ni by Cu by up to 30 atomic %, are of particular interest for their narrow hysteresis, large transformation plasticity, high shock absorption capacity, and basic SME.

Ceramic Particulate/SMA Matrix Composites

In an SMA matrix, dispersed second-phase particles may precipitate or form during solidification or thermal (mechanical) processing, thereby creating a native composite. The martensitic transformation characteristics and properties of the composites can be modified by controlling the particles, as demonstrated in Ti-Ni(Nb), Cu-Zn-Al, and Cu-Al-Ni-Mn-Ti alloys. Alternatively, the presence of a ceramic second phase within the SMA matrix may lead to a new composite with decreased density and increased strength, stiffness, hardness, and abrasion resistance. Compared with common ceramic/metal composites, a higher plasticity may be expected for this composite because the stress-induced martensitic transformation may relax the internal stress concentration and hence hinder cracking. Previously, Al_2O_3 particle-reinforced Cu-Zn-Al composites were prepared with the conventional casting method, and this kind of composite was suggested to be suitable for applications requiring both high damping and good wear resistance. Using explosive pressing of the powder mixture, a TiC/Ti-Ni composite was prepared. In the sintered TiC/Ti-Ni composite it was found that the bend strength, compression strength, and stress intensity factors were significantly higher than those for TiC/Ni and WC/Co composites. With increasing TiC content, the hardness and compressive strength increases, while the ductility and toughness decrease.

More recently, Dunand et al.[71] systematically investigated the Ti-Ni matrix composites containing 10 vol % and 20 vol % equiaxed TiC particles, respectively. The composites were prepared from prealloyed Ti-Ni powders with an average size of 70 μm and TiC particles with an average size of about 40 μm, using powder metallurgy techniques. The TiC particles modify the internal stress state in the Ti-Ni matrix, and consequently, the transformation behavior of the composite: Unlike composites with matrices deforming solely by slip, the alternative deformation mechanisms, namely twinning and stress-induced transformation, are expected to be operative in the Ti-Ni composites during both the overall deformation of the matrix and its local deformation near the reinforcement, thereby resulting in the pseudoelasticity and rubber-like effect.

Magnetic Particulate/SMA Matrix Composites

Giant magnetostrictive materials $(TB_yDy_{1-y})_x Fe_{1-x}$ (Terfenol-D) provide larger displacements and energy density, and superior manufacturing capabilities, as compared to ferroelectrics. However, their applications have been limited by the poor fracture toughness, eddy current losses at higher frequencies, and bias and prestress requirements. More recently, composite materials based on Terfenol-D powders and insulating binders have been developed in Sweden. These composites broaden the useful range of the Terfenol-D material, with improved tensile strength and fracture toughness, and potential for greater magnetostriction and coupling factor. Most recently, a design concept was proposed to embed Terfenol-D particles within an SMA matrix to create a ferromagnetic shape memory composite with the combination of the characteristics of SMAs and magnetostrictive materials. The Terfenol-D particles will be elongated by about 0.1% when applying a magnetic field. The generated force is high enough to induce the martensitic transformations in the matrix at appropriate temperatures. Therefore, the orientation and growth of the martensite plates may be controlled by the magnetic field and by the distribution and properties of the Terfenol-D particles embedded in the matrix.

As an alternative, the high passive damping capacity of the magnetic powders/SMA matrix composites may be utilized. It is known that Cu-Zn-Al SMAs have high-damping capacity at large strain amplitudes due to thermoplastic martensitic transformations, but their stiffness is inadequate for some structural applications. The ferromagnetic alloys, including Terfenol-D, Fe-Cr, Fe-Cr-Al, and Fe-Al, are known to have relatively high strength as well as high damping capacity in the range of small strain amplitudes. In principle, the combination of Cu-Zn-Al matrix and ferromagnetic alloy inclusions should yield high damping capacity over a wide range of strain amplitudes and higher stiffness than that of the monolithic Cu-Zn-Al alloys.

Accordingly, three kinds of MMCs were fabricated from prealloyed Cu-26.5 wt% Zn-4.0 wt% Al powders (as a matrix) and rapidly solidified Fe-7 wt% Al, Fe-20 wt% Cr, and Fe-12 wt% Cr-3 wt% Al alloy flakes (30 vol%), respectively, by powder metallurgy processing. The interfaces between the Cu-Zn-Al matrix and the flakes in the consolidated composites were delineated and were free of precipitates or reaction products. In all three composites, the damping capacity with the strain in the range from 1.0×10^{-4} to 6.0×10^{-4} was found overall to show substantial improvements. In particular, the Fe-Cr flakes/Cu-Zn-Al composite demonstrated the highest overall damping capacity and exhibited an additional damping peak at strain 165×10^{-6}.

Material Forms

SMAs are manufactured in many of the conventional forms expected of metal alloys: drawn round wire, flat wire, tubing, rolled sheet, and sputtered thin films. Additional forms include shaped components, centerless ground tapered wires and tubing, alternate core wire (Ni-Ti filled with a conductive or radiopaque material), PTFE-coated wire, stranded wire, and embedded composites. Ni-Ti-X alloys are the most readily available at present in all of these forms.

The processing of SMA material is critical for the optimization of shape memory behavior. Many adjustments can be made to optimize the properties of a material form for a particular application; however, most efforts are made to optimize a balance of strain recovery, ductility, and tensile strength. SMAs such as Ni-Ti are typically melted with extreme purity and composition control, hot-worked to bars or plates, cold-worked to their final form, and subjected to specialized thermomechanical treatments to enhance their shape memory properties.

Furukawa Electric Co., Ltd. has developed an Ni-Ti SMA foil that is as thin as 30 µm. SMAs are brittle and fracture with ease, making rolling quite difficult, so sheet materials with thicknesses of less than 100 µm have hardly been commercialized. This is the first development of an alloy foil as thin as 30 µm.

This SMA is produced from a mixture of Ni and Ti, but normally is cracked easily when under tension. It also becomes hard right after formation, and the reliability is only about 20% compared with 70–80% of the Cu-Al alloy type.

Research into the causes of the alloy cracking succeeded in developing technology to prevent the alloy cracking, by which it became possible to tension and roll the alloy. At the same time, the rolling conditions were optimized, by which the alloy rollability was improved to 50%. As a result, it is now possible to manufacture Ni-Ti SMA plates and strips thinner than 1 mm up to a width of 50 mm, and thicker than 1 mm up to a width of 130 mm.

The SMA consisting of Ni and Ti enables the shape recovery temperature to be set within 0–100°C. The development of this new SMA, in succession to the conventional types of wire material SMA, enables the alloy to be used in a broad range of applications such as the manufacture of electronic components and machine parts.

Design Constraints and Considerations

When assessing a potential design challenge, designers are often anxious to develop a solution using the unique and exciting properties of SMAs. It is critical, however, for designers to understand the complexity of SMA behavior. As a general rule, if conventional materials and designs can be applied to a solution to yield an acceptable and desirable result, the use of SMAs to provide an alternative solution will amplify complexity and cost. SMAs are best utilized when their unique properties are necessary for design success — when conventional materials cannot meet the demands of the application.

The design of SMA applications requires more than traditional design techniques and textbook methods. Due to the many unique properties of SMA materials, several considerations specific to SMA design must be addressed and accounted for. The majority of issues that should be addressed prior to designing an application using SMAs are discussed below.

General Guidelines

1. Recoverable Strain. The expected recoverable strain of SMA material must be within the limitations of the alloy of choice. For example, Ni-Ti may recover 8% strain for a single cycle application, but less than 4% for higher cycle applications. Recommended strain limits are 6% for Ni-Ti and 2% for Cu-Zn-Al for lower cycle applications, and 2% and 0.5% for higher cycle applications, respectively. Applications requiring two-way shape memory should expect a maximum strain recovery of about 2%.

2. High-Temperature Stability. Alloy stability must be considered when an application requires or expects high operating temperatures. Ni-Ti alloys tend to be the most stable of all SMAs at elevated temperatures and can withstand exposure to temperatures up to approximately 250°C before previously memorized shaping may be deleted. For Cu-Zn-Al, this maximum temperature is around 90°C.

3. Fatigue. SMA fatigue can be defined as a degradation of any or all of its functional properties. Affected by application cycle quantity, frequency, temperature range, stress, and strain, SMAs may fatigue by fracture, decreasing recoverable strain, shifting transformation temperatures, or decreasing recovery stress.

4. Cost. Most SMAs are inherently more expensive than conventional materials due to the higher cost both of raw material components

and processing methods required. The compositional control necessary for the raw forms of SMA material requires special furnaces
and processes, the sequence of cold working and annealing to
ensure optimal SMA properties is extensive, and the special tooling and fixturing required for producing the many forms and
shapes of the materials all increase the cost of using SMAs.

5. Computer Modeling Capability. Finite-element analysis (FEA),
often used in conventional material design, has also been used to
model the behavior of SMAs. The analysis is difficult, however,
and should not depend on standard material templates and sub-
routines because the functional properties of SMAs rely on non-
standard factors, such as composition and processing history.
Highly specific and complex modeling techniques must account
for the state of the SMA material once formed in its trained shape,
and they then must incorporate the nonlinearity of the stress-
strain curve, the property dependence on temperature, and the
difference between loading and unloading stress behavior. The
illustration in Figure 4.5 demonstrates an FEA model of a superelastic coronary stent, achieved with custom modeling subrou-
tines to predict mechanical properties with a high degree of
accuracy.

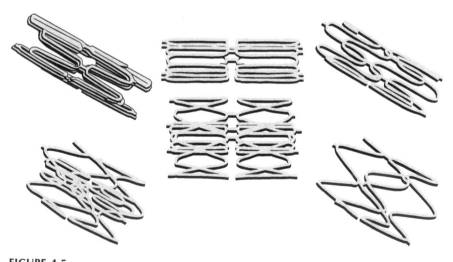

FIGURE 4.5
Finite element analysis model of self-expandable Ni-Ti stent; displays a quantified mapping
of stress and strain amplitudes in both expanded and compressed positions (courtesy of
Pacific Consultants, LLC).

Material Processing Techniques

In recent years, Cu-based SMAs have been widely developed because of their more competitive cost and easier fabrication process than for Ni-Ti SMAs. Among the Cu-based SMAs, the CAN system has better thermal stability.[86] For the preparation of SMAs, the conventional casting method has difficulties in controlling the grain size. Coarse grains will weaken the mechanical properties of alloys. It has been reported that mechanical alloying (MA) and powder metallurgy (P/M) with hot isostatic pressing (HIP) can be used to fabricate Cu-based SMAs.[87–91] P/M can reduce the hot working processes in fabricating the near-net shape products and usually give better control of grain size.

Cu-Al-Ni-based SMA

MA produces prealloyed powders, which can shorten the sintering time. However, no published work has been done on the preparation of CAN SMA by MA and the conventional P/M with cold compaction. In this study, high-energy planetary ball milling was applied to convert the Cu-Al-Ni elemental powder mixture to prealloyed powders. Conventional P/M with cold compaction was used to produce bulk Cu-Al-Ni-based SMA from the prealloyed powders by MA.

Tang et al. conducted a series of investigations[92] that found:

1. MA can be applied to prepare Cu-Al-Ni-based prealloyed powders. A single phase of the structure with lattice parameter close to that of Cu was produced after MA for 40 h of the elemental powder mixtures.
2. The grain size of sintered compacts is refined with longer milling times and is smaller than 2 μm.
3. The microhardness of green and sintered compacts increases with longer MA times, but the microhardness of sintered compacts decreases with greater open pore porosity.
4. The density of the compacts is affected by the hardness and the size of the powder particles. The highest density of green and sintered compacts was obtained front the powder mixtures with MA for 15 h.
5. Conventional P/M with cold-compaction can be used to produce CAN with prealloyed powders by MA.

Fabrication Techniques[93,94]

Joining

Because there are more than 100 different welding, brazing, and soldering processes available for the joining of materials,[95] the important decision to make is, which joining process should be used? This depends on many factors such as the properties to be achieved in the joint region, the properties to be achieved in the base metal close to the joint region (heat-affected zone, or HAZ), the size of the assembly, and the thermophysical properties of the materials being joined.

Ideally, the joints produced using processes such as welding and brazing should have properties identical to those of the base materials. Further, the joining process should not alter the properties of the base materials being joined. Most often due to the inherent characteristics of the joining processes, some properties vary widely from one region of the joint to the other regions of the joint. During the joining process, the joint region, the area close to the joint region, or sometimes the whole assembly is subjected to thermal cycles. As a result of thermal cycles experienced by the assembly, the microstructure and properties in different regions of the joint may be different. It is also most likely that the joint will develop a complex residual stress distribution, further reducing the life of the joint.[96] The extent of variation and what properties may vary depends on the joining processes, base materials, and filler metals used for joining.

Accepting the fact that there will be property variations across the joint, a compromise is made as to what properties of the joint should be matched with that of the base material and also what properties of the base metal are to be preserved. Depending on the service requirements of the joint assembly, a decision is made as to which properties (e.g., strength, toughness, fatigue, corrosion resistance) are matched. While one property is being matched, other properties may vary across regions of the joint. In the case of SMA, it is logical to expect that the "property" or the joint to be matched with that of the base material would be SME.

Fundamental Difference

The properties to be matched between the base metal and joint region are chemical composition and microstructure, which dictate the shape memory response. This is where, fundamentally, the joining of SMAs is different from that of conventional structural alloys such as steels, aluminum, and titanium alloys.

In structural alloys, matching of strength, toughness, fatigue, and corrosion resistance can be achieved despite differences in chemical composition and microstructure in the base metal and joint regions.[95,97–101] The flexibility of using different compositions and microstructures in the joint region, base material, and HAZ increases the probability of being able to produce a successful joint. In joining structural materials, often the filler metals used during welding are different in chemical composition as compared to the base metal, but satisfactory joints still can be produced.[95,97–101] Some typical examples are welding of steels,[95] where the carbon content of the filler metal is kept low, as compared to the base metal, to avoid the formation of martensite, which may cause cracking.

In the case of welding of Al-Zn-Mg alloys, superior strength, fracture toughness, and fatigue properties are achieved by using Al-Mg filler metals.[98,99,101] In the case of titanium weldments, even though the microstructures of the different regions of the weldment vary, satisfactory mechanical properties are achieved.[95,100]

The point to be noted from the above examples is that successful joints are obtained despite different chemical compositions and microstructures of weld metal, heat-affected zone, and base metal regions of the joints. In the case of SMA, this is not the case; matching of chemical composition, microstructures, and transformation temperatures is very important in addition to matching the mechanical properties and corrosion resistance.

Joining of SMAs using conventional welding processes is extremely difficult and challenging because it is not possible to control the microstructural development and chemical composition of the weld region as precisely as required by SMA due to the inherent characteristics of the processes itself.

Fusion Welding of SMAs

Fusion welding using processes such as gas tungsten arc and laser and electron beam welding have been tried to join Ni-Ti alloys. Some of the problems encountered during the fusion welding process were embrittlement due to the solubility of oxygen, hydrogen, and nitrogen; loss of superelasticity and SME in the HAZ; and formation of intermetallic compounds, such as Ti_2Ni or $TiNi_3$, which are brittle and have no SME.[10,11,102–105] Akari et al.[103,105] joined 3-mm-thick Ni-Ti sheet using a 10 kW CO_2 laser. Even though the laser weldments showed good SME and superelasticity, the tensile strength of the weldment was found to be low. Fracture occurred at the center of the weld metal, and this has been attributed to the coarse grain size of the fusion zone. Based on previous work, it appears that fusion welding of SMAs such as Ni-Ti would be difficult.

Edison Welding Institute[10,106] initiated a study for welding Nitinol to other ferrous metals such as stainless steels. This is a particularly attractive metal combination because it allows a manufacturer to use the more expensive

Nitinol only for the critical elements of a design, such as the nose bridge and temple portions of a pair of eyeglasses. Alternative joining methods have been investigated and used for Nitinol-to-stainless steel joints such as soldering, brazing, mechanical fastening, and organic adhesives. None of these are straightforward solutions and all have serious shortcomings in terms of strength, application, or impact on the Nitinol performance.

Successful fusion welding, particularly laser welding, of Nitinol to itself has been well established and documented over the last 10 years. With proper procedures and high-quality material welding, Nitinol welding has become a relatively routine process. However, fusion welding Nitinol to other metals, including stainless steels, has proven much more difficult. This is due to the tendency for Ti to form brittle intermetallic phases with most other metals. In the case of ferrous metals, the intermetallic phases TiFe and $TiFe_2$ form. Nevertheless, welding Nitinol to stainless steel continues to be a desirable material combination for many applications including medical devices.

Various welding approaches have been considered including resistance welding, friction welding, and using a transition material. Some of the more promising approaches have been solid-state forge welding techniques such as friction and resistance welding. In solid-state welding, a bond is formed through the application of heat and pressure, but without large-scale melting of the base metals. The absence of liquid prevents or minimizes the formation of brittle intermetallic phases and can result in a sound weld. These processes are typically limited to particular joint geometries or part dimensions. Consequently, despite the apparent feasibility of forge welding, its practical application appears to be somewhat limited.

Through an internal research program EWI[106] has developed a more accessible and universally applicable fusion welding process for joining Nitinol to stainless steel. Successful laser welding of wire-to-wire butt joints has demonstrated the successful results of this work. Weld tensile strengths were increased from 0 to as much as 621 MPa. This strength increase was accompanied by a dramatic reduction in the weld metal hardness from 900 down to 450 Vickers hardness. Functionally, the weld strengths were sufficient to allow extensive superelastic bend of the Nitinol at the weld. This approach is not limited to just Nitinol-to-stainless steel joints. Similar improvements have been demonstrated on joints between carbon steel and Ti.

Solid-State Welding of SMAs

Solid-state welding processes such as friction welding, resistance welding, and diffusion welding are inherently attractive for joining SMAs because they have potential for developing joints with little, if any, microstructural changes. Welds can be obtained at relatively low temperatures as compared to fusion welding and without melting the metals being joined.

EWI has conducted a research program to investigate the resistance weldability of Nitinol to itself and tungsten.[10] Resistance-welded Nitinol showed excellent superelasticity without any postweld treatment. In the resistance welding process, some amount of localized melting may occur as the fusion nugget forms at the interface. The regions of localized melting may solidify under very high cooling rates, which may produce some nonshape memory microstructural features. Proper selection of postweld treatments may obviate the formation of nonshape memory microstructural features. Additional research is required in this area before conclusions can be drawn on the applicability of the resistance welding process to join SMAs.

Sound joints in 6-mm-diameter bars of Ni-Ti using friction welding[11] have been produced. An upsetting value of at least 127.8 MPa was used to weld them, and with a postweld heat treatment, there was little variation in transformation temperatures of the weld with those of base metal. The most important result of the experiment was that the shape memory characteristics of base metal and welded joint were similar.

Friction welding seems to be a viable process for joining SMAs, but the geometrical limitations of the parts to be joined may be a problem in applying this process. However, with the recent advances in linear friction welding process technology, friction welding can be used for joining of nonaxisymmetric parts.[94] In linear friction welding, frictional heat is generated through the linear translation of two workpiece interfaces through a small amplitude at a suitable frequency. Another variation of the friction welding process is inertia friction welding, in which a high axial force is used during joining. The rapid thermal cycle and high axial force associated with inertia friction welding will cause the expulsion of heat- and deformation-affected, plasticized metal out of the weld interface.[107] Inertia friction welding may produce sound joints by sufficient expulsion of oxide layers and heat-/deformation-affected regions, which are detrimental to the SME.

Diffusion welding is a process that produces a weld by the application of pressure at elevated temperature with no macroscopic deformation.[94] A filler metal may be inserted between the faying surfaces. This is a potential process for joining Ni-Ti.

Other solid-state welding processes that could be tried to join SMAs are percussion welding and ultrasonic welding.

Alternative Processes for Joining

A patented soldering process[108] for joining Nitinol to other materials uses a halogen-containing flux that removes titanium oxide from the surface of Nitinol and deposits a metal coating to prevent further oxidation. A material such as Sn-Ag solder is flowed onto the tinned surface to displace the activated flux. Electively, another member is applied to the molten solder to form a joint. There is no claim that SME is retained in the joint, even though it is claimed that the base material SME will be retained. It appears this type of joining can be used only when mechanical continuity is required and where the SME is not required at the joint.

Another patented process supports the component to be welded within a collet such that a substantial portion of the component projects beyond the collet. A heat source renders the projecting portion molten, after which Nitinol is embedded into the molten mass. After cooling, the joints were found to have good pull strength and low brittleness. With this process, it appears that Nitinol can be joined to stainless steel at least from the mechanical point of view, which is the requirement for medical guidewire fabrication.

Transient Liquid Phase Bonding

A variety of advanced material systems, such as nickel-based superalloys and intermetallic compounds, are joined with transient liquid phase (TLP) bonding. In the TLP bonding process, filler metal is clamped between the contacting metal surfaces, and the whole assembly is heated to the bonding temperature. The filler metal forms a liquid at that temperature, and the resulting liquid film resolidifies isothermally as the joint is held at that temperature. Because, in the joining of SMAs, it is important to produce a joint with the same microstructure as the base material and to retain the SME, TLP bonding is a strong candidate for joining SMAs.

An Auburn University research group has tried to join Ni-Ti to Ni-Al.[102] During this investigation, it was observed that a layer of Ni_2AlTi was present at the bond line. Some problems encountered with TLP bonding of Ni-Ti to Ni-Al are formation of secondary phases, which may not exhibit SME, and diffusion of Ti and Al from the substrate to the bond line. With proper selection of interlayers and postbond treatments, these problems could be obviated. This process shows promise for joining SMAs, and its successful use depends on the optimizing of the parameters for a given alloy system.

Work To Be Done

SMAs are extremely difficult to join using conventional joining processes because it is not possible to precisely control the chemical composition, microstructural development, and transformation temperatures as required to maintain SME.

Solid-state welding processes such as friction welding and resistance welding are promising. Other solid-state processes such as inertia friction welding and percussion arc welding are worth trying.

Transient liquid phase bonding with postbond treatments and soldering with special fluxes appear to be viable alternatives to join SMAs.

Superplastic Forming

Superplasticity is a phenomenon in which a material exhibits a much larger elongation at an elevated temperature. In other words, the superplastic material exhibits excellent ductility under appropriate conditions, providing superior advantages for material forming.

Among a number of superplastic forming techniques are bulge forming, vacuum forming, and superplastic forming (SPF) diffusion bonding, all of which have been widely applied in the aerospace industry. In the field of Cu alloys, a few workers have studied the tensile properties of superplastic Cu alloys;[109] however, few studies have been done on the forming properties of thin Cu alloy sheets. Hsu and Wang[94] conducted a study on the SPF behavior of Cu-Zn-Al-Zr SMAs.

This investigation examined the effects of temperature and pressure on the pressure forming of Cu-26.5 wt% Zn-4wt% Al-1wt% Zr alloy sheet under free bulging conditions using argon gas. The effects evaluated were the dome height measured at the dome pole, the specific thickness (the ratio of the actual thickness to the initial thickness), and the thinning factor (the ratio of the actual thickness to the average thickness).

The results showed

1. The dome height as well as dh/dt increase with increasing temperature or pressure.

2. A larger effective strain rate in specimens bulged by a higher pressure at a constant temperature can explain why dh/dt for higher forming pressure is larger than at a lower pressure.

3. The specific thickness decreases with increasing fractional height. Furthermore, the specific thickness at the pole decreases with increasing temperature or pressure.

4. The gradient of the thickness strain varies with the fraction height and gradually increases with increasing temperature or pressure.

5. The thinning factor decreases with increasing fractional height. This decrease becomes more pronounced with an increase in either the forming temperature or the pressure.

Forming

A shape memory polymer (SMP) has recently been developed by Mitsubishi Heavy Industries.[110] This development allows for a material used in a child's bicycle helmet to be remolded at home so that you can adjust its size as the child grows; secondly, an eating utensil can be reshaped by the physically handicapped to meet each user's special needs. The patented "smart" polyurethane is easy to process. Under certain conditions that can be customized, SMP can transform its shape and hardness, then return to its original state, if so desired. Shaping takes place when the material is heated, while quick cooling enables it to retain its shape. If reheated, the material "remembers" its original shape and returns to it.

The material comes in pellet, solution, or liquid form for easy compounding. Conventional equipment can be used to injection-mold, extrude, coat, or cast SMP. Semifinished products are available, including foam and microbeads. The material also recently passed critical biocompatibility lab tests for use in medical devices.

Machining

SMA materials are infamously difficult to machine. Tool wear is rapid for conventional machining methods such as turning, milling, drilling, and tapping.[111] Currently, the most successful machining techniques include surface grinding, abrasive cutting, EDM, and laser cutting. Component shaping must be considered as well; memory training of SMA shapes are performed at high temperatures, typically around 500°C. Unlike most conventional materials that may be cold-formed, SMAs must be rigidly clamped in a desired shape and exposed to these elevated temperatures.

Design for Assembly

Designing for assembly is also an important manufacturing consideration. Fastening SMAs to other materials presents additional challenges, such as bonding and joining.

Because SMAs are designed to exhibit strains of up to 8% and other materials have a strain limit of less than 1%, when the two are rigidly joined the conventional material may break during operation. This often causes problems with plated or painted SMA materials, as the coating material on the SMA will often crack and flake during the strains of operation.

Micromachining and Fabrication of SMA Microdevices

Silicon-Based Microelectromechanical Systems (MEMS)
Fabrication Processes[112,113]

Silicon technologies and related processes were major breakthroughs in microelectronics as well as in sensors, actuators, and microsystems. Silicon-based processes provide a unique method for large-scale production and miniaturization in the development of microactuators. These production methods are massively parallel and allow batch processing. However, these fabrication processes have a few limitations:

- Silicon microstructuring technologies are planar.
- The technological investment is very high.
- These methods are usually confined to structures of limited aspect ratios.

Ni-Ti and Ni-Ti-X (where X = Cu, Hf, Pd) alloys can be deposited on various substrates such as Si, silicon dioxide (SiO_2), or Ti. However, the Si substrate is usually avoided because of the possible formation of silicide (SiNi) during crystallization of Ni-Ti thin films. The most common process for producing Ni-Ti thin films is deposition by DC or RF magnetron sputtering. An in-depth description of this process as well as the properties of thin film produced can be found in Reference 113. Other processes such as deposition by flash evaporation,[114] laser ablation,[115] and multilayer processing[116] can also be used to produce thin films.

Ni-Ti thin films are usually etched by using $Hf/HNO_3/H_2O$ solutions. According to Machino[115] results, a solution of $Hf:HNO_3:H_2O = 1:1:4$ allows selectively etching of an Ni-Ti film deposited on an Si substrate.

Different fabrication methods can be used that depend on the type of actuator. When the actuator is moving freely, an intermediate layer, the so-called sacrificial layer, is required.

Figure 4.6 presents a method of producing a freely moving micromechanical element represented by a spring.[117] In the first step, a polyamide layer is deposited by spin coating and is subsequently cured. The second step is Ni-Ti thin film deposition on the polyimide layer. In the next step, a photoresist thin film is spin-coated over the Ni-Ti thin film. Then, the photoresist is baked, patterned, and developed to expose specific portions of the Ni-Ti layer selectively. Then the Ni-Ti film is etched. Finally, the structure is freed through reactive ion etching of the sacrificial polyimide layer.

Others[118] have proposed depositing the Ni-Ti film just before removing the sacrificial layer. A sacrificial Cr layer is deposited on an SiO_2 substrate. A polyimide layer is spin-coated on the Cr layer. Then, a second Cr layer is deposited on the polyimide. The upper Cr layer is patterned by photolithography and wet-etched. The polyimide layer is vertically etched by oxygen plasma using the previously patterned Cr layer as a mask. Then, the Ni-Ti layer whose thickness is less than that of the polyimide layer is deposited. The final step consists of removing the polyimide layer by wet etching ("liftoff") and releasing the mobile part by wet etching the Cr layer. Because the Ni-Ti layer is never etched, the aspect ratio of the structure is theoretically not limited.

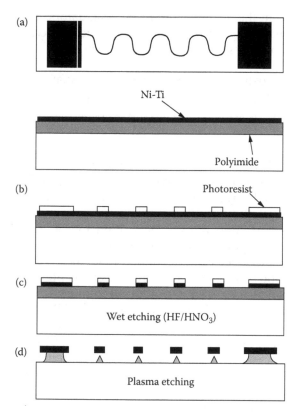

FIGURE 4.6
Thin-film microactuator fabrication process: (a) deposition of the layers, (b) photoresist deposition and patterning, (c) wet etching of SMA layers, and (d) plasma etching of the polyimide layer [after Walker et al.].[17]

Several applications such as microvalves use membrane actuators or biomorphs of Ni-Ti on SiO_2. This structure can be fabricated by using the process proposed by Wolf et al.[119] An Si wafer that has an oxide layer on both sides (SiO_2) is coated by a photoresist and softbacked. The back side photoresist is patterned and results in direct exposure of the Si in defined parts, which creates window-like structures. Then, the Si is etched and almost all of the volume "behind the window" is removed. The front layer is then removed using a buffered oxide etchant. The next step is depositing the Ni-Ti layer. Finally, the rest of the Si is removed. When biomorph-like structures are desired, the front SiO_2 layer is not removed, and Ni-Ti is deposited directly on it.

Laser Machining

A laser can be used for milling and cutting SMA elements. Nd-Yag lasers are usually used. The part is fixed on a two-axis linear stage under a fixed focusing objective. A rotational stage is sometimes added for tube cutting. The laser is focused on the part, and an additional gas flow is usually used to drain off the molten material. The cutting precision depends on the object size and the material. It is possible to cut sheets whose thicknesses range from a few millimeters down to 0.005 m.

The spot size will vary according to the part thickness. For Ni-Ti, results have shown that the minimum spot size ranges from 0.02 mm for 0.01-mm-sheet thickness to 0.08 mm for a 1-mm-sheet thickness. For small to medium production volume and prototyping, laser cutting is an efficient method for Ni-Ti material.

SMA Applications

SMAs have been engineered for applications and devices since the first discovery of the SME in the 1930s. The majority of this design activity was initiated by the discovery of Nitinol in 1962, and since then more than 10,000 patents have been issued for applications using SMAs.[120]

Applications utilizing each of the unique properties of SMAs have been designed, prototyped, and marketed throughout the world.

The following examples of these applications may include some discussion regarding design choices, material limitations, and SMA behavior. These examples are categorized by industry to demonstrate the varied and widespread use of SMA applications.

Aeronautics/Aerospace

The fields of aeronautics and aerospace pioneered many of the initial product ideas and applications incorporating SMAs. SMA materials are used in this industry to take advantage of properties such as high power-to-mass ratios and ideal actuation behavior in zero-gravity conditions. Designs utilizing these properties replace heavier, more complex conventional devices for weight reduction, design simplicity, and reliability.

Cryofit® Hydraulic Pipe Couplings — SMA couplings were the first successful commercialized application of SMAs (Figure 4.7). In 1969, Raychem Corporation introduced shrink-to-fit hydraulic pipe couplings for

FIGURE 4.7
Shape memory devices. *Clockwise from top left:* memory card ejector mechanism for laptop computers; Cryofit® hydraulic pipe couplings: Cryocon® electrical connector; fire safety lid release for public garbage receptacles (courtesy of Shape Memory Applications, Inc.).

F-14 jet fighters built by Grumman Aerospace Corporation. This coupling is fabricated from an Ni-Ti-Fe alloy with a martensitic transformation temperature below -120°C and is machined at room temperature with an inner diameter approximately 4% smaller than the outer diameter of the piping it is designed to join. When cooled below –120°C with liquid nitrogen, the coupling is forced to a diameter 4% greater than the pipe diameter for an overall internal strain of about 8%. When warmed above its transformation temperature, the coupling diameter decreases to form a tight seal between the pipes.[116]

This shape memory application of constrained recovery continues to be a commercial and financial success. Despite the difficulties of cooling the couplings to liquid nitrogen temperatures for expansion storage, the aerospace industry has welcomed the many advantages over traditional pipe-joining techniques, such as welding or brazing. Installation is simple, less costly, and not reliant on high levels of operator skill. The replacement of couplings and hydraulic lines is straightforward, and the possibility of annealing and damaging the hydraulic lines as with welding or brazing is eliminated.[119]

Newly developed stainless steels,[122] containing from 4.4 to 6.3% Ni, promise to expand the market for shape memory materials, says Chenxu Zhao at Xi'an Jiaotong University in P.R. China. These steels are more corrosion-resistant, easier to work with, and, moreover, much less expensive than SMAs currently in use.

Researchers have found that certain steels (FeNiC, FeMnSi, FeMnSi-CrNi) undergo a non-thermoelastic transformation but still exhibit good SME. These materials rely instead on a stress-induced transformation to restore their shape. The reversible strain is less than that of traditional SMAs (3% compared with 8% for Ni-Ti and 6% for Cu-based alloys) but strong enough to warrant further research into commercial applications. Zhao says that pipe couplings are the most practical use for shape memory stainless steels. Couplings that shrink instantly to provide an effective seal over smaller diameter pipes could have a welcoming market at the right price. Traditionally, this market has been limited to high-end applications (fighter jets and blood-clotting filters).

Zhao further states that the lower-cost shape memory stainless steels have great potential for industrial applications in this field, because 2% reversible strain is sufficient for pipe joint applications.

One Chinese company has already capitalized on this discovery. Shanghai Tianhe Shape Memory Material Corp. claims to have developed a stainless steel coupling that, when heated to 250°C, shrinks to fasten two separate pipes together.

Shape memory stainless steels could also prove to be useful in marine environments, where corrosion is a concern.

Frangibolt® Release Bolts — Shape memory bolts were developed by TiNi Alloy Company to replace conventional exploding bolt devices in aerospace release mechanisms. The bolts are used to attach spacecraft accessories during launch and to release them after launch with an activated heating element.[123] A martensitic, shape memory cylinder is compressed and assembled to a notched bolt, and when activated by an electrical heater the cylinder increases in length and delivers a force greater than 22 KN to fracture the bolt at its notch.[124] Used successfully aboard the spacecraft Clementine in 1994, these release bolts have improved upon designs for conventional explosive mechanisms by eliminating the risks of off-gassing, accidental activation during shipment, and potential spacecraft damage during explosions.

Mars Sojourner Rover Actuator — An SMA wire was used to actuate a glass plate above a small solar cell on the Rover unit during the Pathfinder/Sojourner mission to Mars. A material adhesion experiment performed during the mission used the actuator to replace large, heavy motors and solenoids. The small, simple length of Ni-Ti wire heated and contracted when the Rover applied power, pulling a glass plate away from the solar cell to allow a comparison of sunlight intensity with and without the plate. The rate of dust collection was then determined, and the resulting data will be used to design cleaning methods for future missions to Mars.[125]

Self-Erectable Antenna — A prototype space antenna was constructed by Goodyear Aerospace Corporation. Designed to fold compactly at room temperature, the device would unfold to a large, extended antenna shape when heated with solar energy.[112,126] Although this did not become a commercial success, the concept is feasible and the prototype has served as a model for similar designs pursued within the aerospace industry.

Smart Airplane Wings — Composite structures embedded with SMA wires can be used to change the shape of an airplane wing. The embedded wires may be activated to constrict and improve vibration characteristics of the wing, heated to change their effective modulus to reduce vibration, or activated to alter the shape of the wing for optimal aerodynamics. All of these properties can be used to produce an adaptive airplane wing, altering as environmental conditions change to improve efficiency and noise reduction.

O'Handley of MIT and Ullakko[127] from the Helsinki University of Technology in Finland reasoned that an alloy of Ni, Mn, and Ga might respond to a magnet field with a change of shape. When the researchers exposed a sample of the new alloy to a magnetic field about 2 or 3 times stronger than that generated by a refrigerator magnet, the material became rubbery and expanded. When the researchers removed the field, a spring pushing on the alloy caused it to revert to its original shape By alternating the magnetic field at high speed, the researchers could repeat the cycle several thousand times per minute. "It sort of looks like a heart beating very quickly," says O'Handley.

Adding magnetic control, researchers believe, will make SMAs much more practical to use. "There's a great deal of promise in these [magnetically controlled] materials," says the University of Minnesota's Thomas Shield, who also studies SMAs. O'Handley, for example, envisions using the alloy to make lightweight mechanical parts in airplane wing flaps. A lump of the new material wrapped in a magnetic coil could raise and lower a flap on cue, replacing today's heavy hydraulic systems that require a central pump and multiple oil lines running to the flaps.[128]

Leading-Edge Deicing[129]

Icing conditions can occur as aircraft fly through clouds with supercooled water droplets. These water droplets can freeze immediately upon impact or flow back along the surface as a liquid film before freezing. Accumulation of ice can greatly degrade aircraft aerodynamic performance. Small amounts of ice can decrease lift, increase drag, and significantly change pitching moment, thus potentially causing loss of control.

Current deicing systems either prevent ice accumulation via heating or break the ice layer after it is formed using surface deformation. Turbofan aircraft often use hot bleed air from the engine for anti-icing, whereas on propellers and helicopter rotors, resistive electric heaters are commonly used. However, such systems have high input requirements: High bypass-ratio turbofans have limitations on the amount of available bleed air, while resistive heating requires excessive electrical power input. Pneumatic boots consist of a flexible material that is glued to the leading edge in strips, leaving unbonded passages. Pressurized air is forced into these passages, causing the boot to swell outward and the ice to be shed. Although the input requirements for a boot system are much less than those for a thermal heating system, the boot system may only be effective in a narrow range of operating conditions. Such a system may also have high maintenance costs due to deterioration and erosion of the boot material with time.

Deficiencies with current deicing systems have pushed researchers to investigate new techniques, such as electroimpulse deicing and electro-expulsive and eddy-current repulsive systems. Because a key goal with these systems is low-power use, the surface deformation method is typically employed. Because SMAs can generate significant amounts of strain, their use may be especially appropriate when large deformations are required.

Figure 4.8 reflects the tooling used to fabricate a continuous-wire SMA deicing device. The wire is placed continuously onto the specimen by looping it around the two threaded rods. These rods provide the wire tension that is required during the composite material curing process.

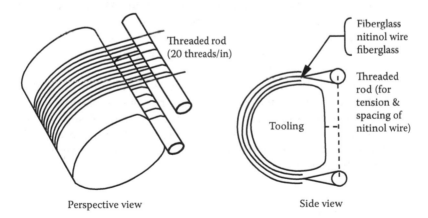

FIGURE 4.8
New tooling to fabricate continuous-wire SMA deicing device. Schematic diagram of the new tooling.

After successful fabrication of a continuous-wire SMA deicing device, initial static icing tests were successfully performed with reduced electrical power. However, the length of Nitinol wire looped around the threaded rod was longer than originally planned. With additional modifications to deal with the longer Nitinol wire, this should further reduce the power input required for deicing.[129]

Helicopter Smart Materials[130]

SMAs may have helicopter applications, because the material has the capability of changing its crystalline configuration as a function of temperature. It is the same characteristic that allows a blacksmith to make a tough cold chisel out of a soft steel rod.

The change is from a crystal configuration known as martensite at room temperatures to austenite at higher temperatures. In the case of the cold chisel, the temperature changes required can be over 649°C, but SMA can be made to change its characteristics over a much smaller range.

Engineers at the University of Maryland wish to get a high-speed shaft through its resonance speed without using dampers. Tail-rotor shafts in some helicopters are designed to operate above their "first critical frequency" as a weight-saving measure, but they require dampers to safely pass through that frequency as the rotors are brought up to speed. (Although it should be noted that these dampers are very simple and do not weigh much.)

These engineers proposed to eliminate dampers by increasing the stiffness just before going through the critical frequency and then returning it to normal when safely on the other side. The analysis shows that with a composite tail-rotor shaft based on, for example, the dimensions of a U.S. Apache helicopter and containing appropriate SMA wires, the resonate frequency could be increased by about 15%.

Another application might be to vary the stiffness of such components as stabilizers in flight test to find the right characteristics to avoid resonance with blade-passage excitation.

Space System Vibration Damper — Comprised of composite materials using prestrained, embedded SMA wire or ribbons, vibration dampers can reduce unwanted motion in various space systems. A sensor detects vibration in the system and sends a signal to activate the embedded composite, which then alters the structural dynamics to damp or cancel the existing vibration.[114]

Consumer Products

SMA devices and components have been used in high-volume consumer products for more than 25 years. Although many consumers who use

FIGURE 4.9
Deformation resistant eyeglass frames (courtesy of Marchon Eyewear, Inc.).

these products are unaware of their SMA components, there is a growing public awareness of SMAs due to recently marketed items that advertise their merits.

Flexon® Optical Frames — Superelastic eyeglass frames are one of the most widely known uses for SMAs. They are frequently advertised in television commercials and can be found at most optical frame retailers. The components of eyeglass frames that are most susceptible to bending, the bridge and temples, are wire forms of Ni-Ti; the remainder of the frame is comprised of conventional materials for adjusting purposes and cost savings. Due to the high strain recovery capability of Ni-Ti, these frames are highly deformation- and kink-resistant (see Figure 4.9).

Portable Phone Antennae — The growing demand for portable phones has resulted in a high-volume application for superelastic Ni-Ti material; most cell phone antennae produced today are Ni-Ti wires muted with polyurethane. The superelasticity resists permanent kinking and withstands the abuses of user handling during the lifetime of portable phones.

Greenhouse Window Opener — An SMA that has a small temperature hysteresis is used as an actuator to open and close greenhouse windows of predetermined temperatures for automatic temperature control. The opener is a spring-loaded hinge that has a Cu-Zn-Al shape memory spring and a conventional metal biasing spring. The SMA spring is compressed by the biasing spring at temperatures below 18°C, and the window is closed. The SMA spring activates around 25°C, overcomes the force of the bias spring, and opens the window. This actuator design relies on reduced thermal hysteresis using a biasing force. As the SMA spring cools to 18°C, although not sufficiently cool to completely transform to its softer martensitic phase, it is transformed enough to accommodate deformation via stress-induced martensite.

Recorder Pen Mechanism — A shape memory pen driver was designed in the early 1970s to replace conventional pen-drive mechanisms, which used a galvanometer to actuate a pen arm. The replacement used Ni-Ti wires pretensioned in a driver unit and actuated by heat from an induction coil in response to input signals. The new design reduced the number of moving parts, improved reliability, and decreased costs. The new recorder pen units were first introduced in 1972; by 1980 more than 500,000 units were produced.[131,132]

Nicklaus Golf Clubs — Superelastic SMA golf club inserts were developed for Jack Nicklaus golf clubs. The damping properties of the inserts hold the golf ball on the clubface longer and provide more spin and greater control for golfers.[133]

Brassiere Underwires — Superelastic Ni-Ti shapes that conform to the user's body are ideal for underwire applications, because they are a unaffected by the temperatures and external forces from repeated washings. Wires are shaped in predetermined configurations, using either round wire or flat ribbon. The product is a commercial success in Asia, but the increased cost compared to that of conventional underwires has reduced the overall sales throughout markets in North America and Europe.

Residential Thermostatic Radiator Valve — SMA actuators have been used to regulate the temperature of residential radiators. An actuator expands when the room temperature increases, overcomes a biasing spring force, and closes a radiator hot water valve. Assisted by the biasing spring, the SMA temperature hysteresis can be as low as 1.2°C.[131] The thermostatic valve can be adjusted via a knob that alters the compression of the biasing spring — the more compression it exerts, the higher the temperature required for the SMA actuator coil to activate and close the hot water valve.[132]

Rice Cooker Valve — SMA valve mechanisms have been successfully employed to improve the performance of rice cookers. Comprised of an SMA spring and a bias spring, the mechanism is inserted into the top lid of a rice cooker. The valve is open while rice cooks and steam is generated, but when the rice is finished cooking, the SMA spring cools and the bias spring closes the valve to keep the rice warm. A Ni-Ti-Cu alloy is used for the SMA spring for its low-strain, high-cycle fatigue properties. Although its recovery force decreases with repeated cycling, this application has demonstrated repeatability for over 30,000 cycles, which corresponds to several daily operations over 10 years.[134]

Robotic Doll — SMA actuator wires were designed to move the arms and legs of a doll to display human characteristics. The application is technically feasible and prototypes were successful; however, the power requirements to activate the wires were too great. The required battery changes were sufficiently frequent to limit the market acceptance of the product.

Miscellaneous Products — An SMA-actuating air-conditioning set of louvers to deflect air up or down depending on temperature was introduced in Japan. Also introduced were coffee makers with temperature-control valves that initiated the brewing process when water started to boil.[135] Other products include superelastic fishing lures, superelastic SONY Eggo™ headphones for the minidisc Walkman®, and novelty items, such as a magic teaspoon with a memory. The teaspoon is given to someone to stir a hot drink and when the spoon is exposed to the hot liquid, it immediately transforms to a bent position.

Toy Miniature Motors — A maker of tiny linear actuators could displace millions of miniature motors in toys, cameras, disk drives, and automobile controls. Though the actuator has electric movers and shakers that are about the size of a paper clip, their active part is a very thin wire with a nanoscale inner structure. The Ni-Ti alloy forms nanosize crystals that combine to make strands 50 μm across, and the electric current rearranges them by heating.

The SMA wire is heated past a critical temperature, and it contracts by as much as 40%, pulling in a linear direction: something motors, with their rotary motion, can do only with the help of bulky crankshafts. They have about 1000 times the energy density of muscle and 4000 times that of an electric motor. In practice, it is only a 400-fold increase in the energy-to-weight ratio because the little wires are bound to a much heavier heat sink that cools them in preparation for their next contraction. It is apparent that a 400-fold jump is quite significant.

The real opportunity is for automobiles. However, because car-driven cycles take many years, while fad-driven toy cycles take only months, the more immediate showcase for the technology is in toys like Luke the robot. Luke can move his head, eyes, and jaws, simultaneously or separately, and in any order, all without making the slightest sound. In contrast, his companion, a motor-driven Furby, just repeats the same, hardwired routine, whirring and clicking, all while consuming 5 times as much electricity. True, not many toys need the high motive force. But consider Luke as a proof of concept for a larger and more lucrative market: The typical luxury car has 170 motors.

Commercial/Industrial Safety

Antiscald MemrySafe® Valve — An SMA valve has been designed to shut off a faucet's hot water source if water temperatures become sufficiently hot (above 49°C). The valve reopens when the water temperature cools to safer temperatures, protecting the user from scalding water.

Firechek® Valve — A safety device employing an SMA actuator is often used in industrial process lines to shut off gas supply in the event of fire.

Exposure to high temperatures activates a valve and cuts off the pneumatic pressure that controls flammable gas cylinders and process line valves.[136]

Circuit Breakers — SMAs have been used in circuit breaker designs to replace conventional bimetals. Due to the high forces required for large circuit breakers, a series of levers must be employed to amplify the forces available from bimetals. High-temperature Cu-Al-Ni alloys have been used in this application for their high-temperature activation and low hysteresis. Simple cantilever beam designs increase force and stroke and eliminate the need for levers.

Proteus® Safety Link Device — A chain link has been fabricated from Cu-Zn-Al to change shape at high temperatures and act as a release mechanism. The release may activate sprinkler systems or trigger fire doors to close, depending on the application.[137]

Telecommunication Line Fuses — Cu-Zn-Al shunts are coupled with high-sensitivity fuses throughout Europe to protect communication systems from lightning strikes. During normal operation, the fuse heats up more rapidly than the SMA, and the shunt remains inactive. Under heavy usage, however, the shunt increases in temperature and activates to bypass the fuse and protect it from a critical burnout temperature.[137]

Safety Trash Lid Mechanism — An SMA device has been designed to smother accidental fires in public trash receptacles. The device holds a trash lid in the open position at normal operating (ambient) temperatures, but when heated by a fire within the trashcan, the shape memory component releases a latch and the lid drops to extinguish the fire.

Medical

The medical industry is rapidly accepting the use of SMAs in a wide variety of applications. From simple pointed needles to complex components implanted in the bloodstream, Ni-Ti has been adopted by the industry for its ability to offer unique and ideal solutions to traditional medical challenges. Well known for its excellent biocompatibility and corrosion resistance, Ni-Ti has been used in many successful medical devices and is now widely accepted throughout the medical industry.

The majority of SMA medical applications use the superelastic property of Ni-Ti, and many of them are in the expanding field of minimally invasive surgery. Due to the high strain recovery of Ni-Ti, components can withstand extreme shape changes for minimal profiles during delivery and then expand to larger devices within the body. Many of these SMA devices have eliminated the need for open-heart surgery and thereby reduce patient risk and decrease hospital recovery periods.

Orthodontic Dental Arch Wires — Dental arch wires, one of the first medical applications that used SMAs, were first introduced in 1977 to replace stainless steel arch wires for straightening teeth. The wires were initially used in the martensitic condition, cold-worked, and deformed around the teeth. They exhibited sufficient springback properties for this application, although later superelastic (austenitic) forms of Ni-Ti wire were introduced to improve product performance. The superelastic arch wire is now designed to exploit the plateau region of SMAs stress-strain curve, which provides nearly constant stress on the teeth as the wire recovers its shape and straightens the teeth.

Mitek Homer Mammalok® — A superelastic needle wire localizer was introduced in 1985 that is used to locate and mark breast tumors to make surgical removal less invasive. The needle is used as a probe to pinpoint the location of a breast tumor first identified by mammography. Surgeons find it difficult to discern the tumor from surrounding tissue, so the probe highlights the correct location for the surgical procedure.[138]

Mitek Suture Anchors — Mitek anchors, fabricated from a Ti or Ni-Ti body that has two or more arcs of superelastic Ni-Ti wire, are secure, stable suture holders used to reattach tendons, ligaments, and soft tissues to bone. The anchors are placed in a hole drilled into a patient's bone and are locked in place by Ni-Ti arcs. These were introduced in 1989, and these anchors are used in shoulder surgery to fasten sutures to bone as well as in many other orthopedic applications, such us ligament anchors used for reattaching the anterior cruciate ligament of the knee.[139]

Guidewire Cores — Ni-Ti wires have been snaked through the tortuous pathways of the human body to guide and deliver other tools and devices for interventional procedures. These superelastic guidewires are optional for use in minimally invasive surgery where procedures are performed through a small portal in a major artery and offer superior flexibility, kink resistance, and torqueability for optimal steering and ease of operation.[140]

Guidewires, which have been used to thread the catheter and deliver the stent, also benefit from Nitinol's superelastic properties. Stringing a catheter is like trying to push a wet noodle through the vein, so a guidewire that can resist bending and kinking by springing back into shape has a significant advantage.

Stents — SMA stents are becoming increasingly popular in the medical industry. These structural, cylindrical components, designed to prop open and support human blood-vessel walls, ducts, and other human passageways, are implanted to prevent collapse or blockage and to patch lesions. Ni-Ti materials are used in place of more conventional metals for coronary artery stenting but are most often used in a peripheral location such as the carotid artery, esophagus, or bile duct. Several shapes and forms of Ni-Ti stents are displayed in Figure 4.10.

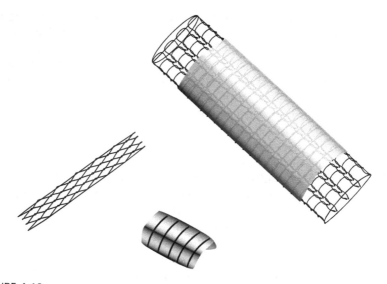

FIGURE 4.10
Nitinol stents. *Left to right:* Superelastic stent (5 mm OD X 40 mm long), laser-cut from Ni-Ti tubing; Ni-Ti ribbon set in a coil configuration; Ultraflex™ esophageal knitted stent with fabric covering (courtesy of Shape Memory Applications, Inc.).

Stents currently on the market and in development use various functional properties of SMAs: the superelasticity of austenite, the thermal shape memory, and the low effective modulus of martensite. Many of these SMA stents use a combination of superelastic and shape memory properties. For example, a stent may be chilled in icewater for transformation to martensite, compressed in the martensitic state, covered with a protective sheath for a minimal profile, and then delivered into the body through a small portal. When in place, the sheath is retracted, and the stent warms to body temperature to recover its original shape. Once recovered, or transformed to austenite, the superelastic properties of the scent result in gentle and constant radial forces on the vessel wall. Stents that are martensitic at room and body temperature must be compressed on a delivery balloon for expansion once delivered into the body. The Paragon Coronary Stent developed by Vascular Therapies is an example of a martensitic stent that is marketed for its even symmetrical expansion and flexibility during delivery.

A new and different kind of stent, unlike coronary stents, sits close to the surface of the skin, where it can be easily squeezed or crushed. But using Nitinol, researchers have developed stents that can be bent and compressed and will still pop back into shape.

Besides the aorta, ailing body parts that may benefit from Nitinol stents include the esophagus, the urinary tract, and the liver. One of the world's largest manufacturers of shape memory medical devices recently introduced a Nitinol stent for treatment of life-threatening obstructions in bile ducts.

The market for Ni-Ti "stems" and related devices that shore up blood vessels is growing as doctors and their patients seek less invasive medical techniques, while engineers improve stent delivery systems.

Stems are tiny wire mesh tubes used to prop open damaged or clogged arteries. In most cases, the stem is pulled over a balloon catheter, which is then threaded up to the blockage site. When the balloon is inflated, the stem expands, forming a scaffold to hold the artery open and allow more blood flow. The stent remains in the artery permanently. After the balloon is deflated, the catheter is withdrawn.

But a new procedure to repair aneurysms in the abdominal portion of the aorta, the body's main artery, simplifies stent placement by using a self-expanding device. Abdominal aneurysms, like worn spots on a tire, sometimes rupture. The stent graft procedure is emerging as a much less invasive alternative to abdominal surgery, the traditional way to repair the aorta.

In this case, a catheter is inserted into the femoral artery of the patient's leg, then threaded up to the aorta. Next, the stent is directed through the catheter on the tip of a wire to the section of the aorta with the aneurysm, where it expands in response to body heat.

Surgeons at the Mayo Clinic in Rochester, MN, have performed over 100 aneurysm repairs using this procedure. Two types of stems used by the Mayo surgeons were made from Nitinol. A stem graft system received approval from the FDA in 1999.

Although stents are made of various materials, including stainless steel, Nitinol is emerging as a favorite for some applications because it can change shape in response to thermal or mechanical changes and is highly flexible. A Nitinol stent that has been cooled into a compact shape and inserted into a blood vessel will gently return to its expanded shape when exposed to the body's warmth, reducing any trauma associated with stem placement.

Among the properties that render Nitinol useful for these medical applications are high loading plateau stress (450 MPa at 3% strain), a low permanent set (0.2% at 6% strain), and transformation temperatures of 5 to 18°C.

Simon Nitinol Filler — An SMA vena cava blood clot filter was invented by Dr. Morris Simon of Nitinol Medical Technologies, Inc. The design is an Ni-Ti wire form shaped like an umbrella frame or wire basket. For delivery in the body, the filter is chilled below its transformation temperature and collapsed into a small insertion tube. The filter is cooled by a flow of cold saline solution while inserted into the patient and then expands when exposed to body temperature. Its recovered, umbrella-shaped form is designed to catch blood clots in the patient's bloodstream to prevent a pulmonary embolism.[141]

Amplatzer® Septal Occluder — Occlusion devices are designed to serve as Band-Aids to cover and heal holes in the heart without requiring open-heart surgery and are typically fabricated from traditional materials. The Amplatzer Occluder is an SMA device comprised of an Ni-Ti wire frame that is woven and shaped into two flat caps connected by a short tubular section. The device is deployed through a portal in the femoral artery of a patient and placed at the center of the hole to be patched. A sheath is then retracted, and the two flat caps spring to shape and clamp on either side of the hole, thus closing the hole and providing a tight seal.[143]

Orthopedic Devices — SMA materials are used for their superelasticity and shape-memory properties in a variety of orthopedic devices designed to accelerate bone and cartilage formation under constant compressive stresses. Prestrained SMA plates are used to treat bone fractures; when attached on both sides of a fracture and screwed together, they provide a compressive force to heal the fracture areas. Staples are used to heal fractures as well. An Ni-Ti staple in the martensitic state is positioned so that its legs can be driven into bone sections across the fractured area to be closed. When heated, the legs of the staple bend inward and pull the bone sections together, creating a compressive force.[144] Ni-Ti spacers are also used to assist spinal reinforcement in surgical procedures. In this procedure, a spacer is inserted as a compressed ring and then is heated to expand and force open a gap in the vertebrae. Bone chips are placed in the gap and fusion occurs over time to create solid bone. SMAs in solid rod form have been used to treat spinal curvature. Martensitic rods are predeformed to conform to a patient's original spinal curvature, wired to the spine, and then gradually recovered to move the spine to its corrected position.[114]

Minipump — Biomedical engineers at Case Western Reserve University have built a prototype drug pump the size of a contact lens. The miniature implant could monitor its own flow rate to ensure a steady stream of medicine, such as insulin for diabetics. The closed-system micropump employs a flow sensor to measure and adjust the pumping rate for more accurate pumping. The prototype consists of a rectangular silicon chamber with one of the outer walls made of two thin layers of Ni-Ti alloy sandwiched around a layer of silicon. The alloy forcefully changes shape when heated to around 60°C. To operate the pump, a staggered pulse of electrical current passes directly through the alloy, setting up a cycle of heating and cooling that causes it to flex. This forces the chamber to expand and contract. The expansion pulls fluid into the chamber through an intake valve, and the contraction expels the fluid through an exhaust valve.

Hydroxyapatite Coatings on Ni-Ti SMAs — Implant material represent one of the fastest growing areas in the health care industry. Both the medical as well as the economical significance of metallic implant materials to solve traumatisms and osteoarticular diseases has been well documented during the last several years.[145] In the past decade, coated metallic materials having an enlarged surface area, which facilitates joining between the prosthesis and the osseous tissue and increases the integrity of the implant, are being widely applied despite uncertainties in the long-term stability of the coating. Coating with the intent of increasing corrosion resistance has no practical significance, because Ni-Ti SMAs inherently have high corrosion resistance to the environment in which they will be placed.

Plasma spray techniques provide the opportunity to join dissimilar materials without creating a substantially modified interface and allow the creation of roughened surfaces. Roughening allows bone to grow into the interstices of the surface, resulting in a higher surface area of contact as well as mechanical interlocking between the implant and bone tissue. The mineral phase of the bones has a chemical composition and a crystallographic structure that corresponds to hydroxyapatite. It has been shown that porous sulfate, carbonate, and phosphate layers break down and dissolve in time. Under favorable circumstances, they are replaced by new bone tissue.[146]

It seems that the pure hydroxyapatite (HAP) $Ca_{10}(PO_4)_6(OH)_2$ will remain stable (bioinert). When HAP is mixed with organic substances or modified ($Ca_3(PO_4)_2$ — tricalcium phosphate), the implant becomes biodegradable.[147] A new generation of intelligent composite biomaterials based on the Ni-Ti SMA with ceramic coating was investigated by Filip et al.[148] The "passive part" of the implant was coated, which guarantees the anchoring effect in the bone structure, and the active part of shape memory implants provides mechanical work, which guarantees a stable anchoring or fixation of bone fragments. The common problem of ceramic coatings is connected with a relatively weak metal/ceramic bonding. In contrast to usual materials like the Ti-6Al-4V, Co-Cr, or AISI 316L steels, both the coating process as well as the application of such intelligent implants are connected with phase transformations in the Ni-Ti metallic substrate.

The microstructure of HAP ceramic coatings deposited by plasma spraying on Ni-Ti SMA substrates showed, with respect to metal/ceramic interface strength, excellent coatings of HAP on Ni-Ti substrates. From the point of view of adherence, these coatings possess higher strength than Epoxy 1200 adhesive (~30 MPa). The observed excellent metal/ceramic interface strength was attributed to the formation of chemical bonding according to the reaction $Ca_{10}(PO_4)_6(OH)_2 + 2TiO_2$ $3Ca_3(PO_4)_2 + CaTi_2O_5 + H_2O$ and the energy dissipation due to stress-induced martensite formation (SIM) or martensite reorientation (RE) during stressing the composite and failure. The applied power parameters of the plasma spraying influence the porosity and fusion between particles of the ceramic coatings.

FIGURE 4.11
Suture retrieval loops designed to recover their shape once deployed from a 6 fr. cannula
(courtesy of Shape Memory Applications, Inc.).

Lengthening Limbs — Researchers at RPI have developed a treatment
for scoliosis and to lengthen limbs. Traditional surgery carries the risk of
paralysis. The new technique noninvasively heats the Ni-Ti SMA wire,
which causes actuators along an implanted spinal rod to move the verte-
bral column, applying corrective force in a controlled fashion.

Miscellaneous Instruments and Devices — Thousands of other medical devices have been developed to exploit the unique behavior of SMAs.
A few additional typical SMA products include coronary angioplasty
catheters, which can be used where metallic tubes will not be successful;
endodontic files; aneurysm clips; retrieval baskets; surgical needles;
retractable grippers;[149] and suture retrieval loops (see Figure 4.11).

Automotive

SMA applications for the automotive industry are challenging for two primary reasons: the extreme range of operating temperatures expected during use and the market demand for low-cost components. Most automotive devices are expected to perform during exposure to the temperature extremes of climates throughout the world. The success of SMA automotive devices has been limited due to the deterioration of shape-memory properties over time caused by exposure to high temperatures.[131] The successful SMA devices have exploited the benefits of lightweight, simple solutions. For example, SMA actuators are used as replacements for thermostatic bimetals and wax actuators and provide single-metal components in place of complex systems. The simple SMA solutions further improve design performance by activating more quickly because they can be completely immersed in a gas or liquid flow.[135]

Pressure Control Governor Valve — A shape-memory governor valve developed by Raychem Corporation and Mercedes-Benz AG was introduced in 1989 Mercedes cars to improve the rough cold-weather shifting of automatic transmissions. The valves employ Ni-Ti coil springs to counteract the effects of increased oil viscosity. An Ni-Ti coil is immersed in the transmission fluid. At low temperatures, it is martensitic and is forced by a steel bias spring to move a piston. This activates a mechanism to reduce pressure and ease shifting. At higher temperatures, the Ni-Ti spring is much stronger and pushes the bias spring in the opposite direction to optimize shifting pressure at the warmer temperatures.[135]

Due to the ideal performance parameters of the application, this governor valve design is one of the few technical and economic successes in the automotive industry. Operating temperatures are within the limit of the material, required force output is low, and expected life is less than 20,000 cycles. These conditions reduce the possibility of fatigue and degradation of SMA properties during the life of the product.[150]

Toyota Shape Memory Washer — Ni-Ti Belleville-type washers were developed by Toyota Motor Corporation and were used in Sprinter/Carib cars to reduce vibration and rattling noise at elevated temperatures. Automotive assemblies such as gearboxes are often combinations of many dissimilar metals. During the temperature increase in standard operation, a difference in the thermal expansion rates of the metals causes assemblies to loosen and rattle. The washers were designed to change shape at high temperatures using forces up to 1000 N, which is sufficient to tighten the assembly and reduce the undesired rattling noise.[135]

Shock Absorber Washer — This component, an application similar to the governor valve, is also designed to counteract the high viscosity of oil at colder temperatures. SMA washers are placed in shock absorber valves to alter their performance effectively in cold and hot temperatures.[135]

Automotive Clutch Fan — This SMA device, developed as a selective switching mechanism for air-cooled engines, requires the activation of a Cu-based SMA coil to control the operation of a clutch fan. The coil is activated at a temperature close to 50°C and engages a clutch to power the engine fan. The fan speeds up to cool the engine until the temperature is reduced, at which point the SMA coil is forced to disengage the clutch via a set of four steel leaf springs that serve as a biasing force. The device is designed to reduce engine noise and fuel consumption when the car is idle.[131]

Industrial/Civil Engineering

Many SMA solutions have been designed and implemented to satisfy some of the rigorous, large-scale demands of civil engineering projects and miscellaneous industrial applications. Although the constraints of SMA properties often restrict use in industrial applications (the limited acceptable temperature range of operation, for example), many creative SMA solutions and design alternatives have successfully improved or replaced traditional industrial designs and devices.

Pipe Couplings — Cryogenic couplings developed by Raychem Corporation for the aerospace industry have been adapted for use in deep-sea operations and are also used in the chemical and petroleum industries. The advancement of an Ni-Ti-Nb alloy for its wide hysteresis has helped to expand the use of these shape memory couplings. When the components are machined at room temperature and then chilled in liquid nitrogen for expansion, they will not recover their shape until temperatures reach approximately 150°C. This allows storing and transporting expanded couplings without using liquid nitrogen. Once applied to the piping and heated to transform to austenite, they maintain their strength at temperatures colder than –20°C.[150]

Structural Elements — Superelastic SMA materials can be used to increase strength and energy dissipation in a building. A project is currently under way to reinforce the Basilica of St. Francis in Assisi, Italy, after severe damage during earthquakes in 1997. SMA wires will be placed in series with horizontal conventional steel ties to connect the walls to the Basilica's roof. The superelastic behavior of the wires will allow ductile, high-strain motion to occur without breakage during an earthquake and will recover the strain by using a gentler, lower force stroke (exploiting the lower plateau of the stress-strain curve). It has been estimated that the SMA structural design will withstand an earthquake at least 50% stronger than if the Basilica were reinforced with conventional steel bars.[151]

Whittaker et al.[152] conducted a study whereby they developed the technical base for the design of SMA energy dissipation devices for building structures by (1) characterization of the basic materials behavior for the design of prototype (SMA or SMM) energy dissipators, (2) development of conceptual designs for SMA structural damping devices, (3) detailed analysis of the seismic response of a preselected nonductile concrete building with and without SMM energy dissipators under moderate earthquake shaking, and (4) parametric analyses of a reduced-order model of the preselected building upgraded with SMM energy dissipators possessing different hysteretic characteristics from that used for the detailed analysis. SMAs can be configured to provide an SME or a superelastic effect (SEE); energy dissipation devices based on both SME and SEE were shown to be technically viable. Several prototype SMA energy dissipators were designed and one energy dissipator was fabricated. The vulnerability of one building typical of many in the DoD inventory was mitigated by adding SMA energy dissipators.

Reinforcing or Decommissioning Structures

Large structures — typically tubular steel structures that are common in the oil and gas industries — can now be reinforced without the need for welding or other hot working techniques. The approach involves the use of expanding shims made from SMAs: a group of metal alloys that have the ability to undergo a plastic transformation at lower temperature and then recover their original shape on heating, exerting large forces as they do so. This can be exploited to achieve the desired sharing of load between the existing and the newly developed load, something that is not normally possible. Structures can now be strengthened safely with maximum efficiency and without the need to interrupt operations. In addition, by these SMAs, a variety of highly efficient clamping and jacking techniques are now possible, offering advantages for decommissioning operations.[153]

Rock Breakers — SMAs are used to replace explosives in demolition, which is similar in function to the Frangibolt[R] release mechanisms used in the aerospace industry. A prestrained SMA cylinder is placed within a crevice of a structure, electrically heated to expand, and the recovery forces produced are sufficient to destroy rocks and cement structures. This concept has been employed in Russia to yield demolition forces greater than 100 tons of force.[114] Rock breakers comprised of Ni-Ti rods are also used in Japan. A firm uses rods 29 mm long and 15 mm in diameter, which are compressed and inserted into boreholes in rock using an assembly of wedge-shaped platens. The Ni-Ti rods are heated by attached electric heaters, causing them to expand and break the rock apart. Tests have demonstrated forces as high as 14 tons when they are heated to 120°C.[134]

Power Line Sag Control — SMA materials are often used to prevent a sag or droop in overhead power lines. Using thermal SMAs, sag control in power lines has been successfully tested in Canada, the Ukraine, and

Russia.[154] When temperatures increase because of ambient temperatures or high load along the power lines, the lines tend to sag. Nitinol wires are attached to the lines and deform at the colder, high-tension state, but contract when warm and remove the slack from the lines.

Steam Pipe Sag Control — Similar to the power line application, pipe hangers made from Ni-Ti are used to reduce the sagging of large steam pipes heated by the steam. This reduces the load variation in the system, rather than counteracting the shifts in geometry, as in power lines.[154]

Power Generation System

The Central Research Institute of the Electric Power Industry in Tokyo[155] has been engaged in research to develop a power generation system using an Ni-Ti-based SMA with the objective of effective utilization of low-temperature exhaust heat for power generation. The system does not utilize heat intact but rather converts heat into motive power via an SMA engine, stores the energy, and revolves a generator whenever necessary.

When attempting to utilize the heat generated by power plants and waste incineration plants, the heat loss will be extremely large when transporting the heat over long distances, so heat utilization will naturally be restricted to nearby areas. Therefore, an attempt was made to ultimately convert this heat into electricity, by the use of an SMA that can be activated at low temperature and with small temperature differences. Using this alloy as the energy conversion element will enable motive power to be obtained from untapped low-temperature energy resources.

When a metal is impressed with a strain surpassing its limit elasticity, a part of the strain remains even if the load is removed. When a material is heated to beyond its transformation temperature, SMA recovers its original shape with a large restoration force. The research team is striving to utilize this pseudoelasticity of SMA for power generation.

SMA, when deformed by impressing a force at room temperature, returns to its original state with a large force when heated to over a transformation temperature. Therefore, the difference in this strength caused by elongation and contraction was utilized in an engine using pistons with reciprocal motions. In power generation, the energy of the reciprocal motions was converted into hydraulic pressure for storage, and the generator revolved with a hydraulic motor.

The new power generation system consists of the SMA engine that generates motive power from a temperature difference, a weight-type accumulator that serves as an energy storage unit, a hydraulic motor, a transmission system, and a generator. Inside the engine are two heat-insulated pool-type chambers, each aligned with 26 SMA rods with one terminal fixed into position and the other connected to the hydraulic cylinder pistons. Each shape memory rod is linear with a diameter of 2 mm and length of 1 m. Conversion to motive power is performed by alternately sending hot water of 85°C and cold water of 20–25°C from a tank into the two chambers. The heated SMAs recover their shapes and, in concert, generate a recovery force and shift the hydraulic cylinder pistons. When hot water is fed into the cooled chamber, the same motions are accomplished, and the pistons achieve reciprocal motions by repeated chamber heating and cooling.

The hydraulic energy acquired in this manner is accumulated in the energy storage unit. The system used for storage is a weight-type accumulator that uses high-pressure oil to raise a steel weight weighing 2 tons to convert the oil compression pressure into potential energy. Opening the valve at the cylinder outlet revolves the hydraulic motor and generates electricity with the generator. However, adjusting the valve method of opening enables control of the output.

The research team used SMA with a shape recovery length of 12 mm (strain factor 1.2%), conducted experiments in cycles of 1–2 min, and has already succeeded in generating 100 W of electricity (power generation volume 0.8 Wh) for 30 sec. To enable commercial utilization of this plant, the SMA will have to display a service life of 500,000 cycles at the minimum (see Figure 4.12).

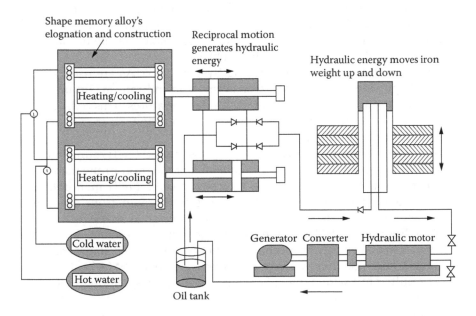

FIGURE 4.12
Mechanism of engine and power generation using SMA.[155]

Transformer Core Compression — SMAs assist in compressing transformer cores, a critical aspect of transformer design. Ni-Ti bolts are pre-strained axially to couple core sheets in large transformers, then heated to contract and provide high compressive forces on the sheets. This SMA solution improves on traditional techniques of tightening with nuts and bolts, where core sheets are placed in a vacuum to withdraw air from between sheets, then removed from the vacuum to install the bolts. Although technically a design improvement, the Ni-Ti solution requires 50.8- to 76-mm-diameter bolts, which is larger than ideal for SMA products.[154] The cost to process and machine the bolts may offset the benefits gained in the assembly process.

Multiwire Tension (MWT) Device — SMA devices are used to increase piping integrity. They prevent crack initiation and propagation by compressing areas of a pipe using the concept of multiwire tension developed by the ABB Nuclear Division in Sweden. A split-sleeve coupling is wrapped with pretensioned SMA wire, the wire ends are fixed, and the coupling is placed over a weld point on a pipe. Then, the assembly is heated so that the wire contracts and the coupling then clamps to a tight fit. MWT techniques are used to improve stress in welded areas, preventing stress corrosion cracking (SCC) and connecting the ends of the pipe should the weld point fail and break.[154]

Indicator Tags — SMAs can be used to indicate high temperature points in a system. Wires used as tags are bent manually at typical operating temperatures, and the tags straighten when temperatures reach a critical level (the material's A_f temperature). Operators note the site of the high-temperature source and can fix the problem.[154]

Sentinel® Temperature Monitoring System — SMAs are used on the low-voltage side of step-down transformers to indicate maximum temperatures and to close a switch at critical temperatures. This SMA mechanism is used to provide information for monitoring, so that operators can prevent overheating.[154]

Injection Molding Mandrels — Centerless ground Ni-Ti round bar and wire are often used to replace conventional, deformation-prone mandrels. Plastics are molded in the shape of a superelastic Ni-Ti mandrel, and during rough handling as the mandrel is withdrawn from the cured polymer, the superelastic material recovers its original shape.

Heat Engines — SMA elements are used in heat engine designs to convert thermal energy to mechanical energy via thermal shape memory behavior. Thermodynamic analyses of ideal, theoretical SMA heat engines have resulted in a wide range of calculated efficiencies, although the most thorough calculations yield maximum thermal efficiencies of only 2–4%.[156] A great number of prototype engines have been constructed using SMA elements that change shape when they pass between hot and cold reservoirs. However, due to loss factors, such as friction, hysteretic effects inherent in the material, and energy input required to maintain a reservoir temperature differential, the practical efficiencies of these heat engines are much too low to serve as low-cost, high-volume energy converters.

The pursuit of a revolutionary SMA heat engine typifies a common occurrence in SMA application design. Intrigued by the potential of SMAs to provide unique anal dramatic design solutions, inventors often pursue creative solutions before completing cost/benefit analyses. Although suitable for small-scale demonstrations, many SMA applications (as in heat engines) do not provide the necessary efficiencies — cost or energy — to warrant replacing conventional designs.

Miscellaneous

There are several other SMA applications that deserve mention.

Elastic Memory Composite — Take an SMP, reinforce it with fiber, and you have created the basis for a device capable of gently lifting 30 times its weight against gravity. With funding from NASA and the Ballistic Missile Defense Agency, Composite Technology Development has developed an elastic memory composite (EMC) that is one fifth the density of SMAs. EMC is a thermally activated material that can be heated, deformed, and then cooled to maintain a new shape. It recovers its original form with the subsequent application of heat and applies force more gradually than SMAs, decreasing risk of damage from shock. The material is currently being qualified to be used in space, but also has varied terrestrial applications such as use in composite pipes where a fully developed cross-section could be heated and rolled onto a spool, enabling the laying of pipe lengths of 300 m.

SMA Thermal-Conduction Switches[157] — Variable-thermal-conduction devices containing SMA actuators have been proposed for use in situations in which it is desired to switch on (or increase) thermal conduction when temperatures rise above specified values and to switch off (or decrease) thermal conduction when temperatures fall below those values. The proposed SMA thermal-conduction switches could be used, for example, to connect equipment to heat sinks to prevent overheating, and to disconnect the equipment from heat sinks to help maintain required operating temperatures when ambient temperatures become too low. In comparison with variable-conductance heat pipes and with thermostatic mechanisms that include such components as bimetallic strips, springs, linkages, or louvers, the proposed SMA thermal-conduction switches would be simple, cheap, and reliable.

The basic design and principle of operation of an SMA thermal-conduction switch is derived from an application in which thermal conduction from hot components to a cooling radiator takes place through the contact area of bolted joints. The thermal conductance depends on the preload in each joint. One could construct an SMA thermal-conduction switch by simply mounting an appropriately designed SMA washer under the bolt head. As the temperature falls below (or rises above) the SMA transition temperature, the SMA washer would contract (or expand) axially by an amount sufficient to unload (or load) the bolt, thereby shutting off (or turning on) most of the thermal conduction through the joint contact area. SMA washers with various transition temperatures can be made to suit specific applications.

Pneumatic Valves[158] — Replacing the electromagnetic coil in a pneumatic valve with an Ni-Ti SMA Nitinol cuts valve cost 60%. Why? Compared to a solenoid valve, the Nitinol valve consumes less power; has virtually no audio, electrical, or magnetic emissions; can modulate flow proportionally; requires loose manufacturing tolerances; has fewer parts; and is smaller in size. It is also lightweight and nonmagnetic.

How does it work? A heat-activated SMA wire functions as the valve actuator. It contracts when electrical power is applied, relaxes when power is removed. A compression spring holds the seal against the seat when power is off. When power is applied, the Nitinol wire heats up and pulls the seal off the seat, allowing air to flow (see Figure 4.13).

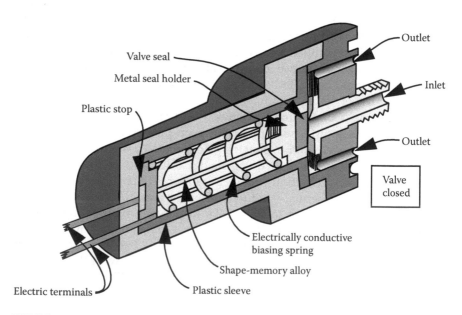

FIGURE 4.13
Terminals, actuator, seal holder, and spring form a continuously closed electric circuit to eliminate sliding contacts and arcing. Applying a pulse frequency modulated electrical current controls position, flow.

Springs for Isolation Devices[159]

There are numerous applications in which metallic isolation devices can offer definite advantages over rubber or other elastomeric mounts and isolators. The advantages are most promising in areas where gasses, oils, corrosive agents, and high temperatures or greatly fluctuating heat levels are present.

Graesser[159] showed and emphasized in his study that a very significant improvement in acoustic transmissibility of lightly loaded springs is realized by use of the Ni-Ti SMA as a spring material. It is judged that very effective isolation mounts can be designed with the SMA material and that springs are but one possible type of mount. Even better designs could be made by incorporating annular leaf spring concepts or SMA wire rope designs, among others. An SMA cable made like wire rope could be doubly effective from a damping perspective, because damping would arise from two sources: SMA intrinsic damping and interfiber friction due to relative motion between cable fibers. Furthermore, an SMA isolation device or mount could be made to include the important features of sensing, stiffness adaptation, or force actuation. A truly "intelligent" mount would integrate all these characteristics with a control algorithm. It is noted, however, that these concepts require much more study, including further research, development, and testing of various design ideas and concepts.

Electronics

The electronics industry has adopted SMA materials primarily for connection mechanisms. Applications in nearly all industries use the electronic activation capability of SMA materials to exploit thermal shape memory behavior (including robotics); however, a few examples are highly specific and unique to the electronics industry.

Cryocon® Electrical Connector — An SMA connector has been designed to attach the braid sheathing of an electrical cable to a terminal plug. The connector is a shape memory ring sheath on a split-walled, collet-shaped tube whose diameter expands when chilled due to the radial forces of the tube. When warmed to room temperature, the ring recovers its smaller diameter shape and clamps the tubing collet prongs together to form a tight electrical connection.[160,161]

ZIF Connector — A zero insertion force (ZIF) electrical connector has been designed to simplify the installation and increase the quality of circuit board connections (see Figure 4.14). A U-shaped strip of Ni-Ti is martensite at operating temperatures and is forced to grip the boards via the bias force of a conventional closing spring. When heated by electrical current, the Ni-Ti strip overcomes the force of the closing spring and opens

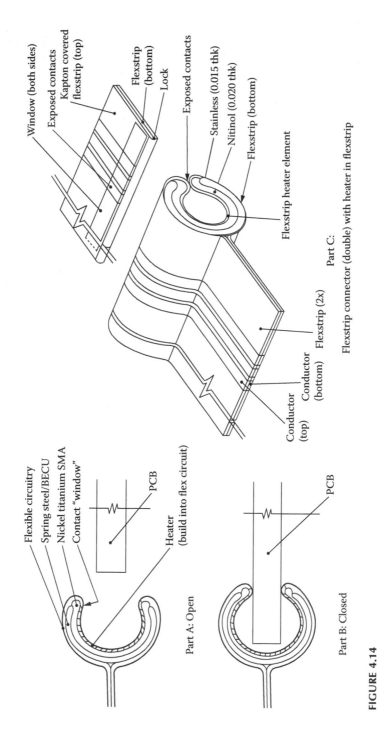

FIGURE 4.14

Schematic diagram of a zero insertion force connector. Reprinted with permission from J. F. Krumme, *Electrically Actuated ZIF Connectors Use Shape Memory Alloys, Connection Technology.* Copyright 1987, Lake Publishing Corporation.[162]

the radius of its U shape, allowing insertion or removal of the boards without force. This combines installation simplicity and maintains optimal, high-force electrical contact between mother and daughter circuit boards.[162]

Microactuators

SMAs have been successfully processed thin film actuators for micro-actuating devices such as tiny valves, switches, and microelectromechanical systems.[163]

Ni-Ti films are sputter-deposited on silicon substrates, and actuators are fabricated by chemical milling and lithographic processes. Devices 1 mm in diameter and 3 μm thick have been fabricated, resulting in tiny actuators such as microvalves for fluid and pneumatic control. The Ni-Ti film actuators can provide up to 3% strain recovery.[164]

The tool consists of connectors/sockets that connect multichip modules and high-density ceramic packages to printed circuit boards. The products enable designers to gain up to 40% more board space and provide 100 signal lines to the inch.

The new connector technology actually incorporates two well-proven technologies: photolithography to etch the contacts onto flexible circuits, allowing a greater number of contacts in a given area, and SMA technology, represented by an Ni-Ti element within the connector. When a low-voltage power supply heats the alloy, it opens the connector's spring, allowing the mating printed circuit board to be inserted with zero force.

Ni-Ti Thin Films

Ishida et al.[165] prepared thin films of Ti-51.3 at% Ni by sputtering. The sputter-deposited thin films were solution-treated at 973°K for 1 H and then aged at various temperatures between 573 and 773°K for three different times of 1, 10, and 100 H. After the heat treatment, the shape memory behavior was examined with a thermomechanical tester. The aging effects on the shape memory characteristics, such as the critical stress for inducing slip deformation, the maximum recoverable strain, and the R-phase and martensitic transformation temperatures, were evaluated on transmission electron microscopic (TEM) observation of the microstructure in the age-treated thin films. In all the age-treated thin films, the presence of Ti_3N_4 precipitates was confirmed. When the precipitate diameter was less than 100 nm, the shape memory characteristics were very sensitive to the microstructure. The aging effects on the shape memory characteristics of the Ni-rich Ni-Ti thin films were found to be almost consistent with those reported in bulk specimens, and, thus, the shape memory behavior of the sputter-deposited thin films of Ni-rich Ni-Ti can be controlled in the same manner as that of bulk specimens.[165]

Bendahan et al.[166] claimed that one of the major barriers in the development of applications using Ni-Ti SMA films is the difficult control over their chemical composition. In fact, the transformation temperatures of Ni-Ti are very sensitive to the alloy composition. They showed that optical emission spectroscopy (OES) can be used to monitor the composition of Ni-Ti films during sputter deposition. The parameter controlling the composition of the films is the product of sputtering gas pressure and target-to-substrate distance. They found a linear relationship between the ratio of optical emission intensity of Ni and Ti atoms and the Ni concentration in the films.

Using OES, they were able to control the composition of the Ni_xTi_{1-x} thin films in the whole range of SME ($0.485 < x < 0.535$). This facilitated the development of applications of the Ni-Ti films.

Miyazaki, Hashinaga, and Ishida[167] prepared Ni-Ti and Ni-Ti-Cu thin films by R.F. magnetron sputtering. The compositions were determined by electron probe X-ray microanalysis, the Cu content ranging from 0 to 13.7 at%.

The transformation and the corresponding deformation behavior were measured by DSC, X-ray diffractometry, and constant stress thermal cycling tests. The following results were obtained:

1. Ni-rich Ni-Ti binary alloy films show an aging effect with the transformation temperatures varying in a wide range, while Ni-poor Ni-Ti binary alloy films do not show such an aging effect.

2. The Ni-poor Ni-Ti films show a single-stage transformation between the parent B2 and the martensitic phase, while the Ni-rich Ni-Ti films show a rhombohedral phase (R-phase) transformation prior to the martensitic transformation.

3. Ni-Ti-Cu films with Cu content less than 9.5 at% show a single-stage transformation and the corresponding deformation behavior between B2 and martensitic phases, while a Ni-Ti-Cu film with Cu content of 9.5 at.% shows a two-stage transformation and the corresponding deformation behavior between B2, orthorhombic, and martensitic phases.

4. The lattice constants of the parent B2, R-phase, and martensite in the Ni-Ti binary alloy films were determined by X-ray diffractometry, while in an Ni-Ti-Cu ternary alloy film with Cu content of 9.5 at.%, the lattice constants of the parent B2, orthorhombic and martensitic phases were determined.

5. The transformation hysteresis associated with the transformation B2 ↔ O showed a strong dependence on Cu content, and it decreased from 27 to 11°K with increasing Cu content from 0 at% to 9.5 at%.

Toyama Prefectural Industrial Technology Center, Hokuriku Electric Industrial Co., Ltd., Takagi Seiko Co., Ltd., and Tanaka Engineering Co., Ltd.[168] have jointly developed an SMA film that changes its shape gradually in response to changes in the surrounding temperature.

Conventional SMAs contain fixed amounts of Ni and Ti, which causes them to assume their memorized shapes suddenly on reaching a specific temperature. In contrast, this new film has a gradient function, changing its composition and shape gradually as the temperature changes.

The research team focused its attention on the fact that the deformation temperature differs depending on the composition of Ni and Ti. In producing an alloy film, the force applied to the substrate was changed continuously and the heat-treatment temperature, conventionally set at over 1000°C, was lowered to 700°C. The result is a film whose composition in the direction of its thickness undergoes a gradual change.

With the new alloy film, the temperature difference between the point at which deformation commences to the point at which it is completed is 40°C, about twice the difference used in the making of conventional alloy films of uniform composition. The shape of this new film can also be controlled by adjusting a current passed through it and controlling the heat.

The research team now plans to apply the film's ability to change continuously to the fabrication and commercialization of superprecision equipment such as medical equipment for treating thrombosis, micropincettes, microvalves, and miniature temperature sensors.

Microrobotics and Microdevices

We usually speak of microdevices when the resolution of the motion and the dimensions of the parts are smaller than the precision and dimension usually achievable in a workshop. In a resolution-of-motion vs. size-of-components representation, microrobotics is typically located in a region defined by resolutions ranging from 10 μm to 1 nm and dimensions ranging from 10 mm to 1 μm. These boundaries are rather a trend than a definition.

Microdevices that integrate other functions such as controlling integrated circuits are usually called MEMS. The MEMS are also often associated with silicon-based technologies and processing. However, according to the definition, MEMS should not be restricted to these special processes. Therefore, as a clever definition, the term "microdevices" is used, rather than MEMS, to qualify small mechanisms, integrated or not.

Technology Innovations LLC, West Henrietta, NY, and Innovation on Demand, Inc., Los Gatos, CA, have developed a concept for a wireless microrobot that can be operated by focused beams of energy. Also known as microactuators, the microrobots are made of Ni-Ti SMA from TiNiAlloy

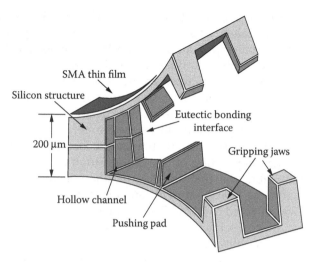

FIGURE 4.15
The Lawrence Livermore National Laboratory's microgripper (200 microns = 0.008 inch).[169]

Co., San Leandro, CA. They were made by sputtering the SMA onto substrates that may be 2 µm wide by 10 µm long.

The microbot has a gripper for 100 nm objects and "legs" for mobility. Purple dots are targets heated by the e-beam; they then conduct the heat to the SMA "muscles" in the legs and gripper to generate the programmed movement. By changing shape, the actuators can grip nanoscale objects or even move. The technology could be used for construction and control of medical devices such as valves and stents, microsurgical instruments, and miniaturized manufacturing molds.

In microrobotics, Lee et al.[169] developed one of the smallest SMA microgrippers. This gripper's dimensions are $1 \times 0.2 \times 0.38$ mm³. The design principle is a kind of "bimorph" of SMA/Si materials. The Si layer acts as a bias spring. The mechanism consists of two identical jaws actuated by Ni-Ti-Cu. The thin films are deposited on both external sides of the gripping jaws. The jaws are made of Si and were shaped by a combination of precision sawing and bulk micromachining of Si. The upper part of the gripper is bonded to the lower part by selective eutectic bonding. The Si gripper cantilevers act as bias springs. When heated, the Ni-Ti-Cu films bend the Si cantilever, and when cooled, the cantilever deflects back and stretches the shape-memory film. A pushing pad is designed to assist in releasing the gripped object by pushing forward as the gripper is opened. An IC-fabricated thin-film resistor heater pad applies heat to the microgripper. Experimental results have shown a gripper opening to 110 µm, when fully actuated, and an estimated gripping force of 13 mN. The time response was estimated at 0.5 sec. (See Figure 4.15.)

Using Si-based processes, Buchaillot et al.[118] developed an XY linear stage made of Ni-Ti thin films. This linear stage consists of four leaf springs operating in the same plane as the substrate. Each actuator has two parallel leaf springs and is initially stretched during mounting to induce a martensitic reorientation.

Actuators are mechanically connected to each other and are designed to avoid any coupling between degrees of freedom. Therefore, the basic actuating principle is an antagonistic design, where each actuator deforms its counterpart. The whole structure is a few millimeters square.

Many other reversible actuators have been proposed in the literature.[170] Among these designs, Kuribayashi et al.[171] in 1993 proposed a millimeter-sized robotic arm. The all-around effect was used to produce a reversible effect in an Ni-Ti alloy beam. Three of these microcantilevers were combined to realize a SCARA-type microrobot that has three degrees of freedom. Later, Kuribayashi and Fujii developed a microcantilever made of an SMA thin film prestressed by a polyimide layer.[172] The polyimide layer provides reverse motion when cooling.

Fluidic Applications — The pioneers in the field of fluidic application research were Busch and Johnson.[173] In 1989, they patented a microvalve that uses an SMA thin film that has been processed by MEMS technologies. The valve consists of an Si orifice die, an actuator die that has a poppet controlled by an Ni-Ti SMA microribbon, and a bias spring. An electrical current heats the actuator (the Ni-Ti microribbon). When no electric current passes through the actuator, the bias spring pushes the poppet against the orifice and closes the valve. When an electric current is applied, the Ni-Ti actuator contracts and lifts the poppet from the orifice, opening the valve. The whole device is less than 1/2 cm². The displacement of the poppet is more than 100 µm and produces a force of ∫ newton. In a rough approximation, flow through the valve is proportional to the current applied to the actuator. The current and an appropriate feedback may be used to control the flow.[173] (See Figure 4.16.)

In another realization, Kohl et al.[174] designed microvalves by integrating SMA bending actuators fabricated from laser-cut, cold-rolled Ni-Ti sheets (see Figure 4.17).

The valve consists mainly of a PMMA housing that has an integrated valve seat, a polyimide membrane, a polyimide spacer, and an SMA microdevice. The SMA microactuator is used for deflection control of the membrane, which opens or closes an opening in the valve seat. The actuator is made of "stress-optimized" cantilever elements whose actuation direction is perpendicular to the plane of the substrate.

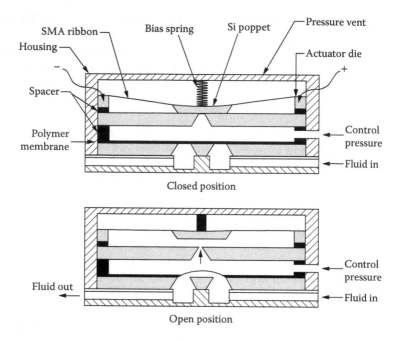

FIGURE 4.16
The microvalve developed by Johnson et al.[173,174]

Specialized Actuator Applications[175-180]

The Boeing Co. is developing and will be testing the enhancement of performance of tilt-rotor aircraft (applicable to V-22) to include redefining the static blade twist distribution and enabling the rotor blades to be retwisted during flight by the use of SMA actuators inside the rotor blades.

NASA-LRC[176] has developed and devised an improved method of designing and fabricating laminated composite-material (matrix/fiber) structures containing embedded SMA actuators. Structures made by this method have repeatable, predictable properties, and fabrication processes can readily be automated.

Such structures, denoted as SMA hybrid composite (SMAHC) structures, have been investigated for their potential to satisfy requirements to control the shapes or thermoelastic responses of themselves or of other structures into which they might be incorporated, or to control noise and vibrations. Much of the prior work on SMAHC structures has involved the use of SMA wires embedded within matrices or within sleeves through parent structures.

The disadvantages of using SMA wires as the embedded actuators include (1) complexity of fabrication procedures because of the relatively large numbers of actuators usually needed; (2) sensitivity to actuator/matrix

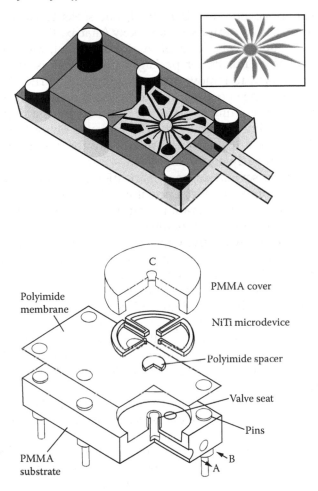

FIGURE 4.17
The microvalve developed at the Forshung Zentrum in Karlsruhe.[174]

interface flaws because voids can be of significant size, relative to wins; (3) relatively high rates of breakage of actuators during curing of matrix materials because of sensitivity to stress concentrations at mechanical restraints; and (4) difficulty of achieving desirable overall volume fractions of SMA wires when trying to optimize the integration of the wires by placing them in selected layers only.

In the present method, one uses SMA ribbons instead of SMA wires. This reduces the number of actuators that must be embedded, thereby making it possible to simplify fabrication processes and to exert better control over the locations and volume fractions of actuators.

In a typical application of this method, one seeks to fabricate one or more structure(s), each comprising an epoxy/matrix/fiber laminate containing one or more embedded SMA ribbons. First, SMA ribbon, as received from the manufacturer packaged on a spool, is removed from the spool and treated to remove the packaging strain. Then by use of a tensile testing machine operating in stroke-control mode, the SMA ribbon is stretched to the amount of prestrain required by the design of the structure. To save time, several parallel lengths of ribbon, separated by spacers and held by grips designed specifically for the purpose, can be stretched simultaneously on the machine.

Figure 4.18 shows several stages of fabrication of a panel-type structure following elongation of the SMA actuators. Fabrication of the SMAHC laminate proceeds by incorporating lengths of the SMA ribbon at the specified locations during the lamination process. Depending on the design of the specific structure, either the SMA ribbons can be laid up between laminae, or else precisely dimensioned portions of the laminate can be removed and SMA ribbons inserted in the resulting voids to incorporate the SMA ribbons within the laminae. Upon completion of the layup, the free ends of the SMA ribbons are constrained within the knurled mechanical grips to maintain the preset elongation during elevated temperature cure. The entire assembly is then vacuum-bagged and subjected to autoclave cure. Subsequent to consolidation, the free ends of the SMA ribbons are released from the grips, and the SMAHC structure is machined to final dimensions.

At NASA-JSC[177] a shape memory ribbon was wrapped around a cylinder to build up length to form a cylindrical shape memory rotary/linear actuator.

A compact actuator generates rotary or linear motion with a large torque or force, respectively. The original version of this actuator is designed to pull a wedge that, until pulled, prevents retraction of the proposed extended nose landing gear of the space shuttle. The original version is also required to fit into a volume that is severely limited by the size of the landing-gear assembly. The basic actuator design could be adapted to other applications in which there are requirements for compact, large-force actuators with similar geometries.

The transducer portion of this actuator is a ribbon of an Ni-Ti SMA. A component made of such an alloy undergoes a pronounced deformation to a "remembered" shape (in the present case, the ribbon becomes shorter) when its temperature rises through a transition valve, causing a transformation in its metallurgical structure from a martensitic phase to an austenitic phase. The component resumes its previous shape (in this case, the ribbon lengthens) when its temperature falls below a lower transition temperature (there is hysteresis in the transformation). In this case, the transition temperatures are somewhat above room temperature.

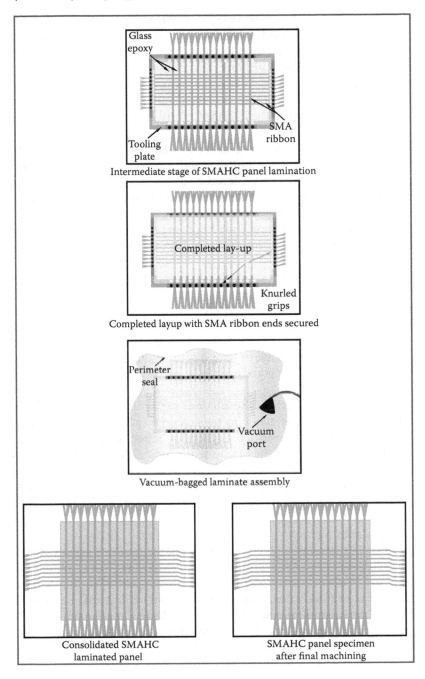

Intermediate stage of SMAHC panel lamination

Completed layup with SMA ribbon ends secured

Vacuum-bagged laminate assembly

Consolidated SMAHC laminated panel

SMAHC panel specimen after final machining

FIGURE 4.18
SMA ribbons were incorporated into an epoxy/fiberglass laminate panel. The panel was then machined to obtain SMAHC beam specimens for testing.[176]

To obtain sufficient length in order to obtain sufficient stroke, the shape memory ribbon is wrapped three times around a cylinder. One end of the ribbon is attached to the cylinder, the other to an object that does not move, relative to the axis of the cylinder. The ribbon is heated by passing an electrical current through it. When the temperature of the ribbon rises above the first transition temperature, the resultant shape-memory shrinkage exerts a torque, causing the cylinder to turn. In the original application, the rotary motion of the cylinder is converted to linear motion by use of ramps, attached to the cylinder, that rolls along wheels. In a different application, the rotary motion of the cylinder could be the desired output.

The primary difficulty encountered in initial tests was that the sliding friction engendered by wrapping the ribbon three times around the cylinder was so large as to impair actuation. The friction was reduced to an acceptably low level by placing the ribbon on nonconductive roller assemblies to enable the ribbon to move around the cylinder by rolling instead of sliding.

Electralock microclamps powered by "programmed metal" technology have reportedly been developed by Jergens Inc., Cleveland, OH. The clamps are made of an SMA that allows the metal actuator to expand and contract during crystalline phase transformations.

By eliminating the support equipment needed for hydraulics, setting up and maintaining the power clamps is much simpler. When operators need to move the clamping fixture, they just unplug it from a standard electrical socket.

The technology is based on the phase transformation of an Ni-Ti alloy as it moves from the martensitic to the austenitic phase. This transformation has been refined to produce a trainable, repeatable, and predictable result. The actuator within the clamp works when the alloy is electrically heated, expanding the material and releasing the 340 kg clamping pressure.

The clamp's portability is said to make it ideal for palletized fixturing applications. They remain securely clamped when disconnected, and the waterproof construction and 110 V power requirements provide high flexibility within a shop.

Figure 4.19 illustrates one of several related pin-puller mechanisms that harness the recovery characteristics of an SMA. Like some other SMA-actuated mechanisms described above,[178] this pin puller is suitable for use in place of a pyrotechnically actuated pin puller and can be operated under remote control to release a door or deploy a stowed object, for example. Unlike pyrotechnically actuated pin pullers, these can be reused.

The SMA component in this pin puller serves as a trigger to release potential energy stored in the driver spring. This spring biases the driver toward retraction (downward in the figure). However, under normal (non-actuation) conditions, the pin is set in the extended (uppermost) position and the ball-detent latch prevents retraction.

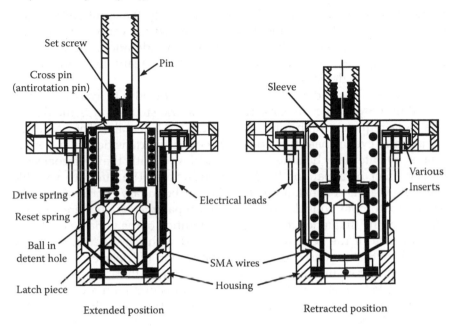

Extended position   Retracted position

FIGURE 4.19
In this SMA-actuated pin puller, the SMA wire serves as a trigger that releases potential
energy stored in the driver spring.[178]

A component made of an SMA undergoes a pronounced deformation
to a "remembered" shape, as described by Reference 177. The difference
here is that the SMA component is a Nitinol wire. In this application, to
initiate retraction of the pin, electric current is made to flow along the wire,
thereby heating the wire above its transition temperature and causing it
to shrink to its "remembered" (unstretched) length. The shrinkage of the
wire pulls the latch piece upward, against the downward bias of the reset
spring. As a result, the balls are no longer forced to protrude through the
detent holes. Thus, the balls no longer block the retraction of the pin.

Later, the mechanism can be reset by pulling on the pin to extend it.
Provided that the SMA wire has cooled enough to recover its stretched
length, the mechanism will remain reset because the reset spring biases
the latch piece into its lowest position, where it forces the balls to protrude
through the detent holes.

Biwa et al.[179] examined theoretically as well as experimentally the
dynamic behavior of a bias-type actuator using Ni-Ti SMA. Some basic
characteristics of SMA actuators were examined by formulating an analytical model of a bias-type actuator driven by electric input, based on the
thermomechanical constitutive equations. Their model proved promising
for describing the behavior of the actuator observed experimentally, in

terms of its dependence on the bias-spring rigidity, the input voltage, and the cycle period. With the identification method for various parameters involved in the model to be given further consideration, the proposed model is expected to offer a useful analytical means for designing a number of SMA devices.

D. C. Lagoudas and A. Bhattacharyya[180] modeled a thermoelectrically cooled thin SMA layer linear actuator, as a first step towards the design of a high-frequency, high-force, large-strain SMA actuator. The SMA was subjected to cyclic phase transition between the martensitic and austenitic phases by alternate heating/cooling, achieved with the thermoelectric Peltier effect of a pair of P/N semiconductors. The effect of variable actuating load and constant load applied as boundary conditions on the SMA actuator was considered. The primary parameters of interest in this work were the frequency response and evolution of the variable load. The performance of the actuator was compared with various commercially available actuators based on energy conversion efficiencies and energy output per unit volume of active material. Results of the analysis indicated that thin SMA layers under partial phase transformation are capable of delivering frequencies of about 30 Hz at peak stresses of about 145 MPa.

Ullakko[181] examined magnetically controlled SMAs as a new class of actuator materials. Materials that develop large strokes under precise and rapid control exhibit a great potential in mechanical engineering. Actuators made from those kinds of materials could replace hydraulic, pneumatic, and electromagnetic drives in many applications. Piezoelectric and magnetostrictive materials exhibit rapid response, but their strokes are small. In SMAs, strokes are large, but their control is slow due to thermomechanical control. Magnetic control of the SME was recently suggested by Ullakko for a principle of new kinds of actuator materials. These materials can develop strains of several percent, and their control is rapid and precise. Actuation of these materials is based on the reorienting of the twin structure of martensite or the motion of austenite-martensite interfaces by applied magnetic field.

In References 182 through 186 Ullakko demonstrates the magnetically induced motion of the austenite-martensite interfaces in an Fe-33.5 Ni alloy.

Ullakko and Handley recognized the advantages of an Ni-based alloy that changed shape in response to a magnetic field.[187]

Ni_2MnGa is a known SMA that, when distorted, will snap back to its original shape in response to temperature increases. Recently, the material has been shown to react just as strongly to magnetic fields.

The ability to control the SME with magnetism may be a significant leap forward in the development of so-called smart materials. That is because the magnetic response is quicker and more efficient than the traditional temperature-induced response.

Researchers first recognized the potential of the magnetic SME in Ni_2MnGa about a decade ago. A few years later, a significant field-induced strain was observed by a group at the Massachusetts Institute of Technology (MIT) led by Robert O'Handley and Kari Ullakko. Work has continued at MIT to develop magnetic shape memory (MSM) materials for use in sonar transducers and vibration mitigation systems for the U.S. Navy.

So far, the university consortium has been able to boost the strain to 6% at room temperature from about 0.2% at $-8°C$. This has important implications for the material's commercial applications in pumps and valves.

They found that by changing the composition slightly (to 50%Ni, 29% Mn, and 21% Ga), they could get even better results at room temperature. They showed that by applying a static stress to the alloy, it could be reset and actuated quasistatically many times over.

More recent breakthroughs have allowed the MIT group to demonstrate strain at frequencies of up to 500 Hz and characterize the material's ability to do work (i.e., strain under load). Ni_2MnGa has shown 30 times greater strain at room temperature relative to other magnetically controlled materials.

The magnetic SME differs from the classic temperature-induced or thermoelastic effect. Although the latter requires a transformation from the martensite phase to the austenite phase, the former occurs entirely within the martensite phase. The shape changes when twin structures oriented favorably relative to the magnetic field grow at the expense of the other twin structures in the material.

Companies hope to take advantage of this unique feature by developing applications that convert magnetic field energy into mechanical motion. These would include simple electromechanical devices that could replace complicated machines in everything from home electronics to automobiles.

Currently in production are MSM sensors as well as actuators capable of stroke lengths of up to 5 mm and forces up to 2 KN.

Future Trends and Prospects

SMA applications continue to gain acceptance in a variety of industries throughout the world. SMAs are introduced to an increasing number of students who study engineering and metallurgy and to the general public via the growing number of SMA products available to consumers. Although property values and design techniques for SMAs are not as readily available or as thoroughly standardized as those of conventional materials, current trends indicate a steady course toward complete characterization of SMA materials. The medical industry has already prompted the standardization of material production, testing, and mechanical behavior; efforts include the formation of an ASTM standard for using Ni-Ti in medical devices. Spearheaded by supplier representatives of both the medical device and SMA material supply industries, a standard has been completed and approved. This new standard represents a significant milestone in the continual effort to demystify SMAs and augment their use in engineering and design.[188–193]

Scaling down things leads to new design approaches. This rule applies to SMA microdevices. Successful breakthroughs have been made in this field of research, but there is still a further need to understand the behavior of small devices on the microscale. It is well known that the mechanical behavior of a thin film of a few microns is not the same as the behavior of a sheet that is a few hundred microns thick. As W. Nix mentioned,[192] thin film materials are, for instance, much stronger than their bulk counterparts. This may be due to the fine grain sizes commonly found in thin films, but single-crystal thin films are also much stronger than bulk materials. Many issues remain unresolved or little addressed for SMA thin films:

- How thin can a functional SMA thin film be? In other words, where is the size limit of the SME?
- If a micron-sized structure is built, how smooth will the motion be? Will we observe a discontinuous effect?
- How does the scaling law apply to fatigue properties?

Many new advances can be expected if, in the future, shaping and chemistry are performed simultaneously. Moreover, new degrees of miniaturization can be taken into account if bottom-up, rather than top-down, approaches are favored. One of the major tools in this field can be the ever-increasing varieties of scanning microprobe technology in order to modify surfaces and to initiate design procedures (e.g., by electrochemical processes on a nanometer scale).[194,195]

Final Remarks on Future Prospects

Further prospects must be seen in the direction of new materials, new geometrical design, as well as enhanced intelligent behavior. Here, only selected aspects of the wide range of attempts in various fields of application could be mentioned. The current generation of smart material systems is driven by the engineering perspective and must be realized with the materials available, to terms of actuators, sensors, etc. The next practical step seems to be the combination of sensor/actuator functions with a control system by the deposition of electroceramic coatings on integrated circuit Si chips.[196]

Research in nanoactuation systems is still in its early stages.[197] However, initial steps towards the realization of nanomolecular machines were, for example, the attempts to reconstruct a biological motor, a flagellar motor of bacteria in vitro, which is simple because it consists of only a few molecules and requires minimal energy to produce relatively large forces. Thus, it operates with high speed and efficiency.[198]

There is a growing demand in fields such as anthropomorphic robotics and microrobotics for muscle-like actuators with high power-to-weight ratios and a large degree of compliance. Muscles are very unconventional actuators, because they are neither pure force nor pure motion generators. They behave rather like springs with tunable elastic parameters. Such a combination of behavior should be obtained in the future potentially by piezoelectric polyvinylidene polymers or by polymer gels that undergo chemomechanical conversion and yield large strain and force densities comparable to those in muscles. Also, conducting polymers have already been shown to possess very high force-generation capabilities.[199] Currently, some of the materials and systems come close to one or two of the desired advanced properties near the level of intelligence. The examples range from electroactive polymers that act as "sleeping" security shields (capacitive sensors) to protect data stored in EPROMs,[200] up to the temperature-responsive PIPAAM polymers that are bioactive and can be separated from reactive products with small temperature increase so that direct recycling systems would be possible.[201]

As a longer term vision, it will be necessary to follow nature's design paradigms, with control at the molecular, nano, micro, and macro levels of synthesis and assembly of actively self-repairing systems in functional shapes.[202] From this viewpoint, our learning stems only from failures as a consequence of our inability to learn from structures during their life.

Scientists and engineers have made advances in developing a large variety of structures and systems, which take advantage of a few of nature's capabilities. They can change their properties, shape, and color, and to some extent, variations of load paths are already possible that account for damage and repair. However, long-term vision should open the perspective upon human-made intelligent material systems that become more and more natural and are capable of "living" in the sense of being able to learn, grow, survive, and age with grace and simplicity.

For example, the Office of Naval Research has applications for ships, submarines, and other vehicles and structures.

These hybrid composite materials would use plates of SMA Ni-Ti to sandwich shape-memory, superelastic foams and rods embedded with hollow glass beads to elicit even greater superelastic performance.

SMPs

Finally, as mentioned previously and discussed briefly, SMPs have the ability to store and recover large strains by the application of a prescribed thermomechanical cycle. The typical protocol is to apply a specified initial deformation at an elevated temperature, cool the predeformed material under constraint to a lower temperature where the shape is fixed (storage), and then heat the material to recover the original shape (recovery).[203]

Of particular interest to the medical community is triggering shape recovery at body temperature, with polymers that provide variable stiffness and recoverable force levels that can be tailored depending on the application. Shape recovery at body temperature permits the deployment of complex packaged shapes into the body environment without local heating.

SMP Advantages

Compared with alternative active materials such as SMAs, SMPs have several advantages, such as low density, large recovery strain, controllable activation temperature range, and low-cost curing and molding processes. The capacity of SMPs to store large strain deformations for deployment in restricted environments has resulted in the recent development of SMP-based medical devices.

Because many medical applications involve bending and large displacement, and because shape change can be achieved in flexure at much more modest strain levels, the work in Reference 203 considered the thermomechanical response in three-point flexure. During cooling, the thermal contraction was not as severely constrained, and the stress needed to maintain the fixed shape could decrease to zero as the temperature was reduced.

A typical three-point flexure predeformation and recovery cycle for SMPs can be described in three steps:

> Step 1: The first step involves a high-strain deformation, called predeformation.
>
> Step 2: The second step is a constraining procedure in which the SMP is cooled while maintaining the fixed shape.
>
> Step3: In the final step, the SMP structure is subjected to a prescribed constraint level and heated toward the glass transition temperature. Unconstrained recovery implies the absence of external stresses and the free recovery of the induced strain. Constrained recovery implies the fixing of the predeformation strain and the generation of a gradually increasing recovery stress.

Because the shape memory cycle goes through the glass transition range, the thermomechanical behavior of SMPs depends on both time and temperature. In other words, the cooling/heating rate can have a significant impact on the SMP stress recovery. The peak recovery stress is affected by the cooling rate. At a fixed strain, the stress will relax as a function of time.

Other studies have shown that when predeformed at a temperature above the glass transition temperature, the evolution of the recovery stress as a function of temperature is gradual and saturates at a stress equal to the stress imparted at the predeformation.

On the other hand, when predeformed at a temperature below the glass transition temperature, the stress recovery response as a function of temperature demonstrates a sharp increase to a peak value. It then shows a decrease towards a saturation stress value governed by the predeformation stress or strain at a higher temperature above T_g.

Higher cooling rates during constraint reduce the temperature necessary for complete shape fixity, but raise the peak recoverable stress. Lower healing rates during recovery decrease the temperature at the onset of recovery and increase the peak recoverable stress.

References

1. Schetky, L. Mc., Shape-Memory Alloy Applications, in J. H. Westbrook and R. L. Fleischer (Eds.), *Intermetallic Compounds*, Vol. 2, Practice, New York: John Wiley & Sons, 1994. pp. 529–558.
2. Wayman, C. M., Shape Memory Alloys, *MRS Bulletin*, 18 (4), 49–56, 1993.
3. Haasters, J., Salis-Soho, G. V., and Bensmann. C., The Use of Ni-Ti as an Implant Material in Orthopedics, in T. W. Duerig, K. N. Melton, D. Stöckel, and C. M. Wayman (Eds.), *Engineering Aspects of Shape Memory Alloys*, Butterworth-Heinemann, 1990, pp. 426–444.
4. Hodgson, D. E., Shape Memory Alloys, *NiTi Smart Sheet*, http://www.sma-inclSMAPaper.html.
5. Bellouard, Y., Microrobotics and Microdevices Based on Shape Memory Alloys, in M. Schwartz (Ed.), *Encyclopedia of Smart Materials*, Vol. 2, New York: J. Wiley and Sons, Inc., 2002, pp. 620–644.
6. Development of High Temperature Two-Way Shape Memory Alloys, Final Report, October 1991–March 1995, U.S. Army Res. Office, Res. Triangle Park, NC, October 17, 1995, p. 27.
7. Wayman, C. M. and Duerig, T. W., An Introduction to Martensite and Shape Memory, in T. W. Duerig, K. N. Melton, D. Stöckel, and C. W. Wayman (Eds.), *Engineering Aspects of Shape Memory Alloys*, Butterworth-Heinemann, 1990, pp. 452–469.

8. Delaney, L., Chandrasekaran, M., Andrade, M., and Van Humbeeck, J., Nonferrous Martensites, Classification, Crystal Structure, Morphology, Microstructure, *Proceedings Int. Conf. Solid-Solid Phase Transformation*, Met. Society of AIME, pp. 1429–1453, 1992.

9. Maki, T. and Tamura, T., Shape Memory Effect in Ferrous Alloys, *Proc. ICOMAT-86*, Japanese Institute of Metals, 1986, pp. 963–970.

10. Wnag, G., Weldability of Nitinol, Edison Welding Institute publication, 1995, available for private circulation only.

11. Shinoda, T., Tsuchiya, T., and Takahashi, H., Friction Welding of Shape Memory Alloy, *Welding International*, 6(1), 20–25, 1992.

12. Gu, Q., Van Humbeeck, J., and Delaey, L., A Review of the Martensitic Transformation and Shape Memory Effect in Fe-Mn Si Alloys, *Journal de Physique IV*, Col. C3, 4, C3-135–140, 1994.

13. Maki, T., Ferrous Shape Memory Alloys, in K. Otsuka and C. M. Wayman (Eds.), *Shape Memory Materials*, Chapter 5, Cambridge Univ. Press, 1998.

14. Sato, A., Sama, K., Chishima, E., and Mori, T., Shape Memory Effect and Mechanical Behaviour of an Fe-30Mn-1Si Alloy Single Crystal, *J. Physique*, 12 C4, 797–802, 1982.

15. Sato, A., Chishima, E., Soma, K., and Mori, T., Shape Memory Effect in Transformation in Fe-30Mn-1Si Alloy Single Crystals, *Acta Metall.*, 30, 1177–1183, 1982.

16. Robinson, J. S. and McCormick, P. G., Shape Memory in an Fe-Mn-Si Alloy, *Mat. Sci. Forum*, 56, 649–653, 1990.

17. Sato, A., Shape Memory Effect and Physical Properties of Fe-Mn-Si Alloys, *MRS Int'l. Mtg. on Adv. Mats.*, 9, 431–445, 1989.

18. Murakami, M., Otsuka, H., Suzuki, B. G., and Matsuda, S., Complete Shape Memory Effect in Polycrystalline Fe-Mn-Si Alloys, *Proc. of Int. Conf. on Mart. Transf. (ICOMAT-86)*, Jap. Inst. of Metals, 1986, pp. 985–990.

19. Jonsson, S., Review on Fe-Based Shape Memory Alloys, Swedish Inst. for Metals Research, Stockholm, *NTIS Alert*, Vol. 96, No. 4, February 15, 1996, p. 37.

20. Lecis, G., C., Lenardi, C., and Sabatini, A., The Effect of Mn-Depleted Surface Layer on the Corrosion Resistance of Shape-Memory Fe-Mn-Si-Cr Alloys, *Metallurgical and Materials Trans. A-Physical Metallurgy and Materials Science*, 28(5), 1219–1222, 1997.

21. Chung, C. Y., Chen S. C., and Hsu, T. Y., Thermomechanical Training Behavior and Its Dynamic Mechanical Analysis in an Fe-Mn-Si Shape Memory Alloy, *Materials Characterization*, 37(4), 227–236, 1996.

22. Hamers, A. A. H. and Wayman, C. M., Shape Memory Behavior in Fe-Mn-Si Alloys, *Scripta Metallurgica Et Materialia*, 25(12), 2723–2728, 1991.

23. Liu, D. Z., Liu, W. X., and Gong, F. Y., Engineering Application of Fe-Based Shape Memory Alloy on Connecting Pipe Line, *Journ. of Physique III*, Vol. 5 (ICOMAT-95, Proceedings), 1995, pp. 1241–1246.

24. Van Humbeeck, J. and Delaey, L., A Comparative Review of the (Potential) Shape Memory Alloys, in E. Hornbogen and N. Jost (Eds.), *The Martensitic Transformation in Science and Technology*, Oberursel, Germany: DGM, Verlag, 1989, pp. 15–25.

25. Van Humbeeck, J., Chandrasekaran, M., and Stalmans, R., Copper-Based Shape Memory Alloys and the Martensitic Transformation, *Proc. ICOMAT-92*, Monterey, CA: Monterey Institute of Advanced Studies, 1993, pp. 1015–1025.
26. Tadaki, T., Cu-Based Shape Memory Alloys, in K. Otsuka and C. M. Wayman (Eds.), *Shape Memory Materials*, Chapter 3, Cambridge University Press, 1998.
27. Ahlers, M., Martensite and Equilibrium Phases in Cu-Zn and Cu-Zn-Al Alloys, *Progress in Materials Science*, Vol. 30, J. W. Christian, P. Haasen, and T. B. Massalski (Eds.), 1986, pp. 135–186.
28. Wu, S. K. and Ming, H., Cu-Based Shape Memory Alloys, in T. Duerig et al. (Eds.), *Engineering Aspects of Shape Memory Alloys*, U.K.: Butterworths Scientific, 1990, pp. 69–88.
29. Rapacioli, R. and Chandrasekaran, M., The Influence of Thermal Treatments on Martensitic Transformation in Cu-Zn-Al Alloys, *Proc. Intl. Conf. on Martensitic Transformations* Cambridge, MA: MIT Press, 1979, pp. 596–601.
30. Rodriguez, P. and Guenin, G., Thermal and Thermomechanical Stability of Cu-Al-Ni Shape Memory Effect, in E. Hornbogen and N. Jost (Eds.), *The Martensitic Transformation in Science and Technology*, Oberursel, Germany: DGM, Verlag, 1989, pp. 149–156.
31. Van Humbeeck, J., Delaey, L., and Roedolf, D., Stabilisation of Isothermally Transformed 18R Cu-Al-Ni Martensite, *Proc. Int. Conf. on Mart. Transf. (ICOMAT-1986)*, Japan Institute of Metals, 1986, pp. 862–867.
32. Abeyaratne, R. and Knowles, J. K., On the Kinetics of an Austenite-Martensite Phase-Transformation Induced by Impact in a Cu-Al-Ni Shape Memory Alloy, *Acta Materialia*, 45(4), 1671–1683, 1997.
33. Kainuma, R., Takahashi, S., and Ishida, K., Thermoelastic Martensite and Shape-Memory Effect in Ductile Cu-Al-Mn Alloys, *Metallurgical and Materials Transactions A — Physical Metallurgy and Materials Science*, 27(8), 2187–2195, 1996.
34. Xiao, T. and Johari, G. P., Mechanisms of Internal Friction in a Cu-Zn-Al Shape-Memory Alloy, *Metallurgical and Materials Transactions A — Physical Metallurgy and Materials Science*, 26(3), 721–724, 1995.
35. Wei, Z. G., Peng H. Y., and Zou, W. H., Aging Effects in a Cu-12Al-5Ni-2Mn-1Ti Shape Memory Alloy, *Metallurgical and Materials Transactions A — Physical Metallurgy and Materials Science*, 28(4), 955–967, 1997.
36. Romero, R., Somoza, A., and Jurado, M. A., Quenched-In Defects and Martensitic-Transformation in Cu-Al-Be Shape Memory Alloys, *Acta Materialia*, 45(5), 2101–2107, 1997.
37. Elst, R., Van Humbeeck, J., and Delsaey, L., Grain Refinement of Cu-Zn-Al and Cu-Al-Ti by Ti-Addition, *Material Science & Technology*, 4, 644–648, 1988.
38. Elst, R., Van Humbeeck, J., and Delaey, L., Evaluation of Grain Growth Criteria in Particle-Containing Materials, *Acta Metall.*, 36, 1723–1729, 1988.
39. Van Humbeeck, J., High-Temperature Shape Memory Alloys, *Transactions ASME*, 121, 98–101. 1999.
40. Hornbogen, E., The Effect of Variables on Martensitic Transformation Temperatures, *Acta Metall.*, 33, 595–601, 1985.
41. Nishiyama, Z., *Martensitic Transformation*, Academic Press, 1978.

42. Adnyana, D. N., The Effect of Grain Size on the M_S-Temperature in a Grain-Refined Copper-Based Shape Memory Alloy, *Proc. Int. Conf. on Mart. Transf.*, The Jap. Inst. of Met., 1986, pp. 774–779.

43. Jianxin, W., Bohong, J., and Hsu, T.Y., Influence of Grain Size and Ordering Degree of the Parent Phase on M_S in a CuZnAl Alloy Containing Boron, *Acta Metall.*, 36, 1521–1526, 1988.

44. Elst, R., Van Humbeeck, J., and Delaey, L., The Evolution of Martensitic Transformation Temperatures During Annealing of Supersaturated -Cu-Zn-Al-Co Alloys, *Proc. Int. Conf. on Mart. Transf.*, The Jap. Inst. of Met., 1986, pp. 891–895.

45. Gu, N. J., Peng, H. F., and Wang, R. X., Influence of Training Time and Temperature on Shape-Memory Effect in Cu-Zn-Al Alloys, *Metallurgical and Materials Transactions A -Physical Metallurgy and Materials Science*, 27(10), 3108–3111, 1996.

46. Saburi, T., Ti-Ni Shape Memory Alloys, K. Otsuka and C. M. Wayman (Eds.), *Shape Memory Materials*, Cambridge University Press, 1998, chap. 3.

47. Treppman, D., Hornbogen, E., and Wurzel, D., The Effect of Combined Recrystallization and Precipitation Processes on the Functional and Structural Properties in NiTi Alloys, *Journal de Physique IV*, Colloque C8, *Suppl. au Journal de Physique III*, Vol. 5, *Proc. ICOMAT-95*, R. Gotthardt and J. Van Humbeeck J (Eds.), December 1995, pp. 569–574.

48. Wayman, C. M., Transformation, Self-Accommodation, Deformation, and Shape Memory Behavior of NiTi Alloys, *MRS Int'l. Mtg. on Adv. Mats.*, 9, 63–76, 1989.

49. Saburi, T., Structure and Mechanical Behavior of Ti-Ni Shape Memory Alloys, *MRS Int'l. Mtg. on Adv. Mats.*, 9, 77–91, 1989.

50. Todoriki, T. and Tamura, H., Effect of Heat Treatment after Cold Working on the Phase Transformation in TiNi Alloy, *Trans. of the Jap. Inst. of Metals*, 28, 83–94, 1987.

51. Nishida, M., Wayman, C. M., and Honma, T., Precipitation Processes in Near-Equiatomic TiNi Shape Memory Alloys, *Met. Trans.*, 17A, 1505–1515, 1986.

52. Wayman, C. M., Phase Transformations in NiTi Type Shape Memory Alloys, *Proc. Int. Cont. On Mart, Transf.*, 1986, The Jap. Inst. of Metals, pp. 645–652.

53. Otsuka, K. and Wayman, C. M. (Eds.), *Shape Memory Materials*, Cambridge University Press, 1998.

54. SMA Inc., Company data sheet.

55. Gere, J. M. and Timoshenko, S. (Eds.), *Mechanics of Materials*, CBS Publisher, 1986.

56. Li, D. Y. and Liu, R., *Wear*, 225-229, 727–783, 1999.

57. Dueng, T., Pelton, K. N., Stöckel, D., and Wayman, C. M. (Eds.), *Engineering Aspects of Shape Memory Alloys*, Butterworth-Heinemann, 1990.

58. Lin, P. H., Tobushi, H., and Tanaka, K., Influence of Strain-Rate on Deformation Properties of TiNi Shape-Memory Alloy, *JSME Intl. J. Ser. A- Mech. Mater. Eng.*, 39(1), 117–123, 1996.

59. Eckelmeyer, K. H., The Effect of Alloying on the Shape Memory Phenomenon in Nitinol, *Scripta Met.*, 10, 667–672, 1976.

60. Honma, T., Matsumoto, M., Shugo, Y., and Yamazaki, I., Effects of Addition of 3D Transition Elements on the Phase Transformation in TiNi Compound, *Proceedings ICOMAT-79*, pp. 259–264.

61. Huisman-Kleinherenbrink, P., On the Martensitic Transformation Temperatures of NiTi and Their Dependence on Alloying Elements, Ph.D. thesis, University of Twente, The Netherlands, 1991.

62. Kachin, V. N., Martensitic Transformation and Shape Memory Effect in B2 Intermetallic Compounds of Ti, *Revue Phys. Appl.*, 24, 733–739, 1989.

63. Kachin, V. N., Voronin, V. P., Sivokhe, V. P., and Pushin, V. G., Martensitic Transformation and Shape Memory Effect in Polycomponent TiNi-Based Alloys, *ICOMAT-95 Proceedings*, Part I and II, Lausanne, France, R. Gotthardt and J. Van Humbeeck (Eds.), Editions de Physique, *Journal de Physique*, IV, 5(8), 760–770, 1995.

64. Saburi, T., Takagaki, T., Nenno, S., and Koshino, K., Mechanical Behaviour of Shape Memory Ti-Ni-Cu Alloys, *MRS Int'l. Mtg. on Adv. Mats.*, 9, 147–152, 1989.

65. Horikawa, H. and Ueki, T., Superelastic Characteristics of Ni-Ti-Cu-X Alloys, *Trans. Mat. Res. Soc. Jpn.*, 18B, 1113–1116, 1993.

66. Melton, K. N., Proft, J. L., and Duerig, T. W., Wide Hysteresis Shape Memory Alloys Based on the Ni-Ti-Nb System, *MRS Int'l. Mtg. on Adv. Mats.*, 9, 165–170, 1989.

67. Yang, J. H. and Simpson, J. W., Stress-Induced Transformation and Superelasticity in Ni-Ti-Nb Alloys, *ICOMAT-95 Proceedings*, Part I and II, Lausanne, France, R. Gotthardt and J. Van Humbeeck (Eds.), Editions de Physique, *Journal de Physique*, IV, 5(8), 771–776.

68. Pu, Z., Tseng, H., and Wu, K., Martensite Transformation and Shape Memory Effect on NiTi-Zr High-Temperature Shape Memory Alloys, Florida International University, Miami, U.S. Army Res. Office, Research Triangle Park, NC, Rept. DAALO3-91-G-0245, October 17, 1995, p. 9.

69. Wu, K. H., Development of High-Temperature Two-Way Shape Memory Alloys, Florida International University, Miami, U.S. Army Res. Office, Research Triangle Park, NC Rept. DAALO3-91-G-0245, January 10, 1991 to March 31, 1995, p. 109.

70. Yang, D. Z., Pu, Z., and Wu, K. H., A Metal-Based Intelligent Composite with SMA Materials, *Proc. Fourth Intl. Conf. Adaptive Structures*, November 2–4, 1993, Cologne, Germany, pp. 405–417.

71. Fukami-Ushiro, K. L. and Dunand, D. C., NiTi and NiTi-TiC Composites, PT3: Shape-Memory Recovery, *MIT*, November 26, 1995.

72. Van Humbeeck, J., Shape Memory Materials: State of the Art and Requirements for Future Applications, *J. Phys. IV*, Colloque C5, *Suppl. au Journal de Physique III*, 7, November 1997, pp. C5-3–12.

73. Collings, E. W., The Physical Metallurgy of Titanium Alloys, in *ASM Series in Metal Processing*, 1984.

74. Ivasishin, O. M., Kosenko, N. S., and Shevskenko, S. V., Crystallographic Features of "-Martensite in Titanium Alloys, *ICOMAT-95 Proceedings*, Parts I and II, Lausanne, France, R. Gotthardt and J. Van Humbeeck (Eds.), Editions de Physique, *Journal de Physique*, IV, 5(8), 1017–1022, 1995.

75. Baker, C., The Shape Memory Effect in a Ti-35wt% Nb Alloy, *Metal Sci. J.*, 5, 92–100, 1971.
76. Sasano, H. and Suzuki, T., Shape Memory Effect in Ti-Mo-Al Alloys, *5th Proc. Intl. Conf. of Titanium*, 3, 1667–1674, 1984.
77. Hamada, T., Sodeoka, T., and Miyagi, M., Shape Memory Effect of Ti-Mo-Al Alloys, *Proc. Intl. Conf. of Titanium*, 2, 877–882, 1988.
78. Sugimoto, T., Ikeda, M., Komatsu, S., Sugimoto, K., and Kamei, K., Shape Recovery and Phase Transformation Behaviour in Ti-Mo-Al-Sn-Zr Alloys, *Proc. Intl. Conf. of Titanium*, 2, 1069–1074, 1988.
79. Sohmura, T. and Kimura, H., Shape Recovery in Ti-V-Fe-Al Alloy and Its Application to Dental Implant, *ICOMAT-86*, NARA, pp. 1065–1070.
80. Fedotov, S. G., Chelidze, T. V., Kovernistyy, Y. K., and Sanadze, V. V., Phase Transformations During Heating of Metastable Alloys of the Ti-Ta System, *Phys. Met. Metall.*, 62, 109–112, 1986.
81. Duerig, T. W., Abrecht, J., Richter, D., and Fisher, P., Formation and Reversion of Stress-Induced Martensite in Ti-10V-2Fe-3Al, *Acta Met.*, 30, 2161–2172, 1982.
82. Europaïsche Patentanmeldung 0062365, anmeldetag November 3, 1982.
83. Lei, C. Y., Wu, M. H., Schetky, L. Mc., and Burstone, C. J., Development of Pseudoelastic Beta Titanium Orthodontic Wires, *Proc., Wayman Symposium*, May 1996, K. Inoue et al. (Eds.), TMS Publications, 1998, pp. 413–418.
84. Yang, D. and Wei, Z., Dalian University of Technology, In M. Schwartz (Ed.), *Encyclopedia of Smart Materials*, Vol. 1, New York: John Wiley & Sons, 2002, pp. 551–558.
85. Barrett, D. J., Metal Matrix Composite Reinforced with Shape Memory Alloy, Dept. of the Navy, Washington, D.C., *NTIS Alert*, 96(21), November 1, 1996.
86. Sugimoto, K., Kamei, K., and Nakaniwa, M., *Engineering Aspects of Shape Memory Alloys*, T .W. Duerig (Ed.), London: Butterworth-Heinemann Ltd., 1990, pp. 89–95.
87. Jean, R. D., Wu, T. Y., and Leu, S. S., *Scripta Metall.*, 25, 883–888, 1991.
88. Igharo, M., and Wood, J. V., *Powder Met.*, 28, 131–139, 1985.
89. Janssen, J., Willems, F., Verelst, B., Maertens, J., and Delaey, L., *J. de Physique*, 43, 809–812, 1984.
90. Nakanishi, N., Shigematsu, T., and Machida, N., *Proc. Conf. International Conference on Martensitic Transformations*, Monterey, CA, July 1992, Monterey Institute for Advanced Studies, pp. 1077–1082.
91. Kim, Y. D. and Wayman, C. M., *Scripta Metall.*, 24, 245–250, 1990.
92. Tang, S. M., Chung, C. Y., and Liu, W. G., Preparation of Cu-Al-Ni-Based Shape-Memory Alloys by Mechanical Alloying and Powder-Metallurgy Method, *J. Mater. Proc. Technol.*, 63(1–3), 307–312, 1997.
93. Potluri, N. B., Joining of Shape Memory Alloys, *Welding J.*, 39–42, 1999.
94. Hsu, C.-C. and Wang, W.-H., Superplastic Forming Characteristics of a Cu-Zn-Al-Zr Shape-Memory Alloy, *Mater. Sci. Engrg. A- Structural Materials Properties Microstructure and Processing*, 205(1–2), 247–253, 1996.
95. *AWS Welding Handbook*, Vol. 4, Materials and Applications, Part 2, Miami: American Welding Society, 1998.

96. Suresh, S., Potluri, N. B., Raju, A. S., and SambaSiva R. C., Experimental Investigation on the Distribution of Residual Stresses in Weldments Associated with Fabrication-Related Processes, *Proc. Intl. Conf. Weld. Technol.*, University of Roorkee, India, 1988, pp. 125–132.

97. Potluri, N. B., Ghosh, P. K., Gupta, P. C., and Shukla, D. P., Fatigue Behavior of Thermit Welded Medium Manganese Rail Steel, *Proc. Intl Symp. Fatigue and Fracture in Steel and Concrete Structures*, SERC, Madras, India, 1991, pp. 1055–1065.

98. Hussain, H. M., Ghosh, P. K., Gupta, P. C., and Potluri, N. B., Properties of Pulsed Current Multipass GMA-Welded Al-Zn-Mg Alloy, *Welding J.*, 75(6), 209-s to 215-s, 1996.

99. Potluri, N. B., Ghosh, P. K., Gupta, R. C., and Reddy, Y. S., Studies on Weld Metal Characteristics and Their Influence on Tensile and Fatigue Properties of Pulsed Current GMA Welded Al-Zn-Mg Alloy, *Welding J.*, 75(2), 62-s to 70-s, 1996.

100. Murthy, K. K., Potluri, N. B., and Sundaresan, S., Fusion Zone Microstructure and Fatigue Crack Growth Behavior in Ti-6Al-4V Weldments, *Mater. Sci. Technol.*, 13, 503-310, 1997.

101. *AWS Welding Handbook*, Vol. 3, Materials and Applications — Part 1, 1996, Miami: American Welding Society.

102. Gale, W. F. and Guan, Y., Microstructural Development in Copper-Interlayer Transient Liquid Phase Bonds between Martensitic NiAl and NiTi, *J. Mater Sci.*, 32, 357–364, 2000.

103. Hirose, A. and Araki, N., Welding Shape Memory Alloys, *Metals*, 59(8), 61–68, 1989.

104. Nishikawa, M., Welding Shape Memory Alloys, *Bull. Metals Soc.*, 24(1), 56–60, 1985.

105. Araki, T., Hirose, A., Uchihara, M., Kohno, W., Honda, K., and Kondh, M., Characteristics and Fracture Morphology of Ti-Ni Type Shape Memory Alloy and Its Laser Weld Joint, *Materials*, 38(428), 478–483, 1989.

106. EWI Develop a Technique for Welding Nitinol Shape Memory Alloy to Stainless Steel, *EWI Insights, Technology Update*, www.ewi.org, 2001, p. 5.

107. Nishikawa, M., Welding Problems of Shape Memory Alloys, *Metals*, 53(6), 36–41, 1983.

108. Hall, T., Joint, a Laminate, and a Method of Preparing a Nickel-Titanium Alloy Member Surface for Bonding to Another Layer of Metal, U.S. Patent 5,354,623, 1994.

109. Zheng, Z. Y., Xie, Z. W., Yuan, S. G., and Chen, Z. P., *Proc. Intl Meet Adv. Mater.*, Materials Research Society, Pittsburgh, PA, 1989, p. 117.

110. Chamberlain, G., Exotic Materials Enter the Marketplace, *Design News*, p. 23, March 27, 1995.

111. Jackson, C. M., Wagner, H. J., and Wasilewski, R. J., 55-Nitinol — The Alloy with a Memory: Its Physical Metallurgy, Properties, and Applications, *NASA Report SP-5110*, Washington, D.C., 1972, p. 74.

112. Feynman, R., There's Plenty of Room at the Bottom, Annual Meeting of the American Physical Society, December 29, 1959, Caltech Engineering and Science, February 1960.

113. Lang, W., Silicon Microstructuring Technology, *Material Science and Engineering:* Reports R17, No. 1, pp. 1–55, 1988.
114. Miyazaki, S., Ishida, A., Martensitic Transformation and Shape Memory Behavior in Sputter-Deposited Ti-Ni-Base Thin Films, *Materials Science and Engineering*, A273-275, 1999, pp. 106–133.
115. Makino, E., Uenoyama, M., and Shibata, T., Flash Evaporation of TiNi Shape Memory Thin Film for Microactuators, *Sensors and Actuators*, A 71, 187–192, 1998.
116. Eucken, S., Shape Memory Effect in Alloys Produced by Meltspinning, *Progress in Shape Memory Alloys*, S. Eucken (Ed.), DGM Informationsgesellschaft, pp. 239–277.
117. Walker, J. A., Gabriel, K. J., and Mehregany, M., Thin-Film Processing of TiNi Shape Memory Alloy, *Sensors and Actuators*, A21, 243–246, 1990.
118. Buchaillot, L., Nakamura, S., Nakamura, Y., Ataka, M., and Fujita, H., Constitutive Parts of a Shape Memory Alloy Titanium-Nickel Thin Film Catheter, *Proc. of 2nd Intl Conf. on Shape Memory and Superelastic Technologies*, Pacific Grove, CA, 1997, pp. 183–188.
119. Wolf, R. H. and Heuer, A. H., TiNi (Shape Memory) Films Silicon for MEMS Applications, *J. Microelectromechanical Syst.*, 4(4), 206–212, 1995.
120. Schetky, L. Mc., The Present Status of Industrial Applications for Shape Memory Alloys, *Proceedings: Shape Memory Alloys for Power Systems*, Palo Alto, CA, 1994, pp. 4.1–4.11.
121. Lehnert, T., Tixier, S., Böni, P., and Gotthardt, R., A New Fabrication Process for Ni Ti Shape Memory Thin Films, *J. Mater. Sci. Eng. A*, 273–275, 713–716, 1999.
122. *Nickel*, 16(4), 6, 2001.
123. Brailovski, V., Trochu, F., and Leboeuf, A., Design of Shape Memory Linear Actuators, *Proc. 2nd Intl. Conf. on Shape Memory and Superelastic Technologies*, Pacific Grove, CA, 1997, pp. 227–232.
124. Morgan, N. B., The Stability of NiTi Shape Memory Alloys in Actuator Applications, Ph.D thesis, Cranfield University, U.K., 1999.
125. Stöckel, D., *Proceeding Actuator-92*, 1992.
126. Jackson, C. M., Wagner, H. J., and Wasilewski, R. J., *NASA Report SP-5110*, Washington, D.C., 1972, pp. 74, 78, 79.
127. Stikeman, A., Attractive Shapes, *Techno. Rev.*, July/August 2001, p. 32.
128. Phillips, E. H., Memory Alloys Key to "Smart Wing," *Av. Wk. & Sp. Technology*, July 22, 1996, p. 68.
129. Myose, R. Y. et al., Leading Edge Deicing, *Aerospace Eng.*, April 2000, pp. 26–29.
130. Prouty, R. W., Smart Materials and Helicopters, *Rotor & Wing*, May 1996, p. 62.
131. Wayman, C. M., *J. Met.*, June 1980, pp. 129–137.
132. Schetky, L. McD., *Sci. Am.*, 241(5), 79, 81, 1979.
133. http://www.techtran.smfc.nasa.gov/new/memmetal.html.
134. Tamura, H., *Proc.: Shape Memory Alloys Power Syst.*, Palo Alto, CA, 1994, pp. 8-3, 8-5.
135. Stöckel, D., *Adv. Mater. Process.*, October 1990, pp. 35, 38.
136. Market Forecast, *High-Tech Materials Alert*, June 21, 1996, pp. 1–4.
137. Moorleghem, W. V., *Proc: Shape Memory Alloys Power Syst.*, Palo Alto, CA, 1994, pp. 9-1, 9-3.

138. O'Leary, J. P., Nicholson, J. E., and Gatturna, R. F., in *Engineering Aspects of Shape Memory Alloys*, London: Butterworth-Heinemann, 1990, p. 477.
139. Mitek Surgical Products, Inc., Company brochure, 1995.
140. Stice, J., in *Engineering Aspects of Shape Memory Alloys*, London: Butterworth-Heinemann, 1990, p. 483.
141. http://www.nitinolmed.com/products/.
142. http://www.nidi.org/nickel/0602/5b.htm.
143. http://www.agamedical.com/patients/index.html.
144. Haasters, J., in *Engineering Aspects of Shape Memory Alloys*, London: Butterworth-Heinemann, 1990, pp. 426–444.
145. Silver, F. H., *Biomaterials, Medical Devices and Tissue Engineering*, London: Chapman & Hall, 1994.
146. Poulmaire, D., Ducos, M., Setti, Y., and Hypolite, M. P., in S. Blumm-Sandmeier, H. Eschnauer, P. Huber, and A. R. Nicoll (Eds.), *2nd Plasma-Technik Symp., Proc.*, Vol. 3, June 1991, Lausanne, Switzerland Plasma Technik AG, Wohlen, Switzerland, 1991, pp. 191–199.
147. Søballe, K., *Acta Orthopaedica Scandinavica*, Suppl. 255, 64, 1993.
148. Filip, P., Melicharek, R., and Kneissl, A. C., Hydroxyapatite Coatings on TiNi Shape-Memory Alloys, *Zeitschrift Für Metallkunde*, 88(2), 1997.
149. Shape Memory Alloys: Medical and Dental Applications, *NTIS Alert*, 96(5), 33, 1996.
150. Stöckel, D., *Proc.: Shape Memory Alloys Power Syst.*, Palo Alto, CA, 1994, pp. 1–9.
151. http://www.alphagalileo.org/, accessed September 20, 1999.
152. Whittaker, A. S., Krumme, R., Sweeney, S. C., and Hayes, J.,R., Structural Control of Building Response Using Shape-Memory Alloys, Phase I, Construction Engrg. Res. Lab., U.S. Army, Champaign, IL, *NTIS Alert*, 96(8), 28–29, 1996.
153. http://www.yet2.com/app/utility/external/indextechpak/29617, accessed July 7, 2003.
154. Cederstrom, J., *Proc: Shape Memory Alloys Power Syst.*, Palo Alto, CA, 1994, pp. 6-2 to 6-7.
155. Power Generation System Using Shape Memory Alloy, *JETRO*, May 1997, p. 2.
156. Wollants, P., De Bonte, M., Delaey, L., and Roos, J. R., Nitinol Heat Engine Conf., Silver Spring, MD, 1978, pp. 6.10 to 6.22.
157. Shape-Memory-Alloy Thermal-Conduction Switches, *NASA Tech Briefs*, February 1999, p. 64.
158. Locking Key, *Design News*, March 3, 1997, p. 70.
159. Graesser, E. J., Effect of Intrinsic Damping on Vibration Transmissibility of Nickel-Titanium Shape-Memory Alloy Springs, *Metallurgical and Materials Transactions A — Physical Metallurgy and Materials Science*, 26(11), 1995.
160. Funakubo, H. (Ed.), *Shape Memory Alloys*, New York: Gordon and Breach, 1987, pp. 201, 206.
161. Cydzik, E., in *Engineering Aspects of Shape Memory Alloys*, London: Butterworth-Heinemann, 1990, pp. 149–157.
162. Krumme, J. F., *Connection Technol.*, 3(4), 41, 1987.
163. http://www.sma-mems.com/t.film.htm.
164. Johnson, A. D. and Busch, J. D., *Proc. 1st Intl. Conf. Shape Memory Superelastic Technol.*, Pacific Grove, CA, 1994, pp. 2–304.

165. Ishida, A., Sato, M., and Takei, A., *Metallurgical and Materials Transactions A — Physical Metallurgy and Materials Science*, 27(12), 1996.

166. Bendahan, M., Seguin, J. L., and Canet, P., NiTi Shape-Memory Alloy Thin Films — Composition Control Using Optical-Emission Spectroscopy, *Thin Solid Films*, 283(1–2), 1996.

167. Miyazaki, S., Hashinaga, T., and Ishida, A., Martensitic Transformations in Sputter-Deposited Ti-Ni-Cu Shape-Memory Alloy Thin-Films, *Thin Solid Films*, 28(1–2), 1996).

168. Shape Memory Alloy Film Changing Gradually with Temperature, *JETRO*, October 1995, pp. 19–20.

169. Lee, A., Ciarlo, D., Krulevech, P., Lehew, S., Trevino, J., and Northrup, A., *Sensors and Actuators*, A54, 755–759, 1996.

170. Yoshida, H., Composite Structure with Temperature-Dependent Shape Reversal, *96-03-001-03, JETRO*, 23(12), 18–19, 1996, Nat. Insti. of Materials & Chemicals Research, AIST.

171. Kuribayashi, K. and Fujii, T., A New Micro SMA Thin Film Actuator Prestrained by Polyimide, *Proc. Intl. Symp. on Micromechatronics and Human Science*, Piscataway, NJ: IEEE, 1998, pp. 165–70.

172. Busch, J. D. and Johnson, A. D., Shape-Memory Alloys Micro-Actuator, U.S. Patent, WO 91/00608, June 27, 1989.

173. Johnson, A. D. and Martynov, V. V., Applications of Shape-Memory Alloy Thin Film, *Proc. 2nd Intl. Conf. on Shape Memory and Superelastic Technologies*, Pacific Grove, CA, 1997, pp. 149–154.

174. Kohl, M., Strobanek, K. D., and Miyasaki, S., Stress-Optimised Shape Memory Microvalves, *Sensors and Actuators*, A72, 243–250, 1999.

175. Bellouard, Y., Clavel, R., Gotthardt, R., Bidaux, J.-E., and Sidler, T., A New Concept of Monolithic Shape Memory Alloys Microdevices Used in Micro-Robotics, *Proc. Actuators*, H. Borgmann (Ed.), 1998, pp. 502–505.

176. Fabricating Composite-Material Structures Containing SMA Ribbons, *NASA Tech Briefs*, February 2003, pp. 57–58.

177. Cylindrical Shape-Memory Rotary/Linear Actuator, JSC, *NASA Tech Briefs*, October 2002, p. 30.

178. Latch-Release Pin Puller with Shape-Memory-Alloy Actuator, *NASA Tech Briefs*, July 2002, pp. 72–72.

179. Biwa, S., Yamada, K., and Matsumoto, E., Analysis of Bias-Type Using Shape-Memory Alloy Based on Its Thermomechanical Constitutive Description, *JSME Intl. Series A — Mechanics and Material Engrg.*, 39(4), 526–532, 1996.

180. Lagoudas, D. C. and Bhattacharyya, A., Modeling of Thin Layer Extensional Thermoelectric SMA Actuators, Texas A&M, *NTIS Alert*, 97(3), 1997.

181. Ullakko, K., Magnetically Controlled Shape-Memory Alloys — A New Class of Actuator Materials, *J. Mater. Eng. Perform.*, 5(3), 405–409, 1996.

182. Ullakko, K., Jakovenko, P. T., and Gavriljuk, V. G., High-Strength Shape Memory Steels Alloyed with Nitrogen, *Scr. Metall. Mater.*, 34, 6, 1996.

183. Ullakko, K., Magnetic Control of Shape Memory Effect, International Conference on Martensitic Transformations ICOMAT-95, August 20–25, 1995, Lausanne, Switzerland, Ecole Polytechnique Federale de Lausanne (Abstract).

184. Ullakko, K., Large-Stroke and High-Strength Actuator Materials for Adaptive Structures, *Proc. 3rd Intl. Conf. on Intelligent Materials and 3rd European Conf. on Smart Structures and Materiels,* June 3–5, 1996, Lyon, France, INSA de Lyon, France and University of Strathclyde, U.K., p. 6.

185. Ullakko, K., Jakovenko, P. T., and Gavriljuk, V. G., New Developments in Actuator Materials as Reflected in Magnetically Controlled Shape Memory Alloys and High-Strength Shape Memory Steels, *Proc. of Symp. on Smart Structures and Materials,* February 26–29, 1996, San Diego, CA, Intl. Society for Optical Engineering, p. 9.

186. Ding, Z. and Lagoudas, D. C., Solution Behavior of the Transient Heat Transfer Problem in Thermoelectric Shape Memory Alloy Actuators, Texas A&M, *NTIS Alert,* 97(3), 1–2, 1997.

187. Magnetic Shape Memory Alloy, *Nickel,* 17(3), 7, 2002.

188. Van Humbeeck, J., Nonmedical Applications of Shape Memory Alloys, *Mater. Sci. Eng.,* A273-275, 134–148, 1999.

189. Duerig, T., Pelton, A., and Stöckel, D., An Overview of Nitinol Medical Applications, *Mater. Sci. Eng.,* A273-275, 149–160, 1999.

190. Van Humbeeck, J. and Staimans, R., Shape Memory Alloy Systems and Functional Properties, in M. Schwartz (Ed.), *Encyclopedia of Smart Materials,* New York: John Wiley & Sons, Vol. 2, 2002, pp. 951–964.

191. Bellouard, Y., Microrobotics, Microdevices Based on Shape-Memory Alloys, in M. Schwartz (Ed.), *Encyclopedia of Smart Materials,* New York: John Wiley & Sons, Inc., Vol. 1, 2002, pp. 620–644.

192. Nix, W., Yielding and Strain Hardening of Thin Metal Films on Substrates, *Scripta Materialia,* 394/5, 545–554, 1988.

193. Mucklich, F. and Janocha, H., Smart Materials — The IQ of Materials in Systems, *Zeitschrift Für Metallkunde,* 87(5), 357–364, 1996.

194. Bard, A. J., Denault, G., Lee, C., Mandler, D., and Wipf, D. O., *Acc. Chem. Res.,* 23, 357–363, 1990.

195. Thundat, T., Nagahara, L. A., Oden, P. I., and Lindsay, S. M., *J. Vac. Sci. Technol.,* A8, 3537–3541, 1990.

196. Newnham, R. E., *MRS Bulletin,* 4, 24–33, 1993.

197. Drexler, K. E., *Nanosystems, Molecular Machinery, Manufacturing and Computation,* New York: John Wiley & Sons, 1992.

198. Aizawa, S. I., *Proc. Bionics Design Intl .Workshop,* Tsukuba, Japan, 1992.

199. DeRossi, D., in C. R. Rogersand G. G. Wallace (Eds.), *Proc. 2nd Conf. Intellig. Mat. ICIM '94,* Lancaster, Basel: Technomic Publ. Comp., 1994, pp. 44–51.

200. Unsworth, J., in C. R. Rogersand G. G. Wallace (Eds.), *Proc. 2nd Conf. Intellig. Mat. ICIM '94,* Lancaster, Basel: Technomic Publ. Comp., 1994, pp. 23–32.

201. Ogata, N., in C. R. Rogersand G. G. Wallace (Eds.), *Proc. 2nd Conf. Intellig. Mat. ICIM '94,* Lancaster, Basel: Technomic Publ. Comp., 1994, pp. 13–22.

202. Clark, D. T., in C. R. Rogersand G. G. Wallace (Eds.), *Proc. 2nd Conf. Intellig. Mat. ICIM '94,* Lancaster, Basel: Technomic Publ. Comp., 1994, pp. 3–12.

203. Liu, Y., Gall, K., Dunn, M. L., McCluskey, P., and Shandas, R., Shape Memory Polymers for Medical Applications, Dept. of Mech. Engrg., University of Colorado, Boulder, CO, *AM&P,* December 2003, pp. 31–32.

Bibliography

Adler, P., Manufacture of Nitinol Tubing, *AM&P*, August 2003, p. 53.

Bai, Y.-J, Liu, Y.-X., Geng, G.-L., and Li, L., Microstructural Origin of Degradation of Shape Memory Effect during Martensite Aging of CuZnAlMnNi Alloy, *J. Adv. Mater.*, to be published, 2005.

Dunand, D. C., Mari, D., Bourke, M. A. M., and Goldstone, J. A., Neutron Diffraction Study of NiTi during Compressive Deformation and after Shape-Memory Recovery, Intl. Conf. on Martensitic Transformations, Lausanne, Switzerland, *NTIS Alert*, 96(3), February 1, 1996.

Fukami-Ushiro, K. L. and Dunand, D. C., NiTi and NiTi-TiC Composites — Shape-Memory Recovery, *Metallurgical and Materials Transactions A — Physical Metallurgy and Materials Science*, 27(1), 193–203, 1996.

Goldberg, D., Xu, Y., and Murakami, Y., Characteristics of Ti50Pd30Ni20 High-Temperature Shape-Memory Alloy, *Intermetallics*, 3(1), 35–46, 1995.

Han, X. D., Zou, W. H., and Wang, R., Structure and Substructure of Martensite in a Ti36.5Ni48.5Hf15 High-Temperature Shape-Memory Alloy, *Acta Materialia*, 44(9), 3711–3721, 1996.

Hurlbut, B. J. and Regelbrugge, M. E., Evaluation of a Constitutive Model for Shape-Memory Alloys Embedded in Shell Structures, *J. Reinforced Plastics and Composites*, 15(12), 1249–1261, 1996.

Inagaki, H. and Inoue, K., Effect of Mn on the Shape-Memory Effect in Fe-Mn-Si-Cr-Ni Alloys, *Zeitschrift Für Metallkunde*, 85(11), 790–795, 1994.

Inque, H., Miwa, N., and Inakazu, N., Texture and Shape-Memory Strain in TiNi Alloy Sheets, *Acta Materialia*, 44(12), 4825–4834, 1996.

Investigation of Power Generation System Using Shape Memory Alloys, New Energy Dev. Org., Tokyo, Japan, *NTIS Alert*, 86(8), 1–2, 1996.

Lam, C. W. H., Chung, C. Y., and Ling, C. C., Removal of Martensite Stabilization in Cantim Shape-Memory Alloy by Post-Quench Aging, *J. Mater. Proc. Technol.*, 63(1–3), 600–603, 1997.

Lee, C. H. and Sun, C. T., The Nonlinear Frequency and Large-Amplitude of Sandwich Composites with Embedded Shape-Memory Alloy, *J. Reinforced Plastics and Composites*, 14(11), 1160–1174, 1995.

Lin, P. H., Tobushi, H., and Tanaka, K., Deformation Properties of TiNi Shape-Memory Alloy, *JSME Intl. J. Ser. A- Mech. Mater. Eng.*, 39(1), 108–116, 1996.

Pons, J. and Portier, R., Accommodation of Gamma-Phase Precipitates in Cu-Zn-Al Shape-Memory Alloys Studied by High-Resolution Electron-Microscopy, *Acta Materialia*, 45(5), 2109–2120, 1997.

Shape Memory Alloys: Medical and Dental Applications, NERAC, Inc., Tolland, CT, *NTIS Alert*, 97(7), April 1, 1997.

Shield, T. W., Leo, P. H., and Grebner, C. C., Quasi-Static Extension of Shape-Memory Wires under Constant Load, *Acta Materialia*, 45(1), 67–74, 1997.

Xu, Y., Shimizu, S., and Suzuki, Y., Recovery and Recrystallization Processes in Ti-Pd-Ni High-Temperature Shape-Memory Alloys, *Acta Materialia*, 45(4), 1503–1511, 1997.

Yoshida, H., Creation of Environmentally Responsive Composites with Embedded Ti-Ni Alloy as Effectors, *Adv. Comp. Materials*, 5(1), 1–16, 1995.

Yungang, T. and Baoqi, T., Smart Composite Damage Assessment System Based on the Neural Network, National Air Intell. Center, *NTIS Alert*, 96(14), 18–19, 1996.

5

Nanostructured Materials (N$_S$M)

Nanoscale science and engineering are the wave of scientific discovery and have relevance in almost all disciplines and areas of the economy mostly focused on fundamentally new concepts in nanomanufacturing and instrumentation.

Like any emerging technology, there are questions about societal implications. Societal implications are quite broad — it implies increasing productivity, changing health care, and expanding the limits of sustainable development of the Earth. With existing technologies, there is a limit to the number of people who can live on the Earth in a sustainable way. If we improve that technology, moving to processes with no waste, or with a small amount of materials, this limit will be extended. Also, like any new technology, there are social effects like potential negative impacts if, by accident, materials are released into the environment. All of this has to be taken into account.

We are moving ahead because all the indications so far show that the advantages of nanotechnology by far outweigh the potential unexpected consequences. For example, nanoparticles can eventually target cancer cells and visualize illnesses before they happen. At the same time, nanoparticles may have effects that are not yet fully determined. The positive aspect is that nanoparticles already exist all over the Earth. With nanotechnology, we can address current issues that cannot be solved otherwise.

At the same time, we have to develop nanotechnology in a responsible way. We have to pay sufficient attention to environmental and health issues, as well as to ethical and economic issues.[1]

It has been said that nanotechnology is as complex as the space program. The main advantage to the space program was developing knowledge about the universe. Moving to nanotechnology not only is exciting, but also will change the way we live on Earth. Both bring a lot of excitement. But one thing that is different is that in moving to the nano world, you have significant implications for living conditions on Earth and effects on health that are probably broader than any other technology developed so far. The science base of nanotechnology is developing now.

There are several phases. The first phase, which we are already in, is dealing with nanostructures and starting to have systematic control of this. The next phase, in 3 or 4 years, will be when we start to have active nanostructures such as a transistor, an actuator, an active drug-delivery device. It will be another several years before we have nanosystems.

We are moving in this direction — the only question is how far. Although the space program had a lot of knowledge development, nano-technology will have much more, besides an equal interest in knowledge and understanding once we develop the tools, and that will mark a major change in the economy.

One cannot predict exactly the time involved; many developments will be made by breakthroughs. We cannot make an exact timeline for each event. One has to keep in mind that, basically, we are still at the beginning of the road. Most of the activity in nanotechnology is empirically based — it is experimental, stemming from another field prior to nano.

The long-term vision for nanotechnology has three or four components. It is not important what is happening on the "small" end. What is important are the new phenomena, the new processes, the new functions that happen at the intermediate-length scale. It is not important to work in one field — what is important is the exchange of tools, methods, archi-tectures, and applications among different disciplines. And it is important that you have the ability to transform with a small amount of energy or materials, and to do things that are not possible at any other scale, and to do them more economically.

This is robust development; it is not science fiction or hype. The devel-opment is real and is penetrating into many fields. Expansion is expected into four new areas: food and agricultural systems, energy conversion and storage, regenerative medicine and nano-/biomedicine, and environ-mental research and education, according to Dr. Michail C. Roco.[2]

The question that now arises is, "How many atoms create a solid?" It depends on the type of material and property studied. In metal clusters, although a crystalline structure is already formed at about 100 atoms, the macroscopic melting point is only reached for clusters with 1000 atoms. In semiconductor crystals, up to 10,000 atoms are necessary to show optical excitations similar to that of the bulk. Nanocrystals are located in the regime between molecular clusters and crystals of micrometer size. They are characterized by an interior component, structurally identical to a corresponding bulk solid and a distinguished, substantial fraction of atoms on the crystallite surface. The physical and chemical properties of solid matter changes as the particle size decreases well below the micrometer regime because of quantum size effects and changes in the crystallo-graphic structure.

The scientific and technological development of nanostructures has two major approaches: superlattices and quantum well devices, and three-dimensional ultrafine, polycrystalline microstructures. The latter, also called *nanocrystalline materials*, are defined as polycrystalline solids with

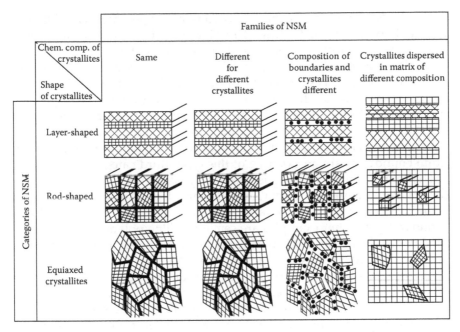

FIGURE 5.1

Classification scheme for NsM according to their chemical composition and the dimensionally (shape) of the crystallites (structural elements) forming the NsM.[3]

grain sizes of a few nanometers.[2] Grains, pores, interface thicknesses, and defects are of similar dimensions.

N_SM are solids composed of structural elements — mostly crystallites — with a characteristic size (in at least one direction) of a few nanometers. The various types of N_SM may be classified according to their chemical composition and the shape (dimensionality) of their structural constituents (see Figure 5.1).

For the sake of simplicity, we shall first focus on N_SM formed by nanometer-sized crystallites. According to the shape of the crystallites, three categories of N_SM may be distinguished: layer-shaped crystallites, rod-shaped crystallites (with layer thickness or rod diameters in the order of a few nanometers), and N_SM composed of equiaxed nanometer-sized crystallites. Depending on the chemical composition of the crystallites, the three categories of N_SM may be grouped in four families. In the simplest case (first family, Figure 5.1), all crystallites and interfacial regions have the same chemical composition. Examples of this family of N_SM are semicrystalline polymers (consisting of stacked crystalline lamellae separated by noncrystalline regions; first category in Figure 5.1) or N_SM made up of equiaxed Cu crystals (third category).

N_SM belonging to the second family consist of crystallites with different chemical compositions (indicated in Figure 5.1 by different hatchings). Quantum well (multilayer) structures are probably the most well-known examples of this type (first category). If the compositional variation occurs primarily between crystallites and the interfacial regions, the third family of N_SM is obtained. In this case, one type of atoms (molecules) segregates preferentially to the interfacial regions so that the structural modulation (crystals/interfaces) is coupled to the local chemical modulation. N_SM consisting of nanometer-sized Cu crystals with Bi atoms segregated to the grain boundaries are an example of this type (third category).

The fourth family of N_SM is formed by nanometer-sized crystallites (layers, rods, equiaxed crystallites) dispersed in a matrix of different chemical composition. Precipitation-hardened alloys are examples of this group of N_SM. Other examples are semicrystalline polymer blends solidified under shear. The N_SM considered so far consisted mostly of crystalline components. However, in addition, N_SM are known in which one or all constituents are noncrystalline. For example, semicrystalline polymers consist of alternating (nanometers thick) crystalline and noncrystalline layers. Other examples are partially crystallized glasses. Spinodally decomposed glasses represent N_SM in which all constituents are noncrystalline. Finally, crystalline or noncrystalline materials containing a high density of nanometer-sized voids (e.g., due to the α-particle irradiation) are N_SM of which one component is a gas or vacuum.[4-7]

Structure and Properties of N_SM[7]

The structure of an N_SM consists of nanometer-sized, equiaxed crystallites formed by only one kind of atom (Figure 5.1, first family/third category). An N_SM of this kind consists structurally of the following two components (see Figure 5.2): the crystallites (atoms represented by open circles) and the boundary regions (dark circles). The atomic structure of all crystallites is identical. The only difference between them is their crystallographic orientation. This is not so in the boundary regions where two crystallites are joined together. In the boundary regions, the average atomic density and the coordination between nearest neighbor atoms deviate from the ones in the crystallites. The presence of two structural components (crystals and boundaries) of comparable volume fractions and with typical crystal sizes of a few nanometers is crucial for the properties of N_SM. The properties of a crystalline solid are known to depend on (1) the size of the crystalline regions and (2) the atomic structure of the solid characterized by the average atomic density and the coordination between nearest

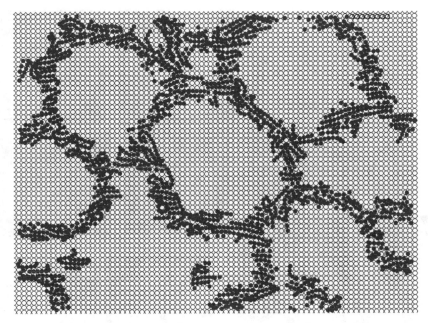

FIGURE 5.2
Computed atomic structure of an NsM. The computations were performed by modeling the interatomic forces by a Morse potential. The black (boundary) atoms are atoms the sites of which deviate more than 10% from the corresponding lattice sites.[3]

neighbors. It is exactly these two parameters that vary if one compares a single crystal of macroscopic dimensions with an N_SM with the same chemical composition. In N_SM, the crystal size is reduced to a few nanometers, and, moreover, the average density and coordination between nearest neighbors is changed. Hence, the properties of N_SM have been proposed to differ from the ones of a single crystal.[8]

Nanocrystals and Nanocrystalline Materials

Nanocrystals

Engineers at Sandia National Laboratory (SNL) are studying seashells and diatoms that pull calcium and silicate ions from cool ocean water to build hard crystalline tissues that protect them. The purpose is to view these structures (see Figure 5.3) to uncover ways to build similar structures using low-temperature, environmentally benign methods. Currently, they are using low temperatures and chemical concentrations in an aqueous

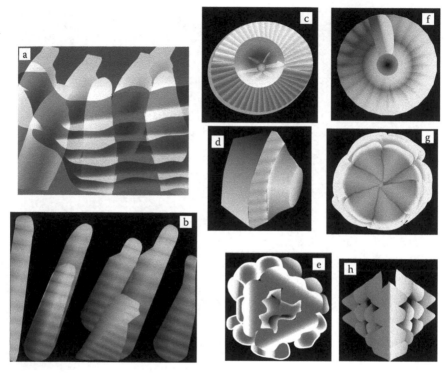

FIGURE 5.3

Researchers have made nanostructures that are strikingly similar to those in nature. For example, (a) is an acre from red abalone, (b) includes synthetic ZnO crystals, and (c) is a diatom. Images (d) to (h) are man-made silica crystals.

environment rather than the high temperatures, high concentrations, and organic solvent method more widely used to build nanocrystals. So far, the team has been able to control where and how crystals form in the early phases of their studies.

At another government agency, Argonne National Laboratory (ANL) researchers have developed a process that has resulted in the synthesis of nanocrystals: particles small enough to contain only thousands, instead of millions, of atoms.

The developed process using a physical vapor synthesis (PVS) plasma reactor is based on ANL's technique for making nanophase materials from the assembly of physical vapors into nanocrystals using a natural convection, gas-phase condensation process.

That technology is the basis for the physical vapor synthesis reactor, which uses a forced-convection process. The end result is a nanocrystal powder whose particles have weak agglomeration and clean surfaces: attributes that make the powder industrially useful.

The PVS plasma reactor is being used to make powders that serve as cosmetic pigments and sunblocks.

Scientists at Purdue University have worked for years to find a way to mass-produce tiny, unusually hard nanocrystals; they should look no farther than their local machine shop. These researchers say they found that metal shavings produced by lathes, drills, and milling machines are filled with nanocrystals, which hold the promise of superstrong mechanical parts and composite materials.

The researchers said these shavings, now routinely discarded or melted down for reuse, could be a source of the crystals at a hundredth of the cost of producing them using current methods.

Stretching or otherwise deforming a metal with intense strains breaks up this crystal structure, shrinking some crystals and making them stronger and harder than normal-size crystals.

Such strains occur when a hardened metal die is forced into a spinning metal rod at a 90° angle during the lathing process, according to engineering Professor Srinivasan Chandrasekar of Purdue University, West Lafayette, IN.

Nanocrystals of various materials have been shown to be 100 to 300% harder than the same materials in bulk form. They can help make superstrong metal parts or be added to plastics to make new types of composite structures. But current methods are expensive. In fact, nanocrystals can cost as much as $100/lb to make. By using the leftover chips, Purdue engineers say they have created a $1/lb process for producing these materials in large quantities.

In the future, metal nanocrystals from chips maybe used to economically produce longer-lasting bearings and new types of sensors and components for computers and electronics. But in the meantime, researchers must determine whether the nanocrystals found in scrap chips retain their desired properties after standard processing steps. Early lab findings have shown that they do.[9]

The engineers have measured increased hardness in nanocrystals of copper, tool steel, stainless steel, two other types of steel alloys, and iron and their data has demonstrated that these materials are nanocrystalline and that they have enhanced mechanical properties.

Further research will be needed to determine whether the nanocrystals contained in scrap chips retain their desired properties after standard processing steps. Those steps include milling the chips to make powders and then compressing and heating the powders to make metal parts. Nanocrystals currently produced in laboratories have been subjected to such processes, and they have retained their nanocrystalline properties.

Researchers at Oak Ridge National Laboratory (ORNL) have synthesized a wide range of metal and semiconductor nanocrystals embedded in materials like sapphire and crystalline silicon.

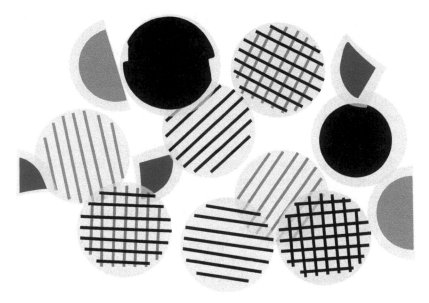

FIGURE 5.4
Brilliant colors come from nanocrystals formed in sapphire and fused silica. Encapsulated gold nanocrystals produce purple or red, while silver nanocrystals produce yellow.

The researchers hope to predict and control the size, structure, and optical/physical properties of nanocrystalline composites produced by high-dose ion implantation. Because of the unique ways these materials absorb and emit light, they could be used to develop faster, smarter computers and better flat-panel displays (see Figure 5.4).

University of Georgia researchers have discovered a way to grow tiny crystals, measured in mere billionths of a meter, which may prove useful in making both semiconductor chips and catalysts. Dubbed nanocrystals, these flyspeck crystals can contain the absolute minimum number of atoms (27) for cubic structures composed of two elements such as carbon and a metal.

The nanocrystals are produced by vaporizing titanium or other metals by zapping it with a laser, then mixing the metal vapor with methane, a gas that contains carbon. As the research proceeds, it could ultimately be possible to grow metal-carbon structures that would be far smaller than transistors on microchips. As catalysts, titanium-carbon nanocrystals should be vastly more efficient than anything now known. That is because the key to catalytic processes is the amount of surface area exposed to the reaction, and shrinking the size of catalyst molecules increases their surface-to-volume ratio. One must remember one thing: The work is very fundamental and it is not sure when, if ever, it will become practical.

Researchers at Rensselaer[10] have created large symmetrical crystals that rarely occur in nature. These crystals could be harder than conventional engineering materials. The accidental discovery was made during attempts to make superconducting nanostructures with a simple technique used to create carbon nanotubes.

Pulickel Ajayan and Ganapathiraman Ramanath, faculty members in materials science and engineering, used boron carbide, a common engineering material, in the high-temperature experiment. In the ashes, they discovered large crystals with fivefold crystallographic symmetry.

Nanosized fivefold symmetrical, or icosahedral, crystals are fairly common, but these larger micronsize crystals with fivefold symmetry are rare in nature because their smaller units cannot repeat their pattern infinitely to form space-filling structures. As the nuclei of these crystals grow, the strain on the crystals increases. This causes them to revert to their common bulk crystal structures.

Ajayan says that the inherent structure of boron carbide, which has icosahedral units in the unit cell, allows the crystals to grow to micron size without the strain. These crystals are unique due to their high symmetry, and because of the hardness inherent to the crystal structure, one could anticipate a better material for engineering, specifically coatings.

Some metals, such as certain aluminum alloys, reach superplasticity when heated to temperatures as high as 982°C. At that point, the metal stretches like plastic, enabling it to be molded into a strong, detailed part. Although the technology has won acceptance in the aerospace industry, it remains relatively unused by producers of autos and consumer goods. The problem: Superplasticity requires a long time to fabricate a part, and the temperatures are too high to form some components. Researchers at the University of California say they have overcome these drawbacks by achieving superplasticity at temperatures as low as 232°C. And they claim their technology works with a wide range of materials, including nickel and ceramics. The key, says Amiya Mukherjee, professor of materials science, was switching from materials made of microcrystals to those made of nanocrystals. (Microcrystals range from 1 to about 20 μm in diameter; nanocrystals are about 7000 times smaller.) The finished nanostructured materials also were much stronger than microstructured ones, Mukherjee reports.[11]

Nanocrystalline Materials

Using a powerful electron microscope to view atomic-level details, Johns Hopkins researchers have discovered a "twinning" phenomenon in a nanocrystalline form of aluminum that was plastically deformed during lab experiments. The finding will help scientists better predict the mechanical behavior and reliability of new types of specially fabricated metals.

At the microscopic level, most metals are made up of tiny crystallites, or grains. Through careful lab processing, however, scientists in recent years have begun to produce nanocrystalline forms of metals in which the individual grains are much smaller. These nanocrystalline forms are prized because they are much stronger and harder than their commercial-grade counterparts. Although they are costly to produce in large quantities, these nanomaterials can be used to make critical components for tiny machines called microelectromechanical systems (MEMS), or even smaller nanoelectromechanical systems (NEMS).

But before they build devices with nanomaterials, engineers need a better idea of how the metals will behave. For example, under what conditions will they bend or break? To find out what happens to these new metals under stress at the atomic level, Johns Hopkins researchers[12] conducted experiments on a thin film of nanocrystalline aluminum. Grains in this form of aluminum are 1000th the size of the grains in commercial aluminum.

Two methods were employed to deform the nanomaterial or cause it to change shape. The researchers used a diamond-tipped indenter to punch a tiny hole in one piece of film and subjected another piece to grinding in a mortar. The ultrathin edge of the punched hole and tiny fragments from the grinding were then examined under a transmission electron microscope, which allowed the researchers to study what had happened to the material at the atomic level. The researchers saw that some rows of atoms had shifted into a zigzag pattern, resembling the bellows of an accordion. This type of realignment, called deformation twinning, helps explain how the nanomaterial, which is stronger and harder than conventional materials, deforms when subjected to high loads.

"This was an important finding because deformation twinning does not occur in traditional coarse-grain forms of aluminum," said Michael Chen, associate research scientist at Johns Hopkins University, Baltimore, MD. "Using computer simulations, other researchers had predicted that deformation twinning would be seen in nanocrystalline aluminum."

By seeing how the nanomaterial deforms at the atomic level, researchers are gaining a better understanding of why these metals do not bend or break as easily as commercial metals do.

"This discovery will help to build new models to predict how reliably new nanoscale materials will perform when subjected to mechanical forces in real-world devices," says Chen.

Scientists[13] at NRL have been able to produce nanocrystalline cobalt-copper with magnetic coercivities as high as 370 Oe by reducing metal acetates in a polyol. This greatly extends the usefulness of the polyol process, which was developed in France in the early 1980s to produce micronsized powders of individual metals. Subsequent work stretched this technique to include synthesis of submicron powders. This new breakthrough permits creation of far smaller bimetallics. By combining single magnetic domains (cobalt nanograins) in a nonmagnetic matrix (copper), researchers hope to produce materials that could be used in high-density magnetic storage.

The chemical route to nanocrystalline Co-Cu has several other advantages. It eliminates boron impurities found in similar materials made by aqueous borohydride reduction of constituent salts. Chemical techniques also give researchers better control over particle composition, size, and distribution. It should also permit effective stabilization against agglomeration and cost-effective production in volume.

The powders are made by suspending cobalt (II) acetate tetrahydrate with copper (II) acetate hydrate in ethylene glycol. Refluxing the mixture at 180 to 190°C for 2 h causes the Co-Cu particles to precipitate. The powders themselves consist of agglomerated nanoscale crystallites. The face-centered cubic (fcc) copper appears to nucleate first and serves as a template for growth of metastable fcc cobalt. Annealing those powders with low cobalt contents increased both saturation magnetization and coercivity.

J. Y. Ying and colleagues[14] conducted studies of nanocrystalline CeO_{2-x}, Al_2O_3, and ZnO using Photoacoustic Fourier-Transform Infrared Spectroscopic (PA-FTIR). They evaluated nanocrystalline oxides because they are novel materials with crystal sizes of <10 nm. They give rise to unique surface structures and size-dependent properties that are useful to advanced catalytic, ceramic, and semiconducting applications.

During the PA-FTIR surface structural characterization of nanocrystalline CeO_{2-x}, Al_2O_3, and ZnO, researchers explored several structural features unique to nanocrystalline oxides: oxygen vacancies, surface adsorbates, surface phonons, grain boundaries, and quantum confinement. These components may be engineered through synthesis and processing parameters for a better understanding and utilization of the unique properties of nanocrystalline oxides. They further found that the relationships between surface, interface, bulk crystalline structure, and microstructure are emphasized for structural tailoring towards catalytic, ceramic, electronic, and optical applications.

Xiao et al.[15] examined TiAl, which has potential of being a strong, lightweight structural material. Nanocrystalline TiAl was synthesized under controlled conditions by sputtering and condensation in Ar-carrier gas and subsequent cold compaction in vacuum. The grain sizes ranged from 3 to 35 nm, with most of them measuring <10 nm.

Structure and chemical composition of individual grains were investigated by x-ray diffraction, conventional transmission electron microscopy (TEM), high-resolution electron microscopy, and microchemical analysis.

No ordered intermetallic phases were found. The majority of the grains consisted of disordered hexagonal phase with a composition close to Ti-55% at.% Al.

Mazumder et al.[16] conducted research efforts in the synthesis by laser ablation deposition of $NbAl_3$ in nanocrystalline forms, as well as process characterization and modeling. Nanocrystalline powders and films were produced under a range of process conditions. Multilayer structures, alternating Al and $NbAl_3$ films, were also fabricated. Process yields ranged over 0.1–0.3 µg/laser pulse. Variations in yield reflect different gas dynamics at different process conditions. Mean powder particle sizes ranged over 5–10 nm, increasing with gas pressure and decreasing with laser power.

Temperatures and expansion velocities were also measured via the optical diagnostic; axial expansion velocity correlates well to mean particle size.

Ying[17] also examined TiN, which is potentially very useful in demanding engineering applications due to its hardness and wear resistance. However, due to low sintering activities in microcrystalline powders, TiN has been primarily used as a coating, rather than a monolithic ceramic material. Ying outlines processing techniques by which bulk, dense, nanostructured TiN materials may be produced. A valved filter collection device added to a novel forced flow reactor used for synthesizing nanocrystalline TiN powders allows for the collection of these powders without exposing them to air. Subsequent careful powder handling procedures allow the production of dense (99%), nanostructured (grain size 250 nm) TiN through a pressureless sintering process at 1400°C.

Capacitors in which the main dielectric layers are made from sintered nanocrystalline $BaTiO_3$ have been fabricated and tested in an initially successful and continuing effort to increase energy densities, breakdown potentials, and insulation resistances beyond those of prior commercial capacitors that contain coarser-grained sintered $BaTiO_3$. This development is based on the premise that the relevant physical properties of $BaTiO_3$ grains vary with their sizes in such a way that smaller grains are better suited for use as dielectrics in capacitors.

The variations in question can be summarized as follows:

- Capacitance and energy-storage density: For reasons too complex to be explained in the limited space available, hysteretic switching of ferroelectric domains in $BaTiO_3$ gives rise to a loss of capacitance and thus a loss of incremental energy-storage density with increasing applied potential. It had been conjectured that this detrimental effect of ferroelectric-domain switching could be minimized by reducing grain sizes to the nanocrystalline range (<100 nm). Thus, it should be possible to store more energy, especially near the upper limit of applied voltage for a given capacitor.

- Breakdown potential and energy-storage density: The breakdown potential of $BaTiO_3$ or another ceramic dielectric material is related to its mechanical strength, which is approximately inversely proportional to the square root of the size of its smallest internal flaw. Inasmuch as the flaw size cannot be smaller than the grain size, it is expected that, along with mechanical strength, the breakdown potential should increase with decreasing grain size. The expected increase in the breakdown potential would contribute, along with the expected increase in capacitance, to an increase in achievable energy-storage density.

- Insulation resistance: The insulation resistance of a capacitor is quantified by measuring the direct current that it passes when charged to a steady potential. A simplified electric model of a grainy dielectric material is that of grain-boundary and grain-interior elements in series. In a nanocrystalline (grain sizes < about 100 nm) dielectric, more inherently resistive grain boundaries are present per unit thickness than are present in a coarser-grained version of the same material, and thus one expects the insulation resistance to be greater.

In preparation for testing these concepts, multilayer capacitors that contained sintered nanocrystalline dielectric layers were fabricated. The nanocrystalline dielectric materials were formulated to satisfy an Electronics Industries of America (EIA) standard, called X7R, that specifies acceptable ranges of dielectric properties as functions of temperature. Each grain of the X7R-compliant $BaTiO_3$ has a duplex microstructure comprising a lightly doped ferroelectric core surrounded by a heavily doped paraelectric shell. (The dopants are Bi, Nb, Zn, and Mn.)

Table 5.1 presents results of tests of capacitors made from one of the nanocrystalline $BaTiO_3$ formulations and of commercially available capacitors made from coarser-grained $BaTiO_3$. These results clearly indicate the superiority of the nanocrystalline $BaTiO_3$ as the dielectric material. On the basis of these results and of other observations made during the tests, it appears that in comparison with capacitors made from coarser-grained $BaTiO_3$, capacitors made from nanocrystalline $BaTiO_3$ can operate more reliably at high temperatures and high voltages, can be made smaller and lighter for a given capacitance value, and can have higher energy-storage densities and higher capacitances for a given case size.

TABLE 5.1

The Nanocrystalline-BaTiO$_3$ Capacitors Were Tested along with Commercial BaTiO$_3$-Dielectric Capacitors and Found to Be Superior with Respect to Insulation Resistance, Dielectric-Breakdown Electric Field, and Energy-Storage Density

Property		Capacitors Made from Nanocrystalline BaTiO$_3$	Commercial Capacitors Made from Coarser-Grained BaTiO$_3$
Grain Size		<100 mm	0.5 µm
Relative Permittivity		1.815	2.498
Insulation Resistance	at Temperature of 25°C	1.240 GΩ	132 GΩ
at Applied Potential of 200 V	at Temperature of 200°C	730 MΩ	138 MΩ
Dielectric Breakdown	Potential/Thickness	863 V/8.75 µm	744 V/17.3 µm
	Electric Field Equiv. to Potential/Thickness	98.6 V/µm	43.0 V/µm
Energy-Storage Density at Half of Average Breakdown Potential		3.20 J/cm^3	1.86 J/cm^3

Nanocrystalline Processing

Initially came the examination of microstructure in nanocrystalline materials of which the type, crystal structure, number, shape, and topological arrangement of phases and defects such as point defects, dislocations, stacking faults, or grain boundaries in a crystalline material was reported by Gleiter.[18]

Earlier[19] he had reported that nanocrystalline materials are polycrystalline solids with grain sizes below 100 nm. Grains as well as pores, interfaces, and other defects are of similar dimensions. This nanospecific microstructure (nanostructure) leads to chemical and physical size effects. The coincidence of methods for the synthesis of large quantities of such materials, new methods of analysis (e.g., HRTEM or STM), and the original idea of Gleiter that with the compaction of nanoparticles, novel materials with interesting microstructure and properties can be expected, provided the basis for the broad research field nanomaterials, which has increased enormously over the last 2 decades.

The synthesis of ultrafine powders is only the first important step in the production of nanocrystalline materials. Depending on the application (e.g., ion conducting membranes in fuel cells, sensors, or structural ceramics), different microstructures have to be obtained. Nanoporous structures are, for example, required for applications in ultrafiltration, adsorption, or catalysis. Ultrafine-grained structures in dense ceramics are desirable because smaller grain sizes result, for example, in improved mechanical properties and superplastic behavior necessary for net shape forming.[20-24] Nanocrystalline ceramics may also be used as intermediate products where the nanostructure is lost after the final processing steps (Allen et al.[25]). In order to exploit these nanoscale properties, nanocrystalline ceramics (<100 nm) of high density (>98%) with homogeneous microstructures are needed. In advanced ceramic processing, ultrafine powders are employed to reduce sintering temperature and time to obtain a fine-grained structure.

Sintering

Nanosized powders can be synthesized by various techniques, but the consolidation into fully dense ceramics without significant grain growth still remains a challenge. The biggest limiting factor for sintering of many ultrafine powders to full density is the presence of agglomeration (e.g., Yan[26]). During compaction, these agglomerates behave as large entities introducing a wide variation in pore size. Due to different rates of pore closure during densification, larger pores will develop into a porous microstructure.[27] The pores can only be removed at high sintering temperatures and long sintering time, resulting in exaggerated grain growth. In the processing of conventional materials, optimized firing cycles, hot pressing methods, or dopants are used to promote densification over sintering.[26-28] It has been shown that these methods also work for nanocrystalline materials; for example, a two-step sintering process and doping with Nb results in a reduction of grain size in nanocrystalline yttria,[30] as well as hot isostatic pressing in case of nanocrystalline TiO_2.[30] For nanocrystalline titania[31] and alumina[31] sintered at low temperature and high pressure, the grain growth is limited by low temperature and pressure-induced phase transformations. Sintering is the single most important step during powder processing and determines the properties of the final product. The characteristics of the starting powder has a profound influence on the processability and microstructure of the final product (see Table 5.2).

Interest in nanocrystalline zirconia ceramics with average grain sizes below 100 nm has increased during the past few years as their properties (e.g., sinterability, mechanical properties, or superplastic behavior[33]) are often significantly different and considerably improved compared to conventional zirconia ceramics with coarser grain structures. Many of the sintering studies on zirconia-based ceramics are concerned with nanopowders produced by wet-chemical methods such as precipitation of hydroxides from salts or alkoxide hydrolysis.

TABLE 5.2

Characteristics of a Powder with Ideal Characteristics for the Formation
of Dense, Nanocrystalline Ceramics[26]

Powder Characteristics	Effect
Small particle size	Low sintering temperature and times
Narrow size distribution	No anomalous grain growth, homogeneous microstructure
No agglomeration	High final density at small grain size, homogeneous microstructure
Equiaxed particle shape	Easy particle rearrangement, high final density
High purity or controlled, homogeneously distributed dopant	Grain growth inhibitor

The wet-chemical procedures require little capital investment and
enable the production of relatively large quantities of powder. However,
one drawback of this technique is that most of the as-synthesized powders
are amorphous, requiring calcination at temperatures up to 600°C prior
to compaction, although it is known that microstructural changes
(changes in pore size distribution and even densification) can start taking
place at lower temperatures in nanocrystalline ceramics.

Zirconia-based nanopowders synthesized by gas phase processes, for
example, inert gas condensation (IGC), consist of crystalline, loosely or
nonagglomerated, ultrafine particles well below 20 nm.[34-36] Therefore, IGC
powders usually show improved sinterability compared to the wet-
chemical nanozirconia powders. Nanocrystalline ZrO_2 prepared by IGC
reached full density at a final grain size of 60 nm by vacuum sintering at
975°C, while a relative density in excess of 99% with grain size of 45 nm
can be obtained by sinter-forging at 950°C under an applied pressure of
300 MPa.

CVS

Nanocrystalline powders have been produced by CVS of pure zirconia,
zirconia doped with alumina, zirconia coated with alumina,[37-39] and zirconia doped with yttria.[40]

As-synthesized CVS-ZrO_2 powder is nonagglomerated with a crystallite
size of about 5 nm; it has a narrow size distribution and high crystallinity.
Upon uniaxial compaction, a transparent green body of ultrafine, uniform
microstructure and narrow pore size distribution with pore diameters
below the grain size distribution is formed. A transparent, fully dense
ZrO_2 ceramic with a grain size of 60 nm is formed after sintering in
vacuum at 950°C for 1 h. Sintering and densification temperatures are
lower in case of vacuum sintering.

Powders of nanocrystalline zirconia doped with 3 to 30 mol% alumina have been synthesized by CVS. The microstructural development of the doped samples depends strongly on the alumina content. Sintering of zirconia samples with 3 and 5 mol% Al_2O_3 at temperatures of 1000°C for 1 h results in fully dense, transparent ceramics with grain sizes of 40 to 45 nm and homogeneous microstructures.

Ultrafine particles with a narrow size distribution and low agglomeration as produced by CVS of zirconia are necessary for the production of green bodies with small and narrowly distributed pores, which in turn is a requirement for good sinterability. The grain growth during sintering of zirconia is effectively suppressed by doping with alumina or yttria.

The CVS process enables the control of the distribution of the elements at the time of synthesis at the molecular level. CVS offers the potential to design the microstructure on the nanometer scale and tune properties like sinterability or dispensability. The combination of advanced methods of structural analysis can resolve the details of the ultrafine particles in powders and sintered materials, provide information for improved models of the CVS process, and lead to optimized processing methods and nanostructured products.

Powder quality has been shown by Winterer[41] to be a crucial factor for the production of a particular "nanomicrostucture." CVS allows the control of powder quality on the molecular level. However, the homogeneous distribution of dopants is a challenge because the number of dopant atoms at the percentage level in a particle that is 5 nm or smaller has to be controlled on the order of a few atoms per particle. Segregation and precipitation occurs at the local, molecular level, and often impurities are also involved.

In summary, the following four approaches for generating N_SM have been applied so far: (1) The first approach involves a two-step procedure consisting of the production of isolated, nanometer-sized crystallites with uncontaminated free surfaces followed by a consolidation process (with or without elevated temperatures). Various methods have been devised to produce the isolated nanometer-sized crystallites (e.g., inert gas condensation, precipitation from solutions, or decomposition of precursor compounds). Depending on the details of the production process, the nanometer-sized crystallites obtained may be chemically identical or chemically different, or they may be coated at their free surfaces. Surface coatings have been achieved by PVD, by CVD, or by exposing the crystallites to sonic reactive atmosphere (e.g., oxygen). (2) The second approach involves the introduction of a high density of crystal defects (e.g., dislocations, grain boundaries etc.) into a formerly perfect (or nearly perfect) single crystal by heavy deformation (e.g., by ball milling, extrusion, shear, wear, or high-energy irradiation). (3) The third approach is based on crystallization from unstable states of condensed matter. So far, crystallization from glasses or from undercooled melts and precipitation from supersaturated solid or liquid solutions have been utilized. (4) N_SM may be generated by depositing atoms (molecules) on suitable substrates (e.g., by CVD, PVD, electrochemical methods, or precipitation reactions from dilute solutions). If chemically different atoms (molecules) are deposited simultaneously or consecutively, nanocomposites belonging to the second, third, or fourth family (see Figure 5.1) are obtained. Numerous variants of these production methods have been developed in recent years to account for the chemical peculiarities of specific alloy systems and for other factors such as reactivity, production rate, and grain size. In fact, the development of technologically optimized production procedures seems crucial for technological applications of N_SM when large quantities are required at low costs.

Nanocomposites

Processes to make new ceramic and metallic materials are many and depend on the final desirable structure and macroscopic shape to be obtained. For example, SiC can be used as an abrasive, refractory, structural material or a semiconductor with a negative temperature coefficient. Processes developed to produce such materials are extremely diversified. A process is selected based on the required properties and structural shape of the parts to be obtained. As an example, the preparation of carbon fibers, which need specific polymer precursors for spinning followed by heat and oxidation treatments, had been adapted to the preparation of SiC fibers. Other processes of transformation, descended from physics or chemistry domains, are CVD and chemical liquid deposition. They are able to present improved mechanical, thermal, or electrical properties, not solely depending on a peculiar chemical composition but on the arrangement of the crystalline phases or the size of the grains. Nanocomposites constitute a new class of extremely diversified materials that have appeared in the last 2 decades.

Ceramics

Researchers Niihara et al.[47] and Sajgalik proposed a classification into two and three main systems:

- Micronanocomposites in which the matrix is constituted by micrometric crystals embedded into a nanometric phase. Ceramic composites can be divided into two types: microcomposites and

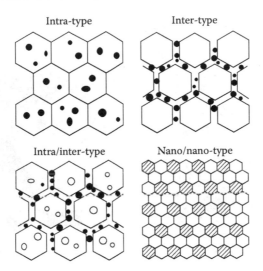

FIGURE 5.5
Schematic drawing of ceramic nanocomposites.

nanocomposites. In the microcomposites, microsize second phases such as particulate, platelet, whisker, and fiber are dispersed at the grain boundaries of the matrix. The main purpose of these composites is to improve the fracture toughness. On the other hand, Niihara et al.[47] distinguishes three different configurations: intragranular, intergranular, intra/intergranular (see Figure 5.5).

- Nano/nanocomposites in which the nanometric grains of the matrix and the second phase are spread uniformly (see Figure 5.6).

What makes a nanocomposite especially interesting is that at least one of its phases has dimensions in the nanometer range (10–100 nm). In this range, chemical and physical interactions have critical length scales, and if the nanoscale building block is made smaller than this, the corresponding fundamental properties can be changed. An example of this is light scattering, where by controlling the size of pores and nanocrystals in the range of 8 nm, transparent ceramics have been produced in the visible domain. Other enhanced properties have been developed (superplasticity, magnetoresistance, low-temperature densification, enhanced and finer homogeneity, etc.).

In the intra- and intergranular nanocomposites, the nanosized particles are dispersed mainly within the matrix grains or at the grain boundaries of the matrix, respectively. The aim of intragranular particles is the dislocation, generation, and pinning during the cooling down from fabrication temperatures or the in situ control of size and shape of matrix grains. The former role of nanodispersoids is important for the oxide ceramics such

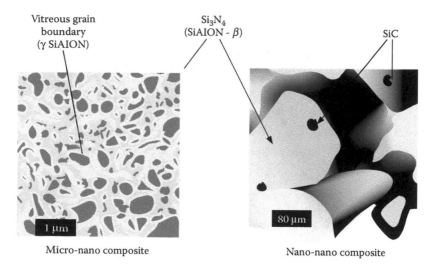

Micro-nano composite   Nano-nano composite

FIGURE 5.6
Ceramic/ceramic micro-nano and nano-nano composites' aspects.

as Al_2O_3 and MgO, which become ductile at high temperatures, while the latter role is significant for the nonoxide ceramics such as Si_3N_4 and SiC with strong covalent bonding even at high temperatures. The intergranular nanodispersoids must play important roles in the grain boundary structure control of Si_3N_4 and SiC, which gives the improvement of high-temperature mechanical properties. In other words, the aim of nanodispersoids is to improve the mechanical properties at both room and high temperatures. On the other hand, the nano/nanocomposites are composed of the dispersoids and matrix grains both of nanometer size. The primary purpose of this composite is to add the new functions such as machinability and superplasticity, like metals, to ceramics.

Based on the linear fracture mechanics of brittle materials and the micro- and nanostructure in Figure 5.5, the inter- and intragranular nanodispersoids will not give the remarkable improvement of fracture toughness, although other mechanical properties such as hardness, strength, and creep resistance are expected to be significantly improved by the inter- or intragranular nanocomposite technology. Therefore, it is necessary to combine the nanocomposite with the microcomposite for designing the supertough and superstrong ceramics (i.e., the nanocomposites reinforced with particulate, whisker, platelet, and long fiber in the microsize).

Sintering

The sintering methods are most promising for the structural ceramics. The optimum sintering processes for nanocomposites were determined by

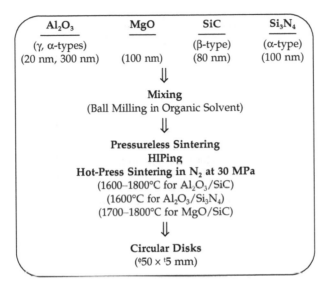

FIGURE 5.7
Fabrication processes for Al_2O_3/SiC, Al_2O_3/Si_3N_4, and MgO/SiC nanocomposites by sintering methods.[47]

transmission electron microscopic observation of composites sintered at various conditions and by the in situ observation of sintering processes using high-voltage and high-temperature TEM.[48] Finally, the ceramic-based nanocomposites were successfully prepared by the usual powder metallurgical techniques such as pressureless sintering, hot-pressing, and HIPing in the Al_2O_3/SiC, Al_2O_3/Si_3N_4, Al_2O_3/TiC, Y_2O_3/SiC, MgO/SiC, mullite/SiC, B_4C/SiC, B_4C/TiB_2, SiC/amorphous SiC, and Si_3N_4/SiC systems. The Si_3N_4/SiC nanocomposites were prepared by the liquid-phase sintering developed for the monolithic Si_3N_4 ceramics, and the solid-state sintering was applied for the other composite systems. Figure 5.7 shows the fabrication processes of typical oxide-based nanocomposites. The mullite/SiC nanocomposites in this work were prepared by the pressureless reaction sintering of natural kaolin-Al_2O_3-SiC powder mixtures. In these oxide-based composites, the commercially available powders were selected as the starting materials. But the special amorphous Si-C-N precursor powder converts to the nanosized Si_3N_4 and SiC powders by the high-temperature heat-treatment during sintering.[49] As the sintering additive, the 8 wt% Y_2O_3 was mixed with the amorphous Si-C-N powders with various carbon contents. In this fabrication process, the starting materials are only different from the processes of the monolithic Si_3N_4 ceramics.

In the ceramic/metal system, the nanosized metal-dispersed Al_2O_3 nanocomposites were successfully prepared by sintering of Al_2O_3 and fine

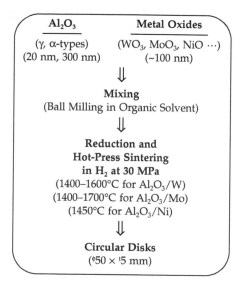

FIGURE 5.8
Fabrication processes for ceramic/metal nanocomposites by reduction/sintering methods.[47]

metal powder mixtures and by reduction/sintering of Al_2O_3 and metal oxide mixtures. Figure 5.8 shows the fabrication processes of Al_2O_3-based metal nanocomposites such as Al_2O_3/W, Al_2O_3/Mo, Al_2O_3/Ti, and Al_2O_3/Ni.[50–52]

K.N.H. Uchida of Mitsushita Electric and Professor Niihara of Osaka University[53] have developed new nanocomposites in which nanometer-sized, microfine molybdenum particles are uniformly dispersed in a matrix of yttria/partially stabilized zirconia (Y-PSZ or Y-TZP). The addition of nanomicroparticles of molybdenum to comparatively strong and tough PSZ formed the composite, and the material manifested strength and toughness equivalent to those of superhard alloys and corrosion resistance nearly equal to that of ceramics.

Matsushita Electric has used the composite for electric tools' cutting edges and barber's equipment, including electric razors.[53]

The research group in Reference 53 selected molybdenum for the metal component because of its high melting point, approximately 2600°C. In superhard alloys, metallic properties, as opposed to ceramic, are dominant, because their metal component, cobalt, with a low melting point of 1500°C, tends to cover the ceramic grain boundaries. Therefore, superhard alloys have excellent toughness, but poor corrosion resistance. By using a high-melting metal, the group was able to overcome the poor corrosion resistance, the weakness of superhard alloys. Thus, molybdenum was chosen because its readily wettable, microfine powder was commercially available.

The mechanical properties of the newly developed composite was strongly influenced by the amount of added molybdenum. The composite's bending strength increased as the molybdenum addition increased from 0–30 vol%, above which the strength did not change significantly. The composite's toughness, on the other hand, hardly increased until the molybdenum amount reached 30 vol%, and rapidly increased as soon as the molybdenum amount exceeded 30 vol%. The explanation: "As long as the molybdenum amount is 30 vol% or less, the composite's strength improvement is governed by the nanoparticles."[53]

When the nanoparticles precipitate inside zirconia particles, the structure of zirconia becomes finer, and the composite's strength increases. On the other hand, as more molybdenum is added, molybdenum in the grain boundaries begins to grow to become sub-µm-sized particles; hence, greater toughness. The composite's toughness becomes high because molybdenum in the boundaries can prevent the progress of any cracks that have been generated in the sinter, in addition to the PSZ's phase-transition reinforcing mechanism.

The composite's mechanical properties, when 50 vol% molybdenum was added, were determined as follows: Bending strength was 1800 MPa and toughness was 17.5 $MPa/m_{1/2}$. These values were 1.8 and 4.2 times, respectively, the values with no molybdenum added. According to the group: "We have never seen a nanocomposite that has simultaneously manifested bending strength of 1000 MPa and toughness of 10 $MPa/m_{1/2}$." The group confirmed the excellent corrosion resistance of the composite with a saltwater spray test; thus, the group concluded that the material should have no problem in an ordinary environment.[53]

Other Processes

Solid-State Thermolysis of Polymer Precursors

Under controlled atmosphere, pyrolysis of polymers, such as polysilazane and polycarbosilane, is an efficient process for preparation of ceramics. The in situ crystallization of these materials permits the preparation of nanocrystalline materials by a completely powder-free process. The structure and composition of the grain boundaries that originate from these crystallization processes strongly depend on the molecular structure of the initially used precursors. Such processes are particularly interesting in the preparation of fibers.

Powder-Making

To obtain nanocomposites, it is necessary to prepare nanopowders so as to increase the reactivity of the different phases during the sintering process. This is due to a high surface-to-volume ratio (generally diameter <100 nm). To obtain such powders, which have to present ideal characteristics (nanometric in size, spherical in shape, monodisperse, low agglomerated, high or controlled purity), it is necessary to develop new processes. These new methods must be innovative and need to be adapted to industrial production.

Such recently developed processes deal with reactions such as

- Solid phase: Evaporation-condensation, laser ablation, ball milling (or mechanical alloying), self-propagating high-temperature synthesis (SHS).

- Liquid phase: Sol-gel, spray-drying of solutions, aerosol pyrolysis.[54]

- Gaseous phase: Chemical vapor condensation at low pressure, plasma and laser synthesis.[55–58]

Explosive Energy

The Munitions Directorate of the U.S. Air Force Research Laboratory (AFRL)[59] recently demonstrated the capability to produce nanodimensional explosive composites that enhance explosive performance with the added feature of improved handling safety. Researchers fabricated the nanocomposite material using a novel synthesis method that allows for the formation of high-energy agglomerate particles containing nanoaluminum fuel coated with nanocrystalline high explosive. This concept (see Figure 5.9) allows intimate mixing of the reacting explosive coating with the high-density, high-energy fuel core, thus taking advantage of the shorter reaction pathway.

These explosive agglomerate particles exhibited improved safety properties during initial safety handling testing. The improved safety handling

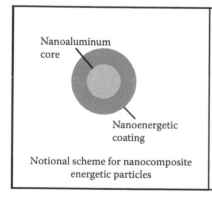
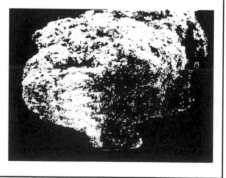

FIGURE 5.9
Concept of composite nanoparticle and actual electron microscope photo of a composite nanoparticle aggregate.

properties are directly attributable to the inherent strength properties of the nanosized components. Researchers characterized the nanocrystalline structures of the nanocomposite materials using a powder x-ray diffractometer, which enables evaluation of structures in the nanometer length scale. By applying this analytical technique to the material, researchers can understand the structure of the materials present and validate their condition after processing.

Initial dynamic testing of these nanometric materials produced encouraging results. Researchers measured increases in detonation velocity and pressure as predicted for these small-length scale, large surface area nanometric materials. Shock sensitivity measurements indicate greatly improved response to high-pressure shock loads.

The slow energy release properties of using high-energy-containing metal-oxidizer, solid-state systems historically obstructed the promise of delivering high energy in high-density form. Even though these systems contain up to 4 times the energy content of traditional explosive formulations, it is only recently, with the development of methods for producing nanosized metal particles, that this energy could be released on a useful time scale. Nanotechnology has given researchers the ability to increase the surface areas and reduce the reaction path lengths of these solid particles by three orders of magnitude. These increases dramatically improve the reaction rates and bring these solid-state particle systems into the family of high-energy-density systems that can deliver useful explosive power.

Polymer/Clay

Polymer/clay nanocomposites promise a veritable explosion of thermoplastic polymer applications because they are lightweight materials that rival metals in stiffness and strength while providing enhanced gas-barrier characteristics, superior dimensional stability, and high heat-distortion temperatures: all with low mineral loading and virtually no loss in impact resistance.[60] However, the material has a significant drawback: the high cost and difficulty of dispersing nanometer-scale smectite clay particles in plastic matrices. This has delayed the onset of a materials revolution, with the result that nanocomposites today are found in only a few niche products.

Now, a manufacturing method recently developed at ANL[60] promises to drastically reduce nanocomposite manufacturing costs by eliminating several processing steps and avoiding the need for expensive coupling agents in most cases.

In current technology, uniform dispersion of nanometer-scale clay particles in polymer matrices has always been problematic. Smectite clays have extremely large surface areas (e.g., 750 m^2/g), which means that particle behavior is dominated by a complex interplay among surface-acting chemical forces. Additionally, smectite clay surfaces are naturally hydrophilic and must be made hydrophobic to become compatible with olefinic resins.

The most common approach to this challenge is based on 50-year-old technology, which relies on surfactants that permit organoclays to function as rheological control agents in oil-based paints, inks, and greases. Dependence on these agents has greatly boosted manufacturing costs and negatively affected chemical resistance, gas-barrier characteristics, and other end-product properties.

The ANL manufacturing approach begins by dispersing clay particles in water and mixing them with surfactants (see Figure 5.10). After draining off the water, the surface-treated organoclay slurry is filtered through a high-pressure filter press to reduce water content in the resulting cake to approximately 40–60%.

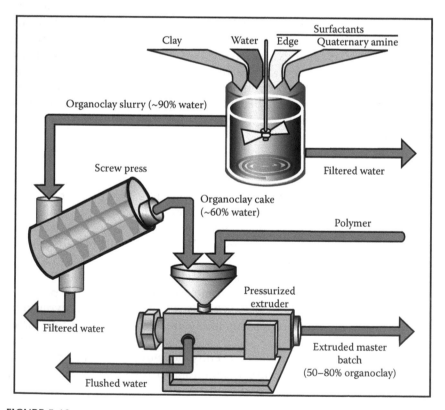

FIGURE 5.10

Argonne process technology includes proprietary surfactant chemistry, which enables the organoclay to be simultaneously dewatered by a flushing operation while preparing a nanocomposite concentration.[61]

Each organoclay particle is made up of a stack of clay platelets. The basic challenge in making a nanocomposite is to separate the clay platelets from one another to such an extent that polymer molecules may be intercalated between the platelets. Clay particles in which this has been achieved are said to be "exfoliated." The next step begins the exfoliation process by mixing the organoclay cake with the polymer in a pressurized extruder, where no coupling agents whatsoever are used.

The remaining water is flushed from the filter cake and removed from the extruder in the liquid state, thereby avoiding the expense of providing heat for evaporation. This step replaces the energy-intensive drying and crushing steps of conventional manufacturing methods. The resulting nanocomposite master batch is an intercalated (but not fully exfoliated) material that has anywhere from 50 to 80% organoclay. The material is not fully exfoliated because of the high clay loading, but it has been primed by incorporation of polymer particles between the clay platelets.

These polymer particles are key to the process. Normally, polymers are incorporated very slowly, with the rate-controlling step involved with establishing a foothold for polymer entry. However, the new process enables this first step to proceed very rapidly, resulting in a master batch that is a highly concentrated, intercalated, cost-effective, pelletized product.

A fully exfoliated nanocomposite material may then be produced by simply mixing the master batch pellets with a suitable polymer in a conventional extruder or a Banbury-type mixer. The organoclay concentrate can be handled in much the same way that die concentrates are handled today by conventional plastics processing technology.

Transparent nanoclay dispersions have been produced in a variety of thermoplastic polymers, such as polyethylene, polypropylene, and ethylene-propylene copolymers, as well as in the elastomers polyisoprene and polybutadiene. The flushing technique is also suitable for low-molecular-weight materials such as oligomers, plasticizers, and monomers.

Most ANL research has concerned polymer/clay nanocomposites based on polyolefins and elastomers. Less work has been done with fluoropolymers, even though some commercial interest has been expressed in this area. Thermoset polymers are also attractive candidates for the process. One potential problem, of course, is that organoclay surfactants tailored to facilitate exfoliation in the monomers may not necessarily be suitable for the polymerized material. This does not appear to be a problem with epoxy resins. The organoclay master batch process has been successful for epoxy oligomers, which are the basis for epoxy resins.[61]

In a recent development program at NASA-LRC,[62] a novel class of polymer/clay nanocomposites has been invented in an attempt to develop transparent, lightweight, durable materials for a variety of aerospace applications. As their name suggests, polymer/clay nanocomposites comprise organic/inorganic hybrid polymer matrices containing platelet-shaped clay particles that have sizes of the order of a few nanometers thick and several hundred nanometers long. Partly because of their high aspect ratios and high surface areas, the clay particles, if properly dispersed in the polymer matrix at a loading level of 1 to 5 wt%, impart unique combinations of physical and chemical properties that make these nano-composites attractive for making films and coatings for a variety of industrial applications. Relative to the unmodified polymer, the polymer/clay nanocomposites may exhibit improvements in strength, modulus, and toughness; tear, radiation, and fire resistance; and lower thermal expansion and permeability to gases while retaining a high degree of optical transparency.

The clay particles of interest occur naturally as layered silicates. In order to fully realize the benefits of a polymer/clay nanocomposite, it is necessary that the clay particles become fully exfoliated (delaminated) and uniformly dispersed in the polymer matrix. Concomitantly, it is necessary to maintain the exfoliation, counteracting a tendency, observed in prior formulations of polymer/clay nanocomposites, for dispersions of exfoliated clay particles to collapse back into stacked layers upon thermal treatment. One reason for the difficulty in achieving and maintaining exfoliation and uniform dispersion is the incompatibility between the silicate particle surfaces (which are hydrophilic) and the polymer matrix (which is hydrophobic).

Figure 5.11 depicts a process for making a polymer/clay nanocomposite film according to the invention. In one of two branches of the first step, a hybrid organic/inorganic matrix resin in a sol-gel form is prepared. The organic precursor of the hybrid is a compound or oligomer that contains

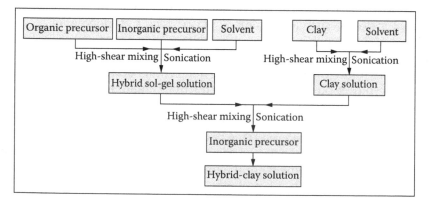

FIGURE 5.11

This flow diagram shows the major steps of a process for making a film of an organic/ inorganic hybrid polymer/clay nanocomposite.[62]

both a cross-linkable functional group (e.g., phenylethynyl) and an alkoxy-silane group. The inorganic precursor of the hybrid is also an alkoxysilane. Both precursors are mixed with a solvent to form the sol-gel matrix resin. In the other branch of the first step, a clay solution is prepared by initially dispersing layered clay particles in the same solvent as that used to form the sol-gel matrix resin. To achieve intercalation of the solvent into the stacked layers, the mixture is subjected to high-shear mixing and ultrasound. Suitable clays to provide compatibility to the organic polymer include chemically modified organophilic cation-exchanged smectite-type clays and synthetic clays having hydroxyl functional groups on the edges or elsewhere on the surfaces of the particles.

In the second step of the process, the clay solution is added to the hybrid sol-gel solution and the resulting mixture subjected to high-shear mixing and ultrasound. The hydroxyl groups of the organic/inorganic hybrid react with hydroxyl groups on the surfaces and the edges of the exfoliated clay particles, forming covalent or hydrogen bonds that enhance exfolia-tion in the presence of high shear.

The third step involves a film casting process. For example, to make an unoriented film, one begins by simply casting the solution onto a glass plate or other suitable clean, dry surface. The solution is allowed to dry to a tack-free film in ambient, desiccated air, then further dried and cured in flowing heated air. During this thermal treatment, the remaining silanol groups of the hybrid undergo condensation reactions, forming a molec-ular network that prevents the reunion of the exfoliated particles into stacked layers. In addition, the organic matrix can be consolidated with further cross-linking among the cross-linkable functional groups during thermal cure. It may be possible to prepare oriented films and fibers by using shear, drawing, and fiber spinning processes.

Polymer/Silica

Optically transparent materials are needed for the low-cost packaging of photonic components and modules. Critical properties required for those packaging materials include controlled refractive index, low cure shrink-age, and a coefficient of thermal expansion matched to that of the associ-ated photonic components.

Conventional polymeric optical packaging materials tend to have high shrinkage and large CTEs, limiting their applications. One way to reduce the CTE and minimize associated problems is to incorporate glass parti-cles into the polymer. Addition of a glass filler, however, usually causes light transmittance to be lost — a result of the difference between the refractive index of the particle and that of the epoxy matrix.

Some researchers have reported that when the refractive index difference is reduced to the order of 10^{-3}, optical transparency can be achieved.[63–65]

Recently, nanometer-sized silica particles have been used instead of the typical micron-sized particles.[66] Here, Shi, Zhou, and Edwards[67] introduced a new type of optical nanocomposite for photonic packaging and device applications. The CTE and the light transmittance of these optical nanocomposites were determined as a function of filler fraction.

The matrix of the nanocomposites was an epoxy-resin system. The resin used in the experiments was a clear-grade diglycidal ether of bisphenol A (BADGE). The hardener was substituted hexahydrophthalic anhydride; a tertiary amine salt was used as a catalyst. The nanometer-order silica filler had a narrow particle-size distribution with a most populated size of approximately 25 nm. The maximum particle size was 50 nm. The nanoparticles were predispersed in BADGE.

The bulk material was prepared by mixing the epoxy resin and the predispersed nanometer-order silica with a stoichiometric quantity of hardener and catalyst. The resulting composite was deaired and frozen at –40°C until use.

When samples of the composites were placed on characters printed on paper, the characters were legible, suggesting that the composites transmit light well in the visible wavelength region. A spectrophotometer was used to determine the light transmittance over a range of 300 to 900 nm. Generally, addition of fillers reduces the light transmittance; however, the magnitude of the reduction is wavelength dependent. The filler addition caused light transmittance to be reduced more for shorter wavelengths.

The spectral transmittance of the nanocomposites as a function of filler content slopes more steeply at shorter wavelengths; as the wavelength becomes longer, the particle size of the filler becomes much smaller than the wavelength, causing much less reduction in light transmission. Another interesting experimental finding was worthwhile to mention: As the filler content was increased beyond 10%, no additional reduction in light transmission was observed.

Adding a nanometer-order silica filler to an epoxy matrix is a feasible way to make optical nanocomposites for packaging applications, especially in light of the discovery that increasing filler content beyond 10% by weight does not seem to further drop the light transmittance. There are many other opportunities in this field, such as the *in situ* synthesis of nanoparticles.[68, 69]

Laminated

Haseeb, Celis, and Roos[70] designed, developed, and constructed an automated electrochemical setup for the deposition of nanocomposites consisting of alternate stacks of very thin laminates of a few nanometers in thickness. The setup used two electrochemical cells based on an impinging electrolyte jet, which provided uniform mass transfer at the substrate during deposition. The setup avoided the mutual contamination of the baths by using a novel gas curtain system and intermediate rinsing. Using the setup, Cu/Ni nanocomposites with individual laminate thickness of 25 nm were deposited. Structural investigations revealed the presence of distinct layers of uniform thickness. Exposure of the substrate to pressurized air used in the gas curtain system did not cause any disruption in the morphology as revealed by SEM.

Extrusions

D. Powell, NASA-GSFC,[71] has had a task to develop more formally some of the composites to be employed on the Titan Orbiter Aerorover Mission (TOAM). The research was targeted at replacing graphite epoxy with a very lightweight and flexible material that exhibits nearly the strength of steel, but also produces radiation shielding and electromagnetic shielding. This research took carbon nanotubes and oriented them in a polymer substrate, and then cross-linked them chemically with that substrate.

The results produce the strength characteristics of the carbon nanotubes with enhanced electrical and thermal conductivity. Also, there is a significant weight reduction per unit strength and, obviously, added flexibility, which currently is not available in graphite epoxy.

"This enabling technology — it's good for everything from car fenders to parachutes to bulkheads on airplanes. It's a basic material that is very strong, lightweight, and flexible," according to Powell.

The Hubble Space Telescope application involves employing the first useful nanotechnology in space. This was putting an improved thermal interface on the Space Telescope Imaging Spectrometer (STIS) instrument, which was one of the four instruments on Hubble, Powell says. The mission need was to extract more heat than the current radiator allowed in the STIS instrument, but engineers could not afford any forceful contact as well as could not add any hardware that would, in a significant way, contact the STIS instrument.

Powell decided, once again, to employ carbon nanotubes, orienting them in an aligned array, sort of like the bristles of a toothbrush in which they are all aligned vertically and standing in very closely packed formation. What this allows NASA to do is increase the net thermal contact area with the existing radiator. The thermal interface accomplishes this by pressing the carbon nanotubes into the surface defects of the existing radiator — the little crevices, joints, and valleys micrometers high — reducing the net thermal surface area. It is sort of like shoving your toothbrush up against a stucco ceiling — you could touch more places than if you had a flat blade.

Nanomanufacturing

Like many things bearing the prefix "nano," nanomanufacturing — also known as nanofabrication, nanomachining, and nanoprototyping — is pushing the boundaries of what the laser industry has typically considered materials processing. In the process, it is providing some of the first true commercial applications for lasers and optics in the nanoscale realm.[72]

Laser processing is a well-established technique for micromachining, thin-film synthesis, and microdevice prototyping, but it is also showing unique capabilities for nanoscale synthesis and processing of materials. Much of what has been done so far in nanomanufacturing (typically defined as <100 nm) leverages existing processes such as photolithography or takes concepts such as laser sintering into new dimensions for applications in microelectronics, optical communications, solar panels, optical storage, biofluidics, and semiconductor manufacturing.

In fact, the basic lithographic concepts used in semiconductor manufacturing should carry over to nanomanufacturing, especially with the never-ending push to create smaller and smaller form factors on semiconductor chips and integrated circuits.

As a matter of fact, this transition is already happening. Laser nanopatterning and nanolithography by tip-enhanced laser irradiation have been reported on since the late 1990s. By combining laser processing and scanning-probe microscopy, researchers have etched lines with widths as small as 30 nm.

Currently, "The focus is on patterning to carry current lithography approaches to the micro- and nanoscales using long-exposure lasers, such as the excimer, which would be useful for parallel patterning of entire masks at the same time," claims Haris Doumanidis, program manager for the National Science Foundation's Nanomanufacturing Program (Arlington, VA). "While the applications of lasers and energy sources in general are for two-dimensional patterning of surfaces to create features with some functionality, this same process can be structured into three dimensions through layering."

Nanoscale optical lithography techniques have already found their way into the commercial sector. Engineers, for example, are using wafer-based nanofabrication techniques to create a new class of optical components: nano-optics, which feature physical structures far smaller (<250 nm) than the wavelength of light used to create them.

The fine-scale surface structures of nano-optics interact with light according to novel physical principles, yielding new arrangements of optical processing functions with greater density and more robust performance. In addition, the small dimensions of nano-optics allow multilayer integration, yielding complex optical components "on a chip" with a broad range of applications and the ability to quickly modify component design, giving a whole new meaning to the concept of rapid prototyping.

Like its counterparts in semiconductor manufacturing, the nanopattern transfer process involves several steps: constructing a mold inscribed with the complement of the desired nanopattern, transferring the pattern to a resist layer on a prepared substrate, and selectively removing the resist with reactive ion etching to transfer the nanopattern to the target material layer on the substrate. Various techniques, including holographic lithography and electron beam lithography, can be used to create the desired negative image of a nanostructure.

Many nano-based R&D efforts are recognizing that, as with many laser-based applications, the materials being manipulated are a critical consideration in the choice and design of the instrumentation.

In line with this trend, researchers at the University of California at Berkeley[72] have developed a laser approach for curing gold nanoparticles that saves time and money in the production of integrated circuits. Using an argon-ion laser in conjunction with drop-on-demand ink-jet technology to deposit liquified gold on a substrate, Costas Grigoropoulos and colleagues in the mechanical engineering department at the University of California at Berkeley have demonstrated the ability to print two-dimensional micropatterns on thin films in about 5 min. The nanoparticles are suspended in a solvent and directly heated by the laser; as the solvent evaporates, molten gold nanoparticles agglomerate, resulting in the formation of a continuous gold line The radiation of the argon-ion laser at 488 nm was shown to be entirely absorbed within a depth of about 1 μm below the surface of the gold nanoparticle solution.

According to Grigoropoulos, working with nanoparticle suspensions offers many advantages compared to conventional subtractive integrated circuit processes, including saving expensive materials by depositing material only at desired places on a substrate and working with gold at room temperature rather than in a molten state.

"The thing you cannot forget is that the properties of the nanoscale, such as the melting point and boundaries between phase transformation, are different," Grigoropoulos says. "The research we have been pursuing is on the lower melting point of metal nanoparticles, which will enable the sintering of these nanoparticles and the writing of electronic conducting patterns at temperatures that are relatively low and compatible with plastic substrates."

"The main issue with using laser energy for processing at the nanoscale is diffraction limits," Doumanidis says. "The wavelength is usually at 100 μm of scale, which is too large for processing at the nanoscale. But we are coming up with new ways to work around this problem, such as near-field optics, multiphoton techniques, and plasmonic patterning."

Nanocomposite Applications and Directions

Automotive

A new process could make nanocomposites feasible for parts such as body panels. The technology uses sound waves to increase the compatibility between microscopic reinforcement materials and plastic resin used to make nanocomposite parts.

Mixing solid microscopic particles of smectite clay with plastic resin creates the so-called nanocomposites. Automotive manufacturers prefer the tiny clay particles over larger talc, mica, or glass-fiber fillers, which often leave part surfaces bumpy and cause parts to crack more easily in the cold. Nanocomposites are stronger and lighter than other plastic composites and may replace steel, aluminum, and conventional plastics in body-panel applications.

But developing nanocomposites has been mired by the clay's lackluster interaction with automotive plastics such as polypropylene and polyethylene. Automotive scientists found that bombarding clay particles with sound vibrations during mixing causes the filler to better disperse throughout the design matrix, improving a part's strength without using costly agents. The method disperses clay particles as single platelets throughout the matrix.

For example, various auto manufacturers have used plastic door panels for years. But, to let the material expand and contract, larger than normal gaps are placed between doors and body side panels.

Nanocomposite parts are cheaper and lighter, with better impact characteristics and lower coefficients of linear and thermal expansion. Using nanocomposite parts will allow manufacturers to get around things like big gaps for clearances between doors. Mass production of nanocomposite parts using this technology is likely within the next 5 years.

General Motors, for example, recently began using a nanoengineered material in the running boards for its sport utility vehicles and pickup trucks. The material is composed of clay that is ground to the nanometer scale and combined with a polymer that gives it the toughness and lightness critical to the auto industry. Even though the industry has been using nanoscale carbon fibers to strengthen its tires for years, GM's innovation, caught the rest of the industry by surprise.

Even traffic lights at most intersections are now built with solid-state nanoscale materials that make them brighter and 10 times more energy-efficient, claims Terry Michalske, director of the Center for Integrated Nanotechnologies at SNL in Albuquerque, NM.

Coatings

Thermal spraying is a widely used industrial process for applying protective coatings to materials surfaces. An attractive feature of the process is its ability to produce coatings, ranging in thickness from 25 μm to several millimeters, of almost any desired material. Historically, thermal spray deposition of metal, ceramic, and composite coatings was developed for applications in aircraft gas turbine engines. During the past 10 years, however, the range of applications has rapidly expanded into other areas, including land-based gas turbines, diesel engines, automobiles, surgical implants, and wear parts.

In thermal spraying, powders are fed into a combustion flame or plasma arc spray gun, where they are rapidly accelerated by the high-velocity gas stream exiting from the gun nozzle. During the short residence time in the flame or plasma, the particles are rapidly heated to form a spray of partially or completely melted droplets. The large impact forces created as these particles arrive at the substrate surface promotes strong particle–substrate adhesion and the formation of a dense coating.

Recent research at the University of Connecticut[73] has demonstrated the feasibility of thermal spraying of n-WC/Co powders to produce high-quality coatings.[74] A procedure has also been devised to reprocess as-synthesized powders into sprayable powder agglomerates, suitable for use in standard powder feed systems, irrespective of whether they have been prepared by chemical or physical methods. Important differences in the thermal spraying of "micrograined" and "nanograined" WC/Co powders can be appreciated from Figure 5.12. Micrograined particles experience surface melting only, which contrasts with homogeneous or bulk melting of nanograined particles. Thus, when the particles impinge on the substrate, the semisolid nanograined particles flow more freely, thereby forming a much denser coating, which has a completely uniform nanocomposite structure. Because of the rapid kinetics of WC nanograin dissolution in the liquid Co, the relative amounts of these two phases can be predicted using the equilibrium phase diagram. Thus, a controlling factor is the degree of superheat above the pseudobinary eutectic in the WC-Co system. The higher the particle superheat, the lower its viscosity, and hence the higher the deformation rate when it collides with the substrate. Under appropriate conditions, the semisolid particles should display thixotropic behavior (i.e., the dynamic viscosity should decrease with increasing shear rate). The turbulent flow created in the impacting particles should be useful in disintegrating particle agglomerates and thus promoting structural homogeneity in the deposited coating.

Tests have shown a much higher hardness in the nanograined composite coating, provided that precautions are taken to avoid decarburization in

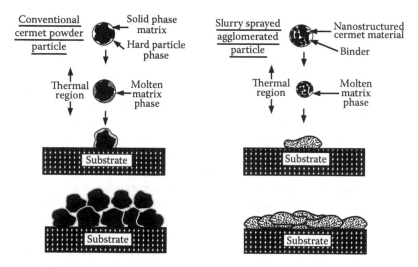

FIGURE 5.12
Comparison between thermal spraying of conventional and nanostructured WC/Co powders.

the flame or plasma. One method of accomplishing this is to use low-pressure plasma spraying, where the powder particles are naturally protected from oxidation in the plasma flame.[75] High-density coatings, with reproducible high hardness values, can be achieved by this means. In high-cobalt alloys, the resulting "splat-quenched" coating consists of nanodispersed WC grains in an amorphous Co-rich matrix phase. The problem of decarburization in thermal spraying of WC/Co powders is less acute in high-velocity oxy-fuel (HVOF) spraying, because of the lower particle temperatures and shorter particle residence times, compared with plasma spraying.

Very thin diamondlike nanocomposite (DLN) coatings may have a variety of tailorable properties in various fields. Everything depends upon the exact composition of the coating and upon the addition and amount of other elements, alloying elements, or doping material. The substrate to be coated can take various forms ranging from mechanical parts to plastic films. The electrical resistivity can be continuously varied over at least 18 orders of magnitude from a purely dielectric to a metallic state, resulting in various applications as resistor coatings, conductive coatings, and dielectric coatings.

DLN coatings comprise two interpenetrating amorphous networks. The first network is a diamond-like carbon (a-C:H) network and the second is a glass-like (a-Si:O) network. The second network conveys high stability to the first network. Other properties include a high resistance against corrosion and erosion, a high transparency, and a high resistance against scratches.

There is a growing demand for abrasion-resistant polymer systems for coating applications in sensors, optics, textiles, and numerous consumer goods. One such system, a carbon filament-wound epoxy composite sensor coating that is subjected to severe in-use wear, is processed by continuous dip-coating of the carbon fiber reinforcement prior to filament winding. During processing, the resin viscosity in which the fiber is dipped must be less than 3000 cP (viscosity) at 80°C to prevent resin degradation.

Conventional epoxy materials generally do not provide enough wear resistance. To solve these problems, a polymer dispersion incorporating nanocrystalline alumina was developed.

Gas Phase Condensation (GPC)

The nanocrystalline alumina is produced by a technology known as gas phase condensation.[76] GPC technology is at the forefront of some of today's most advanced nanophase materials.

Used for synthesizing inorganic and metallic powders, GPC is based on the evolution of a physical vapor — from the evaporation of elemental or reacted material — followed by immediate condensation and reaction of the vapor into small nanometer particles. To maintain nanometer-sized particles and weak agglomeration, the aerosol is rapidly cooled and diluted to prevent extensive sintering (to form hard agglomerates) and coalescence growth (aggregates).

Many applications require the nanosized particles to be dispersed in a fluid. Although every dispersion application is unique, the powder surface must always be rendered compatible with the dispersing fluid. If a coating is used, individual powders should be coated with a minimal layer of material without causing aggregation and agglomeration.

To meet these requirements, a coating process was recently developed that encapsulates nanocrystalline particles with a durable coating. The particle size distribution generated by the GPC process is preserved after coating. The coating can be engineered to allow dispersion of nanoparticles in organic fluids with dielectric constants ranging from 2 to 20, as well as water. Tailored coatings enable the value of individually coated and dispersed nanoparticles to be realized. Steric stabilizers or specific chemical groups can be incorporated into the coating to affect greater dispersion stability or specific (targeted) chemical reactivity, respectively.

For the carbon filament-wound epoxy composites, a nanocrystalline Al_2O_3 powder (57 m^2/g) was rendered compatible with the composite polymer matrix, Shell 862, by coating the powder. In addition, the coating was chemically modified to enable covalent incorporation into the Shell 862 resin. The coated Al_2O_3 was dispersed into the Shell 862 resin. The resin dispersion was activated by mixing it with Curing Agent W (26.2 parts of Agent W per 100 parts of Shell 862).

The composite was manufactured by continuous dip-coating of a graphite fiber into the activated dispersion followed by winding around a mandrel. The composite was cured under vacuum at 149 to 177° C for 1 h.

MEMS

Lucent Technologies Bell Labs, Arryx, and SiWave[77] have been developing a "top-down" approach, which focuses on stretching the limits of existing technologies by engineering devices with ever-smaller design features, for example, nanosized optical switching systems known as microelectromechanical systems. With the "bottom-up" approach, scientists attempt to build nanodevices one molecule at a time, in much the same way that living organisms synthesize large molecules. Scientists are trying to harness natural processes to create organic materials that have never existed in nature before.

Hewlett-Packard's Quantum Science Research Lab revealed a bottom-up storage lab project that had created the highest density of electronically switchable memory to date. The project used molecules known as rotaxanes to create a memory device capable of storing the equivalent of a full-length feature film in the area of a human fingernail. These rotaxane structures are trapped at the intersection of platinum and titanium wires only 40 nm wide. IBM, in its Millipede project, uses 1024 small pivoting arms to make memory indentations just 10 nm wide in a polymer material.

Medical

An example of the top-down and bottom-up methods converging can be found on the medical front, in implantable biosensors designed for drug delivery in diabetics. The bottom-up technology is used to encapsulate and deliver the drugs, while the top-down approach produces the electronic signaling device that activates the release of the drug.

Textile/Clothing

Nano-Tex[78] has prepared engineered fabrics that use tiny fibers to repel stains and wick moisture from sweaty athletes. Other companies, like DuPont and Corning, as well as research institutions, like Cornell University, the University of North Carolina at Charlotte, and the University of Texas at Austin, have been exploring the use of nanotechnology to make fabrics more durable, lighter, and better able to absorb dyes and textures.

Nano-Care, the stain-repellent fabric from Nano-Tex, incorporates billions of tiny fibers, each about 10 nm (that is 0.0000004 in.) long, that are embedded within traditional cotton or linen. The waterproof fibers, which Nano-Tex calls "nanowhiskers," make the fabric dense, increasing the surface tension so drops of liquid cannot soak through — just like raindrops on a freshly waxed car. The company says this Nano-Care treatment will withstand 50 home launderings before its effectiveness is lost.

Nano-Care and other similar nanotech treatments differ from fabric coatings like 3M's Scotchgard in that the nanoscale fibers are actually embedded into the products. In contrast, Scotchgard is applied to the finished goods, typically as a spray, and only protects what can be sprayed.

Nanotechnology can also be used in textiles to increase the ability of synthetic fabrics — like those made of the plastic polypropylene — to absorb dyes. Most polypropylenes resist dyeing, making them unsuitable for consumer goods like clothing, tablecloths, or floor and window coverings, With one experimental technique, a special dye-friendly clay is ground into nanometer-sized particles, which are given a light positive electrical charge to prevent them from clumping together. The clay is then mixed into raw polypropylene stock before it is extruded into threads, which are turned into fibers and woven into fabrics. The result is a composite material that can absorb a dye without weakening the fabric. And it does not even require nanoscale robots (see Figure 5.13).

Other developments include nanofabrics that do not trap odors, as well as a product called Nano-Touch, a cottonlike sheath to wrap around synthetics. Others are manufacturing Nano-Dry dress slacks that whisk perspiration away from the skin.

Micro Springs

Tiny, near-microscopic springs created using semiconductor lithography techniques are poised to enter the marketplace, offering chip designers dramatic improvements in chip-to-package connections and chip testing capabilities.

A consortium made up of the Xerox Palo Alto Research Center (Palo Alto, CA), NanoNexus Inc. (Fremont, CA), and the Georgia Institute of Technology (Atlanta) are developing high-density, integrated circuit test probes and chip packaging.

The springs are batch-fabricated on silicon wafers using a MEMS technique. Narrow thin-film metal fingers — with pitches (spacing from center to center) as fine as 6 μm — are laid down over a substrate using conventional lithography. By careful control of the metal deposition process, researchers build in mechanical stress between the top and bottom of the metal strip. When the substrate is partially etched away, the stress causes

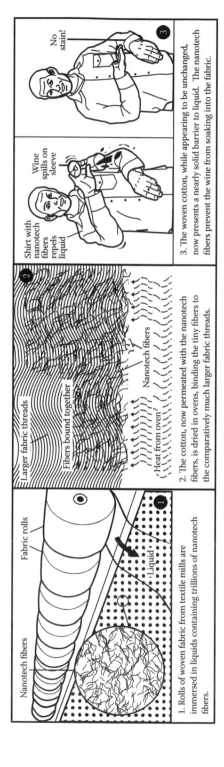

FIGURE 5.13
How stain-repellent nanotech fabric is made.

one end of the ribbon to curl up away from the surface, forming a tiny spring contact.

The technique makes possible very dense chip-to-package connections: 10 times denser than today's best solder connections. This is important because as lithography advances allow ever more complex chips with smaller and smaller features, the problem of connecting the chip to the outside world gets harder and harder. Conventional soldered connections are reaching their limit. But micro-spring contacts are potentially a generation or more ahead, say the researchers. The springiness of the springs also means that they absorb the mechanical strains set up in coupling chips to substrates with a different thermal expansion factor. And, unlike solder, they are lead-free.

Additionally, companies are developing the technology as "Nano-Springs"™ for chip and wafer test probes, because the novel contacts easily can scale to meet shrinking size and precision requirements that cannot be achieved with any other technology, and do it economically. Recently demonstrated was the ability to test contacts to gold bumps at a 52-μm pitch — typical of the contacts for display-driver chips in cell phones — and is pushing the technology to pitches as low as 35 micrometers and other metals.

Computers

Hewlett-Packard Labs scientist R. Stanley Williams[79,80] and his group of researchers face a monumental task: trying to make computers whose functionality rests on the workings of molecules.

To do so will mean reinventing the transistor. Although silicon and other inorganic semiconductors have always been the basic building blocks of microchips, it turns out that organic molecules can also have some potentially useful electrical properties. Indeed, over the last few years, researchers have learned to synthesize molecules that can function as electronic switches, holding binary 1s or 0s in memory or taking part in logical operations. And molecules have one significant advantage: They are really small.

Such work is critical to the future of computing, because conventional chip fabrication technology is on a collision course with economics. Today's best computer chips have silicon features as small as 90 nm. But the smaller the features, the more expensive the optical equipment needed to manufacture them. A state-of-the-art fabrication plant for silicon microchips now costs some $3 billion to build. A chip in which silicon transistors are replaced with molecular devices, on the other hand, could in principle be fabricated through a simple chemical process as inexpensive as making photographic film. A circuit with 10 billion switches could eventually fit on a grain of salt; that is a thousand times the density of the transistors in today's best computers. A computer built from such circuits could search billions of documents or thousands of hours of video in seconds, conduct highly accurate simulations and predictions of weather and other physical phenomena, and do a much better job of imitating human intelligence, perhaps even communicating with us through natural conversation.

Miscellaneous

Scientific clamoring about the cutting edge of nanotechnology — along with its odd lexicon of words like *nanotubes, nanorobots,* and *nanochips* — has overlooked the real, practical stuff. Nanotech in everyday products might seem like an oxymoron, but there are several examples already available in powders, crystals, and slurries. And you do not have to look much further than your apartment to find them.

It is also not surprising that the definition of *nanotechnology* is still murky. As chip makers push for a top-down definition — ever-shrinking blueprints of larger equivalents — materials scientists, looking from the opposite vantage point, argue that this discipline is based upon designing and engineering systems molecule by molecule. Although the science of the small has only recently made headlines in newspapers, scientists have been quietly developing and refining the properties of nanomaterials since the 1950s, when nanoporous molecular sieves, called zeolites, were first used to separate crude oil into useful components. Even stained glass windows, host to a variety of nanocrystals, owe their bright colors to ever-smaller molecular interactions (see Figure 5.14).

Nanoparticles

There are numerous methods used to produce nanoparticles. For example, Ted Kamins[79] has used a CVD reactor. He has grown the nanowires required by molecular electronics, as an alternative to using nano imprint lithography. So far, Kamins has synthesized wires as small as 10 nm in diameter by exposing "nanoparticles" of various materials to a mixture of gases in the deposition reactor. In the ensuing reaction, long chains of silicon grow up around the particles, producing what looks under the electron microscope like a forest of needles.

A new process developed at the University of Illinois produced a family of supersmall nanoparticles that show promise for electronic display and

ANTIBACTERIAL DRESSINGS
First widespread
commercial use: 1998

For years, silver has been used as an antibacterial agent. But in a nanocluster form,
it kills bacteria much faster and (mysteriously)
reduces inflammation. Nucryst
Pharmaceuticals embeds
nanocrystalline silver in burn
and wound dressings
prescribed by doctors or
used by clinics.

EASY-CLEAN BATHROOMS
First widespread
commercial use: 1999

Nanogate, a German company, develops
a thin film coating composed of nanoparticles. The gaps between them are
too tiny for most other particles to fit
between, so dirt doesn't stick to the surface. Also, nanoparticles don't refract
light, so the coating is transparent and
the original color of the ceramic bathtubs, shower stalls, and sinks is visible.

POLISHING MICROCIRCUITS
First widespread
commercial use: 2001

As computer chips diminish in size, they become
increasingly sensitive to ever so small, nano-scale
scratches that occur during the manufacturing
process. Cabot Microelectronics develops slurries containing nanoparticles that can polish away nano-scale
scratches on the circuits. Using a mixture of chemical
and mechanical abrasion, the carefully engineered
particles smooth the silicon layers without
introducing fresh scratches.

SELF-CLEANING GLASS
First widespread commercial use: 2001

British manufacturer Pilkington develops a
nanoglass that cleans itself – with the help of
water and sunlight. Its nanothin surface absorbs
a broad range of light frequencies from the sun
and uses the energy to break organic dirt into
small, soluble molecules. As the nanosurface
is hydrophilic, or water-loving, rainwater easily
sheets down the window, carrying the decomposed dirt with it.

TRANSPARENT SUNSCREEN
First widespread commercial use: 2001

Topical sunscreen ointments work by absorbing harmful ultraviolet
rays, thus protecting the skin. Zinc oxide, which is white because its
crystals scatter light, is an excellent UV absorber. But consumers want
transparent sunscreen. And that's what the Korean-Australian joint venture Advanced Nano Technologies has devised: a transparent nanopowder. Its 25-nanometer crystals are smaller than the wavelength of visible
light and so lose their whiteness. The crystals' ability to absorb sunlight
means they could be incorporated into nanotech-based, potentially
transparent, solar-power devices.

EXTRA-BOUNCY TENNIS BALLS
First widespread commercial use: 2001

A substance that successfully traps air in a tennis ball and yet remains pliable is a challenge to find. Tennis ball maker Wilson Sporting Goods uses the nanomaterial company
InMat's blend of nano-size clay flakes and micro-size rubber spheres as the coating for
its high-end official Davis Cup Double Core ball. The tiny clay flakes form a molecular barrier that traps air molecules, while the intermingled rubber spheres ensure flexibility. There are plans for the coating to be used in ultra-airtight tires for
cars, impermeable protective gloves, and non-glass beer bottles.

NANOPANTS
First widespread commercial use: 2001

Eddie Bauer, Gap, Levi Strauss, and many other
clothing suppliers weave nanofibers, produced by the research company Nano-Tex,
into their pants and sweaters. By engineering clothing on the molecular level,
says Nano-Tex, it becomes wrinkle resistant, oil repellent, and quick drying.

FIGURE 5.14
Nanotechnology applications.

flash memory material. They may also be useful as ultrabright fluorescent markers for tagging biologically sensitive materials.

To get nanoparticles, researchers gradually immerse a silicon wafer into an etchant bath of hydrofluoric acid and hydrogen peroxide while applying an electrical current. The process erodes the surface layer, leaving behind a delicate network of weakly interconnected nanostructures. The wafer then gets an ultrasound bath, crumbling the nanostructure network into clumps of 1-nm-diameter particles, which are separated into groups of different sizes.

The silicon particles fluoresce in different colors when exposed to ultraviolet light, depending on their size. They also fluoresce when exposed to at least two photons of infrared (IR) light in a technique that can noninvasively penetrate human tissue. Further, researchers have been able to get small aggregates of the nanoparticles to go into laser oscillation. At 6 μ in diameter, the particle clusters are one of the smallest lasers in the world and show that microlasing is an important step towards the realization of a laser on a chip, which could ultimately replace wires with optical interconnects.

Some applications for nanoparticles are already in use as additives in paint fillers in ceramics and for processing coal. Other applications are either in their infancy, in an R&D phase, or being proof-tested before being exposed and used in military, commercial, and industrial products.

The following describes many of the above situations for nanoparticles:

Materials researchers at Iowa State University, working in part under a grant from the National Science Foundation, have demonstrated a novel coating that makes surfaces "smart" — meaning the surfaces can be switched back and forth between glassy-slick and rubbery on a scale of nanometers, the size of just a few molecules.

Possible applications include the directed assembly of inorganic nanoparticles, proteins, and nanotubes, and the ultraprecise control of liquids flowing through microfluidic devices that are finding their way into biomedical research and clinical diagnostics.

The new coating is a single layer of Y-shaped "brush" molecules, according to principal investigators Vladimir V. Tsukruk and Eugene R. Zubarev.[82]

Each molecule attaches to the surface at the base of the Y, which forms a kind of handle for the brush, and extends two long arms outward to form the bristles. The coating can be switched because one arm is a polymer that is hydrophilic, or attracted to water, while the other is a polymer that is hydrophobic, or repelled by water.

Thus, say the researchers, when the coated surface is exposed to water, the molecules collapse into a series of mounds about 8 nm wide, with the hydrophilic arms on top shielding the hydrophobic arms inside. Conversely, when the surface is treated with an organic solvent such as toluene, the surface spontaneously reorganizes itself into mounds that have the hydrophobic arms on top.

Not surprisingly, the two states are very different when it comes to properties such as stickiness and the ability to become "wet."

In future work, the Iowa State team hopes to coax the mounds into an ordered pattern, instead of the current random scatter, which may allow the researchers to make surfaces that are lubricating in one direction and sticky in others.

Researchers have uncovered a new way to control the motion of submicroscopic particles in artificial nanodevices and biological systems. The technique may deliver medications to specific cells or replace wires in molecular-sized electronic devices.

The idea devised by a University of Michigan physicist and other researchers is to add auxiliary particles that interact with the primary target particles. This makes it easier to control and manage how particles of interest flow. Ratchets, or asymmetrically shaped sawtooth substrates, can then make particles (ions or molecules) flow in one direction instead of wandering randomly.

For example, if one species of particle repels the other, the two will drift in opposite directions. However, if they attract each other, the more active species will drag the passive one along. This indirect rectification of particle motion could be used to deliver or remove passive ingredients from a cell or from an artificial molecular-scale device. These tiny shuttles or microscopic conveyor belts might also act as new types of wires, replacing standard wires in nanodevices (see Figure 5.15).

According to the team, injecting an appropriate density of passive particles can control the velocity of active particles and vice versa.

Silicone rubber and other rubber-like materials have a wide variety of uses, but in almost every case they must be reinforced with particles to make them stronger or less permeable to gases or liquids. University of Cincinnati chemistry professor James Mark and colleagues have devised a technique that strengthens silicone rubber with nanoscale particles, but leaves the material crystal clear.

Silicone rubber is often reinforced by tiny particles of silica (the primary component of sand and the mineral quartz). However, those silica particles can cloud the silicone rubber, which is a problem for protective masks, contact lenses, and medical tubing that rely on silicone rubber's transparency.[83]

The technique infuses silicone rubber with nanoparticles up to 5 times smaller than the silica particles formed by comparable methods, while still providing the same level of reinforcement and maintaining the silicone rubber's clarity.

Variations on the technique might also be used to enhance other properties of silicone rubber and similar materials, affecting such traits as

How it works

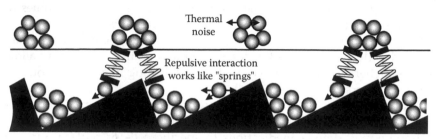

At low temperatures, thermal noise is too weak to overcome the barriers. The red particles move toward the bottom of the grooves, pushing the green particles toward the substrate peaks.

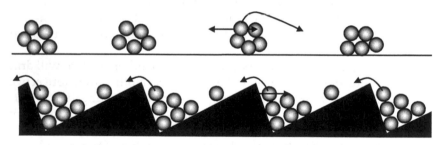

At high temperatures, thermal noise shakes the particles strongly enough to make them jump over the barriers. The opposite spatial asymmetry for red and green balls pumps them in different directions.

FIGURE 5.15
Microscopic conveyor belts.

impermeability to gases or liquids. This could lead to better masks or suits to protect against agents that might be used in terrorist attacks.

The team's technique is an improvement over related methods that use a chemical reaction to create silica particles within the silicone polymers. By generating the required catalyst in place from a tin salt and by restricting the amount of water to only that absorbed from water vapor in the air, the silica particles remain smaller — only 30 to 50 nm across — and are evenly dispersed throughout the silicone rubber. At that size, smaller than the wavelength of ultraviolet and visible light, the silica nanoparticles are essentially invisible.

The development of the ability to harness the large surface area-to-volume ratios and the unique surface properties of nanoparticles in the <10 nm size range, especially in heterogeneous catalysts, is a major driving force in both fundamental research and practical applications of nanoparticle

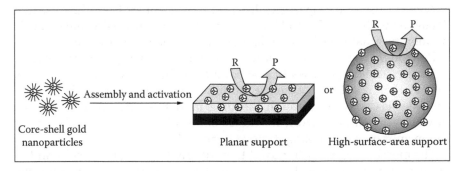

FIGURE 5.16

A schematic illustration of assembly and activation of core-shell nanoparticles as catalysts on planar substrate and high-surface-area support materials. The R-to-P process illustrates a catalytic reaction from reactant to product.

catalysts in fuel cells. It is this size range over which particles undergo a transition from atomic to metallic properties. The study of this unique aspect was indeed inspired by the discovery of the high catalytic activity of nanosized gold towards CO oxidation.[84–87] Gold is traditionally considered catalytically inactive as a practical matter when the size is reduced to a few nanometers. There are two major challenges for developing nanometer-sized particles as catalysts: (1) the control of sizes and compositions and (2) the prevention of the aggregation propensity of nanoscale materials. The recent exploration of core-shell-type nanoparticles, which can be broadly defined as core and shell of different matters in close interaction, including inorganic/inorganic, organic/organic, or inorganic/biological combinations,[88–90] provides intriguing pathways to address these challenges. The use of such nanoparticles as building blocks towards catalytic materials takes advantages of diverse attributes, including size monodisperity, processability, solubility, stability, self-assembly capability, and unique optical, electronic, magnetic, and chemical/biological properties (see Figure 5.16).

In conclusion, gold and its alloy nanoparticles of controlled size and composition can be produced by combining capping-based synthesis and thermally activated processing protocols. Although explorations of gold nanoparticle catalysts have shown great promise in fuel cell catalysis, the question of whether nanostructured gold nanoparticles can be eventually used as practical fuel cell catalysts that are compatible with existing PGM catalysts has not been answered. Such an answer will certainly require in-depth fundamental insights into the full controllability of size and surface composition of the nanoscale gold during the catalytic activation and reaction. The rapid development of gold-based nanotechnology will help enormously in the exciting advent towards gold "catalyst-by-design."

A clever use of nanoparticles let a team of scientists from the Research Triangle Institute (RTI)[91] and North Carolina State University make plastic membranes that pass large molecules faster than smaller molecules.

Previously, membranes became less selective as their permeability rose. "Usually, adding particles to plastics makes them less permeable," says Tim Merkel of RTI. "For example, additives in plastic wrap might be used to allow less oxygen through, keeping food fresh longer. In our case, however, the particles work at a molecular level as 'nanospacers,' opening the membrane and making it more permeable."

The membranes are being used to purify hydrogen for fuel cells and to clean pollution-causing chemicals out of fossil fuels. Currently, only the physical structure changes, but future research will look at making membranes chemically active to pass or block specific chemicals.

BGAs and flip chips are reaching the point where even extremely fine solder paste is too coarse to carry out defect-free board attachment. Fortunately, the next step past solder paste already is in testing: nanoparticle inks. Nanoparticles of metals are generated and suspended within a highly purified organic solvent matrix. This method permits these almost atomic-size level, completely oxide-free particles to be stored and then manipulated easily. For example, their liquid suspensions of particles that resemble dark blue "inks" made from single metals instead of premixed alloys can be mixed together in any desired amounts to generate multi-component alloys. Recently, four inks (tin, silver, copper, and indium) were mixed to provide a complex lead-free alloy. Because the nanoparticles are oxide-free, they have high surface energy. Thus, when the volatile organic solvent is flashed off, the particles immediately bind together or coalesce. Because this can be accomplished at relatively low temperatures compared to the significantly higher soldering temperatures required by the standard lead-free alloys, delicate electronics can be manufactured easily.

A 9% chromium martensitic steel reinforced with carbonitride nanoparticles exhibits time-to-rupture that is two orders of magnitude longer at 650°C than the conventional ASME-P92 alloy, which is the strongest steel currently available for boiler components, reports Japan's National Institute for Materials Science.

The steel has Charpy impact energy of 100 to 150 J at room temperature. The improvement in creep resistance is attributed to a mechanism of boundary pinning by the thermally stable carbonitride particles. According to researcher Fujio Abe, the dispersion of nanometer-sized carbonitride particles is the result of reducing carbon levels to 0.002%.

The steel was designed as the result of a program to determine the optimum composition for steels that can resist creep at high temperatures. The compositions (in wt%) of the steels examined in the study were based on Fe-9Cr-3W-3Co-0.2V-0.06Nb-0.05N, with varying carbon amounts:

0.002, 0.018, 0.120, and 0.160C. The steels were prepared by vacuum induction, melting 50 kg ingots. High-temperature strength was examined by creep tests at 650°C under constant load for up to 10,000 h, with specimens 10 mm in diameter and 50 mm long.

A large number of fine precipitate particles having a size range of 5 to 10 nm are distributed along prior austenite grain boundaries as well as along lath, block, and packet boundaries in the specimen with 0.002%C after tempering heat treatment. The fine particles were identified as MX carbonitrides, where M is a metallic alloying element and X refers to carbon and nitrogen.

Other steels have been developed to withstand creep at high temperatures by distributing fine particles throughout the matrix, but their production is complex and expensive. However, the new steel can be made economically by conventional processing techniques. Its development may lead to the cheaper manufacture of large-scale creep-resistant steel components for high-temperature applications such as those needed for power plants.[92]

During the late 1980s, Japanese scientists discovered that the mechanical properties of ceramics could be improved by adding ultrafine (100–200 nm) SiC particles. Subsequent work at Lehigh University, sponsored by EPRI and the U.S. Office of Naval Research (ONR), has confirmed this discovery and shed new light on the probable mechanism involved. In general, ceramics tend to be brittle: When bent, they break easily, without deformation, along surface defects. Lehigh researchers have found that the presence of SiC nanoparticles apparently facilitates crack healing in the ceramic surface during annealing, thus making the surface stronger. In contrast, annealing a single-phase ceramic material actually makes surface cracks grow. According to the latest results, ceramic-SiC-nanocomposites show a significant increase in high-temperature rupture strength, compared with single-phase ceramics. In addition, the creep rate of the nanocomposites is more than a hundredfold lower. The results also show large increases in abrasion resistance.

In other nanoceramic particle research, Osaka University scientists have developed a photonic crystal with high-k nanoceramic composite particles dispersed in epoxy resin lattices.

The composite particle was developed by mechanochemical bonding (MCB). The particle delivers a dielectric constant that is 50% higher than conventional fine particles, helping reduce lattice intervals by 20%.

These nanoceramic composite particles have been used in the optical modeling system creating a photonic crystal, an artificial grating that is expected to pave the way for the development t of terahertz wave-based sensors, and high-capacity milliwave communications devices.

Engineers at the Georgia Institute of Technology have compiled a "recipe book" for how to create nanoparticles with specific magnetic properties.

The aim is to develop diagnostic and therapeutic tools for medicine. For example, magnetic nanoparticles could carry drugs to a specific area in the body, guided by magnetic fields. Magnetic particles might also improve the contrast of MRI images. And small magnets could serve as tracers, replacing current radioactive methods of tracking drugs in the body.

Researchers are currently trying to overcome several hurdles. One is creating low-energy nanoparticles that allow constant changes in their magnetic states. Because magnetic opposites attract, particles could potentially clump together in the body and clog blood flow. Particles that changed magnetic states rapidly, however, might not congregate or cause problems. Another hurdle is size. Particles must be small enough to escape detection by the immune system, yet large enough to remain in the body and circulate through the bloodstream. They must also be consistent because magnetic properties are a function of size. The Georgia Tech team has produced nanomagnets with size variations of less than 15% and hopes to reduce that even further.

Scientists from Columbia University, IBM, and the University of New Orleans recently announced a new, three-dimensional designer material assembled from two different types of particles only billionths of a meter across.[93]

Precision chemistry methods have been developed to tune the particles' sizes in increments of less than 1 nm and to tailor the experimental conditions so the particles would assemble themselves into repeating 3-D patterns (see Figure 5.17). Designing new materials with otherwise unat-

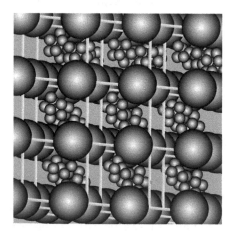

FIGURE 5.17
Sketch of a self-assembled bimodal superlattice of nanocrystals. The superlattice is composed of superparamagnetic iron oxide nanocrystals in a cubic array, with an icosohedron of 13 semiconducting lead selenide crystals at the center of each cube. (The thirteenth lead selenide nanocrystals is at the center of the icosohedron.)

tainable properties, sometimes referred to as "metamaterials," is one of the promises of nanotechnology. Two-dimensional patterns had previously been created from gold nanoparticles of different sizes and mixtures of gold and silver. Extending this concept to three dimensions with more diverse types of materials demonstrates the ability to bring more materials together than previously realized.

"What excites us the most is that this is a modular assembly method that will let us bring almost any materials together," says Christopher Murray, at IBM Research. "We've demonstrated the ability to bring together complementary materials with an eye to creating materials with interesting custom properties."

The scientists chose the materials for the experiments specifically because of their dissimilar yet complementary properties. Lead selenide is a semiconductor that has applications in infrared detectors and thermal imaging and can be tuned to be more sensitive to specific infrared wavelengths. The other material, magnetic iron oxide, is best known for its use in the coatings for certain magnetic recording media.

The combination of these nanoparticles may have novel magneto-optical properties as well as properties key to the realization of quantum computing. For example, it might be possible to modulate the material's optical properties by applying an external magnetic field.

The first step was to create the nanoparticles. The particle sizes were calculated from the mathematical ideal of the structures being created. In addition to fine-tuning the sizes, the particles had to be very uniform, all within 5% of the target size. First, iron oxide particles 11 nm in diameter, were created, and lead selenide particles 6 nm in diameter. There are approximately 60,000 atoms in one of the iron oxide nanoparticles and ~3000 atoms in the lead selenide particles.

Next, the nanoparticles were assembled or, more to the point, had the particles assemble themselves into three different repeating 3-D patterns by tailoring the experimental conditions. Forming these so-called "crystal structures," as opposed to random mixtures of nanoparticles, is essential for the composite material to exhibit consistent, predictable behaviors. Various other materials are known to assemble spontaneously into these structures of close-packed particles, but none have been made of two components in three dimensions and at the length scales. Precision chemistry methods were developed in order to tune the particles' sizes in increments of <1 nm and to tailor the experimental conditions so the particles would assemble themselves into repeating 3-D patterns.

Finally, encapsulated nanoparticles, formed when graphite is present during vaporization with a tungsten-based electric arc, could have applications in magnetic resonance imaging and cancer treatment, say their discoverers, a team of materials scientists and engineers at Northwestern University. Called "buckycapsules," the airtight pellets — which resemble buckytubes — could travel through the body without exposing a person to the capsules' metal core.

Nanopowders

Approaches to Powder Fabrication

CVD

One can make ceramic nanophase powders ranging in purity from 95 to 99.999% using a 5 kW carbon dioxide laser CVD process. Powders produced with lasers have more controlled diameters and a high degree of sphericity. Their fine particle size enhances hot pressing formability of materials. Powders with controlled diameters of <20 nm have a max/min diameter ratio of <2, have a surface area of about 109 m^2/g, and have no aggregation. One company that uses this route recently introduced several nanophase powders including SiN, SiC, SiCN, BN, and SiO.

Nanophase materials have potential in wear components of internal-combustion engines. Nanophase iron nitride powder has superior magnetic properties and is being considered for magnetic storage systems, optomagnetic devices, and bulk magnets.

Laser Synthesis

Infrared laser pyrolysis (IRLP) is a highly versatile method for the production of a wide range of nanopowders including SiC, SiCN, SiCO, BN, and C.

Laser pyrolysis is a versatile method well suited for the production of significant amounts of Si-based nanopowders. Due to their properties (purity limited only by the purity of the reactants, low size dispersion, etc.), the nanopowders were good candidates for applications in ceramics. Nanopowders can be obtained from gaseous or liquid precursors. The powders obtained from the gas phase were crystallized, while the powders obtained from the liquid phase were amorphous. In both cases, the chemical composition of the powders was controlled by the chemical composition of the reactive mixture. Finally, an interesting result was the incorporation of the elements of sintering aids (Al, Y, O) during the synthesis.[94]

A summary of the syntheses performed for materials applications is given in Table 5.3.

In the IRLP process, the reactants are heated by IR laser radiation and decompose, causing clusters to nucleate and grow rapidly. This process is inherently very clean because homogeneous nucleation occurs in a well-defined reaction zone without interaction with chamber walls. The small

TABLE 5.3

CO_2 Laser Synthesized Nanometric Powders

Powders	Chemical Systems	References
Si	SiH_4	Haggerty and Cannon 1981; Cauchetier et al. 1987
SiC	$SiH_4 + C_2H_4$	Sumaya et al. 1985; Cauchetier et al. 1987; Förster et al. 1991
	$SiH_4 + C_2H_2$	Cauchetier et al. 1987; Curcio et al. 1989; Fantoni et al. 1990; Tougne et al. 1993
	$SiH_4 + C_xH_y$ with $x = 4$	Cauchetier et al. 1988
	$SiH_2Cl_2 + C_2H_4$	Suzuki et al. 1992
	$(CH_3)_2Si(C_2H_5O)_2$ vapour	Li et al. 1994a
Si_3N_4	$SiH_4 + NH_3$	Haggerty and Cannon 1981; Kizaki et al. 1985; Symons and Danforth 1987; Buerki et al. 1990
	$SiH_2Cl_2 + NH_3$	Bauer et al. 1989
$SiC + Si_3N_4$ or Si/C/N	$[(CH_3)_3Si]_2NH$ vapour	Rice 1986; Li et al. 1994b
	$SiH_4 + CH_3NH_2 + NH_3$	Cauchetier et al. 1989; Cauchetier et al. 1991
	$SiH_4 + (CH_3)_2NH + NH_3$	Alexandrescu et al. 1991; Borsella et al. 1992
	$SiH_4 + C_2H_4 + NH_3$	Suzuki et al. 1993
	$(CH_3SiHNH)_x$ with $x = 3$ or 4 (aerosol)	Gonsalves et al. 1992
	$[(CH_3)_3Si]_2NH$ (aerosol)	Cauchetier et al. 1994; Herlin et al. 1994; Musset et al. 1997
SiO_2	$Si(C_2H_5O)_4$ (aerosol)	Luce et al. 1994
$SiO_2 + SiC + C$ or Si/C/O	$(RO)_{4-x}Si(R')_x$ with $0 = x = 3$, $R = C_2H_5O$, $R' = CH_3$ and $[(CH_3)_3Si]_2O$ (aerosol)	Armand et al. 1995; Martinengo et al. 1996; Herlin et al. 1996; Fusil et al. 1997
BN	$BCl_3 + NH_3$	Luce et al. 1993; Baraton et al. 1994; Willaime et al. 1995
B_4C	$BCl_3 + CH_4 + H_2$	Knudsen 1987a
	$BCl_3 + C_2H_2 + H_2$	Luce et al. 1993
TiB_2	$TiCl_4 + B_2H_6$	Casey and Haggerty 1987a
	$TiCl_4 + BCl_3 + H_2$	Knudsen 1987b
ZrB_2	$Zr(BH_4)_4$	Cauchetier et al. 1987; Rice and Woodin 1988
TiO_2	$Ti[OCH(CH_3)_2]_4$	Casey and Haggerty 1987b; Rice 1987; Curcio et al. 1991
TiC	$TiCl_4 + C_2H_4$	Alexandrescu et al. 1997
Al_2O_3	$Al(CH_3)_3 + N_2O + C_2H_4$	Borsella et al. 1993
WC	$WF_6 + C_2H_2 + H_2 + SF_6$	Bourgeois et al. 1995

reaction volume and the ability to maintain steep temperature gradients ($10^{6\circ}C/s$) allow precise control of the nucleation rate, growth rate, and residence times. The physical and chemical properties of the particles can be controlled by changing the molecular precursors and the synthesis

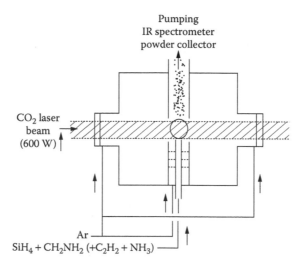

FIGURE 5.18
Schematic laser irradiation cell for gaseous precursors.[95]

parameters (laser power, pressure, etc.). The resulting powders are very fine, spherical, extremely pure, more or less agglomerated, and nearly monodispersed in size. The mean particle size can be adjusted from 10 to 100 nm. All these characteristics (i.e., nanometric size [<100 nm], spherical shape, narrow size distribution, limited agglomeration, and high purity) make laser powders good candidates for use in ceramics (see Figure 5.18).

Since the synthesis of Si and Si_3N_4, the production of ultrafine powders from gaseous reactants has become an innovative application of lasers in chemistry. Although the CO_2 laser synthesis of powders for use in ceramics is a new process compared with plasma or sol-gel processes, it offers considerable promise, and it has been developed in several countries: in Europe (France at the CEA, Italy at the ENEA and the CISE, the Netherlands at the TU Delft, Germany in Jülich), in Japan, and in China.

The laser synthesis technique requires the presence of a species that absorbs laser radiation such as SiH_4. This precursor has been widely used to obtain silicon-based ceramic powders (see Table 5.3).

Production rates at laboratory scale have reached 100 g/h or more for SiC, and the development of the process for a pilot can be estimated to require a production rate of 1–10 kg/day.

Microwave Energy Transfer

In 1995, M. S. Krupashankara[96] began looking into microwaves as an alternative approach for efficiently making production quantities of

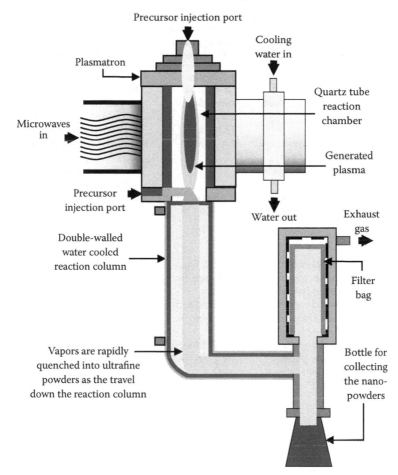

FIGURE 5.19
Coupling microwaves with materials requires tailoring to material electrical and magnetic properties. Precursor residence time in plasma and quench conditions govern the size of the Nanogen.

nanometer-sized constituent materials (see Table 5.4). This later work was aimed at forming lightweight armor plate to protect against projectiles!

Nanometer-sized powders offer high surface area to volume: about 45 times more for 20-nm-diameter iron particles compared to typical powders of the metal at just less than 1 μm. For aerospace and automotive applications, for example, materials made from metal and ceramic nanopowders exhibit superplastic behavior, undergoing elongations from 100 to 1000% prior to failure.

TABLE 5.4

Some Nanopowder Recipes

Nanopowder Synthesis Method	Metal/Ceramic Nanopowder Produced	Process Disadvantages
Ball milling/Mechanical attrition: Powder particles subjected to severe mechanical deformation.	Fe, Co, Ni, NiAl, TiAl, and FeSn	Powder contamination due to steel ball's low production rates, 1–5 gm/hr.
Laser ablation	AlN, MnO_2, TiO_2, Ti	Low production rate, 0.01 gm/hr, and high energy consumption.
Sputtering: A dc or RF source, instead of a laser as in laser ablation, vaporizes the material.	Al, Cu, Mn	Broad particle size distribution but only 6–8% of sputtered material is reported to be <100 nm; >4–5 gm/hr.
Chemical precipitation: Mix two or more precursors and a catalyst to form a gel. Dry the gel and reduce under H_2 to form nanopowder.	SiC, $BaTiO_3$, W-Cu, Mo-Cu	Production rate 10–50 gm/day; agglomeration of powders and oxidation from use of liquid chemical precursors.
Induction plasma: RF generator ionizes plasma. Temperatures up to 10,000°C. Material fed into plasma vaporizes and condenses as nanoparticles on chamber walls.	Fe, AlN, Cu, and metallic borides, nitrides, and carbides.	High energy consumption; efficiency generally under 50%.
Microwave plasma (Nanogen): Microwaves generate plasma, which transfers the heat necessary for chemical reactions.	Fe, Co, Mo, Ni, TiO_2, ZrO_2, Al_2O_3	Coupling of microwave energy to metallic surfaces of particles is difficult; temperatures up to 1,500°C limit the choice of precursors. Information from Materials Modification Inc.

Note: The microwave-plasma Nanogen process (bottom) offers atmospheric-pressure processing for metallics, ceramics, and intermetallics, at higher rates (1 kg/day) than other methods.[96]

Although previous methods could make nanopowders of oxides and carbides readily, due to combining with atmospheric gases, production of pure metal powders was a problem: In some cases, a violent ignition takes place on exposure to air. Prior techniques also had slow production rates of only up to 50 g per day.

The microwave-based Nanogen™ approach developed by Krupashankara and fellow engineer Raja Kalyanaraman "provides a dual advantage: heating the material directly and providing a high-temperature gas envelope to facilitate chemical and thermal processing," Krupashankara notes. The system focuses microwaves (6000 W at 2450 MHz) down a wave-guide. A quartz-walled reaction chamber allows the microwaves to enter and couple with the reactant-laden precursors injected into the chamber. Rapid dissociation, decomposition, and melting or vaporization of the reactants then occurs. Krupashankara says material permittivity (related to dielectric constant) and permeability (magnetic induction divided by field strength) are key properties governing coupling of microwaves to materials in the reaction chamber or plasmatron.

The resulting plasma, at temperatures up to 1500°C, facilitates the formation of particles close to atomic size once the plasma enters a double-walled "reaction column" cooled by water. Controlling residence time of reactants in the plasma insures complete conversion. The time and the quench rate are tuned by varying parameters such as reactant partial pressures, gas flow, and temperature, which also impacts on particle grain sizes and their distribution.

Downstream, a 3M filter bag that can withstand 700°C separates the nanoparticles from the exhaust gas flow, where they fall into a container or bottle below. This direct bottling minimizes surface contamination. Depending on the material, slow introduction of air or a proprietary gas will quench the powder by forming a thin protective coating to eliminate any pyrophoric ignition. Or the powder containers can be sealed under argon or nitrogen for storage. Currently upwards of 1 kg of nanopowders per day can be produced.

"The real use for nanopowders is making various shaped parts," says Krupashankara. They also have the following applications:

- Filters for separating chemicals and purification
- Coatings for strength and wear resistance
- Catalysts to spur reactions
- Sensors for biochemicals
- Color agents for dyes and cosmetics

Mechanical Alloying (MA)

Elemental Cu and NbC powders were mechanically alloyed with either graphite or hexane to form nanocomposite Cu-Nb-C powders. These powders were consolidated by hot-isostatic-pressing at one third of the melting temperature of NbC to form a 10% dense compact, while maintaining a crystallite size well within the nanoregime.[97]

Severe Plastic Deformation

The technique, called *severe plastic deformation*, puts materials and alloys under severe stresses that break down, or refine, grains under loads in excess of 400 tons/in². These high loads are combined with torsion straining or other pressing methods to produce materials of exceptional strength and toughness. Brittle ceramics can become pore-free and malleable under

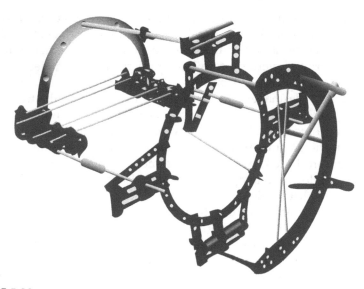

FIGURE 5.20
An orthopedic retainer is one of the first devices made using titanium and nanopowder technology.

certain conditions when formed from these nanopowders. Parts made from nanopowder metals such as copper, iron, silver, aluminum, and titanium are up to 10 times stronger than conventionally structured counterparts.

The process makes high-strength biomechanical and industrial components from ultrafine metal, ceramic, or composite powders. "Parts made with this process exhibit significantly improved performance and durability," says Emil Strumban.[98] "Aluminum components formed from ultrafine powders will be far stronger than conventional large-grained aluminum parts, and steel-cutting tools with ultrafine grains will last several times longer."

This new process that stresses and breaks down microscopic "nanopowders" could produce exceptionally strong orthopedic components like bone implants (Figure 5.20) and stronger parts for cars, airplanes, and spacecraft, and improve the performance of many conventional materials.

In addition to manufactured components, potential applications include new types of filter membranes, cutting blades and drill bits, materials for joining ceramic parts, high-performance bearings, nonsilicon materials for active and passive elements in integrated circuits, and supermagnets to increase the efficiency of electric motors.

High-Energy Ball Milling

Recently, it has been reported that nanocrystalline metals and alloys can be prepared by mechanical working of elemental powders or mechanical alloying of powder mixtures in a high-energy ball mill.[99–102] These claims are mostly based on the analysis of the broadening of x-ray diffraction peak profiles. It is well known that cold-working of metals and alloys produces a broadening of their powder pattern peaks due to dislocations introduced during the deformation process. High-energy ball milling is a form of cold-working where the peak stress can be as high as 50 kbar.[103]

Micronsized powders of Fe, W, and NiAl were mechanically worked in a high-energy ball mill for milling times up to 100 h.[104]

The milling of the W, Fe, and NiAl powders in the high-energy ball mill was an effective way to produce materials whose diffraction patterns exhibited a large broadening of the (hkl) peaks. Even though the sizes of the individual powder particles are still of the order of 1–2 μm, the grain size in each particle is reduced to the nanometer scale.

Densification and Treatment

Nanostructured powders are a novel class of materials whose distinguishing feature is that their average grain or other structural domain sizes are below 100 nm. Within this size range, a variety of confinement effects significantly change the properties of the material. The confinement effects lead to several commercially useful characteristics. From a processing viewpoint, nanostructured powders offer the potential for very high sintering rates at lower temperatures.

A rapid densification technology uses nanostructured powders to produce ceramic devices and components. This technology provides ceramic monoliths and composites that can be used in the automobile, energy, electrochemical, magnetic, structural, biomedical, computing, information-transfer, and pollution-prevention and -control industries.

Nanoscale powders of ceramic, measuring <100 nm, were mixed with <5 wt% of sintering aid, cold-pressed into pellets, and pressureless sintered between 1400 and 1600°C. The sintering environment was evacuated to remove oxygen and maintained in evacuated reducing state.

Micronscale powders of ceramics with the same composition were also processed through the same steps under similar environments.

The densification of a ceramic compact, or sintering, is the process of removing the pores between the starting particles combined with growth and strong bonding between adjacent particles. The driving force for densification is the decrease in surface area and lowering of the surface energy by the elimination of the solid-vapor interface.

In order to reduce the concept of this new technology to practice, the innovators, Yang and Yadov,[105] focused on -SiC and continuous SiC fiber-reinforced SiC composites. Nanoceramic powders were produced and then formed into monoliths. These monoliths were dispersed into the fibers by several alternative methods, isostatically compacted, and hot-pressed to achieve high densities.

Nanosized SiC powders that were synthesized were characterized using x-ray diffraction and TEM. The nanosized SiC powders were further characterized using the B.E.T. method, and surface area was found to be 100 m^2/g. Using 3.2 g/cm^3 as the density of SiC, the average powder size was calculated as 9.4 nm.

The pressureless sintering was carried out in a high-temperature graphite furnace. The graphite heating element is insulated from the water-cooled stainless steel chamber. A EUROCUBE 425 thyristor unit was used to precisely control the furnace temperature and heating/cooling cycle The chamber was first pumped down to ~10 torr using a mechanical pump, then flashed with argon three times. Either an argon or argon-reducing atmosphere was used during the sintering process.

Rapid densification technology will be of great value in the development of new ceramic-matrix-composite technologies of the future.

With 10-nm-sized SiC nanopowders, monolithic SiC and SiC matrix composite samples were pressureless sintered to over 90% of the theoretical density in 240 min at 1450 and 1500°C, respectively. At temperatures such as those presently used in conventional SiC and SiC/SiC densification practices (>2000°C), the densification of nanosized SiC and SiC/SiC is expected to be more than 30 times faster. Beyond processing benefits, the invention offers performance benefits as well. The densified composite samples prepared with nanosized SiC, for example, offer much higher fracture strength (400%) than those prepared with micron-sized SiC as starting powders. This technology breakthrough is applicable to other commercially important carbide, nitride, boride, silicide, and oxide ceramic compound compositions, as well.

Applications

Early experiments in a continuing research program have demonstrated that thermal batteries made from powdered solid electrode and electrolyte materials can be improved by use of smaller (nanometer vs. micrometer size) cathode powder grains. The improvements include the possibility of fabricating and using thinner cathodes, plus increases in mechanical robustness, thermal stability, and overall power density.

A thermal battery is a primary battery that is activated by heating to melt the solid electrolyte to supply electrical power for a limited time. Thermal batteries are highly reliable energy sources with high-power densities and long shelf lives. They are particularly useful for supplying short-term power in expendable weapons (e.g., torpedoes and projectiles) and exploratory spacecraft; likely future commercial applications could

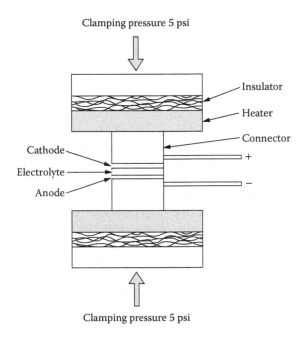

Clamping pressure 5 psi

Insulator

Heater

Connector

+

Cathode

Electrolyte

Anode

−

Clamping pressure 5 psi

FIGURE 5.21
A basic single-cell thermal battery contains electrode and electrolyte disks. Unlike power cells used in many commercial products, this cell must be activated by heating to melt the electrolyte.

include generating emergency power in aircraft and providing startup power for automobiles with weak batteries.

A single-cell thermal battery (see Figure 5.21) contains an electrolyte disk stacked between an anode and a cathode disk. Each disk is made by cold-pressing the appropriate cathode, electrolyte, or anode powder. Heretofore, thermal batteries have been manufactured by techniques that impose lower limits on disk thicknesses needed to ensure adequate mechanical strengths. Accordingly, achievable power densities and other performance characteristics have been limited, and progress toward miniaturization and toward enhancement of activation characteristics and of safety has been impeded.

The anode material used in the experiments was an alloy of 41% Li + 56% Si, supplied as a powder of micronsized particles. The anode disks were formed by pressing this powder in a 2-cm-diameter round steel die at 41 MPa. The electrolyte material included a powder of eutectic salt comprising 45% LiCl + 55% KCl. To strengthen the electrolyte disks, the eutectic salt powder was blended with 35% MgO powder. The blended powder was pressed at 28 MPa to form electrolyte disks.

The cathodes were made from a blend of 68% FeS_2 powder, 30% of the eutectic salt powder, and 2% of SiO_2. The blend was pressed into disks at 28 MPa. To provide a basis for comparison, the FeS_2 powder used to make some cathode disks had particle sizes of the order of 1 μm, while that used to make the other cathode disks had an average particle size of 25 nm. The nanostructured FeS_2 powder was made by ball-milling the micron FeS_2 powder.

For equal weights of blended powder and identical pressing conditions, the cathode disks made from nanostructured FeS_2 came out 23% thinner and thus 30% denser than the cathode disks made from micron-scale FeS_2. The nanostructured disks were found to be more robust than the others by comparison of degrees of shattering in a simple drop test. Thus, it was demonstrated that thinner, more robust cathode disks can be made by use of nanostructured instead of micrometer-scale FeS_2 powder.

In thermogravimetric tests of the thermal decomposition of FeS_2 into $FeS + S$, the nanostructured FeS^2 cathodes were found to be more stable. Finally, the discharge electrical performances of batteries containing nano-structured FeS_2 cathodes were found to be superior to those containing micronscale FeS_2 cathodes. The discharged electrical energy per unit mass averaged over all the cell material (electrodes + electrolyte) was found to be 109 J/g in the nanoscale case and 58 J/g in the micronscale.

Improved flame-retardant additive powders with particle sizes <100 nm are undergoing development. Called "nanostructured flame retardants," these powders can be used as ingredients of flame-retardant coatings or blended into composite polymeric materials to make fabrics and structural components that are flame retardant throughout their thicknesses. The principal ingredient of nanostructured fire retardants tested in preliminary experiments was nanostructured antimony oxide, which was synthesized from micron-particle-sized antimony oxide powder in a subatmospheric-pressure chamber by thermal evaporation and condensation onto a cooled surface.

Because of the high specific surface areas associated with their small particle sizes, nanostructured flame retardants are characterized by relatively high reactivity and correspondingly high effectiveness as flame retardants. The small particle sizes make it possible to blend these materials with other materials at a near-molecular level of fineness and interpenetration.

Unlike older flame-retardant additives with larger particle sizes, nano-structured flame retardants can be incorporated into micron- and submicron-sized fibers; this can be an important advantage in making clothing that stays flame-retardant, as contrasted with clothing made flame retardant by coating materials that can come off with repeated washing. In another potential class of applications, nanostructured flame retardants could be incorporated into microelectronic devices, in which particles of excessive size could adversely effect performance.

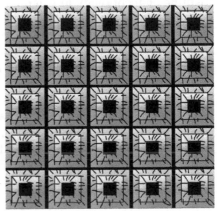

FIGURE 5.22
Nanowires for creating new materials, for example, solid oxide fuel cells.

Nanowires

Scientists at Harvard University have made tiny lasers that can be incorporated into silicon microchips. The lasers, or nanowires, are single wires of the semiconductor cadmium sulfide, which measure only a hundred millionths of a millimeter across (see Figure 5.22). The devices could help make telecommunications and medical technology faster and more compact or even result in new applications.

Although the first nanowire lasers were made from zinc oxide back in 2001 and emitted light in the ultraviolet range, they could only be switched on by being pumped with light from another laser. For most applications, lasers need to be capable of switching on and off electronically.

The Harvard group achieved this necessary electronic control by making a cadmium sulfide nanowire atop a silicon surface. Then, an electrical contact is placed on top of that. At a specific voltage, a current passes from the silicon into the nanowire. Blue-green light is emitted from the ends of the wire. The light color depends on what semiconductor a wire is composed of; for instance, gallium nitride produces blue to ultraviolet light.[107]

Nanobelts

Ultrasmall sensors and inexpensive flat-panel displays are just two of the many potential applications for nanobelts, say researchers at the Georgia

Institute of Technology in Atlanta.[108] Nanobelts are nanometer-scale, thin, flat, and transparent semiconducting metal oxides of Zn, Sn, In, Cd, and Ga, and other commercial-grade materials.

Nanobelts are produced in the lab (from these oxide powders) in an Al_2O_3 tube purged with N_2 or Ar and heated to 1100 to 1400°C. Nanobelts — 30 to 200 nm wide, 10 to 15 nm thick, and up to several millimeters in length — form when the powder evaporates and solidifies on an Al_2O_3 plate in a cooler portion of the furnace. If kept within the proper pressure, temperature, and processing times, they form crystalline belts.

Nanobelts offer significant advantages over existing nanowires and carbon nanotubes, reports the group. For one, nanobelts are chemically pure and need no oxidation protection, as do nanowires. They can also bend 180° without breaking. Nanobelts are structurally uniform and nearly defect-free, each being composed of one crystal with specific surface planes. This is important because defects can destroy quantum-mechanical transport properties. The consistent structure also lends itself to mass production of nanoscale electronic and optoelectronic devices.

Nanofilms

Researchers at the University of Arizona in Tucson have been developing new organic films that could make cloth lights and computerized military garb possible. The organic films for printing on paper, plastic, and textiles could be manufactured on flexible substrates in large quantities.

Researchers aim to print these nanometer-thick films using traditional tools, such as screen printing, ink-jet printing, laser printing, and gravure printing, which reduce costs. These researchers have recently demonstrated that such nanofilms could be printed on cloth.

According to Ghassan L. Jabbour, associate research professor of optical sciences at the University of Arizona, "We now know how to integrate organic materials onto textiles. We haven't solved the whole problem yet, but we understand the pitfalls, what the 'killers' are that prevent these materials from sticking to cloth." The major challenge has been that textiles retain moisture, and organics are moisture sensitive.[109]

Nanocircuits

Nanoscale electronic devices that automatically assemble themselves could be one step closer to reality due to a technique from scientists at the University of Wisconsin Materials Research Science and Engineering Center.[72]

Photolithography today etches circuit elements onto silicon wafers, an approach good for feature sizes to about 150 nm, typical of current-generation chips such as the Pentium 4. But the cost of factories to build the chips rises exponentially with shrinking feature size. It is also unclear if manufacturers can extrapolate current technology much below 50 nm, say experts.

But some experimental techniques build smaller features. For example, a tightly focused electron beam inscribes nanoscale circuit elements on a silicon surface, line by line. But the process takes about a week to do a single chip, impractical compared with photolithography, which can make hundreds of chips per hour.

Another technique that has received attention uses so-called block polymers that spontaneously assemble themselves into periodic structures at the molecular scale. Block copolymers are compounds made of two or more long polymer chains connected at the ends. Unfortunately, the materials tend to organize themselves into spots, cylinders, or broad, swirling patterns, nothing like the precisely ordered structures needed for circuits.

The Wisconsin team has combined these two experimental approaches into what it calls "templated" or "directed" self-assembly. The use of extreme ultraviolet light (which has a much shorter wavelength than the light used in conventional lithography) and some clever optical tricks lay down an alternating pattern of straight, parallel, chemically active stripes just 20 to 30 nm wide. The patterned silicon surface is next washed with a solution containing the block copolymer. In this case, the compound contains two polymers: one chemically attracted to one stripe type, and the other to the opposite stripe type. Changing factors such as stripe spacing and polymer length lets the copolymer organize itself atop the nanoscale stripes.

Researchers concede it is a long way from parallel lines of plastic to fully operational electronic devices. One obvious next step is to grow the block copolymer in nanoscale vertical columns. Such columns would hold one bit of information each. Researchers say the technique may eventually build ultrahigh-density magnetic storage and make possible computers with 4-TB memories. Yet another potential application are nanoscale ICs using copolymer patterns as guides for etching the underlying silicon.

Nanowhiskers

Based on Kear and Strutt's work[73] to fabricate nanostructured whisker-reinforced CMCs entirely from metalorganic precursors, two options have emerged from recent research (see Figure 5.23). Using trichloro-methyl-silane as a precursor compound and a nanodispersed Ni catalyst on a graphite support, SiC whiskers have been grown at practical growth rates of several mm/h at temperatures of 1100–1400°C.[73] Whisker growth occurs by the so-called vapor-liquid-solid (VLS) mechanism, where the silicon source gas reacts with the Ni nanoparticles to form a liquid Ni/Si eutectic; thereafter, whisker growth occurs by transport of the silicon through the liquid droplet to the growing whisker at its base. Because this is in effect a growth-from-the-melt process, a typical SiC whisker displays a high degree of crystalline perfection and exceptionally high strength. Recently, Kear and Strutt[73] have also discovered that CVC-synthesized n-SiC$_x$N$_y$ powders, prepared by thermal decomposition of hexamethyl-disilazane, can also be transformed into whiskers by a simple heat treatment in a reactive gas stream.[110] Thus, heating the n-powders in flowing H$_2$ at 1200°C yields SiC whiskers, whereas heating in flowing NH$_3$/H$_2$ gives Si$_3$N$_4$ whiskers. Rapid growth of the whiskers occurs by two different mechanisms: by a new powder-liquid-solid (PLS) growth mechanism and by the more familiar VLS growth mechanism. An interesting feature of the thermochemical conversion of the n-powders into whiskers is the formation of a three-dimensional random weave, which replicates the shape of the original powder bed from which it is derived. In fact, the whisker preform, which is both strong and resilient, can be removed intact from its container, say a ceramic crucible, and used for subsequent processing. Kear and Strutt have attempted to infiltrate whisker preforms with nano-particle slurries in order to investigate the possibilities for n-CMC sheet fabrication (see Figure 5.23).

Researchers at Washington University in St. Louis have reported that they have made B nanowhiskers by CVD. The particles have diameters in the range of 20 to 200 nm, and the whiskers (also called nanowires) are semiconducting and show properties of elemental B.

In the nano world, the carbon nanotube is king, considered the particle most likely to make new materials, and increasingly valued as potential metallic conductors in the burgeoning experimental world of molecular electronics. However, carbon has its limitations: Its cell wall structure and variable conductivity make it unreliable as a conductor; only one third of those grown have metallic characteristics; the others are semiconductors. And one specific type cannot be predictably grown; instead, a mix of types is grown together.

The researchers at Washington University[111] turned to B, one spot to the left of carbon in the periodic table, to see if it would be a good candidate. If nanotubes could be made of B and produced in large quantities, they should have the advantage of having consistent properties despite indi-vidual variation in diameter and wall structure. The discovery that the

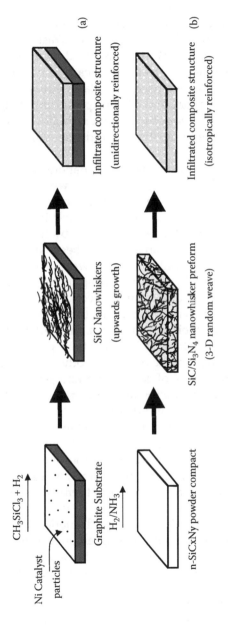

FIGURE 5.23
Schematic diagrams showing two different approaches for synthesizing whisker-reinforced ceramic matrix composites: (a) unidirectionally reinforced, (b) isotropically reinforced.

nanowhiskers are semiconducting makes them promising candidates for nanoscale electronic wires.

"The theoretical papers predicted that B nanotubes may exist and, if they do, should have consistent electrical properties regardless of their helicity. This would be a distinct advantage over carbon nanotubes," claims Carolyn Jones Otten and W. E. Buhro from Washington University. "So, we set out to make these. We had already done some work on BN nanotubes, which are similar in structure to carbon nanotubes but they are electrically insulating. So, we used a similar method to try to make B nanotubes. We grew things that looked very promising: long, thin, wire-like structures. At first we thought they were hollow, but after closer examination, we determined that they were dense whiskers, not hollow nanotubes."

The notion of B nanotubes excites people in nanotechnology more than nanowhiskers because of their unique structure, which could be likened to a distinct form of an element. Carbon, for instance, is present as graphite and diamond, and, recently discovered, in "buckyball" and nanotube conformations. Also, B nanotubes are predicted by theory to have very high conductivity, something researchers have been eager to measure.

The nanowhiskers made by Buhro's group were electrically characterized to see if they were good conductors despite being whiskers rather than tubes. They were found to exhibit semiconducting behavior. However, bulk B can be "doped" with other atoms to increase its conductivity. Often, Buhro's group and their collaborators have attempted to do the same thing with B nanowhiskers to increase their conductivity. Carbon nanotubes have been doped, as have various other kinds of nanowires, and assembled in combinations of conducting and semiconducting ones to make for several different microscale electronic components such as rectifiers, field-effect transistors, and diodes.

Since the early 1990s Buhro and his group have been making many kinds of nanowires and nanotubes that might ultimately be incorporated into nanoelectronic devices. Nanowires and nanotubes have been receiving more current attention as potential transistors, wires, and switches for ultrasmall circuits and devices to be built from them on almost a molecular scale.

"If you want to make electronics smaller and smaller, you have to make the component devices and the wires that interconnect them smaller and smaller," Buhro says. "We are trying to build the scientific infrastructure for electronic nanotechnology and to understand the basic principles involved. We have to find out how these nanowires work and how to connect them into circuits and functional devices. Even when we have that, nobody yet knows how a computer chip will be made that uses these things. That is a wide-open, unsolved problem. But the fundamental science to be done is potentially important and must be done."

Nanocapsules

M. Kusunoki and colleagues[112] at Japan Fine Ceramics Center have discovered that carbon nanocapsules can be produced just by heating carbon fibers. No arc discharge is necessary. The research team performed microscopic investigations of amorphous carbon nanoparticles at high temperatures and ascertained that the particles become graphite at the surface.

Since 1990, when C60, a soccer ball-like carbon molecule, was found in soot, many researchers have examined cages of carbon atoms, or fullerenes. Many possible applications are being pursued in different fields. The most attractive of the fullerene molecules are nanometer-sized tubes and capsules made of carbon atoms because of their properties.

Fullerenes were thought to be produced only in plasma with a strong electrical field and a very high temperature. Most conventional processes use arc discharge The new research shows that nanometer carbon tubes and capsules can be produced in a purely thermal process.

The research team realized that carbon nanocapsules were formed on sintered carbon fibers. Specifically, after heating carbon fibers at 2000°C under tension, carbon nanocapsules were found on the fiber surface. Although some capsules are hollow, most have a shell made of six graphite sheets and contain a single crystal of calcium sulfide (CaS), which seems to come from impurities in the source carbon (see Figure 5.24).

The above team's investigation revealed that amorphous carbon particles grow on the surface of a carbon fiber being heated and then become graphite at the surface and hollow inside at 800°C. Each particle turns into a carbon nanocapsule, either cubic or polyhedral, when the temperature rises to 2000°C.

The capsules are so easy to shake off carbon fibers using ultrasonic waves that the capsules can be obtained in the form of powder. Because so many carbon nanocapsules have been available, their physical properties will be easy to determine. Masses of carbon nanocapsules may serve as an elastic lubricant.

This new method promises to be a source of high-purity carbon. When a material is added as an impurity to source carbon, the method will produce capsules enclosing the material. Possible applications of such capsules are a drug-delivery system (DDS) and delivery of contrast medium for medical imaging.

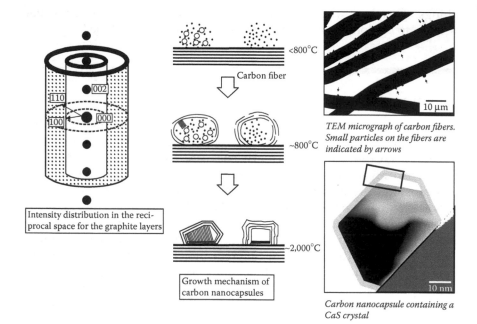

Intensity distribution in the reciprocal space for the graphite layers

<800°C

Carbon fiber

~800°C

~2,000°C

Growth mechanism of carbon nanocapsules

TEM micrograph of carbon fibers. Small particles on the fibers are indicated by arrows

Carbon nanocapsule containing a CaS crystal

FIGURE 5.24
Growing carbon nanocapsules.

Nanospheres

Durable silica nanospheres ranging in size from 2 to 50 nm have been under study by engineers at SNL with an eye toward industrial and medical applications. The spheres are created when researchers take droplets of a homogeneous solution of water, ethanol, silica, and a surfactant and pass them through a reactor or furnace. "As liquid starts to evaporate, the rest of the material self-assembles into a completely ordered particle that maintains its shape when heated," explains Jeff Brinker, a Sandia researcher. They have uniform pores that can absorb organic and inorganic substances, including iron. This means they can be positioned by magnets and the contents released as needed, which might make them well suited as drug-delivery system.

Other engineers believe they could be used as fillers in encapsulants for weapons and tools. The expansion coefficients of encapsulating polymers and the metallic devices they cradle usually differ significantly. So when the temperature changes, the encapsulants expand or contract and stress the devices they are meant to protect. Nanospheres, on the other hand, are porous and would create much less stress when they expand or contract.[113]

Nanofibers

A bone-like material that could be suitable for the repair of bone fractures and the treatment of bone cancer has reportedly been designed by scientists[114] at Northwestern University in Evanston, IL. The molecules are said to self-assemble into a three-dimensional structure that mimics the key features of human bone at the nanoscale level, including the collagen nanofibers that promote mineralization and mineral nanocrystals. Collagen, the most abundant protein in the human body, is found in most human tissues, including the heart, eye, blood vessels, skin, cartilage, and bone, and gives these structures their structural strength.

When the synthetic nanofibers form, they make a gel that could be used as a sort of glue in bone fractures or in creating a scaffold for other tissues to regenerate. When the bone mineral hydroxyapatite is added, the fibers direct the growth of the mineral crystals into an alignment very similar to that of hydroxyapatite around the collagen fibers in natural bone. Because of its chemical structure, the nanofiber gel would encourage attachment of natural bone cells, helping to patch the fracture. The gel could also be used to improve implants or hip or other joint replacements.

The finding also indicates how other materials could be made by self-assembly and spontaneous mineralization that take advantage of an inorganic material growing on an organic material. Such materials could be applied in electronics, photonics, magnetics, and catalysis.

Nanoceramics/Shells

Why does a ceramic coffee cup break much more easily than a seashell? That might seem like a question to ponder during a long, lazy afternoon at the beach.

Seashells, it turns out, have some surprising qualities, and they have been attracting the attention of researchers for more than a decade. "They're extraordinarily tough," says Case Western Reserve University materials science Professor Arthur Heuer, who attributes their crack- and shatter-resistant properties to "an exquisite microarchitecture." Understanding the details of this nanostructure, he adds, could lead to insights into how to make ceramics that are similarly tough and shatter resistant.[115]

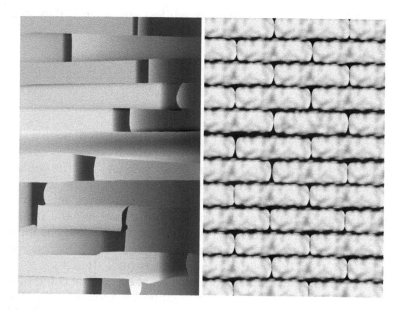

FIGURE 5.25
Researchers at General Electric are studying just why seashells resist cracking and breaking. The microstructure of mother-of-pearl found in red abalone shows layers of platelets (left). On the right is GE's latest version of a synthetic ceramic inspired by seashells.

An abalone shell is made chiefly of calcium carbonate, which is organized into multisided "tablets" that are closely packed in layers. A rubbery polymer glues the tablets together and serves as a cushion between the layers. The shells are unlikely to break or shatter because when a microcrack does form, it propagates along complicated, tortuous paths that, in effect, diffuse the crack. The polymer layers also absorb the damage, so although shells get the equivalent of bumps and bruises, they don't easily break.

GE materials scientist Mohan Manoharan and his team started work on seashells in January 2002. After their study of seashell microstructure was completed, the researchers began attempts to replicate nature's results. Manoharan's team built computer models of shell-inspired materials, starting with models that will consist of just a few layers (see Figure 5.25).

Nanoantennas

Researchers at Purdue University have demonstrated (through mathematical simulations) "nanoantennas," tiny wires and metallic spheres arranged in various shapes. The phenomenon could bring improved medical imagers and detectors for chemical and biological warfare agents able to assay single molecules, or a million times more sensitive than current technology.

Nanoantennas work by moving clouds of electrons in groups or "plasmons." The action lets plasmonic nanomaterials focus tighter than the wavelengths shining on them, something conventional optics cannot do.

Unlike ordinary "right-handed" materials, such as glass, plastic, air, and water, nanoantennas are considered "left-handed" and reverse the normal behavior of visible light and other wavelengths shining on or through them. Nanoantennas arranged in parallel pairs of tiny wires, in theory, produce a negative index of refraction, for instance.

"The phenomenon was first predicted in the late 1960s but all of the work so far has been done with microwaves," says Vladimir Shalaev, professor in Purdue's School of Electrical and Computer Engineering. But pushing the envelope to include visible light could make possible a whole new class of devices, including so-called superlenses for advanced medical diagnostics.

The next step: build nanoantennas and conduct experiments to support the theoretical calculations.[116]

Nanoelectronics

Ultradense integrated circuits with features smaller than 10 nm would provide enormous benefits for all information technologies, including computing, networking, and signal processing. However, recent results indicate that the current very large-scale integrated circuit paradigm based on complementary metal oxide semiconductor (CMOS) technology cannot be extended into this region. The main reason is that below a 10 nm gate length, the sensitivity of silicon field-effect transistor parameters — most importantly, the gate-voltage threshold — to the inevitable random variations of device size grows exponentially. This sensitivity may send fabricating costs skyrocketing and lead to the end of Moore's law sometime during the next decade.

The main alternative nanodevice concept — single electronics, based on controlling the motion of single electrons in solid-state structures — offers some potential advantages over CMOS technology, such as a broader choice of materials. However, the critical dimensions of single-electron transistors for room-temperature operation should be below ~1 nm, far too small for current and even forthcoming lithographic techniques, such as those based on extreme ultraviolet radiation and electron beams.

Hybrid Circuits

Many physicists and engineers believe that this impending crisis may be resolved only by a radical paradigm shift from purely CMOS technology to hybrid semiconductor-molecular circuits, which has been named CMOL, a combination of CMOS and MOLecular. Such a circuit would combine

- A level of advanced CMOS devices fabricated by lithographic patterning
- A few layers of parallel nanowire arrays formed, for example, by nanoimprinting
- A level of molecular devices that self-assemble on nanowires from solution

The CMOL concept combines the advantages of nanoscale components, such as the reliability of CMOS circuits and the minuscule footprints of molecular devices, and the advantages of patterning techniques, which include the flexibility of traditional photolithography and the potentially low cost of nanoimprinting and chemically directed self-assembly. This combination may enable CMOL circuits of unparalleled density — up to 3×10^{12} functions/cm^2 — at acceptable fabrication costs.

For single-molecule components, single-electron devices are the leading candidates because in contrast to field-effect transistors, their operation mechanism does not require high conductivity of the device-to-electrode interfaces. The CMOS layer allows CMOL circuits to circumvent one prominent drawback of single-electron transistors: their low voltage gain. The recent demonstration of single-molecule single-electron transistors by a Cornell University team headed by Paul McEuen and Daniel Ralph and a Bell Laboratories group led by Nikolai Zhitenev offers hope that the first CMOL circuits will be demonstrated within the next 10 years. Hopefully, CMOL circuits will reach the industry in time to preempt the impending Moore's law crisis.

Before that, however, a host of physical, chemical, computer science, and electrical engineering problems must be solved. Notably, researchers need to bring chemically directed self-assembly from the present level of single-layer growth on smooth substrates to the reliable placement of three-terminal molecules on nanowire structures. (Two-terminal devices may be suitable for simple memories but not for more advanced circuits.)

A development and implementation of several molecules suitable for the above arrangement, suggested by Andreas Mayr, professor of chemistry at State University of New York at Stony Brook, essentially combines two well-known single-electron devices, the transistor and the trap. In it, diimide acceptor groups work as single-electron islands, which are sites where an additional electron can be localized. The islands are connected either by oligo-phenylene-ethynylene bridges playing the role of tunnel junctions or by longer chains that do not conduct electrons but stabilize the geometric arrangement. All the bridges and chains are terminated by thiol groups, which serve as alligator clips and should allow molecular self-assembly on gold nanowires.

This molecule works in the following way: When the sum of voltages applied to the top and left nanowires exceed a certain threshold, an additional electron is injected from the right wire into the trap island. The electron's electrostatic field opens the single-electron transistor connecting the top and right wires. This connection would survive a temporary reduction of the applied voltage because it takes time for the trapped electron to escape. As a result, the device functions as a two-input latching switch, essentially a single-bit memory cell controlled by two input signals.[117]

Neuromorphic Networks

Additional studies show that this device discussed above, working together with CMOS layer transistors, may enable the creation of a broad variety of highly functional integrated circuits. A family of distributed crossbar neuromorphic networks, which have been termed CrossNets, seems especially promising. Every CrossNet (Figure 5.26) is essentially a large field of similar tiles (or "plaquettes"), which is now and then interrupted by contacts to "gray cells" located in the underlying CMOS layer. In this biologically inspired architecture, the latching switches (shown as circles with arrows) serve as adaptive "synapses" that connect "axonic" nanowires and perpendicular "dendritic" nanowires. The cells, which are essentially nonlinear differential amplifiers, serve as "somas" or neural cell bodies.

The motivation for using neuromorphic networks comes from the well-known comparison of the performance of present-day digital computers with biological neural systems for one of the simplest tasks: image recognition (or, more precisely, image classification). A mammal's brain recognizes a complex visual image, with high fidelity, in approximately 100 ms. Because the elementary process of neural cell-to-cell communication in the brain takes 10 to 30 ms, this means that the task takes just a few elementary operations. In contrast, the fastest modern microprocessor performing digital number crunching at a clock frequency of a few gigahertz and running the best commercially available software would require minutes (about 10^{11} clock periods) for an inferior classification of a similar image. The contrast is striking indeed.

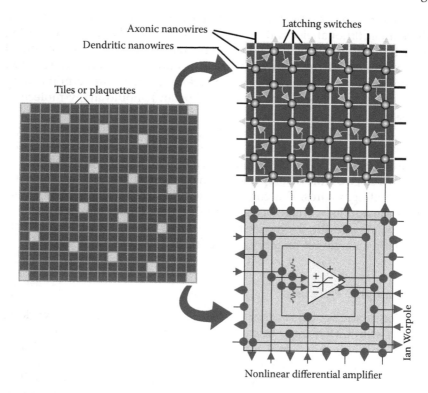

FIGURE 5.26

A nanowire-molecule array, now and then interrupted by contacts to semiconductor cells, forms a distributed crossbar network.

Neuromorphic networks do not require the usual software, but they do need to be trained to perform their tasks. For CrossNets, the main challenge is that external access to individual synapses is impossible.

Success in image recognition can lead to relatively small CrossNet chips that may revolutionize the field of pattern analysis, including such important applications as detection of specific features, such as a tank in a satellite image.

Self-Evolution

The success of the above image recognition and CrossNet chips would pave the way for more ambitious goals. It is plausible that a cerebral-cortex-scale CrossNet-based system — one with ~10^{10} neurons and 10^{15} synapses that would require a ~30×30 cm^2 Si substrate — would be able, after initial training by a dedicated external tutor, to learn directly from its interaction with its environment. In this case, we can speak of a "self-evolving" system.

In computer science, such concepts as evolutionary computation, evolutionary algorithms, and evolving hardware are well known. However, with CMOL circuits, there is the first chance to merge these ideas with the enormous computing power of parallel very large-scale integrated circuits.

Such evolution could lead to self-development of such advanced features as system self-awareness (consciousness) and reasoning. If substantial success along these lines materializes, it will have a strong impact not only on information technology but also on society as a whole.[118, 119]

References

1. Colvin, V., Measuring the Risks of Nanotechnology, *Technol. Rev.*, 71–73, 2003.
2. Roco, M. C., *Nanotech Briefs*, 5–6, 2003.
3. Gleiter, H., Nanostructured Materials — State-of-the-Art and Perspectives, *Zeitschrift Fur Metallkunde*, 86(2), 78–83, 1995.
4. Goddard III, W. A., Brenner, D. W., Lyshevski, S. E., and Iafrate, G. J., Eds., *Handbook of Nanoscience, Engineering, and Technology*, Boca Raton, FL: CRC Press,, 2003.
5. Gonsalves, K. E., Chow, G. M., Xiao, T. D., and Cammarata, R. C., Molecularly Designed Ultrafine Nanostructured Materials, *Mater. Res. Soc. (MRS) Symp. Proc.*, San Francisco, CA, 351, 477, April 4–8, 1994.
6. Siegel, R. W., Synthesis and Properties of Nanophase Materials, *Mater. Sci. Eng.*, A168, 189–197, 1993.
7. Hahn, H. and Padmanabhan, K. A., Mechanical Response of Nanostructured Materials, *Nanostructured Materials*, Elsevier Science Ltd., 6(1–4), 191–200, 1995.
8. Gleiter, H., in N. Hansen, T. Leffers, and H. Lilholt, Eds., *Proc. Second Risø Int. Symp. on Metall. and Mater. Sci.*, Røskilde, Denmark, 1981, p. 15.
9. Metal Chips Ahoy, *Machine Design*, October 10, 2002, p. 47.
10. http://www.rpi.edu/web/campus.News/aug_02/aug_26/ajayan.html.
11. Nanocrystals Hold Key to Lower-Cost Superplasticity Parts, *Design News*, August 2, 1999, p. 20.
12. http://www.jhu.edu/news_info/news/home03/jul03/twinning.html.
13. Chow, G. M., Polyol Process Yields Nanocrystalline Co-Cu Powders, *High-Tech Materials Alert*, November 10, 1995, pp. 3–4.
14. Ying, J. Y., Surface Structure of Nanocrystalline Oxides, NATO Study Insti. on Nanophase Materials: Synthesis, Properties, Applications, Cambridge, MA: MIT, June 1993.
15. Xiao, S. Q., Foitzik, A. H., Welsch, G., et al., Nanocrystalline Titanium Aluminide, *Acta Metallurgica et Materialia*, 42(7), 2535–2545, 1994.
16. Mazumder, J., Duffey, T., Yamamoto, T., and Chung, H., Synthesis of Nanocrystalline Nb-Aluminides by Laser Ablation Techniques, Illinois Univ. at Chicago Circle, *NTIS Alert*, 97(11), 15, 1997.

17. Ying, J. Y., Nanocrystalline Processing and Interface Engineering of Si_3N_4-Based Nanocomposites, Cambridge, MA: MIT, *NTIS Alert*, 97(11), 19, 1997.

18. Gleiter, H., Microstructure, in Cahn, R. W. and Haasen, P. (Eds.), *Physical Metallurgy*, 4th ed., London: Elsevier Science, 1996, p. 844.

19. Gleiter, H., *Progress in Materials Science*, 33, 223, 1989.

20. Betz, U. and Hahn, H., Nanostructured *Materials*, 12, 911, 1999.

21. Hahn, H. and Averback, R. S., *J. Am. Ceram. Soc.* 74, 2918, 1991.

22. Mayo, M. J., in Hori, S., Tokizane, M., and Furushiro, N. (Eds.), *Superplasticity in Advanced Materials*, The Japan Society for Research on Superplasticity, Japan, 1991, p. 541.

23. Hahn, H. and Averback, R. S., *Nanostructured Materials*, 1, 95, 1992.

24. Wakai, E., Sakaguchi, S., and Matsuno, Y., *Adv. Ceram. Mater.*, 1, 259, 1986.

25. Allen, A. J., Krueger, S., Skandan, G., Long G. G., Hahn, H., Kerch, H. M., Parker, J. C., and Ali, M. N., *J. Am. Ceram. Soc.*, 79, 1201, 1996.

26. Yan, M. F., Solid State Sintering, in Schneider, S. J., (chairman), *Engineered Materials Handbook*, Vol. 4, "Ceramics and Glasses," ASM Int., 1991, p. 304.

27. Kingery, W. D., Bowen, H. K., and Uhlmann, D. R., *Introduction to Ceramics*, 2nd ed., New York: Wiley, 1976.

28. Chiang, Y.-M., Birnie, D., and Kingery, W. D, *Physical Ceramics — Principles for Ceramic Science and Engineering*, New York: Wiley, 1997.

29. Chen, I.-W., Wang, X.-H., *Nature*, 404, 168, 2000.

30. Hahn, H., Logas, J., and Averback, R. S., *J. Mater. Res.*, 5, 609, 1990.

31. Mayo, M. J., *Material Design*, 14, 323, 1993.

32. Liao, S. C., Chen, Y. J., Kear, B. H., and Mayo, W. H., *Nanostructured Materials*, 10, 1063, 1998.

33. Hahn, H., *Nanostructured Materials*, 2, 251, 1993.

34. Skandan, G., Processing of Nanostructured Zirconia Ceramics, *Nanostructured Materials*, 5, 111, 1995.

35. Skandan, G., Hahn, H., Kear, B. H., Roddy, M., and Cannon, W. R., *Mater. Lett.*, 20, 305, 1994.

36. Skandan, G., Hahn, H., Roddy, M., and Cannon, W. R., *J. Am. Ceram. Soc.*, 77, 1706, 1994.

37. Srdic, V. V., Winterer, M., and Hahn, H., *J. Am. Ceram. Soc.*, 83, 729, 2000.

38. Srdic, V. V., Winterer, M., and Hahn, H., *J. Am. Ceram. Soc.*, 83, 1853, 2000.

39. Srdic, V. V., Winterer, Möller, M., Miehe, G., and Hahn, H., *J. Am. Ceram. Soc.*, 84, 2771, 2001.

40. Benker, A., Diploma thesis, Darmstadt, 1999.

41. Winterer, M., *Nanocrystalline Ceramics: Synthesis and Structure*, Berlin: Springer, 2002.

42. Gleiter, H., *Progress Mater. Sci.*, 33(4), 223, 1989.

43. Birringer, R., *Mater. Sci. Eng.*, A117, 33, 1989.

44. Suryanarayana, C. and Froes, F. H., *Metall. Trans.*, A 23A, 1071, 1992.

45. Siegel, R. W., *Nanostructured Materials*, 3, 1, 1993.

46. Shull, R. D., *Nanostructured Materials*, 2, 213, 1993.

47. Niihara, K., Nakahira, A., and Sekino, T., New Nanocomposite Structural Ceramics, *MRS Symp. Proc.*, 286, 408–412, 1993.

48. Nakahira, A. and Niihara, K., *J. Ceram. Soc.*, Japan, 100, 448, 1992.

49. Izaki, K., Hakkei, K., Ando, K., and Niihara, K., *Ultrastructure Processing of Advanced Ceramics*, J. M. Mackenzie and D. R. Ulrich (Eds.), New York: John Wiley & Sons, 1980, p. 891.
50. Sekino, T., Nakahira, A., Suzuki, Y., and Niihara, K., *J. Japan Powd. Powd. Met.*, 43(3), 272–277, 1996.
51. Sekino, T., Nakahira, A., Nawa, M., and Niihara, K., *Proc. Aust. Ceram. Soc. '92*, Melbourne, 1992, p. 745.
52. Levin, I., Kaplan, W. D., and Brandon, D. G., Effect of SiC Submicrometer Particle-Size and Content Alumina-SiC Nanocomposites, *J. Am. Ceram. Soc.*, 78(1), 254–256, 1995.
53. JPRS-JST-94-011L, March 25, 1994, pp. 16–18.
54. Novak, B. M., Hybrid Nanocomposite Materials between Inorganic Glasses and Organic Polymers, *Adv. Mater.*, 5(6), 422–433, 1993; *NTIS Alert*, 96(7), 18–19, 1996.
55. Dressler, W., Greiner, A., Seher, M., and Riedel, R., Fabrication of Nanostructured Ceramics by Hybrid Processing, *Nanostructured Materials*, 6(1–4), 481–484, 1995.
56. Prabhu, G. B. and Bourell, D. L., Synthesis and Sintering Characteristics of Zirconia and Zirconia-Alumina Nanocomposites, *Nanostructured Materials*, 6(1–4), 361–364, 1995.
57. Jalowiecki, A., Bill, J., Friess, M., Mayer, J., Aldinger, F., and Riedel, R., Designing of Si_3N_4/SiC Composite Materials, *Nanostructured Materials*, 6(1–4), 279–282, 1995.
58. Rousset, A., Alumina-Metal (Fe, Cr, FeO.8Cr0.2) Nanocomposites, *J. Solid State Chem.*, 11(1), 164–171, 1994.
59. *AFRL Technology Horizons*, March 2003, pp. 29–32.
60. Chaiko, D. J., Leyva, A., and Niyagi, S., Nanocomposite Manufacturing, *AM&P*, June 2003, pp. 44–46.
61. http://www.polyone.com (www.netcomposites.com/news.asp?1510, accessed January 14, 2003, p. 3).
62. Connell, J. W., Smith, Jr., J. G., and Park, C., (NRC), Organic/Inorganic Hybrid Polymer/Clay Nanocomposites, NASA LRC, Hampton, VA, *Nanotech Briefs*, October 2003, pp. 11–12.
63. Sato, H. and Kagawa, Y., *Jap. Soc. Composite Mater.*, Tokyo, 663, 1995.
64. Naganuma, T., Iba, H., and Kagawa, Y., *J. Mater. Sci. Lett.*, 18, 1587, 1999.
65. Kagawa, Y., Iba, H., Tanaka, M., Sato, H., and Chang, T., *Acta Mater.*, 46(1), 265, 1998.
66. Naganuma, T. and Kagawa, Y., *Composites Sci. Tech.*, 62(9), 1187, 2002.
67. Shi, A., Zhou, H., and Edwards, D. E., Optical Nanocomposite is Suited for Photonics Packaging, *Laser Focus World*, October 2003, pp. 93–96.
68. Mikrajuddin, M., Lenggoro, I. W., Okuyama, I. C., and Shi, F. G., *J. Electrochem. Soc.*, 149(5), 107, 2002.
69. Mikrajuddin, M., Lenggoro, I. W., Okuyama, I. C., and Shi, F. G., *J. Chem. B*, 107, 1957, 2003.
70. Haseeb, A. S. M. A., Celis, J. P., and Roos, J. R., An Electrochemical Deposition Process for the Synthesis of Laminated Nanocomposites, *Mater. Manufactur. Proc.*, 10(4), 707–716, 1995.

71. Powell, D., Nanotechnology in Space, *NASA Tech Briefs,* September 2003, p. 19.
72. Kincade, K., "Nano" Takes Its First Real Steps into the Industrial World, *Laser Focus World,* September 2003, pp. 73–78.
73. Kear, B. H. and Strutt, P. R., Chemical-Processing and Applications for Nanostructured Materials, *Nanostructured Materials,* 6(1-4), 227–236, 1995.
74. Strutt, P. R. and Boland, R. F., University of Connecticut, private communication, 1994.
75. McCandlish, L. E., Kear, B. H., Kim, B. K., and Wu, L., *Protective Coatings: Processing and Characterization,* R. M. Yazici (Ed.), Warrendale, PA: The Metallurgical Society, 1990.
76. Bretzman, R. W., Aikens, J., Batilo, F., et al., Nanomaterials and Wear Resistant Polymers, *Ceramic Industry,* June 2000, pp. 39–43.
77. Bruno, L., Nanotechnology, *Red Herring,* March 2003, pp. 50–51.
78. Zeichick, A., The Fabric of Consumer Reality, *Red Herring,* March 2003, p. 54.
79. Tristram, C., Reinventing the Transistor, *Technology Review,* September 2003, pp. 54–62.
80. Yamada, T., Nanoelectronic Devices with Precise Atomic-Scale Structures, *NASA Tech Briefs,* March 2002, pp. 43–44.
81. Biever, C., Behind the Looking Glass, *Red Herring,* March 2003, p. 58.
82. Tsukruk, V. V. and Zubarev, E. R., *Langmuir,* September 16, 2003.
83. Mark, J., Rajan, G., Schaefer, D., Beaucage, G., and Sur, G., Nanoparticles Make Silicone Rubber Clearly Stronger, *J. Polymer Sci. Part B: Polymer Phys.,* August 15, 2003.
84. Haruta, M., *Catal. Today,* 36, 153, 1997.
85. Haruta, M. and Date, M., *Appl. Catal. A,* 222, 427, 2001.
86. Bond, G.C. and Thompson, D. T., *Gold Bull.,* 33, 41, 2000.
87. Bond, G. C., *Gold Bull.,* 34, 117, 2001.
88. Templeton, A. C., Wuelfing, W. P., and Murray, R. W., *Acc. Chem. Res.,* 33, 27, 2001.
89. Crooks, R. M., Zhao, M. Q., Sun, L., Chechik, V., Yeung, L. K., *Acc. Chem. Res.,* 34, 181, 2001.
90. Rolison, D. R., *Science,* 299, 2003, p. 1698.
91. Nanoengineered Membranes, *Machine Design,* September 5, 2002, p. 32.
92. Carbonitride Nanoparticles Strengthen Steel Against Creep, *AM&P,* October 2003, p. 11.
93. www.nsf.gov/od/lpa/news/03/pr0368.htm, accessed July 8, 2003, p. 1.
94. Cauchetier, M., Musset, E., et al., Laser Synthesis of Nanosized Powders, in A. Legrand and S. Senemaud (Eds.), *Nanostructured Silicon-Based Composites,* New York: Taylor & Francis, 2002, pp. 6–23.
95. Cauchetier, M., Croix, O., Luce, M., Baraton, M., Merle, T., and Quintard, P., Nanometric Si/C/N Composite Powder: Laser Synthesis and IR Characterisation, *J. Eur. Ceramic Soc.,* 8, 215, 1991.
96. DeMeis, R., Take a (Nano) Powder! *Design News,* March 1, 1999, pp. 143–144.
97. Murphy, B. R. and Courtney, T. H., Synthesis of Cu-NbC Nanocomposites by Mechanical Alloying, Michigan Technological University, *NTIS Alert,* 95(19), 50, 1995.
98. New "Nanopowders" for Super-Strength Components, *Mater. Perform.,* 36(4), 76, 1997.

99. Hellstern, E., Fecht, H., Fu, Z., and Johnson, W. L., *J. Appl. Phys.*, 65, 305, 1988.
100. Fecht, H., Hellstern, E., Fu, Z., and Johnson, W. L., *Met. Trans.* 21A, 2333, 1990.
101. Oehring, M. and Borman, R., *Mat. Sci. Engr.*, A134, 1330, 1991.
102. Trudeau, M. and Schulz, R., *Mat. Sci. Engr.*, A134, 1361, 1991.
103. Calka, A. and Radlinski, A. P., *Mat. Sci. Engr.*, A134, 1350, 1991.
104. Wagner, C. N. J., Yang, E., and Boldrick, M. S., Structure of Nanocrystalline Fe, W, and NiAl Powders Prepared by High-Energy Ball Milling, *Nanostructured Materials*, 71-2, January-February 1996, Pergamon Press, Conf. Proceedings of TMS-AIME Symp. on the Struct. and Prop. of Nanophase Materials, 1995.
105. Yang, M. L. and Yadav, T., Rapid Densification of Ceramic Monoliths and Composites, *NASA Tech Briefs*, October 1998, pp. 69–70.
106. Au, G., Lei, W., and Yadav, T., Thermal Batteries Made with Nanostructured Material, *NASA Tech Briefs*, April 1999, pp. 28–30.
107. http://link.abpi.net/l.php?20030121A3, Accessed January 21, 2003.
108. Tightening His Nanobelt, *Design News*, June 18, 2001, p. 23.
109. http://link.abpi.net/l.php?20021003A2, Accessed October 3, 2002, p. 1.
110. Chang, W., Skandan, G., Danforth, S. C., Kear, B. H., and Hahn, H., *Nanostructured Materials*, 4, 507, 1994.
111. http://www.netcomposites.com/news.asp?1272, pp. 1–2.
112. Carbon Nanocapsules Grown on Carbon Fibers, *JETRO*, 13, August 1995.
113. Self-Assembling Nanospheres Show Industrial and Medical Promise, *Machine Design*, August 5, 1999.
114. Stupp, S. I., Nanofiber Material Developed to Repair Bone Fracture, *AM&P*, 160(2), 26, 2002.
115. Diop, J. C., Nanoceramics, *Technol. Rev.*, December 2002/January 2003, p. 65.
116. Nanoantennas, *Machine Design*, October 10, 2002, p. 56.
117. Likharev, K. K., Hybrid Semiconductor — Molecular Nanoelectronics, *The Industrial Physicist*, June/July 2003, pp. 20–23.
118. Likharev, K. K., Electronics Below 10 nm, *The Industrial Physicist*, June/July 2003, p. 23, and *Nano and Giga Challenges in Microelectronics;*Amsterdam: Elsevier, 2003, http://rsfq1.physics.sunysb.edu/~likharev/nano/NanoGiga031603.pdf.
119. Likharev, K. K., Mayr, A., Türel, and Muckra, I., CrossNets: High-Performance Neuromorphic Architectures for CMOL Circuits, *The Industrial Physicist*, June/July 2003, p. 23.

Bibliography

Avouris, P. and Appenzeller, J., Electronics & Optoelectronics with Carbon Nanotubes, *The Industrial Physicist*, June/July 2004, pp. 18–21.

Bill, J., Wakai, F., and Aldinger, F., *Precursor-Derived Ceramics*, New York: Wiley-VCH, 1999.

Carbon Nanotube Commercial Manufacturing Breakthrough, http://www.thomasswan.co.uk, p. 1.

Carson, R. T., Savage, H. S., and Rigsbee, J. M., PVD Fabrication of Nanocrystalline Composite Materials, Univ. of Illinois at Urbana-Champaign, Adv. Construction Technology Center, *NTIS Alert*, 96(4), 28, 1996.

Corti, C. W., Holliday, R. J., and Thompson, D. T., Developing New Industrial Applications for Gold: Gold Technology, *Gold Bull.*, 35/4, 111–117, 2002.

Eleskandarany, M. S., Omori, M., and Ishikuro, M., et al., Synthesis of Full-Density Nanocrystalline Tungsten Carbide by Reduction of Tungstic Oxide at Room Temperature, *Metall. Mater. Trans., A-Phys. Metall. Mater. Sci.*, 27(12), 4210–4213, 1996.

Fendler, J. H., Chemical Self-Assembly for Electronics Applications, *Chem. Mater.*, 13, 3196–3210, 2001.

Fluid Nanostructuring, *The Industrial Physicist*, June/July 2003, p. 11.

Gonsalves, K., Nanostructured Bearing Alloy Studies, University of Connecticut, Storrs Institue of Materials Science, *NTIS Alert*, 96(6), 47, 1996.

Goursat, P., Bahloul-Hourlier, D., Doucey, B., and Mayne, M., Processing and Tailoring of Si/C/N-Based Nanocomposites, in A. Legrand and S. Senemaud (Eds.), *Nano Structured Silicon-Based Composites*, New York: Taylor & Francis, 2002, pp. 238–264.

http://f'orbes.z2c.net/rd4/ck/3072-13370-1839-183?m=8-8&e=fef9626194f6; SCIENCE: GREY GOO THEORY.

Kear, B. H. and McCandlish, L. E., Nanostructured W-Base Materials; Synthesis, Processing, and Properties, *J. Adv. Mater.*, 25(1), 11–19, 1993.

Kim, B. K., Ha, G. H., and Lee, D. W., Sintering and Microstructure of Nanophase WC/Co Hardmetals, *J. Mater. Proc. Technol.*, 63(1–3), 317–321, 1997.

Kuzmany, H., Fink, J., Mehring, M., and Roth, S. (Eds.), Molecular Nanostructures, *AIP Conf. Proceedings*, Volume 685, XVII International Winterschool/Euroconference on Electronic Properties of Novel Materials, Kirchberg, Tirol, Austria, March 8–15, 2003. Melville, NY: American Institute of Physics, 2003.

Nanocomposites from Humidifiers, *Machine Design*, May 20, 2004, p. 25.

Nanoscale Technology Yields Better Thermal Insulators, *Machine Design*, April 1, 2004, p. 37.

Nanosized Conveyor Belt, *Machine Design*, June 3, 2004, p. 43.

Niihara, K., *The Centennial Memorial Issue of the Ceramic Society of Japan*, 99(10), 974, 1991.

Nottingham Research Could Lead to New Range of Nanostructures, p. 1, accessed September 12, 2003; http://www.nottingham.ac.uk/public-affairs/press-releases/index.phtml?menu=pressreleases.

Ogino, Y., Yamasaki, T., and Shen, B. L., Indentation Creep in Nanocrystalline 26Fe-Tin and Ni-Tin Alloys Prepared by Mechanical Alloying, *Metall. Mater. Trans., B-Process Metall. Mater. Proc. Sci.*, 28(2), 299–306, 1997.

Ouellette, J., Seeing with Sound, *The Industrial Physicist*, June/July 2004, pp. 14–17.

Park, J. et al., Coulomb Blockade and the Kondo Effect in Single-Atom Transistors, *Nature*, 417, 722–725, 2002.

People: Richard Feynman, http://forbes.z2c.net/rd4/ck/3072-13370-1839-183?m=7-8&e=fef9626194f6.

POLICY: NANO MISCONCEPTIONS; http://forbes.z2c.net/rd4/ck/3072-13370-1839-183?m=14-8&e=fef9626194f6.

Pradhan, S. K., Datta, A., and Pal, M., Synthesis of Nanocrystalline Ni3Cu by Sol-Gel Route, *Metall. Mater. Trans., A-Phys. Metall. Mater. Sci.*, 27(12), 4213–4216, 1996.

Rawers, J., Slavens, G., and Govier, D., et al., Microstructure and Tensile Properties of Compacted Mechanically Alloyed, Nanocrystalline Fe-Al, *Metall. Mater. Trans., A-Phys. Metall. Mater. Sci.*, 27(10), 3126–3134, 1996.

Sajgalik, P., Dusza, J., Hofer, F., Warbichler, P., Reece, M., Boden, G., Kozankova, J., and Sasaki, G., *Mater. Res. Soc. Symp.*, 287, 335. 1996.

Sienko, T., Adamatzky, A., Rambidi, N., and Conrad, M. (Eds.), *Molecular Computing*, Cambridge, MA: The MIT Press, 2003.

Solin, S. A., Magnetic Field Nanosensors, *Scientific American*, pp. 71–77, July 2004, www.sciam.com.

Sporn, D., Raether, F., and Merklein, S., Preparation and Properties of Sol-Gel Derived Nanostructured Thin Ceramic Layers, *Materials Science and Engineering, A — Structural Material Properties, Microstructure and Processing*, 168(2), 205–208, 1993.

Ying, J. Y., Nanocrystalline Processing and Interface Engineering of Si_3N_4-based Ceramics, MIT, Cambridge, MA, *NTIS Alert*, 96(6), 13, 1996.

6

Powder Metallurgy (P/M)

Introduction

Definition

In contrast to other metalforming techniques, P/M parts are shaped directly from powders. The basic P/M process uses pressure and heat to form near-net-shape (NNS) parts. Metal powder is squeezed in a rigid precision die at pressures up to 50 tons/in². After the powder is compressed into shape, it is ejected from the die and fed slowly through a high-temperature, atmospherically controlled furnace. At temperatures below the melting point of the metal, the particles bond, metallurgically fusing or sintering together.[1]

Although P/M is used to fabricate parts from just about any metal, the most common come from iron-based alloys, nickel, tin, copper, or refractory metals such as tungsten or tantalum. Other metals available include aluminum, stainless steel, and nickel, or cobalt-based superalloys. Powder particles are specific in shape and size ranging from 0.1 to 1000 μm.

High-precision P/M parts have been available for years, but limitations in their mechanical properties once restricted their use. A porous, low-density part will not be as strong or as corrosion-resistant as its wrought-metal counterpart. Now, however, this is no longer true for P/M parts that undergo hot forging in closed dies or are produced using hot isostatic pressing (HIP), or those that are metal injection molded (MIM). P/M parts forged, HIPed, or injection molded to 100% theoretical density in production conditions are said to be equal and sometimes better than their wrought-steel counterparts.

For example, relatively complex carbon or low-alloy steel parts in large quantities are ideal candidates for powder forging (PF). And although PF is not a particularly new process, it along with HIP and MIM have sparked the most interest in recent years. Automakers were among the first to realize the merit of fully dense PF parts. The precision-forged components now operate in a wide variety of transmissions, accessory mechanisms, and engines.

Another means of boosting performance is the addition of carbon to P/M steel. Doing so makes the P/M parts heat-treatable. Heat-treating increases hardness, toughness, wear resistance, and strength of the component. Likewise, the addition of alloying elements to the iron-powder mix further enhances the properties of heat-treated steel P/M parts.

With the arrival of the 21st century, P/M is fulfilling its promise as a solid, technologically based industry. The economic and technological fundamentals of the industry have never been better. P/M is advancing in all sectors: conventional "press and sinter," metal injection molding, powder forging, warm forming, hot isostatic pressing, spray forming, and advanced particulate materials.[2]

Materials

P/M techniques can produce alloys that are difficult or impossible to make using traditional casting processes. This chapter on P/M will concentrate on the newer types of materials that have been produced using P/M techniques in lieu of the traditional and conventional steels,[3] coppers, and brasses.

Titanium Alloys

A superelastoplastic titanium alloy with low Young's modulus and high strength has reportedly been developed by Toyota Central Labs Inc., Japan. Its elastic properties enable cold working to 99.9% at room temperature, and strength can be increased to 2100 MPa by applying a simple heat-treatment process. Called *GUM* metal, it is a powder metallurgy beta titanium alloy with a body-centered cubic structure and composition Ti_3 (Ta + Nb + V) + (Zr, Hf, O).

Its elastic coefficient is not constant, and it shows nonlinear elastic deformation, with Young's modulus ranging from 20 to 60 GPa, depending on the amount of cold work. As yield strength increases, the alloy does not show work-hardening, and continuous deformation is possible to any required level.

Properties are based on its unusual nanostructure. No dislocations or twin crystals are observed after cold working; instead, it changes to a marbled structure containing fractal, layered structures with a discrete strain field, and its crystal lattice is curved to a great extent. It is estimated that an unknown plastic deformation mechanism produces its unusual properties, one completely different from those of other metallic materials.

Current applications include spectacle frames and precision screws. Potential applications include a wide range of automotive parts, medical equipment, sporting goods, decorative materials, and aerospace components.[4]

Intermetallics

Mechanical alloying and hot extrusion were studied as a means to dispersion-harden an intermetallic compound based on Ni_3Al-B from elemental powder mixture. The oxide used for the dispersoids was partially stabilized zirconia (PSZ). During mechanical alloying, the microstructure evolved according to the characteristic stages found in other mechanical alloying systems. Completion of the alloying reaction required 16 h, beyond which loss of the crystalline property set in. Experimental observation of the grain refinement during mechanical alloying agreed with a prediction based on an existing model. Compared to V-cone mixing, the mechanical alloying produced a homogeneous distribution of fine dispersoids. The refined grain structure and dispersoids resulted in a high tensile yield strength over a wide range of temperatures.

Table 6.1 shows the chemical composition of the alloy and the characteristics of the powders used in this study band. Figure 6.1 illustrates the flow of the experimental procedure. This work showed that fine PSZ is a potential candidate material for dispersion hardening of intermetallic compounds based on Ni_3Al-B.

In recent years, people have become greatly interested in high-temperature intermetallic compounds, especially TiAl. TiAl has attractive properties, such as high melting temperature and high modulus, low density, and good oxidation resistance. It is one of the most promising high-temperature structural materials. Unfortunately, neither the ductility nor the fracture toughness of these TiAl alloys is particularly high at ambient temperatures. Coupled with poor formability, TiAl alloy is not yet used in aerospace applications. By alloy modification and microstructure control, both ductility and toughness have been improved in recent years.

TABLE 6.1

Nominal Compositions of the Experimental Alloys[5]

Alloy	Composition, wt % (a)			
	Ni	Al	B	PSZ(b)
Ni_3Al	87.4	12.60	—	—
Ni_3Al-B	87.3	12.60	0.10	—
$Ni_3Al-B-PSZ$	85.0	12.35	0.15	2.5

Note: Mean particle size, μm: Ni, 4.5; Al, 5; B, 1; PSZ, 0.3.
(b) Partially stabilized zirconia, ZrO_2 + 3 mol% Y_2O_3.

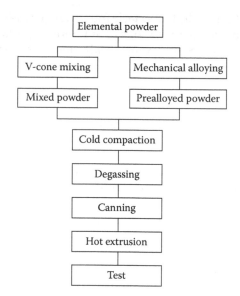

FIGURE 6.1
Flow chart of the experimental procedure.[5]

Yinjiang, Tao, and Lian[6] studied the properties and densification of TiAl. The TiAl alloy and powders with high purity and low oxygen contents were synthesized by infiltration combustion synthesis (ICS, see Figure 6.2). Its densification was studied by both hot isostatic pressing and conventional press-sintering process. The samples, with density of more than 3.75 g/cm³, were obtained by HIP at 196 MPa and 1250°C for 3 h. The microstructure consisted of equiaxed grains and a small volume fraction of lamellar structure. The compressive tests showed the yield strength of TiAl sample was 646 MPa, the compressive strength was 1579 MPa, and compressive ductility is above 26%. The addition of Mn and Cr changed the structure and increased the volume of α_2 phase. The density of the powder compact sintered directly might reach the value of theoretical density.

They determined that the TiAl intermetallic compound could be produced by the following processes: ICS and HIPing or sintering. The mechanical properties of TiAl alloy, sintered by the conventional P/M method or consolidated by HIP from ICS powder, are somewhat superior to those of the TiAl pieces made from the centrifugally atomized powder and SHSed powder.

Composites

An obvious need for structural composites to be used in the temperature interval between that covered by polymer and aluminum matrix composites

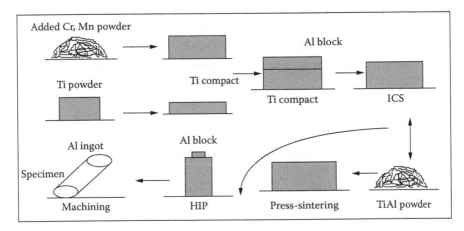

FIGURE 6.2
Flow chart of the production of TiAl produced by ICS.

and that covered by heat-resistant alloys has led to efforts to develop materials with titanium or titanium intermetallic compound matrixes. The main problems to be solved when dealing with titanium matrix composites relate to the chemical reactivity of titanium at high temperatures, but these have been overcome for SiC reinforcement by developing CVD SiC fibers with surface layers enriched in carbon. This has allowed for the production of SiC-Ti composites of high potential, but these are costly as a result of both the high fiber cost and an expensive fabrication process.

Carbon fibers should provide effective reinforcement to titanium matrixes provided appropriate fabrication technologies can be developed. Direct casting is precluded because of the reactivity problems mentioned above. Hence, an alternative fabrication route has been suggested that introduces carbon fibers into a titanium matrix, using a titanium-based eutectic as an interphase.[7]

S. T. Mileiko[8] of the Solis State Physics Institute in Chernogolovka, Moscow District, concluded from his studies that

- Elastic properties of the composites were essentially influenced by the fabrication regime via the formation of a titanium carbide phase at the fiber/matrix interface.
- The content of the titanium carbide phase was optimized.
- In general, the room-temperature fracture properties of the composites were slightly inferior to those of high-strength titanium alloys, although improvements in the fabrication technology can lead to enhancement of the room-temperature strength and fracture toughness.

- Developments have shown that the short-term high-temperature strength of the composites with titanium alloy matrixes surpasses that of titanium alloys at about 700–800°C.

- Specific creep strength appears to be better than that of ordinary titanium alloys.[8]

Copper-Iron-Cobalt Alloys

When used in cutting tools, new alloys of copper, iron, and cobalt have shown an increased blade life of over 30%, according to Pierre-Alain de Chalus[9] and the manufacturer Eurotungstene Poudres, Grenoble, France. Designated Next, the alloys have cobalt contents not exceeding 25% and are said to exhibit an extremely fine microstructure after hot-pressing. Furthermore, the hot-pressing temperature can be reduced by 100 to 150°C, compared with typical cobalt alloys. However, hardness levels reportedly remain about the same (around HBN 320 for Next 100, and HBN 246 for Next 200).

Diamond tools are produced via P/M by hot-pressing to full density a mixture of diamond grit and metal powders in a graphite mold. The function of the metal is to bond and hold the diamond in the tool during the cutting process.

Therefore, the metal or alloy powder selected should have a similar wear rate, to optimize the cutting performance.

Cobalt has traditionally been the bonding material in diamond tools, but it has a history of price instability and high cost. The Next line was designed to overcome these limitations and at the same time provide diamond tools with enhanced mechanical properties. For example, the abrasion resistance of Next alloys is said to be twice that of cobalt, while flexural tests have shown plasticity to be one fourth that of cobalt.

Rhenium

Rhenium and rhenium alloys offer attractive high-temperature properties for use in aerospace applications. However, traditional methods used to make components are time-consuming and have high production costs. Historically, rhenium has been produced in standard rod, bar, plate, and sheet forms, and components were made from these shapes using electrical discharge machining (EDM) to produce the required shapes to specified tolerances. The method works well, but is time-consuming and has a low yield. Therefore, manufacturers are continually looking for ways to improve productivity and reduce costs.

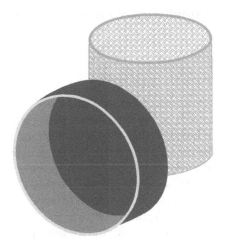

FIGURE 6.3
Typical can mold for cold isostatic pressing.

A recent development in rhenium (P/M) is cold isostatic pressing (CIP) to an NNS using traditional rhenium powder. Rhenium metal powder is packed into a flexible mold having a rigid container to maintain the desired shape (see Figure 6.3), and the can/mold assembly is immersed in water contained within a CIP vessel. In this "wet-bag" process, hydrostatic pressure in the range of 210–410 MPa transfers isostatic pressure to the mold, consolidating the powder into a green NNS rhenium compact. High-performance bipropellant liquid-fuel apogee rhenium-iridium thrusters have been produced using the CIP to NNS method.[10]

Advanced techniques such as vacuum plasma spraying (VPS), direct-hot isostatic pressing of powder (D-HIP), directed light fabrication (DLF), and metal injection molding offer the ability to produce NNS parts in large quantities at lower cost and faster turnaround times using less material. However, these methods require high-density, low-oxygen, spherical rhenium powders in a wide range of particle sizes.

Conventional rhenium powder is an irregularly shaped flake powder having poor flow characteristics and high oxygen concentration (~1000 ppm). A rotating electrode/plasma rotating electrode gas-assisted process (REP/GA-REP/PREP) has been developed[10] and produced spherical rhenium powder (SReP) having low-oxygen content (<50 ppm), an 11 g/cm^3 apparent density, a 12.5 g/cm^3 or higher tap density, and high flow characteristics (4 s/50 g flow rate).

In another development program[10] a consolidation technique of high interest to fabricate small, high-performance rhenium components was powder injection molding (PIM). PIM offers the ability to form intricate shapes to isotropic, near-full density without the need for much secondary machining.

Cermets

Continued R&D has examined hard and ultrahard materials, and cermets designed for components subject to wear (in applications demanding superior fracture toughness to that achieved with monolithic ceramics, or whisker/particle-reinforced ceramics) have also been the subject of research. Particular attention has been given to the study of cermets containing up to 30 v/o metallic binder phase. For the consolidation of these materials, basic research has focused on two aspects: the wetting of the hard dispersed phase by the liquid metal, and the thermodynamic stability of the ceramic particles in contact with the metallic phase during sintering.

A recent development in this area was the fabrication of a family of TiB_2-based cermets. These cermets are constituted of 70–80 v/o of TiB_2 grains (~1–1.5 μm in size) cemented by 20–30 v/o of metallic phase based on Fe-Ni-Ti-Al alloys or stainless steel (Figure 6.4). After pressureless sintering at 1450°C, these cermets achieve 95–97% theoretical density, and after HIPing at 1350°C for 1 h they develop Vickers hardness and K_{IC} values that range from 1450HV and 14 MPa √m for 30 v/o metallic binder, to 1800HV and 11 MPa √m with 20 v/o metal.[11, 12] As witnessed by these hardness/toughness combinations, the materials compare favorably with the high grades of WC-Co hard metals. Interestingly, they have an overall density of ~5.3 g/cm³ and may be machined by EDM. These cemented borides have been used for the fabrication of several wear components

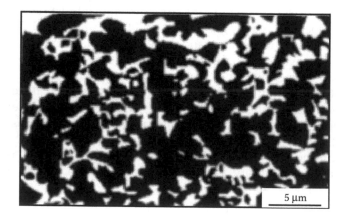

FIGURE 6.4
Microstructure of TiB_2 cement consisting TiB_2 grains (gray), Al_2O_3 particles (black), and Fe/Ni/Ti/Al metallic binder (white). SEM

including positive and negative indexable inserts for turning, tube and wire drawing dies and mandrels, buttons for rock drilling, and sealing rings.

Cemented Carbides

Due to their unique combination of hardness and fracture toughness, materials within the WC-Co system predominate currently in a variety of industrial applications. However, because of the high cost of cobalt, attempts have been made to find an alternative metallic binder phase.

A study has concentrated on the development of metallic binders based on either Fe-Ni[13] or Fe-Mn alloys[14] and a determination of the relationship between microstructural parameters, hardness, and fracture toughness. In both cases, carbon additions were required in order to achieve full density by liquid phase sintering. These additions have to be carefully adjusted to prevent the formation of mixed eta-carbides M6C. In the case of (Fe + Ni)-based binders this results in WC+(Fe-Ni-C) cermets in which the microstructure of the metallic binder phase may be modified by heat treatment in order to increase toughness. For the Fe-Mn alloys, the additions of C, Mo, and Ni in the correct amounts lead to the formation of Hadfield steel as the binder phase.

Tungsten-Based Heavy Alloys

Tungsten-based heavy alloys have been the subject of long-term research at many installations and laboratories. For several years one project at CEIT[15] was aimed at the optimization of the industrial production of conventional W-Fe-Ni heavy metals, of the 90 w/o to 97 w/o W range, for kinetic energy penetrators.

Every stage of the production process, from powder compaction to the final forging operation of the heat-treated pieces, as well as their field performance, was studied. The research required fundamental studies on Ostwald ripening during liquid phase sintering.[16] The aging phenomena during postsintering heat treatments[17] and the effect of microstructure on strength, formability, and toughness of the heave alloys were studied.[18]

The material, which normally contains 90 v/o of brittle tungsten grains, must be able to withstand cold forging operations that raise its strength above 1000 MPa. The ductility required is achieved through strict chemical and microstructural control in order to develop nearly spherical tungsten grains embedded in a ductile fcc matrix.

Significant strength increments, relative to the unforged W-Ni-Fe base alloy, were achieved by means of a very fine dispersion of $Ni_3(Al, Ti)$ precipitates in the fcc matrix, while the concomitant density and ductility losses were kept within acceptable limits.[19, 20]

Processes for Powder Production

The application of P/M techniques has greatly increased in recent years. In many cases, P/M allows for metal parts to be produced at lower costs with improved properties. Another advantage that P/M offers is that it can produce alloys and microstructures that are not possible to produce by standard metallurgical practices such as casting and forging. An important area of P/M is the production of the metal powders used. The physical and chemical nature of the initial powders will determine many of the final properties of the P/M parts produced.

Mechanical alloying (MA) is a method for alloying metals without resorting to external heating or chemical processing. This can be done in almost any of a number of different mills, attritors, or grinders that use the repeated impact of a tool to fracture materials. Ball mills are most commonly used for this purpose. By charging a rotary mill with elemental powders, and tumbling the powders in the presence of a hard media such as hardened steel or ceramic balls, extremely high pressures are generated at the points of contact between media and powders. These pressures are sufficient to cold-fuse individual powder particles together to form larger particles. As the particles grow in size and lose ductility through strain hardening, they will fracture into smaller particles. The process of fusing and fracturing is repeated many times until the original elements are very finely mixed.[21] A very homogeneous powder can be produced at nearly ambient temperature.

MA produces a powder whose average size, shape, distribution, and other attributes can be partially controlled during the process. MA processing is advantageous for the production of alloys whose elements are too reactive or immiscible for ingot metallurgy (melt-alloying) techniques, or direct chemical synthesis routes. Many intermetallic materials fall into this regime, and MA is a relatively easy and safe method for the production of intermetallic powders.[22, 23] It also can produce unusual properties due to the very fine-grained microstructure that typically results.

Status of MA

MA has the potential to produce powders of almost any composition. Since the 1960s, MA has been used to produce oxide dispersion strengthened (ODS) superalloys with increased high-temperature creep resistance. Elemental powders can be mechanically alloyed, eliminating many of the difficulties of preparing powders by fusion techniques. In addition, solid alloys can be quickly and easily reduced to powder. These attributes are very useful for developing new alloys, when modest quantities of unique alloy compositions are required.

MA is also currently used in the production of tungsten-carbide/cobalt (WC-Co) alloys. The milling of a mixture of WC and cobalt powders results in a powder that is typically particles consisting of WC with a coating of cobalt metal. This is usually done in a low-energy ball mill, and while this method combines the two materials, it does not homogeneously mix them as described above.

In recent years, a great deal of research has been performed in the area of mechanical alloying. Similar to rapid solidification processing, many nonequilibrium materials can be produced. Large extensions of solubility limits and nanocrystalline microstructures can be produced. Also, reactive metals can be made into powders without crucible interactions, and bulk materials can be reduced to fine powders.

Many researchers have used MA for the production of intermetallic compounds from elemental metal powders. There is a long list of different compounds that have been fabricated successfully.

Just as for rapid solidification processing (RSP), MA can produce very fine microstructures and amorphous structures. These structure sizes are easily in the nanometer size range. Research efforts in many laboratories are currently trying to exploit these structures and discover their properties. Obviously, these materials would exhibit a great deal of grain boundary strengthening. The potential of these materials is only now being realized. Also, there is a significant body of work that shows sintering temperatures can be dramatically reduced by decreasing the microstructure and particle sizes, as can be achieved by MA. MA produces a highly disordered structure and therefore has a much higher energy than normal materials. During first-stage sintering, this structure should promote neck formation and growth. The small grain size will be very beneficial to the second-stage sintering.

RSP is well known for its ability to produce alloys with solute concentrations dramatically higher than the equilibrium solubility limits. MA can likewise dissolve large quantities of limited-solubility elements.

Solid Solubility Processing

Solid solubility extensions have been achieved in a number of binary alloy systems based on Al, Ti, and Mg. Compared with the RSP method, MA generally leads to higher solid solubilities. In some cases, solid solubility extensions have been achieved by MA in systems (e.g., Ti-Mg) where these are not possible by RS. There have been very few reports of solid solubility

TABLE 6.2

Solid Solubility Extension by Rapid Solidification (RS)
and Mechanical Alloying (MA)

		Solid Solubility, at%		
		Equilibrium at		
Solvent	Solute	Room Temperature	RS	MA
Al	Fe	0.025	4.3	4.5
Al	Mg	18.9	40	23
Al	Nb	0.065	2.4	25–30
Al	Zr	0.083	1.5	9.1
Ag	Ni	~0	...	3.8
Mg	Ti	~0	...	4.2
Nb	Ai	<6	25	60
Ni	Ag	~0	...	9.0
Ti	Al	<11	...	>33
Ti	Mg	<0.2	...	3.6

extensions in intermediate phases. Table 6.2 gives solid solubility extensions achieved by both techniques.

Plasma Processing (PP)

Several investigators have explored the possibilities of using PP techniques to generate very fine-sized metallic, intermetallic, ceramic, or composite materials by high-temperature reaction and rapid quenching.[24–28] The very high temperatures and concentrated enthalpy available in thermal plasma arcs may be used to vaporize most feed materials, and the steep temperature gradients available may allow the nucleation of extremely fine-sized products by rapid quenching.

Physical Vapor Deposition (PVD)

The PVD technique can be regarded as the ultimate "rapid solidification" process, with quench rates estimated at being equivalent to 10^{13} K s^{-1}.[29] Electron beam evaporation and sputter deposition have both been used to produce free-standing PVD alloy films, and grain refinement, solid solubility extension, and amorphization have been observed. Grain size tends to be very fine for the first few micrometers of deposition but usually, depending on the collector temperature, some form of columnar grain growth develops perpendicular to the substrate. Typically, for metals at substrate temperatures less than $0.3T_m$ (where T_m is the melting point in Kelvin), a highly porous columnar structure is formed consisting of crystallites separated by voids. Between approximately $0.3T_m$ and $0.45T_m$ the structure consists of densely packed columnar grains, and at higher temperatures the material recrystallizes to give an equiaxed grain structures.[30] Ultrafine-grained powders have been produced using PVD on cryogenically cooled substrates by repeated removal of the deposit buildup.[29-31]

In general, the greatest solid solubility extensions are found in those systems where the equilibrium solid solubilities are low. Thus, for example, in Ti-Al and Mg-Y, where large equilibrium solubility exists, no extension has been found in vapor-quenched material; Al-Cu and Al-Mn have moderate equilibrium solubility and showed some solubility extension in vapor-quenched material, and Al-Cr, Ti-Mg, and Mg-Ti with very low equilibrium solubilities showed very large solid solubility extensions. Al-Fe is the only alloy that does not follow this pattern, with a maximum vapor-quenched solid solubility of only 5 wt%, despite an equilibrium solid solubility of less than 0.1 wt%.

Novel Plasma Process

A novel plasma melting and rapid solidification (PMRS) process shown in Figure 6.5 has been developed by GTE Products Corp. for the production of rapidly solidified alloy powders based on a number of metals and alloys or a combination of metals and ceramics.[32] Particle size of the obtained powder is said to be similar or smaller to that obtained with gas or water atomized metal powders, but where very fine powders are required an additional step involving microatomization has also been developed by GTE.[32]

In a simplified version of the process shown in Figure 6.6, metal powders are melted in a high-temperature plasma, and the resulting high-velocity spray is rapidly solidified (a) by free fall in inert gases, (b) by quenching into liquid argon or nitrogen, or (c) by impacting against a chilled, moving substrate.

Plasma Melting

The first step in the process involves feeding a coarse feedstock powder of around 45 μm through a plasma flame. The agglomerates are entrained in a carrier gas (argon, helium, nitrogen, and in some cases hydrogen for chemical refining) and fed into the plasma such that melting and solidification occur without interparticle contact. The resulting plasma-melted particles are spherical and dense with excellent flow and packing properties.

In fact, if the feedstock material consists of more than one type of particle, of more than one metal or ceramic component, or any combination of these, the materials will react or alloy during melting to produce prealloyed or composite powder. For example, an aluminum-titanium

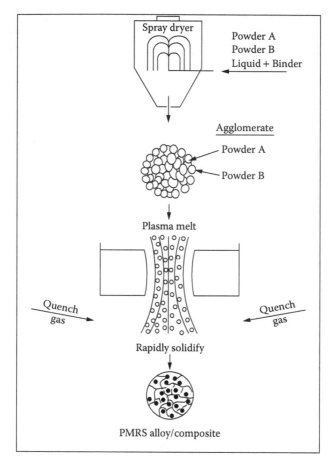

FIGURE 6.5
The plasma melting and rapid solidification (PMRS) powder process developed by GTE.[32]

boride composite powder produced by PMRS, whereby the TiB_2 is on the inside of the particle and the Al on the outside. The novel composite would not be possible to produce by conventional manufacturing techniques. Another example is the WC-Co system, where the microstructure of a spray-dried agglomerated powder after PMRS is said to consist of WC particles, a matrix of Co-W-C, and a few patches of residual cobalt.

The plasma conditions can be varied to accommodate gas flow and diameter; speed can be from subsonic (800 ft/sec) to Mach 2. The point at which powder is injected is also important. The temperature attained by the powder is affected by the agglomerate diameter, the residence time in the plasma, and the flame temperatures that the particles experience.

In the process of melting, the particles are accelerated to nearly the velocity of the plasma gas. The solidified particles are collected in a cooled

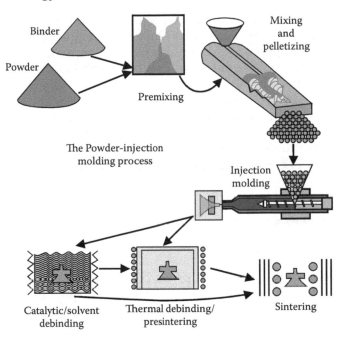

Binder

Powder

Premixing

Mixing and pelletizing

The Powder-injection molding process

Injection molding

Catalytic/solvent debinding

Thermal debinding/ presintering

Sintering

FIGURE 6.6
PMRS simplified version of process.

chamber under inert atmosphere or partial vacuum. Cooling rates are very high after plasma melting because of the high droplet velocity and small droplet size. Cooling can be further enhanced by gas jets designed to impinge upon the molten drops as they exit the plasma flame. However, the particle diameters are smaller and their velocities are higher than for gas atomization, which claims cooling rates of 10^4–10^6 C/sec.

Plasma Discharge Spheroidization (PDS)

Plasma discharge spheroidization (PDS) is a technique that makes ultrafine powders in a cryogenic inert gas environment by localized RF plasma melting of a prealloyed charge. Compared with conventional powder processing techniques, the PDS process produces prealloyed spherical powders 1 to 10 μm in diameter with a tightly controlled particle size distribution. Such ultrafine powders exhibit superior sinterability and improved mechanical properties in a densified state compared to commercial powders.

The oxygen content of these ultrafine powders is higher than that of coarser powders due to their larger surface area. Compared with NiAl ultrafine powders with –325 mesh irregular-shaped powders sintered at

1600°C for 1 h, ultrafine powders can be densified to about 96% of theoretical density compared with about 82% for coarser powders. Mechanical properties of hot-pressed compacts synthesized from ultrafine powders are also superior to those of coarser powders, with higher yield strength and ultimate strength.

The highly sintered powders can be used for net-shape or near-net-shape powder metallurgy applications, particularly in injection molding. The technology was developed under the sponsorship of the U.S. Army Research Laboratories. Ultrafine titanium β and titanium aluminide powders made by PDS have been produced at levels above 50 kg/day.

Spray Compaction

New Al-based alloys have been produced by spray compaction. Due to the extremely high cooling rates achieved during spraying and solidification, the level of the alloying elements can be increased beyond equilibrium. Silicon contents up to 35 w/o have been achieved, leading to excellent wear resistance and very low thermal expansion. Additionally, the production of dispersion-strengthened alloys is possible by this processing route. A combination of these alloying techniques and the resulting materials properties opens up new fields of applications for Al alloys. For example, in the automotive industry, gray cast iron, the dominant material for key components in the engine such as cylinder liners, will be replaced by Al.

Although the principles of spray compaction were developed about 25 years ago, application of this technique to Al alloys requires extensive know-how, keeping in mind the severe safety aspects that must be taken into consideration. Worldwide, only two spray forming plants of this type exist for Al-based alloys.

Other Innovative Advanced P/M Processes

Warm Compaction

The concept of warm compaction, utilizing heated tooling and specific powder during a single compaction step,[33] produces high-density powder metallurgy parts. Generally, it eliminated more conventional processing steps while providing the desired properties of end product.

The development of the Ancordense process, otherwise referred to as warm pressing/warm compacting, has been a major development in P/M work. The industry has always known that the properties of P/M parts are greatly improved as density increases. This is particularly true of dynamic properties such as impact and fatigue strength. Thus, over the years there have been strenuous attempts to increase the compressibility of iron powders, which have resulted in an increase of typical high-density parts from 6.7 Mg m^{-3} perhaps 15 years ago to 7.0–7.1 Mg m^{-3} today. However, to obtain much higher densities it has been necessary either to use prohibitively high pressing pressures, resulting in unacceptable tool costs, or to employ the expensive double-pressing and double-sintering process.

The relative costs of high-density techniques compared with single-press and -sinter technology range from a 40% premium for infiltration, to 60% for double pressing and double sintering, to 100% for powder forging. These high costs have limited the expansion of P/M into the more highly stressed applications that abound in the automotive industry, its largest market.

The warm-pressing process uses the surprisingly simple concept of pressing the powder at an elevated temperature to form the green part. It is reported that a temperature of 143°C is used to achieve densities augmented by about 0.08–0.14 Mg m^3. It has been reported that a linear increase in density was noted up to temperatures exceeding 200°C.

It is claimed that the cost of double pressure, double sintering is only 20–30% greater than conventional processing, whereas properties are equivalent to double pressing and double sintering that costs 60% more than conventional processing. In addition to high density, which can reach 7.4 Mg m^{-3} (it has to be said that this is at a compaction pressure of 770 MPa), it is claimed that the green strength of parts is increased to 17–31 MN m^{-2}, an increment of at least 80% relative to conventional processing at the same density and more than double the green strength at the same compaction pressure. It was claimed that the materials could be machined green, which would give rise to added advantages. Furthermore, somewhat paradoxically, the ejection forces required at a given compaction pressure were in fact lower than those experienced in the conventional techniques, often by ~50%, from which increased tool lives were claimed to result, according to German, Donaldson, and Johnson.

Cold Forming

In the new cold-forming technology, particles (metal, intermetallic, or ceramic) are electronically coated to control the material composition at the particle level and achieve ultimate part with uniform properties. A feed shoe injects powders under pressure into the press die, resulting in more uniform fill. By injecting an activation solution along with conventional or engineered coated particles such as Al/Cu, Cu-coated W and Co, Ni/Al, Ni/stainless steel, and Ni/Ti before pressing, surface oxides are removed and cold welding enhanced. Most parts can be pressed to 100% density in a one-step, ambient-temperature process. Parts that require sintering exhibit higher density, fewer cracks, and higher strength.

Dynamic Magnetic Compaction

In the dynamic magnetic compaction process, very fast-pulsed magnetic pressures are employed to compact complex high-density automotive parts involving metals and alloys, ceramics, and composites.

Laser Manufacturing/Processing

A special-interest program focused on laser manufacturing/processing. Rapid prototyping will be a major benefit of laser processing, which can dramatically shorten developmental cycles for parts. For example, a designer's CAD drawing data can be used to program laser motion and part buildup. The program is designed to move the laser over the surface to build the part one layer at a time.

Laser cladding by layers provides the possibility of fabricating 100% dense metal components. According to Mazunder and associates,[34] computer-controlled five-axis workstations integrated with lasers enable the fabrication of parts of various geometries. Initial work has demonstrated that components with mechanical properties similar to plate materials can be fabricated even from oxide formers such as Al in addition to Cu, Ni, and ferrous alloys inducting H13. Volume deposition of 4.1 cm³/min. was measured for transverse speed of 42.3 mm/sec. for Al. Powder utilization ratio varied from 30 to 90%, depending on the width of the deposit layer (the higher the width, the better the utilization ratio). As-deposited surface roughness was similar to that of a cast structure. Best possible resolution for wall thickness using 6 kW CO_2 laser is around 0.5 mm.

A new near-net-shape metal fabrication process being developed produces fully dense metal components by fusing powdered metals in the focal zone of a laser beam in a single step, with the part being deposited a layer at a time. The process involves the integration of powder flow technology, solidification metallurgy, laser technology, heat flow analysis, CAD-CAM technology; controls, and diagnostics, according to Gary Lewis and co-authors.[35] Al alloys, 300 and 400 series stainless steels, Fe-Ni alloys, Ti alloys, W, Ni-Al, molybdenum disilicide, and P_2O tool steels have been processed.

Eric Whitney and colleagues[36] have developed a system to rapidly prototype near-shape, high-integrity Ti and Ti alloy structurals without molds or dies. Precursor, sieved Ti fines, or fines blended with Al-Va master alloy powders are placed in an argon fluidized powder bed beneath a high-powered CO_2 laser. The laser beam is caused to move in a path representing one cross-sectional slice of the solid article desired. The 14 kW beam fuses the Ti alloy in that slice pattern, and additional powder is subsequently added to the solid target surface by fluidizing the powder bed. High density Ti bars and (2.54 cm) thick wall cylinders have been fabricated. Materials characterization shows minimal contamination in process, and mechanical strengths exceeding ASTM standards for C-2 Ti.

Processes for Part Fabrication

P/M consists of an initial compression of the metal powder followed by the heat-treatment process of sintering. The sintering temperature is normally below the melting point of the main metallic component. In large-scale production, sintering is commonly applied in continuous furnaces; however, due to the higher flexibility, commercial heat treatment companies tend to debinder and sinter in batch furnaces at normal or high pressures.[37]

During sintering, the individual powder particles grow together by diffusion to form a polycrystalline material that gives the compressed part the desired stability. P/M steels are sintered at temperatures between 1100 and 1250°C, while nonferrous metals such as bronze, brass, or aluminum are sintered at far lower temperatures.

Sintering above the melting point is known as liquid phase sintering. For certain steels (namely high-speed steels), the liquid phase is held stable for a certain time, making an exact temperature control necessary. This leads to a rearrangement and further compression of the structure. It's important during liquid phase sintering to prevent distortion of the pressed parts due to high shrinkage.

See References 3, 37, 38, 39, 40, and 41 for debindering and sintering up to 750°C; vacuum and protective sintering up to 1100, 1200, 1600, 3000°C as well as pressure sintering up to 2200°C; microwave sintering;[40,41] and catalytic debindering and sintering of MIM parts (more on this later in this chapter).

Specialty Production Processes: Powder-Injection Molding

Some may remember the original style of orthodontic braces made from welded bands that slid over individual teeth. The braces, prolific up to the early 1970s, gave real meaning to the term "metal mouth." Powder-injection molding (PIM), along with new adhesives, brought dentists smaller brackets that were also more soothing to the eye. Single-piece designs, backed with a waffle pattern to promote adhesion, replaced the two-piece designs with welded mesh backs.

Now, more than 100,000 PIM brackets are produced each day, with production costs reaching $15 million/year. This is just one example of small, complex components used in the medical industry.

Recent developments are focusing on instruments used to perform endoscopic and laproscopic procedures. As a result, engineers must find ways to maintain precision while boosting strength, wear and chemical resistance, and biocompatibility. PIM gives them the ability to form intricate parts with tight tolerances in materials that don't traditionally have the processing ease associated with injection-molded thermoplastics. This technology is reemphasizing metal and ceramic components.[42]

What Is PIM?

PIM is not a new technology, though it did not receive a lot of commercial attention until the 1980s. Recent developments include advanced feedstocks with controlled shrinkage and state-of-the-art processing equipment that precisely control molding, debinding, and sintering.

PIM was first developed in the 1920s and used to mold ceramic spark-plug bodies. In the 1950s it saw limited use in shaping carbide and ceramic components using epoxy, wax, and cellulose binders. Attention to PIM grew in 1979 as two aerospace applications received design awards: a screw seal for a commercial jet and a thrust chamber and injector for a liquid-propellant rocket engine.

PIM begins with fine powder materials consisting of nearly spherical particles ranging in diameter from 0.1 to 20 μm. To prepare them for injection-molding machines, engineers blend powders into a feedstock by compounding them with traditional binders consisting of waxes and thermoplastic resins. Recent developments have given engineers more options including binder systems based on polyacetals and polysaccharides.

The binder's role is to lubricate the powder mixture so it flows easily through molding cavities. As a result, binders also fill the voids between metal or ceramic particles and add integrity to molded shapes so they can be handled during processing. Polymeric binders comprise up to 40% of the mixture by volume. The final step in compounding solidifies the mixture and chops it into easy-to-handle pellets, similar to that of plastic feedstocks.

After compounding their own resin or buying material from a supplier, manufacturers form components at relatively low temperature and pressure in standard plastic injection-molding equipment. Molding temperatures range from 150 to 260°C, which softens binders to a paste-like consistency that will flow through mold cavities.

Once cooled, parts go through a debinding process using heat, chemicals, or even water to remove most of the binder. In thermal debinding, microprocessor-controlled, low-temperature ovens, analogous to convection ovens, sweep air over components and collect condensates. A small amount of binder is left behind so that workers can move parts to sintering furnaces without destroying them. Chemical methods leach binders through a catalytic process (see Figure 6.6.42).

Sintering, the next step, binds particles together at an atomic level by diffusing atoms at temperatures approaching 85% of melting point. Sintering removes the voids left by binders, which boosts density and shrinks parts. Shrinkage is uniform and isotropic, however, so engineers can compensate for it by designing molds approximately 20% larger than final dimensions. Sintering metal parts in a controlled atmosphere or vacuum keeps them from oxidizing. Another method used is single-step debinding/ sintering, where debinding and sintering take place in the same furnace. This further protects surfaces from oxidation because parts aren't handled between steps, leading to improved surface finish, manufacturers report.

Sintered components are typically 96% of theoretical density, with properties approaching that of wrought material. Density is further increased using hot and cold-deformation methods such as HIPing.

Although injection molding is generally associated with high production volumes, PIM is economical for high- and low-volume rates, from as many as 100,000 parts/day to as few as 5000 parts/year. From a production perspective, PIM often eliminates operations such as grinding, machining, drilling, and boring. Manufacturers also save money by recycling feedstock from runners, sprees, and damaged parts, using nearly 100% of raw materials.

Although PIM works for nearly any shape that can be formed by plastic-injection molding, it does have drawbacks and limitations. For example, the process isn't competitive with traditional screw machining or die compacting for simple, axial-symmetric shapes; the cost of materials, tooling, and processing equipment limits the size of components; and manufacturers set limits on thickness from 10 to 50 mm because it lengthens the debinding step.

PIM works best for complex components where the process is competitive with other forming processes such as investment casting and machining.

Design and Material Considerations

The PIM process produces complex, net-shaped components from metals, ceramics, cemented carbides, and cermets. Cost in PIM is largely a reflection of the component size, tooling cost and complexity, mold cycle time, debinding and sintering speed, and other relatively straightforward factors. As with PIM, many shapes are possible, but certain features greatly impact processing ease, yields, and costs. So how do you know a good candidate for PIM? A simple schematic decision tree may answer that question.

Then comes material. PIM materials must be small grain-sized powders that sinter density without extraordinary processing cycles. For ceramics, this usually requires sinter-enhancing additives. For metals, it's best to avoid strong oxide formers as well as reactive, volatile, or toxic metals, such as Be, Hg, Pb, Mn, and Mg. PIM works best for materials with melting points above 1000°C. For lower melting-point materials, die casting is often a better option. Typical PIM materials include stainless and tool steels, Al_2O_3, ZrO_2, Ti alloys, and Ni and Co-based superalloys.

Designs meeting these geometry and material criteria are PIM candidates. Cost becomes the next consideration. The small grain-sized powders can be expensive, so good PIM candidates tend to have high component manufacturing or machining costs that outweigh material costs. Materials account for about 15% of PIM manufacturing costs.

Hard materials prove difficult and expensive to grind or machine. So, applications that benefit most from PIM are those with materials that do not machine easily (tool and stainless steels, Ti, and ceramics) or that have difficult geometries.

Other factors include tolerances and surface finish. For rough surfaces, setup dominates the machining cost. But smooth surfaces require lengthy machining times, which become the dominant cost. Accordingly, there is a region of smooth surfaces between 4 and 5 µm, where PIM is most effective.

There are 10 design factors that justify PIM. The criteria for identifying a PIM candidate component can be captured in a few key considerations. These 10 important design factors help identity good PIM candidates:

- Low mass/volume ratio
- Quantity above 5000/yr, preferably above 20,000/year
- Material difficult to machine
- Moderate to high complexity
- High performance requirements
- Specialized surface features or finishes
- Difficult combination of tolerances

- Consolidation of assembled parts
- Noncritical locations for blemishes, such as gates and ejector pin marks
- Novel compositions or combinations of compositions

Materials and Trends

PIM is used for both metals and ceramics (see Table 6.3). Current medical applications for PIM include surgical tools, fastener hardware for infant incubators and instrument handles, finger and thumb loops, and blades.

Stainless steels such as 316L and 17-4 PH are the most widely used materials for these applications. 17-4 PH is used when components need superior hardness or wear resistance, such as on the cutting edges of instruments, while 316L protects components subjected to more corrosive conditions.

Other PIM materials include low-alloy steels such as Fe-2% Ni and Fe-8% Ni. Low-alloy steels are used for wear resistance and high strength in sealed, oiled environments such as drug-delivery systems, where corrosion is not a problem and there is no direct contact with skin tissue. Ti and Co-chrome alloys, known for their corrosion resistance and biocompatibility, are also under research for use in implantable orthopedic devices.

Ceramics have few applications in the medical industry, besides dental implants, because of their brittleness. Potential applications include ZrO_2 scalpels and surgical tools. With superior wear qualities, ceramics would be ideal for applications such as instrument cutting edges. One barrier to growth in this area, however, is whether the demand for reuse is high enough to justify using such advance technology.

With health care costs under constant scrutiny, medical-device manufacturers must slash costs while remaining profitable. The real driver for PIM

TABLE 6.3

Material Properties for Typical PIM Alloys

Property	17-4 PH Stainless (H-900 condition)	316 L Stainless (as sintered)	Fe-2% Ni (heat treated)	Fe-8% Ni (heat treated)
Yield strength (MPa)	1100–1200	170–200	1300–1400	1300–1350
Ultimate tensile strength (MPa)	1250–1350	500–550	1200–1400	1800–1900
Elongation (%)	4–8	60–80	1	1–3
Density (gm/cm³)	7.60–7.75	7.80–7.90	7.55–7.63	7.55–7.63
Hardness	37–43 HRc	55–65 HRb	55–60 HRc	48–52 HRc

is getting the net shape for less money, while meeting performance requirements such as strength, ductility, and corrosion resistance.

Surface finish is still a barrier to wider use of PIM, as well as either powder metallurgy methods. The medical field is very finicky about quality, especially when it comes to aesthetics and surface finish. There is more concern with surface appearance than other markets such as the automotive industry. Automotive engineers care about tight tolerances, not whether a gear turns black or blue during sintering.

There are some exceptions, however, when device manufacturers take advantage of duller surface finishes, such as the ends of laproscopic and orthoscopic instruments. Very fine surface finishes on instrument tips reflect fiber-optic light back into surgeons' eyes, making it difficult to perform procedures. However, brighter finishes are still required on areas such as thumb loops and handles, where it gives the impression of a clean, sterile surface.

Some manufacturers rely on different material grades to achieve smoother surface finish.

Another Specialty Production Process: Metal Injection Molding (MIM)

Powder-metal parts are as strong as solid steel. That is the allure of MIM technologies, which produce parts with near-theoretical densities (>95%). This results in strength and modulus values comparable to those of wrought metal and gives MIM parts mechanical properties that often equal or surpass those of other metal-forming processes including investment casting, forging, and machining.[43]

MIM is not a new technology. But, most MIM applications materialize to solve a problem in existing parts made some other way. This is changing as materials continue to improve and molders and designers gain experience with the process. More and more parts are specifically designed to be MIM. The ability to make larger parts has also bolstered use of the technology in structural applications ranging from precision medical and aerospace parts to sporting and recreational gear.

Injection molding is usually associated with plastics, where it is known for producing complex, net-shaped parts in quantity. Efficient cycle time is another point in the technique's favor. Additionally, it is often possible to consolidate multiple parts into one injection-molded version that needs no secondary finishing because the process accurately reproduces mold surfaces.

MIM works with an expanding array of metal alloys. Traditional MIM parts compete well in niche markets where parts weigh >200 gm and wall thicknesses range from 0.25 to 6.25 mm. Tolerances are typically held to ±0.3%, and surface finishes of 32 rms (0.8 µm) are common. MIM parts can also be brazed, soldered, and welded as well as plated or ground to size if necessary.

MIM, however, may not be a shoe-in for every small, complex design. Poor candidates may include applications currently using screw machining, zinc die casting, or stamping. Designers must evaluate each part individually, taking into account its complexity and whether there will be secondary machining and finishing operations. If a machined part, for example, needs three or more tool setups and will be made in volumes exceeding 10,000, MIM may provide better economics. The overall design requirements also play an important part in the decision-making process.

Material Basics

MIM is a union of thermoplastic injection molding and conventional powder metallurgy. Binders consisting of wax and polymeric or aqueous systems mix with fine (<20 µm) metal powders in what is called a feedstock. Feedstocks come in various grades of stainless steel, Fe, and Ni alloys. Soft magnetic, W, and Ti are also available.

Historically, metal-parts fabricators developed MIM feedstock in-house. They would formulate mechanical blends of powdered metal and binder systems by trial and error, resulting in "home-brew" feedstocks. As with any injection-molding material, feedstocks that are dependable and consistent can enormously improve processibility and part performance. So, it is not surprising that many home-brew feedstocks are being replaced by commercially precompounded versions with a high degree of consistency and quality assurance.

The materials selected for MIM binder systems are varied, but they all function as flowable carriers for the metal particulates. The feedstocks inject into molds similar to those used for PIM via standard injection-molding equipment. Some MIM feedstock may require modified injection-molding equipment and tooling to accommodate the material's mildly abrasive nature and high viscosity.

Feedstocks are fed into the injection-molding press through a hopper at the back of the machine. The system heats the material, plasticizes it, compresses it via the screw inside the barrel, and then injects it into the mold cavity through the injection screw's shut-off screw tip. The material cools and solidifies in the mold prior to being ejected in what is called a "green" part. Molds incorporating slides and cores produce more complex features than is possible with machining or casting.

Green parts are oversized and somewhat fragile. They are often 10 to 20% larger than the final part size. It is possible to grind off excess material when the part is still in its green state.

Debind and Sinter

The green part undergoes a debinding step to remove most of the carrier. In contrast to conventional powder metallurgy, MIM parts contain up to 40 vol% binder, which needs to be removed. Feedstock materials need to be consistent and offer a predictable and reproducible rate of shrink during further postmold processing.

First-stage debinding typically removes a large portion of the total binder. Density of the part at this point remains below that of the base metal, reflecting both the presence of voids between the metal particulates and the volume of the binder still present. Following first-stage debinding the part is even more fragile and is said to be in a "brown" state.

The debinding method and its duration depend solely on the carrier used. The binder has the greatest effect on cycle times. Polyolefin and wax systems may be debound either thermally, with solvents, or both. Poly-acetal binders must be catalytically debound with nitric acid and a nitrogen-gas atmosphere in a separate debinding oven. Partially hydrolyzed PVOH is soluble in water; remaining polypropylene and plasticizers are thermally debound. Agar-based binding systems undergo debinding by simple air-drying at ambient temperatures (short oven drying may be required with larger or thicker-walled parts).

Debinding is followed by sintering, a high-temperature firing process in a controlled atmosphere to consolidate powdered-metal particles by diffusion. This happens in a sintering furnace at temperatures elevated to just below the melting point of the specific metal. Sintering and densification occurs through multiple processes, including volume diffusion, grain boundary diffusion, and surface diffusion. In some cases, a liquid phase is used to accelerate sintering.

Sintering begins with the molded part undergoing a preheating stage that removes the remaining binder. This stage is necessary to remove any residuals or potential contaminates that could compromise the metal's mechanical properties. A defined sintering schedule typically outlines temperatures, ramp rates, soak or hold times, and cooling rates required for both specific MIM materials and furnaces used. It also defines the atmosphere: hydrogen, nitrogen, vacuum, or combinations thereof.

Successful sintering depends on both the metal being processed and the qualities of the particular furnace. Both should be considered in relationship to each other. After sintering, the metal particulates have been consolidated and densified into a solid mass with nearly the theoretical density of similar wrought metal.

MIM vs. Other Processes

MIM has a number of advantages over traditional processes such as investment casting, press and sinter, forging, and machining (see Figures 6.7, 6.8, and 6.9).[44–47]

Investment casting requires the construction of individual molds for each part produced. Poured rather than injected under pressure, recovered parts yield shapes that — by injection-molding standards — are relatively crude and need secondary finishing operations. Surfaces are rough, and dimensional tolerances inexact. Sinks and voids are common in wall thicknesses over 19.05 mm, and thin walls are difficult to fill. Cast parts need large gates to promote material flow, which may not be easy to hide. The

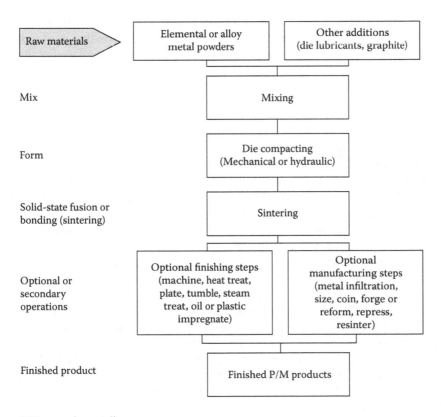

P/M = powder metallurgy

FIGURE 6.7
P/M parts manufacturing: Traditional process.

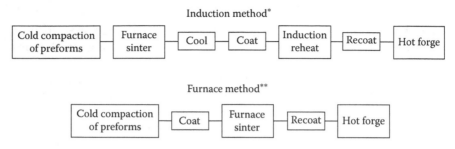

*GM hydramatic, federal-mogul
**ICM/krebsoge, masco tech, toyota

FIGURE 6.8
Commercial powder forging methods.

resulting features could detract from part aesthetics. Individual production steps can include pattern making, tree assembly, investing, stuccoing, and dewaxing. Firing, pouring, and mold knockout as well as finishing steps including deburring, machining, and polishing add more labor to the process.

Press-and-sinter processing, like MIM, uses powdered metal as a raw material. Physically akin to compression molding of plastics, powdered metals are placed in a mold and then ram-compacted to form a fixed shape. Furnace sintering fuses the particulates and slightly increases finished part density. The process produces green parts in quick cycles, and the mold is reusable. But only simple geometries are practical, and mechanical strength is lower than both MIM and wrought metals (see Figure 6.10).[43]

Forging is also best suited for simple shapes. Heated metal is physically hammered into shape under high temperatures and pressures. First, metal ingots are cast and reduced to billets. The billets are heated and placed on formed die halves. Hammering metal against the dies begins to form parts. Heat loss dictates reheating and more hammering. The process is repeated until the final shape emerges. Multiple compaction adds strength. Cross holes are impossible, and it is tough to maintain tolerances and straight edges. Postforging operations include welding, heat-treating, and final machining. Tooling is costly and has a short working life. Cycle times are extended, and overall production costs are relatively high.

Machining of individual parts delivers both exacting tolerances and complex shapes, but there are some limits on part complexity compared to the intricacies of injection molding. Nearly any metal can be machined, and machining normally does not alter the properties of the raw material. Machining centers are well established and accessible. This remains the preferred process for prototyping and short-run production. Machining

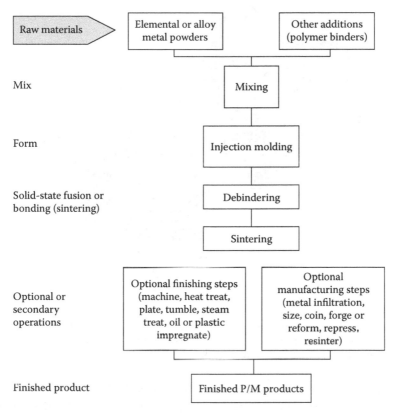

FIGURE 6.9
P/M parts manufacturing: Metal injection molding process.

is, however, labor- and capital-intensive and design limited; production times lag; and costs do not decrease with volume.

Typical Applications

The increased rate of applications for P/M will come through technological developments that provide greater value than other metal, polymer, or ceramic forming processes. It should be noted that P/M is the fastest-growing metal forming business today. Among the P/M metals that are advancing this technology are higher strength, heat- and corrosion-resistant alloys with stainless steels in the forefront.

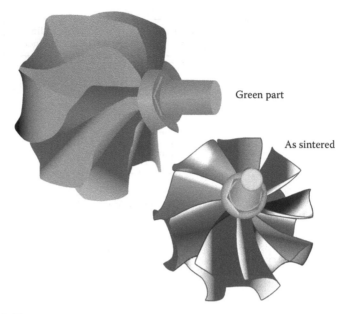

Green part

As sintered

FIGURE 6.10

An automotive turbocharger rotor was produced. The part's complex 3D shape cannot be reproduced by machining and with fewer production steps, MIM is a viable alternative to investment casting for this high-volume part.[43]

Figure 6.11[47] shows the automotive share of the P/M parts market growing to 69% in 1996 as compared to 66.5% in 1995. Typical parts include flanges used on auto exhaust manifolds, multipiece engine bearing races, ring gears with helical geometries, powertrain parking gears and sensors, cylinder liners, main bearing caps, as-sintered connecting rods, and sintered cam lobes.

Using the MIM process, a new 17-4 PH stainless steel endoscopic surgical staple has been produced for the medical industry. The endoscopic linear cutter combines a 45° bilateral articulation with a 360° rotation, providing excellent maneuverability and tissue access for endoscopic procedures such as appendectomies and lung resections.

For the military, the Phalanx Close-In Weapon-System (CIWS),[49] the last line of defense for ships, rapidly fires kinetic energy penetrators at enemy missiles. The major factor affecting the CIWS performance is the mass of the penetrator that remains intact while penetrating its target.

In the past, Phalanx penetrators have been produced from both depleted uranium (DU) and a W-Ni-Fe alloy. Both materials have the high density required for kinetic energy penetrators. DU is useful for penetrating thick armor sections, but it fragments when penetrating missile targets. The

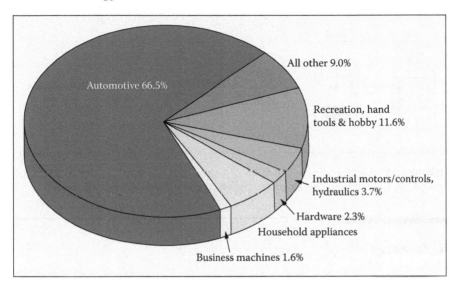

Automotive 66.5%

All other 9.0%

Recreation, hand
tools & hobby 11.6%

Industrial motors/controls,
hydraulics 3.7%

Hardware 2.3%

Household appliances

Business machines 1.6%

FIGURE 6.11
P/M parts market — 1996.[47]

W-Ni-Fe alloy that replaced DU has superior performance, but loses substantial mass at impact velocities greater than 1220 m/s.

Now, for CIWS penetrators[49] is a new W alloy composition and modified process parameters. This material shows a 50% increase in both impact energy and remnant mass over the W alloy presently used in the fleet and over a 100% increase in remnant mass over DU.

High-temperature superconductors (HTS) are being considered for a variety of defense applications, including power generation, surface and subsurface propulsion, and mine sweep systems. In the commercial field, the application of HTS materials may dramatically improve the performance and affordability of magnetic levitation (Maglev) vehicles, magnetic resonance imaging (MRI) systems, superconducting magnetic energy storage (SMES) systems, and metal separation systems. To accelerate the applications of HTS, Concurrent Technologies Corp. recently conducted a program to develop and demonstrate the technology for producing long-length (up to 1 km) HTS wires (see Figure 6.12) and tapes.

This HTS project focused on the Ag-clad, Bi-based compound $Bi_2Sr_2CaCu_2O_x$ (BSCCO), commonly known as Bi-2212. Process models of wiredrawing and rolling were used to help develop manufacturing process specifications. Material properties and constitutive equations were developed to characterize the powder deformation and compaction.

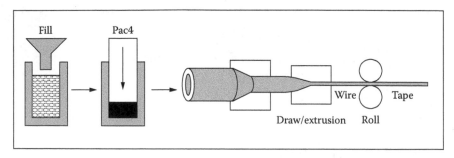

FIGURE 6.12
Process steps in forming long lengths of high-temperature superconducting wires.[49]

References

1. Williams, V. A., Net Shapes Are a Reality, *Production*, March 1983, pp. 44–47.
2. Johnson, P. K., P/M Technology Trends – 1997, *Intl. J. Powder Metall.*, 33(3), 13–19, 1997.
3. Otto, F. J. and Herring, D. H., Gears Heat Treatment – Part 1, *Heat Treating Progress*, June 2000, pp. 55–59.
4. Toyota Central Labs Inc., 41-1, Aza Yokomichi, Oaza Nagakute, Nagakute-cho, Aichi-ken, 480–1192, Japan; www.tytlabs.co.jp.
5. Yoon, E. H., Hong, J. H., and Hwang, S. K., Mechanical Alloying of Dispersion-Hardened Ni_3AlB from Elemental Powder Mixtures, *J. Mater. Eng. Performance*, 6(1), 106–112, 1997.
6. Yinjiang, W., Tao, L., and Lian, Z., Study of TiAl Alloys and Powders Produced by Infiltration Combustion Synthesis (ICS), *J. Adv. Mater.*, October 2000, pp. 52–58.
7. Mileiko, S. T., Rudnev, A. M., and Gelachov, M. V., Carbon-Fibre/Titanium-Silicide-Interphase/Titanium-Matrix Composite, *Compos. Sci. Technol.*, 55, 255–260, 1995.
8. Low-Cost PM Route for Titanium Matrix Carbon-Fiber Composites, *Powder Metall.*, 39(2), 97–99, 1996.
9. de Chalus, P.-A., Copper-Iron-Cobalt P/M Alloys Extend Tool Life 30%, *AM&P*, February 1998, p. 7.
10. Kubel, E., Advancements in Powder Metallurgy Rhenium, *Industrial Heating*, September 2001, pp. 47–51.
11. Barandika, M. G., Sánchez, J. M., and Castro, F., New Developments in TiB_2 Hard Metals, *Met. Powder Rep.*, 49, 10, 1994.
12. Sánchez, J. M., Barandika, M. G., Sevillano, J. G., and Castro, F., Consolidation, Microstructure, and Mechanical Properties of Newly Developed TiB_2-Based Materials, *Scr. Met. et Mat.*, 26, 957, 1992.
13. González, R., Echeberria, J., Sánchez, J. M., and Castro, F., WC-(Fe,Ni,C) Hardmetals with Improved Toughness through Isothermal Heat Treatments, *J. Mater. Sci.*, 30, 3435, 1995.

14. González, R., Iturriza, I., Echeberria, J., and Castro, F., Consolidation of WC Cemented Carbides with Fe-Mn Metallic Binders, *Met. Powder Rep.*, 5(18), 69, 1994.
15. Castro, F., Urcola, J., and Fuentes, M., P/M and Particulate Materials Research, *Intl. J. Powder Metall.*, 33(2), 19–29, 1997.
16. Zubillaga, C., Hernández, F., Urcola, J. J., and Fuentes, M., Experimental Analysis of Tungsten Coarsening in a Heavy Metal, *Acta Metall*, 37(7), 1865, 1989.
17. Zamora, K. O., Sevillano, J. G., and Fuentes Pérez, M., Flow Stress and Ductility of Tungsten Heavy-Metal Alloys, in A. Bose and R. J. Dowding (Eds.), *Tungsten & Tungsten Alloys*, Princeton, NJ: Metal Powder Industries Federation, 1992, p. 281.
18. Ostolaza Zamora, K. M., Sevillano, J. G., and Fuentes, M., Fracture Toughness of Tungsten Heavy Metal Alloys, *Mater. Sci. Eng. A*, 157, 151, 1992.
19. Yáñez, Urcola, J. J., Ostolaza, K., and Sevillano, J.G., Stronger W-Ni-Fe Alloys by Solid Solution or Precipitation Hardening, in A. Bose and R. J. Dowding (Eds.), *Tungsten & Tungsten Alloys — 1992*, Princeton, NJ: Metal Powder Industries Federation, 1992, p. 119.
20. Rodriguez, A. B. and Sevillano, J. G., Viscoplastic Flow of High-Density W-Ni-Fe Alloys during Liquid-Phase Sintering, in A. Bose and R. J. Dowding (Eds.), *Tungsten & Tungsten Alloys — 1992*, Princeton, NJ: Metal Powder Industries Federation, 1992, p. 61.
21. Sundaresan, R. and Froes, F. H., *J. Metals*, 39, 22–27, 1987.
22. Lee, R. Y., Jang, J., and Koch, C. C., *J. Less-Common Metals*, 140, 78–83, 1988.
23. Koch, C. C., Cavin, O. B., McKamey, C. G., and Scarborough, J. O., *Appl. Phys. Lett.*, 43, 1017–1019, 1983.
24. Taylor, P. R. and Pirzada, S. A., *Adv. Performance Mater.*, 1, 33–50, 1994.
25. Taylor, P. R. and Pirzada, S. A., in H.Y. Sohn (Ed.), *Metallurgical Processes for the Early 21st Century*, Vol. 1, Warrendale, PA: TMS, 1994, pp. 65–76.
26. Taylor, P. R., Pirzada, S. A., Marshall, D., and Donahue, S., in K. Upadhya (Ed.), *Plasma Synthesis and Processing of Materials*, Warrendale, PA: TMS, 1993, pp. 215–225.
27. Taylor, P. R. and Pirzada, S. A., U.S. Patent 5369241, November 1994.
28. Taylor, P. R., Manrique, M., and Pirzada, S. A., *Proc. EPD Cong. 1995*, G. Warren (Ed.), Warrendale, PA: TMS, 1995, pp. 49-60.
29. Gardiner, R. W., *Mater. Des.*, 10, 274–279, 1989.
30. Bunshah, R. F. (Ed.), *Deposition Technologies for Films and Coatings*, Park Ridge, NJ: Noyes, 1982, p. 131.
31. Ward-Close, C. M. and Partridge, P. G., *Mater. Lett.*, 11(8–9), 295–300, 1991.
32. Harman, T. R., Novel Plasma Melting Process Produces Rapidly Solidified Metal and Alloy Powders, *METPR* 7, 546, 1986.
33. Donaldson, I. and Hanejko, F. G., Effects of Processing Methods and Heat Treatments on the Mechanical Characteristics of High-Performance Ferrous P/M Materials, *Industrial Heating*, 63(9), 69, 1996.
34. Mazumder, J., Koch, J., Nagarathnam, K., and Choi, J., Rapid Manufacturing by Laser-Aided Direct Deposition of Metals, paper presented at 1995 World Congress on Powder Metallurgy & Particulate Materials, MPIF/APMI, Washington, D.C., June 16–21.

35. Lewis, G. K., Thoma, D. J., Milewski, J. O., and Nemec, R. B., Directed Light Fabrication of Near-Net Shape Metal Components, paper presented at 1996 World Congress on Powder Metallurgy & Particulate Materials, MPIF/APMI, Washington, D.C., June 16–21.

36. Whitney, E. J., Arcella, F. G., and Krantz, D., P/M Laser Manufacturing and Titanium Structures, paper presented 1996 World Congress on Powder Metallurgy & Particulate Materials, MPIF/APMI, Washington, D.C., June 16–21.

37. Irretier, D., Debindering and Sintering of Powder Metallurgical Parts during Batch Processing, *Industrial Heating*, October 1999, pp. 111–116.

38. Herring, D. H., Heat Treating P/M Parts, *Heat Treating Progress*, June 2002, pp. 27–30.

39. Moyer, K. H. and Jones, W. R., Vacuum Sinter Hardening, *Heat Treating Progress*, June 2002, pp. 65–69.

40. Kohl, R., Metal Parts Right Out of the Microwave, *Machine Design*, November 4, 1999, p. 35.

41. Teague, P., Powder Metal Parts Perfected in a Microwave Oven, *Design News*, October 18, 1999, p. 29.

42. Hotter PIM Breathes Life into Medical Products, *Mach. Design*, October 9, 1997, pp. 78–82.

43. Wick, R. and Ruhland, B., Injecting More "Mettle" into P/M Designs, *Mach. Design*, March 20, 2003, pp. 51–54.

44. Puttré, M., Technology Has Steel Shining Like New, *Design News*, December 1, 1997, pp. 69–72.

45. White, D. G., P/M in North-America, *Intl. J. Powder Metall.*, 32(3), 221–227, 1996.

46. Dollmeier, K. and Petzoldt, F., P/M in Germany, *Intl. J. Powder Metall.*, 33(1), 39–43, 1997.

47. Rinek, L. M., Evolving Prospects and Technology Developments for Powder Metallurgy Parts, *SRI Intl.*, D95-1922, 13, 1993.

48. Johnson, P. K., P/M Industrial Trends — New Technologies Propel P/M Growth, *Intl. J. Powder Metall.*, 32(2), 145–153, 1996.

49. Kuhn, H. A. and Dax, F. R., Powder Metallurgy Product and Process — Development at Concurrent Technologies Corporation, *Intl. J. Powder Metall.*, 32(3), 229–238, 1996.

Bibliography

Becker, B. S. and Bolton, J. D., Production of Porous Sintered Co-Cr-Mo Alloys for Possible Surgical Implant Applications: 2. Corrosion Behavior, *Powder Metall.*, 38(4), 305–313, 1995.

Bee, J. V. and Toussaint, E., P/M in Australia, *Intl. J. Powder Metall.*, 33(1), 33–37, 1997.

Chung, H. S. and Kim, B., Powder-Metallurgy at KIMM, Korea, *Intl. J. Powder Metall.*, 32(2), 137–143, 1996.

Compaction and Sintering Characteristics of Composite Metal Powders, *J. Mater. Proc. Technol.*, 63(1-3), 364–369, 1997.

Datta, P. and Upadhyaya, G. S., Effect of Sintering Variables on the Properties of 316L and 434L Stainless Steels, *Industrial Heating*, June 1999, pp. 37–41.

Dunkley, J., Powder Metallurgy & Particulate Materials Technology — 1994, *Powder Metall.*, 37(2), 101–104, 1994.

Froes, F. H., Suryanarayana, C., and Taylor, P. R., Synthesis of Advanced Metals by Powder-Metallurgy Techniques, *Powder Metall.*, 3(1), 63–65, 1996.

Garg, D., Berger, K. R., Bowe, D. J., and Marsden, J. G., Effective Atmosphere for Iron-Copper-Carbon Components, *Industrial Heating*, August 1997, pp. 39–43.

Garner, H. A., Heat-Treated Powder Metal High-Speed Steel Tools Last Longer in Heavy-Duty Nail Production, *Industrial Heating*, June 1996, pp. 40–42.

Harley, M., Polymers for Ceramic and Metal Powder Binding Applications, *Ceram. Industry*, November 1995, pp. 51–55.

Hebelsin, J. C., On an International Scale, What's Hip about HIP? *Design News*, June 9, 1997, p. 196.

James, W. B., Considerations in the Development of Ferrous P/M Alloys for Sinter Hardening Applications, *Industrial Heating*, September 1999, pp. 63–67.

Johnson, J. J. and German, R. M., Pinpointing Parts for PIM, *Mach. Design*, August 7, 2003, pp. 52–53.

Kaufman, S. M., Are Conventional SAE Alloy Steels Coming to Powder Metallurgy? *Modern Metals*, March 1996, pp. 44–46.

Kneringer, G. and Stickler, R., Powder — Metallurgy in Austria, *Intl. J. Powder Metall.*, 32(3), 213–220, 1996.

Mahidhara, R. K., Degradation of RSP/PM Al-8Fe-4Ce during Creep, *J. Mater. Eng. Performance*, 5(2), 256–259, 1996.

Molins, Jr., C., Powder-Metallurgy in Spain, *Intl. J. Powder Metall.*, 32(2), 127–134, 1996.

Pennington, J. N., Exploring Powder Forging, *Modern Metals*, February 1996, pp. 32–35.

Philips, T. and Nayar, H. A., A Troubleshooting Guide for Sintering Furnace Atmospheres, *Industrial Heating*, August 1997, pp. 33–37.

Stoltzfus, J. M., Making Pure Oxide Powders, *NASA Tech Briefs*, January 1997, p. 56.

Tengzelius, J. and Grinder, O., Powder Metallurgy in Sweden, *Intl. J. Powder Metall.*, 32(3), 203–210, 1996.

Upadhyaya, G. S., P/M Trends in India, *Intl. J. Powder Metall.*, 32(2), 117–125, 1996.

7

Nanotubes

Introduction

Imagine a single molecule that is 50,000 times thinner than a human hair and 100 times stronger than steel. Such a molecule is invisible to the naked eye, yet could revolutionize materials technology. Scientists have mastered the synthesis and characterization of components on the microscopic scale; however, the discovery of carbon molecules called nanotubes has sparked new interest in materials on the nanoscale (100 nm or less).

Definition and Structure

A carbon nanotube (CNT) is constructed of a single sheet of graphite rolled into a tube, with fullerene caps on the ends (Figure 7.1).[1] They look like nanoscale cylinders, about 1 nm or so in diameter and a few microns long. Imagine rolling up a sheet of graphite into a tube; that is what we are talking about. Depending on the processing method, nanotubes may have multiple walls with concentric cylinders around the inner tube. However, multiwalled tubes contain imperfections that can limit their properties. As a result, single-walled nanotubes (SWNTs)[1,2] should be of primary interest for materials development; this chapter will also cover multiwalled nanotubes (MWNTs).

These single-walled elongated fullerene molecules contain millions of carbon atoms, each occupying an assigned place, to form defect-free hollow structures just a few atoms in circumference. Nanotubes are many thousands of times longer than their diameters, and they possess exceptional mechanical, electrical, and thermal properties.

These carbon nanotubes are basically hollow crystals of carbon that are stronger than steel and more electrically conductive than copper. CNTs may soon enhance lithium batteries and replace carbon fibers in composites. Aligned CNT arrays can be used to create chemical sensors, nanotweezers, and other devices including field emission products such as flat-panel displays. Designs using nanotube arrays can also be used to greatly reduce

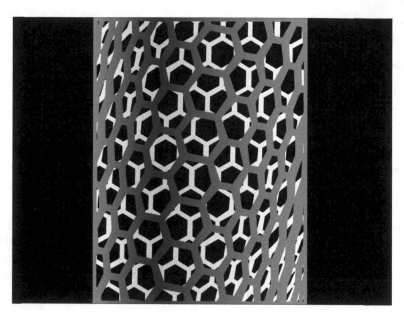

FIGURE 7.1
Schematic of a single layer of carbon nanotube. The cylindrical structure is built from hexagonal honeycomb lattice of sp^2 bonded carbon with no dangling bonds. Imagine taking one layer of graphite and folding it to match the edges that contain the dangling bonds; the cylinder so formed is the basic unit of the nanotube. In multiwalled nanotubes, cylinders of various diameters are arranged concentric to each other with a constant spacing of 0.34 nm between them. In single-walled nanotube ropes, tens of cylinders of near uniform diameter (~1/2 nm) arrange in a triangular lattice.[1]

the cost of optical components such as fiber-optic signal demultiplexing devices.

Because CNTs are carbon hexagons arranged in a concentric manner, they can behave as a semiconductor or metal depending on their diameter and helicity of the arrangement of graphitic rings in the walls.

CNT threads up to 100 m long have been created at the University of Texas at Dallas and Trinity College, Dublin, Ireland, with a toughness 3 times greater than the toughest natural material, spider silk.

A cross-section of the thread measures about 50 µm in diameter and contains hundreds of trillions of tiny CNTs per centimeter in length. To produce the threads, a technique developed by Philippe Poulin, at CNRS in Bordeaux, France, was used.[2-4]

A spinning aqueous solution of CNTs and a surfactant is injected into pipe in which a solution of polyvinyl alcohol flows. The two solutions coagulate to form a rubber-like gel fiber, which is wound onto a mandrel. In a second continuous process, the gel fiber is unwound, washed, and dried to produce a solid polymer fiber of potentially unlimited length. The fiber contains 60% CNTs bound together by polyvinyl alcohol.

The developers have not been able to find any material that is tougher than the CNT composite fibers.[4] The fiber's toughness probably results from structural changes during stretching, and this aligns the nanotubes in the fiber direction.

The threads can also be used as capacitors to store very small electric charges and to power devices embedded in clothes. But improving the electrical properties is the key to future challenges by making them more conductive and more porous.

These are the crucial properties for building sensors out of this material and for making lighter fibers that are still highly conductive.

Growth and Fabrication

The most common methods for producing nanotubes are by laser ablation, electric arc, and chemical vapor deposition techniques. Growth of nanotubes is not well understood. However, it is known that use of metallic catalysts is necessary to sustain nanotube growth. Without this catalyst, fullerenes are formed using the same basic processes.

Engineers at the University of California at Berkeley have found an innovative way to grow Si nanowires and CNTs directly on microstructures in a room-temperature chamber, opening the doors to cheaper and faster commercialization of a myriad of nanotechnology-based devices.

The researchers were able to precisely localize the extreme heat necessary for nanowire and nanotube growth, protecting the sensitive microelectronics — which remained at room temperature — just a few micrometers away, or about one tenth the diameter of a strand of human hair.

The new technique eliminates cumbersome middle steps in the manufacturing process of sensors that incorporate nanotubes or nanowires. Such devices would include early-stage disease detectors that could signal the presence of a single virus or an ultrasensitive biochemical sensor triggered by mere molecules of a toxic agent.

The steps used in creating nanowires and nanotubes are essentially the same, though different chemicals and temperatures may be used. The UC Berkeley researchers, in this case, used an Ni-Fe alloy with acetylene vapor to create CNTs.

The typical nanowire or nanotube production process occurred in a furnace at temperatures of 600 to 1000°C. The procedure began with a 1 cm^2 Si wafer that was coated thinly with a metal alloy. A vapor was then directed towards the substrate, and the metal alloy acted as a catalyst in a chemical reaction that eventually formed billions of nanowire or nanotube precipitates.

The nanomaterials were harvested by being placed in a liquid solvent, such as ethanol, and blasted with ultrasonic waves to loosen them from the wafer surface. Researchers must then ultimately sort through the billions of nanowires or nanotubes to find the few that meet the specifications they need for their sensor applications.

Instead of finding a way to produce nanomaterials separately and then connecting them to larger scale systems, the researchers decided to grow the Si nanowires and CNTs directly onto the circuit board. The challenge was in protecting the sensitive microelectronics that would melt in the tremendously high temperatures needed to create the nanomaterials.

Resistive heating provided the answer. The electrical current flows through the wire to generate the heat, like the wires in a toaster.

The researchers passed the current through the wire to the specific locations on the microstructure where they wanted the nanowires or nanotubes to grow. In one experiment, an area was heated to 700°C, while another spot just a few micrometers away sat comfortably at 25°C. The entire circuit board was placed in a vacuum chamber for the tests. This is the immediate integration of the nanoscale with the microscale.

The experiments yielded Si nanowires from 30 to 80 nm in diameter and up to 10 μm long, and CNTs that were 10 to 30 nm in diameter and up to 5 μm long.

This was a very unique approach and this method allows for the production of an entire nano-based sensor in a process similar to creating computer chips. There would be no postassembly required.

One method above, CVD, uses some hydrocarbon gases such as methane with a catalyst material like iron. A second method, called plasma-enhanced CVD, uses a low-temperature plasmas to grow nanotubes.

Other experiments have shown the feasibility of producing nanotubes in an efficient gas-phase process, sometimes having diameters down to ~0.6 nm. These methods involve CVD in high-pressure CO. This growth technique is a potentially viable means for cost-effective production of large amounts of SWNT. Development of new methods for purification and characterization of nanotubes has given new insight into their growth mechanisms. New data on the morphology and length distributions of SWNT grown by traditional laser-oven methods continue to evolve.

A method has been proposed to develop bimorph actuators and force sensors based on CNTs. The proposed devices could make it possible to generate, sense, and control displacements and forces on a molecular scale, and could readily be integral with conventional electronic circuits. These devices could also enable the development of a variety of novel microelectronic-mechanical systems, including low-power mechanical

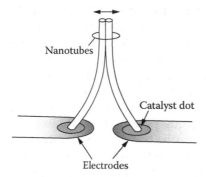

FIGURE 7.2
Carbon nanotubes would grow out from catalytic metal dots on electrodes, eventually becoming attached to each other by van der Waals forces to form a bimorph actuator or sensor.[3]

signal processors, nanoscale actuators and force sensors, and even microscopic robots.

The proposed devices would exploit the dependence of nanotube length on charge injection that has been observed in mats of disordered carbon single-walled nanotubes:[4] The nanotubes become elongated or shortened when biased at negative or positive voltage, respectively. This result suggests that one could produce opposing changes in length in pairs of side-by-side, oppositely biased nanotubes, resulting in a lateral deflection of the unsecured tube ends, as shown in Figure 7.2. Fabrication of such a nanotube bimorph device requires the ability to produce and join the tubes in the desired configuration, with one end of each tube connected to a suitable electrical contact.

The proposed bimorph device could be fabricated by growing two nanotubes by CVD on closely spaced catalyst dots over prepatterned bias electrodes on a substrate. It is likely that during the growth of the nanotubes, the van der Waals attraction would cause the nanotubes to become attached to each other along their sides, as shown in Figure 7.2. Because the electrical conductivity of a nanotube perpendicular to its length is much lower than the electrical conductivity along its length, this configuration should make it possible to maintain a significant differential voltage across the two nanotubes, as needed to cause a differential length change in the pair. Conversely, the application of a lateral external force to the top of the pair should give rise to a voltage between the electrodes, so that this device can also function as a sensitive force detector.

To be able to fabricate nanotube bimorph actuators with the configuration shown in Figure 7.2, it will be necessary to develop the means to control the positions and orientations of individual nanotubes on such substrates as Si wafers. This is likely to entail the use of electron-beam lithography, lift-off, and etching for fabricating catalyst dots 5 to 15 nm wide on prepatterned electrodes. Suitable catalyst materials could include Ni or alloys of Ni, Co, Fe, or Mo. In the contemplated CVD process, suitable precursor and carrier gases (e.g., methane, ethylene, or carbon monoxide plus hydrogen plus ether argon or nitrogen) would interact with the substrate (which would be heated to a temperature between 600 and 950°C), yielding selective growth of nanotubes out from the catalyst dots. There are numerous potential variations on this basic fabrication scheme, including orienting the dots so that the nanotubes grow parallel (instead of perpendicular) to the substrate surface and incorporating other materials to modify the electrical and mechanical properties of nanotube pairs.[3, 4]

An improved process has been developed at Ames Research Center for the efficient fabrication of CNT probes for use in atomic-force microscopes (AFMs) and nanomanipulators. Relative to prior nanotube tip production processes, this process offers advantages in alignment of the nanotube on the cantilever and stability of the nanotube's attachment. A procedure has also been developed at Ames that effectively sharpens the MWNT, which improves the resolution of the MWNT probes and, combined with the greater stability of MWNT probes, increases the effective resolution of these probes, making them comparable in resolution to SWNT probes. The robust attachment derived from this improved fabrication method and the natural strength and resiliency of the nanotube itself produce an AFM probe with an extremely long imaging lifetime. In a longevity test, a nanotube tip imaged an Si_3N_4 surface for 15 h without measurable loss of resolution. In contrast, the resolution of conventional Si probes noticeably begins to degrade within minutes. These CNT probes have many possible applications in the semiconductor industry, particularly as devices are approaching the nanometer scale and new atomic layer deposition techniques necessitate a higher resolution characterization technique. Previously at Ames, the use of nanotube probes has been demonstrated for imaging photoresist patterns with high aspect ratio. In addition, these tips have been used to analyze Mars simulant dust grains, extremophile protein crystals, and DNA structure.

Nanotechnology researchers at the University of Texas at Dallas and Trinity College in Dublin, Ireland, have announced a breakthrough in spinning CNT composite fibers that are tougher than any reported polymer fiber made by humans or nature.[5]

The composite fibers of the CNTs — rolled-up sheets of graphite that are 50,000 times thinner than a human hair — are continuously spun, which results in a material toughness of more than 4 times that of spider silk and 17 times that of Kevlar. The fibers have twice the stiffness and strength, and 20 times the toughness, of the same weight and length steel wire.

"Mankind has largely been unsuccessful in processing untold billions of these invisible nanofibers to make useful articles that exploit these properties. Our spinning method is the first to produce high-performance, continuous fibers of CNTs suitable for potential use in a wide array of applications," states Ray H. Baughman, director of the UTD NanoTech Institute.[4]

Possible applications for the UTD fibers include clothing that could store electrical energy and be used to power various electrical devices, or synthetic muscles capable of generating 100 times the force of the same-diameter natural muscle. The fiber also could be used as a power source for spacecraft on long voyages through conversion of thermal energy to electrical energy, or for "micro air vehicles" the size of an insect that could replace current, much larger military drones used to gather intelligence remotely.

Duke University chemists have developed a method of growing one-atom-thick CNTs 100 times longer than usual, while maintaining straightness with controllable orientation.

Their achievement solves a major barrier to the nanotubes' use in ultrasmall "nanoelectronic" devices. The researchers have also grown checkerboard-like grids of the tubes that could form the basis of nanoscale electronic devices.

The accomplishment involved sprouting the infinitesimally thin SWNT structures, from tiny catalytic cluster of Fe and Mo atoms dotted onto a small rectangle of Si inside a quartz tube.

These growing nanotubes continue to lengthen along the silicon's surface in the direction of the flow of a feeding gas of CO and H_2 that had been quick-heated to a temperature hot enough to melt normal glass. Atoms from the feeding gas were used as molecular building blocks.

The process was described by Duke assistant chemistry professor Jie Liu and associates in the on-line edition of the *Journal of the American Chemical Society (JACS)*.

The researchers claimed lengths initially of more than 2 mm, but now have grown 4-mm-long nanotubes and may get even longer nanotubes in the future.[6]

Another technique forms micropatterns of aligned CNTs by pyrolysis of organic-metal complexes containing both the metal catalyst and carbon source required for the nanotube growth. For manufacturing devices, micropatterns of aligned nanotubes can be produced either by patterned growth of the nanotubes on a partially masked/prepatterned surface or through a contact printing process, where substrate-free nanotube films are transferred to other substrates, such as polymer films, which otherwise would not be suitable for growth of the structures at high temperatures.

It seems likely that two entirely different mechanisms operate during the growth of MWNTs and SWNTs, because the presence of a catalyst species is absolutely necessary for the growth of the latter. Open MWNTs can occasionally be spotted among arc-grown nanotube samples. The simplest scenario could be that by some mechanism, all the growing layers of a tube remain open during growth.[1] Closure of the layer is caused by the nucleation of pentagonal rings due to local perturbations in growth conditions or due to the energetics (stability) between structures that contain hexagons and pentagons.

Successive outer layers grow on inner tube templates, and the large dimensional anisotropy results from the vastly different rates of growth at the high-energy open ends, compared to the unreactive basal planes.

Because the presence of a catalyst is necessary for SWNT growth, any mechanism that accounts for growth should incorporate the role of the catalyst in the growth. There is no consensus as yet on how the SWNTs grow, nor on why the yield of nanotubes increases drastically when a second element (Y) is added; yttrium does not show any catalytic activity when used alone.

Researchers at Rice University and the University of Illinois at Urbana-Champaign have discovered the first method to chemically select and separate CNTs based on their electronic structure. The new process represents a fundamental shift in the way scientists think about the chemistry of CNTs.[7]

Other than low-cost mass production, there is no bigger hurdle to overcome in carbon nanotechnology than finding a reliable, affordable means of sorting SWNTs. In fact, if development efforts can find new technology based on electronic sorting and reliably separate metallic nanotubes from semimetallic and semiconducting varieties, a terrific tool for nanoscience would be available.

The utility of specific CNTs, based upon their precise electronic characteristics, could be an enormous advance in molecular electronics. Until now, everyone has had to use mixtures of nanotubes and, by process of elimination, select the desired device characteristics afforded from a myriad of choices. This could now all change because there is the possibility of generating homogeneous devices.

All SWNTs are not created equal. There are 56 varieties that have subtle differences in diameter or physical structure. Slight as they are, these physical differences lead to marked differences in electrical, optical, and chemical properties. For example, about one third are metals, and the rest are semiconductors.

Although CNTs have been proposed for myriad applications — from miniature motors and chemical sensors to molecule-sized electronic circuits — their actual uses have been severely limited, in part because scientists have struggled to separate and sort the knotted assortment of nanotubes that result from all methods of production.

The researchers and scientists at Rice and Illinois Universities developed a technique for breaking up bundles of nanotubes and dispersing them in soapy water. They applied reaction chemistry to the surfaces of nanotubes in order to select metallic tubes over semiconductors.

To control nanotube chemistry, the researchers added water-soluble diazonium salts to nanotubes suspended in an aqueous solution. The diazonium reagent extracts an electric charge and chemically bonds to the nanotubes under certain controlled conditions.

By adding a functional group to the end of the reagent, the researchers were able to create a "handle" that they can then use to selectively manipulate the nanotubes. There are different techniques for pulling on the handles, including chemical deposition and capillary electrophoresis.

The electronic properties of nanotubes were determined by their structure, so that a way of grabbing hold of different nanotubes by utilizing the differences in this electronic structure was found. Because metals give up an electron faster than semiconductors, the diazonium reagent can be used to separate metallic nanotubes from semiconducting nanotubes.

The chemistry is reversible. After manipulating the nanotubes, the scientists were able to remove the chemical handles by applying heat. The thermal treatment also restored the pristine electronic structure of the nanotubes.

Recently, the consensus has been that the chemistry of a nanotube is dependent only on its diameter, with smaller tubes being less stable and more reactive; however, that is clearly not the case here. Reaction pathways are based on the electronic properties of the nanotube, not strictly on its geometric structure, and this represents a new paradigm in the solution phase chemistry of CNTs.

Properties[8-23]

CNTs can serve multifunctional roles because their tensile strength is at least 10 times stronger and their weight is less than half that of conventional carbon fibers, electrical conductivity is as high as that of Cu, and thermal conductivity is as high as that of diamond. In addition, properties can be tailored through processing to fit a multitude of aerospace, biomedical, and industrial applications. The range of applications for such properties is endless: from chemical sensors and high-strength composites to tiny networks transporting telemetry in a biomimetic fashion. Historically, biological systems have operated on atomic scale principles; now materials engineers will be able to explore this amazing nano world on the same scale.[8]

Mechanical Properties

CNTs have extraordinary mechanical properties. For example, compared to steel, nanotubes have a strength-to-weight ratio of 500. At the same time, nanotubes can be used to make a computer chip, because in addition to these wonderful mechanical properties, they also have very exciting electrical properties.

Historically, the materials that have been used for computer chip applications were impractical for construction of an aircraft. The same with Al or stainless steel: These metals could be used to manufacture an automobile, but they could never be used to make a computer chip. CNTs can be used for both fine applications like computer chips and sensors, and for massive applications in the aerospace and automotive industries.

Absence of detects in individual nanotubes enhances their mechanical properties and their potential as structural reinforcements. Rigorous bonds are formed from covalently linked carbon atoms having three nearest neighbors. Tube ends are sealed, with no dangling chemical bonds to weaken the arrangement. The lightweight structure is assembled as an ideal carbon fiber that has the strongest bonds found in nature. Cohesive strength between carbon atoms and high elasticity compared with that of graphite fibers make nanotubes highly resistant to failure under tension. According to Richard Smalley, "It should be the strongest fiber that you can make of anything — ever. In the strength-to-weight ratio sweepstakes, it should be the ultimate fiber."

Nanotubes are known to have a tensile strength of 50 to 200 GPa, which is at least an order of magnitude higher than conventional graphite fibers. Additionally, predictions of high elasticity and bending stiffness are supported by thermal vibration amplitude measurements of a Young's modulus of over 1 terapascal (TPa). In addition, nanotubes are expected to break at very high strain (5 to 20%), and in dynamic simulations they behave like "superstrings." They narrow down to single carbon chains upon application of tension[2] (see Tables 7.1 and 7.2[24]).

Composite-Reinforced Properties

High modulus and high elastic strain qualify nanotubes as potential reinforcements in composite materials. Microcracks in continuous solids act as stress concentrators, but the failure of one fiber in a composite of loosely coupled nanotubes would result in very little overloading of adjacent tubes; therefore, crack growth would be suppressed. Loads are effectively transferred to the nanotubes, resulting in a composite modulus that is similar to that of an isotropic short fiber composite containing fibers of exceptional modulus and tensile strength.

TABLE 7.1

Strength and Modulus of CNTS[24]

Property	Material	Type	Value	Units	Range	Simulated	Measured	Technique	Equipment	Ref.
Strength, compressive	MWNT		150	GPa			X			31
Bulk Modulus	SWNT		0.191	TPa	0.192–0.19	X		Force-constant		32
Bulk Modulus	SWNT	Ropes	0.022	TPa	0.033–0.015	X		Force-constant		32
Bulk Modulus	MWNT		0.194	TPa	0.194–0.19	X		Force-constant		32
Euler spring const				N/A	4.0–1.6	X		Cerrius-MD		36
Poisson ratio	MWNT		0.269		0.280–0.269	X		Force-constant		32
Poisson ratio	SWNT		0.16			X		Force-constant		30
Shear Modulus	MWNT		0.48	TPa	0.541–0.436	X		Force-constant		32
Shear Modulus	SWNT		0.45	TPa	0.478–0.436	X		Force-constant		32
Strain to failure	MWNT	Outer layer	0.12	Strain	NA		X	SEM	Sem with loading stage	37
Strain to failure	MWNT	In polymer film	0.075				X	Tensile test	Instron	31
Strain to failure	SWNT	Ult. strain at various strain rates			35–28%	X		MD		42

SWNT = single-wall nanotube; MWNT = multi-wall nanotube.

TABLE 7.2
Strength and Modulus of CNTs[24]

Property	Material	Type	Value	Units	Range	Simulated	Measured	Technique	Equipment	Ref.
Strength	MWNT	Outer layer	32.8	GPa	63–20		X	SEM	Sem with loading stage	37
Strength, bending	MWNT		14.2	GPa	22–6		X	AFM	AFM, bending	35
Strength, shear		Nanotube-polymer interfacial shear strength	500	MPa		X		Single fiber fragment model		31
Young's Modulus	MWNT	Full tube		GPa	68–18		X	SEM	Sem with loading stage	37
Young's Modulus	MWNT	Outer layer		GPa	950–270		X	SEM	Sem with loading stage	37
Young's Modulus	MWNT	In polymer film	2	GPa			X	Tensile test	Instron	31
Young's Modulus	SWNT				1.4–3	X		Mechanics		33
Young's Modulus	SWNT	Diamond composite			1.3–1.28	X		Mechanics		33
Young's Modulus	SWNT	Func of geometry			1.2–0.97	X		Cerrius-MD		36
Young's Modulus	SWNT		0.974	TPa	0.975–0.971	X		Force-constant		32
Young's Modulus	SWNT	Ropes	0.56	TPa	0.795–0.43	X		Force-constant		32

SWNT = single-wall nanotube; MWNT = multi-wall nanotube.

The high surface area of nanotubes (compared with graphite fibers) creates a large interfacial region that can have different properties from the bulk matrix. Although interfaces have always been important in composites, the interfacial regions of nanotube composites are so numerous that they dominate the behavior of the bulk material.

Evidence of increased interaction between matrix and fiber is manifest in the fracture surface, which shows nanotubes dispersed in a thermosetting polymer. Dispersion of nanotubes on the microscopic scale is apparent, and some load is transferred between the matrix and reinforcement. One method for increasing this load transfer is to attach molecules to the ends or sides of nanotubes for better bonding. Functionalization at the molecular level brings a new level of design possibilities to composite materials, beyond common methods for increasing interfacial adhesion. It is unknown at this time how the addition of certain functional groups affects the mechanical strength of the tubes. However, the known extraordinary mechanical properties of nanotubes make development of high-strength composites a worthy goal, although its achievement will be very difficult.[1, 2]

Electrical Properties

The most exciting of nanotube properties relates to its electronic band structure. Early calculations revealed that nanotubes could be metallic or semiconducting, depending on their helicity and diameter. The armchair tubes are always metallic, whereas the zigzag and chiral tubes can be either metallic or semiconducting. These predictions have been major driving forces behind the rapid evolution of this field. It is only recently that some of these predictions have been experimentally verified.[1]

Nanotubes also have unique, structure-dependent electrical properties. Depending on how the hexagonal chains are oriented relative to the tube axis (chirality), they may act as metals or semiconductors. A "rollup" vector specifies the oriented tube diameter and describes the number of steps in two directions. The chirality of the tubes is also known to affect their mechanical properties, an important point when researching composite materials.[1]

Assembly of nanotubes in this manner results in a hybridization of electron bonds that differs from both diamond and graphite. Three of four valence electrons around carbon atoms form in-plane bonds, leaving free electrons to travel in one dimension along the tube rather than two dimensions that are available in the graphene plane. A significant increase in carrier electron density is expected in a parallel bundle of nanotubes, resulting in conductivity comparable to a good metal. In fact, recent measurements of conductance in metallic nanotubes shows a stepwise increase with increasing voltage. This indicates quantum-level behavior and shows that nanotubes have potential as molecular wires in devices that rival existing transistors, field emitters, and electrostatic discharge materials. Progress towards molecular electronics is evidenced by the recent demonstration of diodes and field-effect transistors based on SWNTs.[1]

Furthermore, it has been noted that doping with elements such as K or Br enables control of the electrical characteristics of nanotubes. The application of nanotubes for flat-panel displays and commercial lighting may prove possible with the recent demonstration of its field emitting properties under ambient conditions.

SWNTs comprise a well-defined system in terms of electronic properties. Test samples available through the arc process or laser ablation, under optimum conditions, consist of uniform diameter nanotubes, a majority of which have a narrow range of helicity (around armchair configuration).

SWNTs form a network of bundles held together by van der Waals forces. These tube bundles also have been studied for resistance and temperature dependencies. At higher temperatures, true metallic behavior is observed ($dp/dT > 0$). However, there is a minimum above which resistance increases again at low temperature ($dp/dT < 0$). The exact cause of this behavior is not clear.[1]

Thermal Properties

The stricture of nanotubes has a profound influence on heat transport properties in both the parallel and transverse dimensions. Because nanotubes are constructed of a single rolled graphene sheet, the thermal conductivity along the sheet has been measured to be comparable to that of diamond (the highest thermal conductor). These one-dimensional tubes have very high aspect ratios (length to diameter), and most of the heat is expected to be transported along their length. Therefore, a material of highly anisotropic thermal conductivity could be synthesized by suitably aligning the tubes.

Although heat transport in the parallel dimension is expected to rival the best conductor, the rate of heat transfer in the perpendicular direction is much lower.

Other Properties

The electronic and mechanical properties of nanotubes have received the most attention as yet, but there are other characteristics that make nanotubes a material of interest. The hollow structure of nanotubes makes them very light (density varies from ~0.8 g/cm^3 for SWNTs up to 1.8 g/cm^3 for

MWNTs, compared to 2.26 g/cm^3 for graphite), and this is very useful for a variety of lightweight applications from composites to fuel cells. Specific strength (strength/density) is important in the design of structural materials; nanotubes have this value at least two orders of magnitude greater than steel. Traditional carbon fibers have specific strength 40 times that of steel. Whereas nanotubes are made of graphitic carbon, they have good resistance to chemical attack and have high thermal stability. Oxidation studies have shown that the onset of oxidation shifts by about 100°C to higher temperatures in nanotubes compared to high modulus graphite fibers. The oxidation in nanotubes begins at the tube lips, and this leads to the possibility of opening nanotubes by oxidation.[1] In vacuum or reducing atmospheres, nanotube structures will be stable to any practical service temperature.

As described previously, electron transport in nanotubes is unique, and the tubes are highly conducting in the axial direction. Similarly, the thermal conductivity of nanotubes also should be high in the axial direction and should be close to the in-plane value of graphite (one of the highest among materials).[10] Tests to verify the thermal conductivity of nanotube material or nanotube composites have been undertaken. In the case of composites, although the high aspect ratio of nanotubes will aid in improving conductivity, the interface between the nanotubes and the matrix could have a deleterious effect. This is emphasized by the large surface area available for interface formation.

Dialing-Up Properties

Nanotubes, stringy supermolecules already used to create fuel cell batteries and tiny computer circuits, could find myriad new applications ranging from disease treatment to plastics manufacturing to information storage, reports a Purdue University research team.[25]

Scientists led by H. Fenniri have learned to create multiple species of nanotubes that possess unprecedented physical and chemical properties, each of which could lead to a different industrial application. Also unprecedented is the complete control they have over the nanotubes' formation, which allows the team to virtually "dial-up" the properties they wish their nanotubes to possess. The findings could greatly expand the materials available for use on the nanoscale.

"Instead of being limited to building blocks of one size, shape and color, it's as though there is now a yard with many different varieties," according to Fenniri. "This research could give a nanotechnologist a good deal more materials for construction."

Rather than work with carbon or metals, as other groups have done, the Fenniri team has formed nanotubes out of synthetic organic molecules. Although other materials have distinct advantages, they are not as easily managed as the materials the Fenniri team is working with.

"By using synthetic chemistry, Fenniri gained complete control over the formation of nanotubes, and more control in the lab should provide more options to industry," says Professor H. Fenniri of Purdue University.

One way the new nanotubes can be customized is by using them as scaffolding for other materials. The nanotube looks like a spiral-shaped stack of rings; each ring is made of six molecules shaped roughly like pie wedges. On the outside of the spiral, other molecules attach themselves, which hang off the tubes like charms on a charm bracelet. The attached molecules then lend their properties to the outside of the nanotube.

For example, if the component molecules of nylon are attached, the nanotubes can then be turned into very long and flexible fibers that are, nonetheless, very strong.

They could be made into an improved version of nylon. And nylon has a lot more uses than making your socks stretch. One could use these fibers to reinforce everything from boat hulls and aircraft to body armor and parachutes.

Another secret to creating custom-made tubes lies in manipulating a property called chirality, which has to do with the direction the spiral-shaped tubes twist. Nature only twists molecules in one direction — this is why DNA molecules always twist to the right and are described as having right-handed chirality. But the above nanotubes can be made to twist in either direction, creating left-handed nanotubes with abilities that their right-handed cousins often do not have.

Therefore, one can create two nanotubes that are made of the same materials, but that behave differently. Just like a flipped-over puzzle piece does not fit in its hole, a left-handed nanotube can react with different substances than its corresponding right-handed tube.

While experimenting with controlling the nanotubes' properties, Fenniri discovered some unexpected behaviors their nanotubes exhibit. The nanotubes promote their own formation, and such behavior is very reminiscent of living systems, in that they replicate and adapt to their environment. You could imagine that one type of nanotube forms at 25°C, but another type with very different physical and chemical properties can form at 70°C.

These newly discovered principles show the relative ease of manipulating the properties of nanotubes and makes the possibility of many new applications very likely; one can optimistically look forward to the possibility of using nanotubes in disease treatment.

Many drugs destroy infectious bacteria by poking holes in their cellular membranes and leaking out their nutrients, just like pricking a hole in a balloon. These nanotubes could also act in this manner, but in addition, they have the ability to lure the bacteria with a bait that guides them to the cell membrane, where they can start destroying the cell.

Further exploitation of the tubes' dial-up properties could lead to nanotubes that conduct electricity or photons, making them useful in computer memory systems, high-definition displays, biosensors, and drug-delivery systems.

Other Nanotube Materials

Titania nanotubes are 1500 times more sensitive than the next best material for detecting hydrogen gas (H_2) and may be one of the first examples of a material property that changes considerably when reduced to nanometer size, according to researchers at Penn State University, University Park, PA.[26]

H_2 entering an array of titania nanotubes flows around all the surfaces, but it also splits into individual H_2 atoms that permeate the surface of the nanotubes. These H_2 atoms provide electrons that raise electrical conductivity, signaling an increase in the amount of H_2 present.

According to researchers,[26] titania nanotubes can be made "by the mile" and are very cheap as well as sensitive. Furthermore, the material can be used repeatedly, because after the gas is cleared from the tubes, it regains its sensitivity. Tube diameters are 22 and 76 nm.

"The sensitivity comes from the nanoarchitecture, not the surface area," says Penn State professor Craig A. Grimes. They differ in surface area by a factor of 2, but the 22 nm tubes are 200 times more sensitive than the 76 nm tubes.

The researchers suggest that the H_2 molecules are dissociated at the titania surface, and this makes the nanotubes sensitive to H_2. They can monitor H_2 levels from 1 ppm to 4 ppm.

H_2 sensors are used for industrial-quality control in food plants, as weapons against terrorism, and in automobile combustion systems to monitor pollution.

Israeli scientists[27] have created a new type of nanotube built of Au, Ag, and other nanoparticles. The tubes exhibit unique electrical, optical, and other properties, depending on their components, and as such, may form the basis for future nanosensors, catalysts, and chemistry-on-a-chip systems.

The resulting tube is porous and has a high surface area, distinct optical properties, and electrical conductivity. Collectively, the tube's unusual properties may enable the design of future sensors and catalysts (both requiring high surface area), as well as microfluidic, chemistry-on-a-chip systems applied in biotechnology, such as DNA chips (used to detect genetic mutations and evaluate drug performance).

Applying their approach, the scientists have succeeded in creating various metal and composite nanotubes, including Au, Ag, Au/Pd, and Cu-coated Au tubes.

The new nanotube lacks the mechanical strength of CNTs. Its advantages lie instead in its use of nanoparticles as building blocks, which makes it possible to tailor the tube's properties for diverse applications. The properties can be altered by choosing different types of nanoparticles or even a mixture, thus creating composite tubes. Moreover, the nanoparticle building blocks can serve as a scaffold for various add-ons, such as metallic, semiconducting, or polymeric materials, thus further expanding the available properties.

The tubes are produced at room temperature — a first-time achievement — in a three-step process. The scientists start out with a nanoporous Al_2O_3 template that they modify chemically to make it bind readily to Au or Ag nanoparticles. When a solution containing the nanoparticles (each only 14 nm in diameter) is poured through, they bind both to the Al_2O_3 membrane and to themselves, creating multilayered nanotubes in the membrane pores. In step three, the Al_2O_3 membrane is dissolved, leaving an assembly of freestanding, solid nanotubes.

Scientists in Japan at the National Institute of Research (NIR), Inorganic Materials of the Science and Technology Agency,[28] applied a new technique based on laser beam heating under high pressure in a diamond anvil cell (DAC) to synthesize an entirely new type of BN nanotube, and they succeeded in analyzing the nanotube structure and constituent elements using an analytical electron microscope.[28]

A DAC developed by the research institute was used, and monocrystalline plates (about 10 μm thick) consisting of cubic and hexagonal crystal BN were heated. The pressure medium was N_2 gas, and the applied pressure was within the range of 5–15 GPa. A CO_2 laser with a power of 240 W was used to heat the specimens to over 3000°C. The BN specimens were melted or sublimated under these high pressures and high-temperature conditions. Inside the reactive formations obtained were small quantities of BN nanotubes.

Observations were made at a high magnification ratio (400,000–1,000,000 times), which allowed analysis of the compositions of superfine domains of about the same diameter as that of the beam.

The nanotuhe consisted of six layers of telescoped cylinders, the tube diameter was about 8.5 nm, and the tube length about 40 nm. The distance between the cylinders of telescopic structure was about 34 nm, and the tube ends were not open and were closed at an angle of about 120°. The electron energy loss spectrum offered information relating to the specimen's elements and state of bonding, and it was discovered that the nanotube consisted of the two elements of boron (B) and nitrogen (N), without any impurities such as metals. The fine structure of the spectrum showed that the BN nanotube had a graphite-type crystal structure.

By applying the most advanced structural analysis, technology led to the successful discovery of this new type of BN nanotube consisting purely of B and N at a ratio of 1:1.

SWNTs that are encased within an outer sheath of BN nanotubes have been under development at Lawrence Berkeley National Laboratory, Berkeley, CA.[29] The nanotube wire is produced when a BN cylinder is packed with buckyballs, then subjected to a 10 min discharge from an intense beam of electrons. The result is a carbon nanowire conductor enclosed within a BN insulating fiber.

According to physicist Dr. Alex Zettl, "Insulation keeps different wires from shorting to each other or to nearby conductors, and will allow the wires to serve as the basis of coaxial cables or a simple gating configuration for the production of nanoelectronic devices such as transistors."

Zettl and his group made their BN nanotubes by a plasma arc technique, in which a hot electrical discharge is sent between two B-rich electrodes in a chamber filled with pure N_2 gas. This yields an abundance of BN nanotubes in the soot that forms along the chamber walls.

The soot can then be heat-treated to open the tips of the tubes, creating a BN cylinder called a "silo." The silos are packed with buckyballs by sealing the soot in vacuum inside a quartz ampoule along with carbon-60 powder, then heating the ampoule to between 550 and 630°C for 24 to 48 h.

The inside diameter of a BN tube determines the configuration of the carbon buckyballs packed inside. In a 2 nm tube, when individual carbon-60 spheres just barely fit inside the BN cylinder, the buckyballs form a staggered configuration; in a 2.8-nm tube, the buckyballs assemble as rotating triangles; at a diameter of 3.3 nm, the buckyball nanowire is shaped like a corkscrew.

The BN cylinders can also serve as model systems for studying the mechanical, electronic, thermal, and magnetic properties of "dimensionally constrained" configurations of densely packed molecules, a critical need for the development of nanotechnology.

Instead of pure carbon, nanotubes can now be made from the stuff of DNA. Researchers at Purdue University[30] used synthetic forms of guanine and cytosine. Unlike their carbon counterparts, the organic nanotubes are easier to make and have a wider range of properties that can be controlled by dressing up the nanotubes with certain molecules.

Welding Nanotubes

Researchers have discovered how to weld together SWNTs, pure carbon cylinders with remarkable electronic properties, which could pave the way for controlled fabrication of molecular circuits and nanotube networks.

For the first time, SWNTs have been welded together, although MWNTs with junctions previously have been created using growth techniques. The electrical properties of SWNTs surpass those of multiwalled tubes, which is why so many researchers have been anxious to try this experiment, said Pulickel Ajayan, Professor of Material Science at Rensselaer Polytechnic Institute, Troy, NY.[31,32]

"No one knew if junctions could be created," said Ajayan. "SWNTs are perfect cylinders without any defects, but in order to create junctions between them, intertube carbon-carbon bonds need to form. The irradiation and heating process we use creates just enough defects for these bonds to form without damaging their electrical properties."

The researchers used a special high-voltage electron microscope that has the capability to irradiate and produce the heat necessary for the experiment. (See Figure 7.3.)

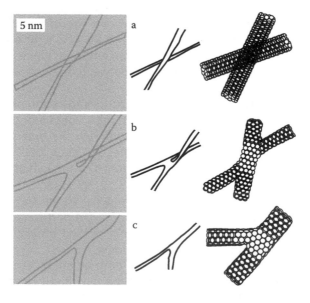

FIGURE 7.3
Joining SWNTs, Pulickel Ajayan, professor of materials science at Rensselaer, and his colleagues in Germany, Mexico, the U.K., and Belgium[31,32] used irradiation and heat to form the welded junctions.

Applications

DNA

You think it is hard keeping your tube socks organized? Try sorting CNTs, those remarkable molecules whose electrical properties make them potential building blocks for everything from ultrasensitive diagnostic devices to transistors 100 times smaller than those in today's fastest microchips. Trouble is, when nanotubes are fabricated, they are a mixed bag; some are electricity conductors, while others are semiconductors. Because a number of practical electronics applications demand nanotubes of uniform conductivity, sorting technologies are needed.

Researchers at DuPont[33] in Wilmington, DE, say they are beginning to solve the problem using another remarkable molecule: DNA. The results are literally visible. A pink-colored vial of nanotubes in solution contains highly conducting nanotubes; other vials, with greenish hues, hold semiconducting ones. "One of the central goals of the field at the moment is to separate nanotubes, because there are applications where having mixtures of semiconducting and metal versions is a real hindrance," says R. Bruce Weisman, chemist and nanotube researcher at Rice University. "If they've got vials of separated nanotubes, that is a big result."

The DuPont researchers found that single-stranded DNA tends to wrap around the nanotubes, forming a stable structure. To enlist this property to sort nanotubes, they engineered DNA to selectively attach to nanotubes with specific conductivities. Then they used standard lab techniques to separate the DNA-nanotube hybrids according to the natural change of the DNA, which is different for different sequences. The attached nanotubes go along for the ride. (See Figure 7.4.)

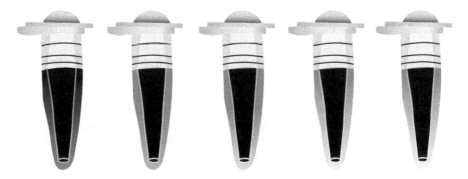

FIGURE 7.4
Unsorted nanotubes in solution appear black (far left). Conducting ones appear pinkish, semiconducting ones, greenish.

Military and Space

What if you could wear lightweight armor that kept you warm — and let you phone home?

Nanotechnologists have come up with a superstrong, flexible fiber that can conduct heat and electricity. It could be made into a modern version of chain mail, the heavy metal mesh worn by medieval knights. If woven from the new fiber, modern chain mail could be light as a cotton shirt, but bulletproof.

Over hundreds of millions of years of evolution, many animals, plants, and natural materials have developed extraordinary properties. Spider silk, for example, is 5 times tougher than steel. (Toughness is defined as the measure of the energy needed to break a fiber.) Some nanotechnologists would like to make synthetic yarn with the same toughness as spider silk (discussed earlier).

Items like antennae, batteries, sensors, and electronic connections could be wired into a lightweight military uniform. As so often happens with military wear, the fiber also could be made into fashionable street wear, but not until the present major obstacle is overcome: the steep price of CNTs — as high as $15,000 an oz. The new fiber will not be widely available until prices drop considerably — and that is not likely for another 5 or 10 years.

A space elevator is a concept that "puts a little 'reach' into the task." Imagine (not immediately) a slender ribbon that stretches from the top of a mid-ocean platform to your destination 100,000 km into space.[34] A space elevator is said to be capable of transforming the economics of space travel, making ventures ranging from space spas to exotic scientific exploration more possible.

Until recently, there has been no material strong enough to make the required cables. But, CNTs might just do the trick.

The space elevator "is no longer science fiction," says David Smitherman of NASA's MSFC in Huntsville, AL. Physicist Bradley C. Edwards agrees. A CNT string half the width of a pencil can support more than 40,000 kg, Edwards notes. That's equivalent to the weight of 20 full-size cars.

Edwards envisions building a space elevator one ribbon at a time, similar to the way bridges were once constructed. In building a bridge across a canyon, for instance, the first step was to catapult or shoot a string from one side of the chasm to the other. Then a larger string was attached to the first string and pulled across. The builder repeated this process until the entire supporting structure of the bridge was in place.

A space elevator must, of course, span a wide gap. The initial string would consist of a flat CNT ribbon 100,000 km long, Edwards says. A conventional rocket would carry a spool of the ribbon to an orbit some 35,000 km above Earth's surface. The scientists have chosen this orbit because it keeps the elevator above the same point on Earth as the planet rotates. Otherwise, the elevator ribbon would drift east or west relative to a fixed point on Earth, and tension on the ribbon would vary.

Edwards' plan is for an elevator shaped like a ribbon that stretches 62,000 miles straight up from a platform floating in the Pacific Ocean off Ecuador. A second platform would hold a powerful laser now being developed that would shoot a strong beam at power cells, similar to solar panels, on the bottom of a mobile plate, pushing the end of the nanotube structure upward. If all goes as planned, the elevator should be carrying cargo, such as satellites, into space within 15 years.

Composites

SWNTs have been used to increase the electrical conductivity of Al_2O_3 by 13 orders of magnitude. Guo-Dong Zhan and colleagues at the University of California at Davis[35] have taken a sample of Al_2O_3, which is a ceramic insulator, and turned it into a fracture-resistant ceramic matrix composite (CMC) with a conductivity that is over 735% higher than the previous record for a nanotube-ceramic composite.

Materials scientists have used nanotubes to improve the tensile strength, conductivity, and thermal properties of various materials. However, combining nanotubes with ceramic materials has proven to be more difficult. Zhan and colleagues first mixed a nanotube-ethanol suspension with Al_2O_3 for 24 hours and then used a spark-plasma sintering technique to fuse the constituents. Unlike other sintering methods, this technique allows consolidation of the mixture at fairly low temperatures, so the nanotubes were not damaged by the process.

The researchers found that the electrical conductivity increased with higher nanotube content and temperature, in contrast to earlier findings. They observed a maximum conductivity of 3375 siemens per meter at 77°C in samples that were 15% nanotube by volume. TEM of the final microstructure revealed that the nanotubes had self-organized into "ropes" held together by van der Waals forces that were entangled within the Al_2O_3 grains. The improved conductivity is a result of these ropes forming a continuous, interlinked electrical pathway throughout the composite. The ropes also make the structure strong and more resistant to corrosion.

The UC Davis research team says the CMC composites could be used in high-performance materials that have to withstand extreme conditions of temperature, mechanical stress, and exposure to chemicals. These materials are widely used in components for the automotive, aerospace, and defense industries. Other potential applications include micro- and nano-electronics, and various medical devices such as implants and prostheses.

Filler-based applications of nanotubes for polymer composites is another area being hotly researched. One of the biggest applications of traditional carbon fibers is in reinforcing polymers in high-strength, high-toughness lightweight structural composites. Epoxy-based MWNT composites have been made and tested, but the results are not very conclusive. Substantial increase in modulus has been reported[1] together with high strain to failure but the strength of the composite is less than expected. The success of nanotube-reinforced composites depends on how strong the interface (between tubes and the matrix) can be made. The atomically smooth surfaces of nanotubes do not guarantee a strong interface. Molecular interlocking of the nanutubes and polymer chains could happen, but it is unclear how such interactions affect strength. Poor dispersion of the samples can create weak regions in the composite where cracks can originate. So far, the failure mode observed in the composites is highly brittle, similar to pure epoxy.

High-conductivity composites (electrical and thermal) using nanotube-filled polymers could be useful, but the problem is getting well-distributed nanotubes in the matrix; heavy settling of nanotubes is seen when larger nanotube epoxy composites are made,[1] probably due to the lack of interaction with the tubes and the matrix. One advantage, however, is the negligible breakdown of nanotubes during processing of the composites. This is a big problem in carbon fiber composites because the fibers are extremely brittle.

Other than structural composites, some unique properties are being pursued by physically doping (filling) polymers with nanotubes. Such a scheme was demonstrated in conjugated luminescent polymer, poly(*m*–phenylenevinylene–co-2.5-dioctoxy–*p*-phenylenevinylene) (PPV), filled with MWNTs and SWNTs.

Other new directions for the use of nanotubes in polymer matrices are being discovered for nonlinear optical properties, membrane technologies, and implant materials for biological applications.[1]

Medical

One program being developed at the National Cancer Institute has been the use of CNT-based biosensors for cancer diagnostics.

Other work has researchers[33] planning to use DNA-nanotube hybrids in prototype sensors for medical diagnostics. For this application, they would marry nanotubes with DNA sequences of a pathogen's DNA; the nanotube would register the binding electrically. Longer term, sorted nanotubes — after being heated to unravel the DNA-nanotube hybrid — could be used as switches or other elements in molecular electronic devices.

Commercial Products and Fiber Optics

The price of big-screen TVs may come down to a point where most people can afford one. This is thanks to CNT technology now in the works at Nano-Proprietary Inc., Austin, TX.[36] The carbon tubes act like electron guns: When subjected to an electric field, they emit electrons. So grouping the tubes around red, green, and blue phosphors creates pixels on a screen, allowing large flat panels only 2 mm thick (see Figures 7.5, 7.6, and 7.7).

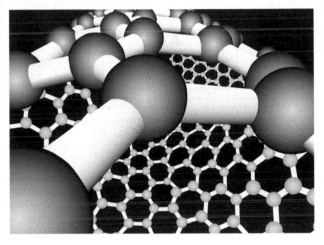

FIGURE 7.5
An artist's model of a carbon nanotube shows a one-layer version. Tubes can be several layers thick. Their great advantage is that in an electric field, carbon atoms easily give up electrons, so they act as little electron guns.[36]

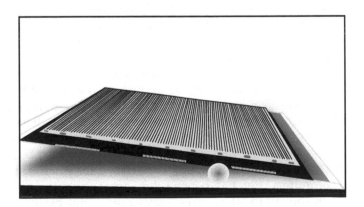

FIGURE 7.6
A prototype carbon-nanotube panel is about 2-mm thick. This proof-of-concept version has a 13.6-in. diagonal, holds 96 X 96 pixels, and can produce 64 levels of gray. A color HDTV version 60 in. or larger could have a switching voltage of less than 100 V.[36]

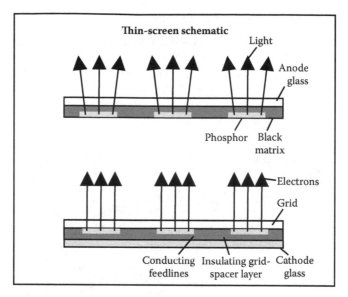

FIGURE 7.7

The display screen Yaniv proposes is made in four simple layers starting with low-cost glass. First to go on are metal electrodes, overlaid with a black layer to support following layers. The black layer is patterned to leave the pixel area exposed. Carbon nanotubes then get deposited on the pixels with a proprietary technology. A metal grid over the black sustaining material finishes the cathode.

"The tubes can be made from single or multiple layers of carbon," says Zvi Yaniv, president of Nano-Proprietary.[36] "They are about 5 to 6 nm in diameter, but relatively long, about 10 to 30 µm. Best of all, they are rich in pi electrons, those that strip off easily in an electric field," he adds.

Yaniv says he has built a gray-scale prototype screen based on nanotubes and that a color version is in the works.

A problem in the display world, explains Yaniv, is that there are many technologies but developers are trying to push each to its limits — even to a point of making a square peg fit in a round hole. For example, liquid-crystal displays are not economical in large sizes — 60 in. diagonal and up. Nevertheless, screen manufacturers are trying to push the technology to larger diagonals because they know HDTV will be the next big market. Plasma, another flat-panel technology, is not good for small displays, as on laptop computers, but has benefits for larger formats. "So already plasmas are in the 50- to 60-in. sizes," says Yaniv. "But plasma is costly and energy hungry," he says. CRTs are out because they are too bulky.

The beauty of CNTs is that each pixel has its own electron gun that is basically the thickness of the glass substrate.

Several Japanese manufacturers have also shown prototypes based on CNTs, understandable because the market for large monitors could exceed $100 billion.

Researchers at CSIRO Molecular Science, Sydney, Australia,[37] believe that advances in nanotube technology may pave the way for a completely new kind of flat television and computer screen. They predict that the new screens will be thinner, more convenient, more energy efficient, and longer lasting than LCD screens. It will also be possible to make the screens flexible.

CSIRO has found a way to control the arrangement of nanotubes, forming them into molecular wires. In screens, these wires work as an intermediary, focusing electrons onto a surface where they react with a fluorescent material to produce light for the display.

Researchers at Rensselaer Polytechnic Institute (RPI)[38] discovered that SWNTs ignite when exposed to conventional photographic flash. According to researchers, SWNTs may find use in light sensors or as remote triggers for explosives.

Before igniting, the nanotubes emit a loud popping sound commonly known as a photoacoustic effect, a phenomenon not previously associated with CNTs. It happens when porous black objects, such as nanotubes, absorb large amounts of light, expanding and contracting the gas surrounding them, releasing sound.

Researchers have experimented to see how light exposure affects the nanotubes. They found that although tubes burn only when oxygen is present, their atomic structure is altered when exposed to flash, even in inert gas. "To our knowledge, no other material emits such a loud sound and ignites spontaneously when exposed to unfocused low-power light," says Ganapathiraman Ramanath, assistant professor of materials science at RPI. "Our work opens up possibly using low-power light sources to create new forms of nanomaterials and will serve as a starting point for developing nanotube-based actuators and sensors that rely on remote activation and triggering," he adds.

Finally, scientists at IBM Research[39] have discovered a new way to get CNTs to emit light, a breakthrough that might one day lead to advances in fiber-optic technology.

At the University of Toronto,[39] meanwhile, researchers have managed to produce light by injecting electrons into a polymer embedded with "quantum dots," microscopic crystals made of lead sulfide. Polymers are being used in research into processor, display, and other technologies.

CNTs have emerged as a candidate to replace Si and metal in chip manufacturing a decade or two down the road. In the more immediate future, nanotubes could be employed to create corrosion-resistant paint or to improve fuel cells or batteries.

The research from these institutions essentially points the way toward another potential application: generating light.

Generating light is not easy or cheap. Current optical equipment does the job, but optical components are difficult to manufacture and, as a result, expensive. By contrast, semiconductors can be mass-produced cheaply. Unfortunately, researchers have tried, and failed, to get Si to generate light effectively.

Besides its potential use in chips, fiber-optic technology also is already used for transmitting information across long-distance telephone lines and in other networks. It carries more information than traditional Cu wires, but it is also more expensive and difficult to install.

The commercial applications will be far away, but definitely this has a potential for great applications for bridging the optical and electrical fields in communications equipment.

The more near-term application for optical is probably for sensors.

In IBM's research, the light appears when a negative charge is applied to one end of the nanotube and a positive charge to the other.

Light is created in this manner now in fiber-optic equipment, but the components have to be "doped," or chemically coated, so that the opposing charges will meet. By contrast, nanotubes are so small — measuring about a nanometer, or a billionth of a meter, in diameter — that they are considered one-dimensional objects. No doping is required.

"When electrons and holes (positive charges) come together, they neutralize each other and become light," said Phaedon Avouris, manager of nanoscale science and technology at IBM Research. "A nanotube is the ultimate in confinement. If you place the electrons in one side and the holes on another, they will find each other."

The light emitted by the nanotubes featured a wavelength of 1.5 μm, the same wavelength used in fiber optics today, noted Avouris. That means arrays of light-generating nanotubes have the potential to be used inside fiber-optic cables to transmit data.

Electronics

Multiterminal CNT junctions are under investigation as candidate components of nanoscale electronic devices and circuits.[40] Three-terminal "Y" junctions of CNTs (see Figure 7.8) have proven to be especially interesting because (1) it is now possible to synthesize them in high yield in a controlled manner and (2) results of preliminary experimental and theoretical studies suggest that such junctions could exhibit switching and rectification properties.

Following the preliminary studies, current-vs.-voltage characteristics of a number of different "Y" junctions of SWNTs connected to metal wires

Materials

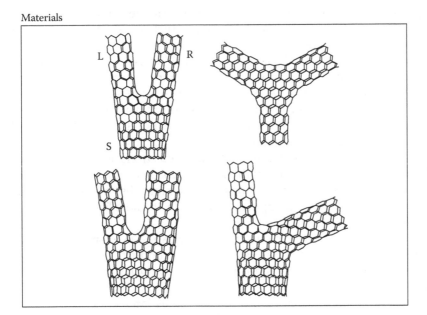

FIGURE 7.8
Symmetric and asymmetric carbon-nanotube "Y" junctions have been studied experimentally and theoretically as candidate electronic switches and rectifiers.[40]

were computed. Both semiconducting and metallic nanotubes of various chiralities were considered. Most of the junctions considered were symmetric.

The results of computation modeling showed that symmetric junctions could be expected to support both rectification and switching. The results also showed that rectification and switching properties of a junction could be expected to depend strongly on its symmetry and, to a lesser degree, on the chirality of the nanotubes. In particular, it was found that a zigzag nanotube branching at a symmetric "Y" junction could exhibit either perfect rectification or partial rectification (asymmetric is current-vs.-voltage characteristic).

A research team at Georgia Tech[41] has managed to carry out a series of experiments that demonstrate ballistic conductance — a phenomenon in which electrons pass through a conductor without heating it up — at room temperature in CNTs up to 5 μm long. The lack of heating lets nanotubes carry extremely large current densities. Researchers measured densities greater than 10 million A/cm^2. A current of that magnitude should have generated temperatures of 20,000°K in the tubes, far above their combustion temperature of 700°K.

"This is the first time ballistic conductance has been seen at any temperature in a three-dimensional system this large," says Walter de Heer, a physics professor at Georgia Tech. "It shows you can constrain current flows to narrow areas without heating, which would be of interest for ultrasmall electronics."

Researchers also found that the nanotubes' resistance is not affected by tube length. "In classical physics, the resistance of metal bar is proportional to its length," says A. Wang, a professor at Georgia Tech's School of Material Science and Engineering. "If you make it twice as long, you will have twice the resistance. But for these nanotubes, resistance is independent of length or diameter."

According to researchers, that is because electrons act more like waves than particles in structures with sizes approaching that of an electron's wavelength. "Electrons pass through these nanotubes as if they were light waves passing through an optical waveguide. Its more like optics than electronics," says de Heer.

This research could prove useful in fabricating ever-smaller electronic devices. The ability of certain structures to conduct relatively large currents without harmful resistance heating would let engineers use very small conductors.

CNTs have been developed to replace Cu as interconnects within integrated circuits, report researchers at NASA Ames Research Center, Moffett Field, CA.[42] Because copper's resistance to the flow of electricity increases greatly as the wire dimensions are reduced, the diameter of Cu conductors has a lower limit.

However, according to researcher Jun Li, "One advantage of CNT interconnects within integrated circuits is that these interconnects have the ability to conduct very high currents, more than a million amperes of current in a 1 cm^2 area without deterioration." In addition, the process does not require the trenches on Si wafers required for Cu. Manufacturers will also be able to add more layers of components to Si chips because of the smaller size and higher conductivity of the CNTs.

The technology involves growing CNTs on the surface of an Si wafer by means of a chemical process. A layer of SiO_2 is deposited over the nanotubes grown on the chip, to fill the spaces between the tubes. Then the surface is polished flat. Scientists can build more multiple, cake-like layers with vertical CNTs that can interconnect layers of electronics that make up the chip.

A big challenge for any transistor is pumping electrons into it from a metal wire. Engineers overcome this so-called Schottky barrier in Si semiconductors by replacing the metal wire with a strand of Si doped with other elements. Now researchers at Stanford and Purdue Universities[43] have found a way around the Schottky barrier in semiconducting CNTs, which are difficult to dope in the required way. The scientists connected wide tubes (3 nm in diameter) to Pd wires, which conduct readily and stick to nanotubes mysteriously well. The nanotubes could then carry about 5 times as much electricity as was previously possible: close to their theoretical ballistic limit (at which electrons travel without ricocheting off other particles). High currents are key to manufacturing high-powered computer chips.

Other physicists have found that semiconducting CNTs have the highest "mobility" of any known material at room temperature. Mobility refers to how well a semiconductor conducts electricity. A semiconducting transistor made from a single CNT showed mobility more than 70 times greater than the Si used today to computer chips.

The researchers had to grow extremely long CNTs, up to 0.3 mm in length, and had to precisely place metal wires on each end of a single tube to make the measurements. The technology holds promise as a replacement for Si chips, if production and substrate issues can be resolved.

Researchers at UC Berkeley and Stanford University[44] have built a device that automates the process of decoding thousands of CNTs on an Si chip. They created a chip with Si metal oxide semiconductor (MOS) circuitry. The chip, called RANT (random access nanotube test chip), contains a network of Si wires and switches that form a circuit (see Figure 7.9).

Containing thousands of CNTs connected to the circuit on a 1 cm² Si chip, the researchers can turn certain switches on and off to isolate the path that leads to an individual nanotube. Not only can they pinpoint which nanotube responds to electrical current passing through the system, they can also tell whether the conductivity can be turned on or off. If they are able to change the conductivity of the nanotube, they know that it is a semiconductor and not metallic.

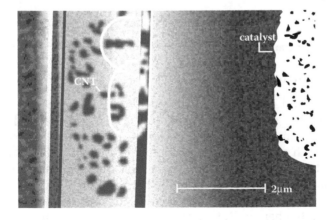

FIGURE 7.9
Magnified view of carbon nanotube grown on silicon MOS circuitry. The bright area on the upper right-hand side is the catalyst island upon which the nanotube was grown.[44]

The development is a step closer to using CNTs for memory chips that can hold orders of magnitude more data than current Si chips — 10,000 times greater, according to some estimates — or for sensors sensitive enough to detect traces of explosives or biochemical agents at the molecular level.

Until now, no group has publicly reported success in directly integrating nanotubes onto Si circuits. This is a critical first step in building the most advanced nanoelectronic products, in which one would want to put CNTs on top of a powerful Si integrated circuit so that they can interface with an underlying information processing system.[44]

Another application that has caught the attention of the scientific and engineering communities is the use of nanotubes as electron emitters.[21,22] Field emission has been observed from arrays of partially aligned MWNTs that have been aligned by pulling a slurry of nanotube dispersion through a ceramic filter. The film of aligned tubes is then transferred onto a substrate, and a voltage is applied across the supporting film and a collector. Such nanotube films act like field emission sources with turn-on voltages of a few tens of volts and electron emission at current densities of a few hundred milliamperes per centimeter squared. The nanotube electron source remains stable over several hours of field emission and is air stable. One of the practical issues that still needs to be resolved is obtaining uniformly aligned tubules, which would guarantee uniform emission.

Due to high aspect ratio, mechanical strength, and elasticity, nanotubes could be used as nanoprobes; for example, as tips of scanning probe microscopes. This idea has been demonstrated successfully and a nanotube tip on an atomic force microscope was used to image the topography of TiN-coated Al film.[1]

Future Potential

In a short period of time, from discovery in 1991 to present day, CNTs have caught the fancy of chemists, physicists, and material scientists. Interest in this material has overshadowed that of fullerenes in recent years, although nanotubes still are not as readily available as fullerenes. The market price of nanotubes (about $400 per gram for good-quality MWNTs and probably twice as much for SWNTs) is still too high, and new synthesis methods based on a continuous process need to appear for this scenario to change. But in pointing to growth mechanisms, so that new fabrication methodologies can be pursued by experimentalists, theory has been rather disappointing. However, it should be noted that the laser method for producing SWNTs reported in 1996 provided a boost by making adequate quantities of pure SWNTs available. Many theoretical predictions were based on modeling SWNTs, and this availability of good-quality SWNTs carried the possibility, for the first time, that some of these predictions could be tested. Indeed, most of the models have been confirmed, which is quite remarkable.

It is hard to tell where the future of nanotubes lies. The most promising and fascinating developments have taken place in exploring the potential of nanotube electronics. Is it realistic to imagine that a few years from now, CNTs will become an integral part of microelectronic circuitry? Important concepts based on nanotube molecular devices have been demonstrated, but the biggest challenge still remains in building nanotube-based architectures to suit existing or future electronic fabrication technology. Manipulating individual nanotubes and placing them in desired locations and configurations have been reported,[1, 45,46] but this is clearly not the approach that needs to be taken if highly complex architectures are to be built from individual elements.

It must be remembered that carbon fibers have existed for more than 3 decades. They were never really considered for electronic applications due to a high density of structural defects. However, carbon fibers have found important applications in composite technology and as electrodes for energy conversion. Nanotubes can be considered to be the ultimate carbon fiber, and it will be surprising if applications for nanotubes are not developed in areas where traditional carbon fibers are abundantly used. One of the problems for nanotube-based composites is the lack of understanding of how the mechanics work around a nanosized inclusion or filler. It will take a lot of dedicated and tedious work before some of the fundamental questions concerning this are answered. One has to differentiate between carbon fibers and nanotubes; the latter is close to a molecular structure, and the properties are governed more from interactions at the atomic level. Also, new areas of composite applications (other than structural) need to be investigated, like properties of polymers, such as photoluminescence, which can be tailored by physically doping with nanotubes. Many new functionalities are emerging from a judicious use of nanotubes in composites.

With extremely small dimensions and mechanical strength, as well as elasticity, one area where nanotubes ultimately may become indispensable is their use as nanoprobes. One could think of such probes being used in a variety of applications, such as high-resolution imaging, nanolithography, nanoelectrodes, drug delivery, sensors, and field emitters.

From the progress made so far, it is highly conceivable that nanotubes may one day become an integral part of our lives through the high technology that it promises.[45,46]

(Millions of Pounds)					Annual Growth
Market	**2003**	**2008**	**2013**	**2020**	**2003–2020**
Packaging	47	120	305	1500	23%
Construction	18	41	220	2000	32%
Electrical & electronics	8	55	210	1400	36%
Motor vehicles	5	40	340	2400	44%
Consumer	2	10	60	1100	45%
Other	40	79	225	2200	27%
Total	120	345	1360	10600	30%
Total plastics compounds	38700	47200	58000	76000	4%
Percentage nanocomposites	0.3	0.7	2.3	13.9	—

FIGURE 7.10
Demand for nanocomposites.[47]

Nanotechnology will change the world — for the better, one hopes. The current and future forecast for U.S. consumption of all types of nanocomposites is shown in Figure 7.10. By 2020, clays and minerals are expected to be the major volume materials, with nanotubes being the highest-value material.

References

1. Ajayan, P. M., Carbon Nanotubes, in H. S. Nalwa (Ed.), *Handbook of Nanostructured Materials and Nanotechnology, Vol. 5; Organics, Polymers, and Biological Materials,* Academic Press, 2000, pp. 329–357.
2. Files, B. S. and Mayeaux, B. M., Carbon Nanotubes, *AM&P,* October 1999, pp. 47–49.
3. Hunt, B., Noca, F., and Hoenk, M., Carbon Nanotube Bimorph Actuators & Force Sensors, *NASA Tech Briefs,* September 2001, p. 33.
4. Baughman, R. H., Carbon Nanotube Actuators, *Sci,* 284, 1340, 1999.
5. *NASA Tech Briefs Insider,* http://link.abpi.net/l.php?20030904A2, September 4, 2003, pp. 1–2.
6. Unusually Long & Aligned Buckytubes Grown at Duke, *Net Composites,* http:// www.netcomposites.con/news.asp?1654, July 5, 2003, p. 1.
7. Chemists ID Process to Sort Carbon Nanotubes by Electric Properties, *The Rice University Weekly Online,* http://www.rice.edu/projects/reno/Newsrel/ 2004/20030911_sorting.shtml, October 24, 2003, p. 1.
8. Dekker, C., Carbon Nanotubes as Molecular Quantum Wires, *Physics Today,* May 1999, pp. 22–28.
9. Yakobson, B. I. and Smalley, R. E., Fullerene Nanotubes: C 1,000,000 and Beyond, *American Scientist,* 85(324), 324–337, 1997.

10. Dresselhaus, M. S., Dresselhaus, G., and Eklund, P. C., *Science of Fullerenes and Carbon Nanotubes*, San Diego, CA: Academic Press, 1996.

11. Ebbesen, T. W. (Ed.), *Carbon Nanotubes: Preparation and Properties*, Boca Raton, FL: CRC Press, 1997.

12. Saito, R., Dresselhaus, M. S., and Dresselhaus, G., *Physical Properties of Carbon Nanotubes*, London: Imperial College Press, 1998.

13. Ebbesen, T. W., *Physics Today*, June 26, 1996.

14. Ajayan, P. M. and Ebbesen, T. W., *Rep. Prog. Phys.*, 60, 1025, 1997.

15. Dresselhaus, M. S., Dresselhaus, G., Eklund, P. C., and Saito, R., *Phys. World*, 33, January, 1998.

16. Iijima, S., Ichihashi, T., and Ando, Y., *Nature*, 356, 776, 1992.

17. Ajayan, P. M., Ichihashi, T., and Iijima, S., *Chem. Phys. Lett.*, 202, 384, 1993.

18. Yakobson, B. I. and Smalley, R. E, *Am. Sci.*, July–August, 324, 1997.

19. Iijima, S. and Ichihashi, T., *Nature*, 363, 603, 1993.

20. Bethune, D. S., Kiang, C. H., de Vires, M. S., Gorman, G., Savoy, R., Vazquez, J., and Beyers, R., *Nature*, 363, 605, 1993.

21. de Heer, W. A., Chatelain, A., and Ugarte, D., *Sci.*, 270, 1179, 1995.

22. Collins, P. G. and Zettl, A., *Appl. Phys. Lett.*, 69, 1969, 1996.

23. Dai, H. J., Hafner, J. H., Rinzler, A. G., Colbert, D. T., and Smalley, R. E., *Nature*, 384, 147, 1996.

24. Harris, C. E., Stuart, M. J., and Gray, H. R., Emerging Materials for Revolutionary Aerospace Vehicle Structures and Propulsion Systems, *SAMPE J.*, 38(6), 33–43, 2002.

25. http:// www.netcomposites.com/news.asp?1394, October 4, 2002, pp. 1–3.

26. Grimes, C. A., Titania Nanotubes Sense Hydrogen 1500 Times Better, *AM&P*, p. 20, October 2003.

27. New Type of Nanotube Made of Gold or Silver, p. 2, February 5, 2004, *Angewandte Chemie*, December 2003.

28. Synthesis of New Type of Nanotube, *JETRO*, April 1997, p. 16.

29. Zettl, A., Boron Nitride Nanotubes, *AM&P*, pp. 25–29, 2002, www.lbl.gov.

30. Malik, T., Nucleotide Nanotubes, *Scientific American*, December 2002, p. 36.

31. Nano-Welding Creates Tiny Junctions, *Rensselaer Polytechnic Institute: Campus News*, p. 1, September 9, 2002.

32. How to Weld Carbon Nanotubes, *Mach. Des.*, January 9, 2003, p. 32.

33. Talbot, D., The Nano Sorter, *Technol. Rev.*, July/August 2003, p. 26.

34. *Science News*, October 5, 2002.

35. Zhan, G.-D. et al., *Appl. Phys. Lett.* 83, 1228, 2003.

36. www.appliednanotech.com, *Mach. Des.*, September 18, 2003, pp. 35–37.

37. Nanotubes for Better TV Screens, CSIRO Molecular Science, Sydney, Australia, *Mach. Des.*, May 2000, p. 56.

38. Pop! Goes the Nanotube, *Mach. Des.*, July 11, 2002, p. 24.

39. Carbon Nanotubes Emit Light, May 6, 2003, netcomposites.com/news.asp?1661, p. 1.

40. Scrivastava, D., Switching Rectification in Carbon-Nanotube Junctions, *NASA Tech Briefs*, October 2003, pp. 48–50.

41. Nanotubes Demonstrate Unique Electrical Properties, *Mach. Des.*, August 20, 1998, p. 51.

42. Li, J., Carbon Nanotubes Replace Copper in Electronic Chips, AM&P, July 2003, pp. 16–17.
43. Minkel, J. R., Barrier-Free Nanotubes, Scientific American, October 2003, p. 36.
44. Nanotube Circuitry, Nano Letters, January 2004, p. 1, http://link.abpi.net/l.php?20040108A2.
45. Rotman, D., The Nanotube Computer, Technol. Rev., March 2002, pp. 36–45.
46. Ross, P. E., Tiny Ventures, Red Herring, August 2002, pp. 56–59.
47. Mapleston, P., Nanotechnology Will Change the World — for the Better, One Hopes, Modern Plastics, July 2004, pp. 32–36.

Bibliography

Bronikowski, M. and Hunt, B., Block Copolymers as Templates for Arrays of Carbon Nanotubes, NASA Tech Briefs, April 2003, pp. 56–57.
Composites: Resins, Filler, Reinforcements, Natural Fibers and Nanocomposites, http://www.netcomposites.com/netcommerce_features.asp?730.
CompositesWeek, 6(27), 4, 2004.
Delzeit, L. D. and Delzeit, C., Improved Method of Purifying Carbon Nanotubes, NASA Tech Briefs, July 2004, p. 46.
Delzeit, L. D., Patterned Growth of Carbon Nanotubes or Nanofibers, NASA Tech Briefs, July 2004, pp. 46–47.
Fan, J. and Wan, M., Synthesis, Characterizations, and Physical Properties of Carbon Nanotubes Coated by Conducting Polypyrole, J. Appl. Polymer Sci., 74, 2605, 1999.
Farley, P., Solar-Cell Rollout, Technol. Rev., July/August 2004, pp. 35–40.
Fischer, J. E. and Dai, H., Metallic Resistivity in Crystalline Ropes of Single-Wall Carbon Nanotubes, Phys. Rev. B, 55, R4921, 1997.
Gao, G., Cagin, T., and Goddard, W. A., III, Energetics, Structure, Thermodynamics, and Mechanical Properties of Nanotubes, Nanotechnology, 9, 183, 1998.
Grobert, N., Novel Carbon Nanostructures, Ph.D. thesis, University of Sussex, 2000.
Halicioglu, T., Stress Calculations for Carbon Nanotubes, Thin Solid Films, 312, 11–14, 1998.
http://www.cmp-cientifica.com.
Hunt, B., Choi, D., Hoenk, M., Kowalczyk, R., and Noca, F., Growing Carbon Nanotubes Aligned with Patterns, NASA Tech Briefs, October 2002, pp. 52–53.
Lower Conductivity in Nanotubes Composites, http://www.netcomposites.com/news.asp?1874, p. 1, accessed November 18, 2003.
Mann, C., Near-Term Nanotech, Technol. Rev., July/August 2004, p. 22.
New Carbon Nanotube Composites with Significantly Improved Electron Emission Properties, http://www.netcomposites.com/news.asp?1414, p. 1, accessed October 21, 2002.
Osawa, E. (Ed.), Perspectives of Fullerene Nanotechnology, The Netherlands: Kluwer Academic Publishers, 2002.
Stix, G., Breaking the Mold, Scientific American, July 2002, pp. 34–35.

Wilson, J. D., Wintucky, E. G., and Kory, C. L., Making Carbon-Nanotube Arrays Using Block Copolymers: Part 2, *NASA Tech Briefs*, January 2004, pp. 42–43.

Xu, C. L. and Wei, B. Q., Fabrication of Aluminum-Carbon Nanotube Composites and Their Electrical Properties, *Carbon*, 37, 855–858, 1999.

Yakobson, B. I., Brabec, C. J., and Bernhole, J., Nanomechanics of Carbon Tubes: Instabilities Beyond Linear Response, *Phys. Rev. Lett.*, 76(14), 2511–2514, 1996.

8

Functionally Gradient Materials

Introduction

In a functionally gradient material (FGM), the composition and structure gradually change over volume, resulting in corresponding changes in the properties of the material. By applying the many possibilities inherent in the FGM concept, it is anticipated that materials will be improved and new functions for them created.

The concept of FGMs was first proposed in 1984 at Japan's National Aerospace Laboratory to prepare thermal barrier materials usable not only for space structures and fusion reactors, but also for future space plane systems. In 1987, a national project supported by the Japanese Science and Technology Agency, Research on the Basic Technology for the Development of Functionally Gradient Materials for Relaxation Thermal-Stress, was begun, composed of 17 government research institutes, universities, and companies.[1-3]

Definition and Design of FGMs

Definition

FGM generally consists of different material components such as ceramics and metals. Therefore, FGM is a composite material having microscopic inhomogenous character. Continuous changes in their microstructure distinguish FGMs from conventional composite materials. The continuous changes in composition result in gradients in the properties of FGM. The differences in microstructure and properties between FGM and conventional composite materials are schematically illustrated in Figure 8.1.[4]

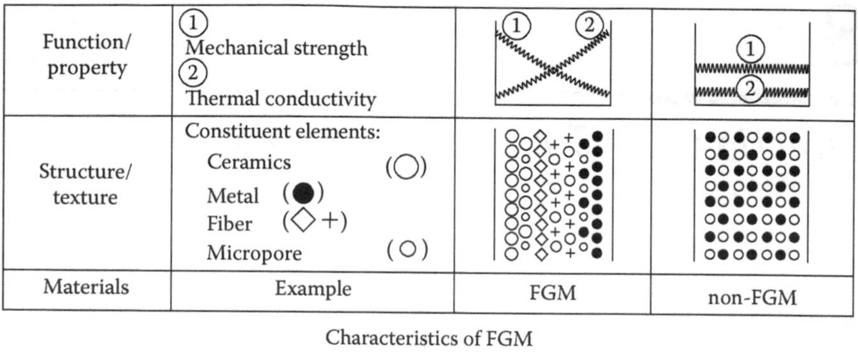

Function/ property	① Mechanical strength ② Thermal conductivity		
Structure/ texture	Constituent elements: Ceramics (○) Metal (●) Fiber (◇+) Micropore (○)		
Materials	Example	FGM	non-FGM

Characteristics of FGM

FIGURE 8.1
Characteristics of FGM.[4]

Design

The design of an FGM must be considered and examined by taking into account the conditions it will encounter in practical application. This involves an effort to select an optimum material combination to ease heat stress, an optimum intermaterial compositional distribution that is suitable for production conditions, and optimum microstructures for the materials involved.

In general, the highest temperature on the surface of a reentry vehicle has been estimated to reach 1800°C.[5] Hence, materials at the surface must withstand temperatures up to 1800°C and a temperature gradient of 1300°C. Therefore, the material must have suitable heat-resistant and oxidation-resistant properties in the surface layer, mechanical toughness of the cool surface, and an effective thermal-stress relaxation throughout the material.

For such material requirements, the unique idea of an FGM was proposed to prepare a new composite by using heat-resistant ceramics on the high-temperature side and tough metals with high thermal conductivity on the low-temperature side, with a gradual compositional variation from ceramics to metals, as illustrated in Figure 8.1. In such a structure, the thermal expansion coefficient can be adjusted by controlling the composition microstructure and mid-porosity ratio between the inner and outer surfaces.

Projects and studies like those above have continued and been carried out by R&D of FGM for the last 18 years and has included specialists in materials design/database, processing, and evaluation.

In designing metal/ceramic FGMs as structural or component materials, the knowledge of the residual stress is important not only to prevent undesirable deformation and cracking, but also to ensure the high reliability of the products. The aim of a study by Fukui et al.[6] was to examine the macroscopic thermal residual stresses in Al-SiC FGM rings theoretically and experimentally. Machining Al-SiC FGM rings is known to be extremely difficult because of distributed SiC particles, and because the heat associated with the turning introduces other thermal residual stresses. Thus, for their experiment the researchers couldn't adopt the well-established experimental techniques known as Sack's method.[7]

Therefore, they adopted a simple and robust cutting method in which the deformation, after the ring was cut at one cross-section, was measured, and then the initial stress state was determined by solving an inverse problem.

Fukui et al. successfully cast Duralcan F3D.20S of Al-20 vol% SiC master composite into Al-SiC FGM rings by the centrifugal method. These rings had a variation of graded composition in the range of 0 to 43 vol% of SiC from the inner to the outer radial planes of the ring.

Excellent agreement between the theory and the experiment was found. The residual stress was found to be generated by a cooling of $\Delta T = 140$ K, which was from half the melting point corresponding stress-free condition to the ambient temperature. The hoop residual stresses in the FGM ring varied in the range of -50 to $+35$ MPa and from tension at the inner surface to compression at the outer surface because of the graded composition. With an increase in wall thickness or composition gradation, the residual stresses were found to increase.

A flowchart, Figure 8.2, for the design and development of FGMs via pulse discharge resistance consolidation with temperature gradient control was prepared by Kimura and Toda.[8] The design process can be illustrated with reference to a model material for thermal stress relief applications, an FGM containing TiAl intermetallic and partially stabilized zirconia (PSZ), having lightness and high strength. The demands placed on the FGM by the service requirements include full density in the layers having higher volume fractions of PSZ and the formation of intermetallics with nanometer grain sizes via low-temperature crystallization of amorphous TiAl.[9–11]

Kimura and Toda[8] developed a novel powder processing route allowing the design and fabrication of FGMs having constituents with greatly different densification rates, giving control of porosity up to full density and of residual stress along a graded composition. This route consisted of pulse discharge resistance consolidation with temperature gradient control obtained through the use of a die having a specially designed outer shape.

Using this method, FGMs graded from nanoscale TiAl intermetallic to fully dense PSZ were successfully prepared. The compacts were fully densified throughout their cross-sections without any discontinuity and

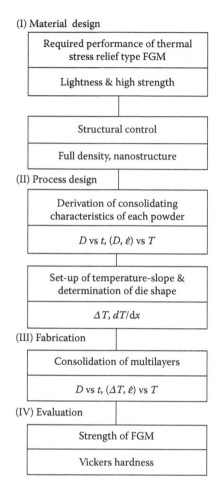

(I) Material design

Required performance of thermal stress relief type FGM

Lightness & high strength

Structural control

Full density, nanostructure

(II) Process design

Derivation of consolidating characteristics of each powder

D vs t, $(D, \dot{\varepsilon})$ vs T

Set-up of temperature-slope & determination of die shape

$\Delta T, dT/dx$

(III) Fabrication

Consolidation of multilayers

D vs t, $(\Delta T, \dot{\varepsilon})$ vs T

(IV) Evaluation

Strength of FGM

Vickers hardness

FIGURE 8.2

Flow chart for fabrication of FGMs by temperature gradient controlled pulse electric discharge consolidation.[9]

had hardnesses increasing from 650 HV in the nanocrystalline TiAl layer to 1560 HV in the PSZ layer, following a mixtures rule between.

Hirano and Yamada developed an inverse design procedure that works as follows.

First, required shape and heat boundary conditions are set in developing an FGM structure, and the selection of suitable materials and a production method is made using the database. Second, the selection of an optimum compositional distribution function is made to control and to represent the required gradient material composition, and the profiles of the FGM are determined by setting the parameters. Third, the selection of an inferred model of material properties is made based on the required microstructure of the FGM. This is achieved by obtaining data on the properties of the basic compositional materials involved from a material property database. At the same time, the shape parameters of the mathematical model of the microstructure are determined. Using property values obtained from an arbitrary material composition created by these processes, the analyses of temperature distribution and thermal stress distribution are conducted. Fourth, an effort is made to derive a compositional distribution profile, and an optimum material combination, to minimize the specific stress (stress/material strength) by adjusting gradient distribution parameters and material combination.

Considering that these requirements involve repeated calculations to attain a global optimization in basic design, a three-dimensional analysis by the finite element method (FEM) is not practical because of its limitations in terms of analysis time. In addition, a method for controlling three-dimensional compositional distribution is still in its infancy of development. Under these circumstances, in the basic design state, it is important to make an effort to attain a one-dimensional optimum compositional distribution in designing the basic shape of an FGM before two-dimensional optimization becomes possible. At the same time, in the detailed design stage, an effort must be made to analyze the nonlinear effects for material shapes that are in practical use, to evaluate thermal fatigue strength, and to optimize localized compositional distribution using the compositional distribution data obtained in the basic design stage.

For example, Obata et al.[13] studied various materials in developing thermal stress-eased superhigh-temperature shields. They selected a combination of SiC, which has good resistance to oxidation at high temperatures and high strength at high temperatures, and graphite (C), which matches well with a C-C composite, a structural material, and has a low Young's modulus and high thermal conduction. An SiC/C FGM had SiC on one side and C on the other, while its composition changed continuously from the SiC side to the C side.

In order to forecast an optimum gradient composition distribution for minimum thermal stress in an SiC/C FGM, imagine an infinite cylinder 10 mm thick and 95 mm in inner radius, one end of which is constrained (Figure 8.3). When the inside temperature is 1327°C and the outside is 29°C with a 1027°C temperature gradient thus generated, the stress status with zero stress in the axis direction can be calculated. It is assumed that the cylinder has an SiC single phase to a depth of 1 mm from the inner surface, a gradient composition (various properties) from SiC to C from 1 mm to 10 mm in the thickness direction, and an SiC/C functionally gradient film with a uniform composition distribution in the inside direction.

The internal stress was obtained for various gradient compositions, and at various temperature gradient conditions. Figure 8.4 shows the stress distribution, temperature distribution, and Young's modulus distribution

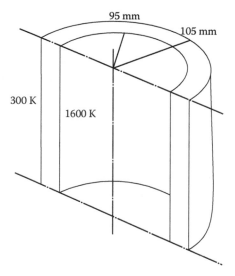

FIGURE 8.3
Cylinder model for calculation of thermal stress in SiC/C functionally gradient composite
(inside: SiC; outside: C).

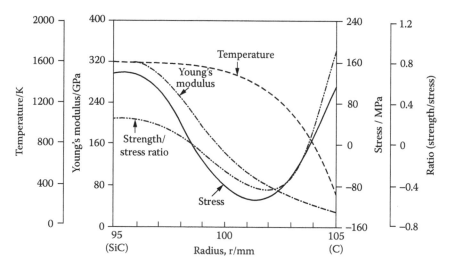

FIGURE 8.4
Calculated distributions of temperature, Young's modulus, stress, and strength/stress ratio
in SiC/C functionally gradient composite.

when the strength-internal stress ratio becomes 1 or lower, while Figure 8.5
shows the gradient composition at that time. It is clear from Figure 8.4
that in an SiC/C functionally gradient film, the composition gradient can
slash the thermal stress to one sixth that of the SiC single phase.

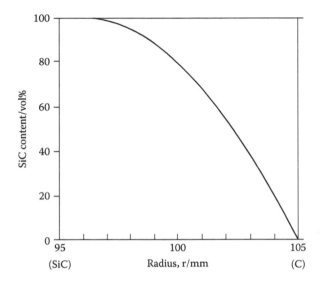

FIGURE 8.5
Suitable compositional distribution in SiC/C functionally gradient composite predicted by calculation.

An SiC/C FGM having a gradient composition shown in Figure 8.5 was synthesized as chemical vapor deposition (CVD) as follows.

The material gas was an $SiCl_4 + C_3H_8 + H_2$ system, and the synthesizing temperature was 1500°C. While keeping the flow of C_3H_8 constant, the flow of H_2, the carrier gas for $SiCl^4$, was controlled to change the ratio of the material gas. This method makes it possible to synthesize an SiC/C functionally gradient film whose composition changes continuously from SiC to C on a graphite substrate. The thickness of the film obtained by synthesis for 100 min was about 1 mm.

Material Combinations

Matsuzaki et al.[14] selected as a cooling side (low–temperature side) material a Ti-Al system intermetallic compound, a material that is attractive for its good high-temperature specific strength (strength/specific gravity), which makes it suitable for use as a high-temperature structural material.

In particular, TiAl has a targeted utilization temperature of 977°C, about 101°C higher than Hastelloy X. The use of this material could contribute to raising the surface temperature of FGMs and, in turn, to lowering the heat transfer rate to the cooling side of these materials.

TABLE 8.1

Candidate Materials for Use in the High-Temperature Side of FGMs

	Stabilized ZrO_2	Al_2O_3	Stabilized HfO_2	TiC	HfC	HfB_2	$MoSi_2$	SiC
Thermal protectivity	++	++	++	—	-	—	+	+
Oxidation resistance	++	++	++	—	-	-	++	++
Oxygen permeation resistance	—	++	—	+	+	++	++	++
Thermodynamic stability	+−	+	+−	++	+−	+−	—	—
Ambient temperature strength	++	+	+	+	+	+	+	+
Elevated temperature strength	-	+	-	+	+	+	+	++

Note: ++: Excellent; +: Acceptable; +−: Under examination; –: Poor; —: Unacceptable.

Table 8.1 lists the characteristics required for heat-side (high-temperature side) application materials.

Al_2O_3 has good chemical, thermal, and dynamic characteristics that make it suitable for use as the FGMs topmost surface material. In particular, its dynamic characteristics should be improved by the introduction of a reinforcing material (such as SiC grains or whiskers). When introducing these reinforcing materials, it is important to ensure thermal stability between the TiAl and these materials.

As the hot side topmost surface of the FGM, Al_2O_3 is used because of its heat-insulation, antioxidation, antierosion, and resistance to hot environment corrosion capabilities. SiC is used to improve the FGM's dynamic strength characteristics. The TiC layer, as an intermediate layer, is also used to prevent mutual reactions between the SiC and TiAl, and it also relieves nonelastic deformations of the FGM in hot conditions. Chemically stable TiC may be used effectively with active silicon-system ceramics or intermetallic compounds to form high-temperature side materials.

The use of multicomponent materials in an FGM makes it possible for these components to assume different functions, but increasing the material component makes material design more complex. However, this means that there is greater freedom for macroproperty control. For example, in this four-layer FGM, freedom in controlling the thermal and mechanical properties of the topmost surface material (Al_2O_3-SiC) is provided by changing the amount of SiC introduced, and changing the amount of TiC and its distribution pattern equally increases freedom in controlling the macroproperties of the entire FGM. The development of this multicomponent FGM leads to the realization of FGM design using the reverse problem analysis design method, the utmost goal in FGM design efforts. This method calls for obtaining an ideal property distribution to achieve a targeted function by calculation and to determine the materials to be used, shape, compositional distribution, and microstructure.

R. Watanabe and his co-workers[15] at Tohoku University attempted to prepare partially stabilized zirconia/type 304 stainless steel FGM by applying powder metallurgical techniques. Powders of PSZ (0.05 µm) and SUS 304 (6 µm) of continuously changing composition were mixed according to the predesigned composition profiles. The stacked powders were pressed, and then compacts were sintered in vacuum.

The results of fracture strength measurements, metallographic examination, and fractography, using the specimen cut from the sintered compacts, indicated that the sintering shrinkage or graded compact could be controlled and that the thermal stress that was generated in the compact during sintering could be reduced, both of which were essential for graded composite with sufficient fracture strength.

In this way, a metal/ceramics-graded composite of partially stabilized zirconia and stainless steel, having a stepwise controlled compositional gradient, was fabricated by a powder metallurgical process.

Powder metallurgical processing with a thermal-stress relief for this type of FGM was also used for PSZ/Mo and PSZ/W systems.[16]

T. Fukushima and his colleagues at the National Institute for Metals established the basic technique for preparing 51-base alloys/ZrO_2 base ceramics-graded coatings by plasma spraying. Here, twin torches were used in air, and the spraying parameters were varied.[17]

N. Shimoda and his colleagues at Nippon Steel Corporation investigated the gradient composition coating process by using a single gun with four ports based on the low-pressure plasma spray (LPPS) method.[18] The ZrO_2-NiCr system was investigated by controlling the supply of ceramic and metal powders to the plasma arc simultaneously in a single gun.

These processes, which cover thin sheet lamination and LLPS, appear to be convenient for large-scale, complex structures and mass production.

Utilizing the characteristics of the Self-Propagating High-Temperature Synthesis (SHS) reaction — extremely high temperature of reaction, very rapid reaction velocity, and remarkably high heating rate — the combustion-consolidation method allows for a continuous change in the metal composition. In a ceramic matrix, the initial arrangement of the constituents in a green body remains, which is due to the rapid wave propagation of the combustion reaction.

N. Sata et al.[19] studied the TiB_2-Cu system and modeled an infinite cylinder to be used as a rocket thrust chamber 105 mm in O.D. and 95 mm in I.D., as shown in Figure 8.6.

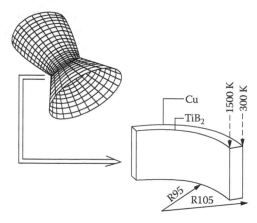

FIGURE 8.6
Thrust chamber model and temperature condition.[19]

The chamber was assumed to be exposed to a temperature difference of 927°C across the wall thickness. The metal-ceramic composite was designed by using thermal and mechanical data. As illustrated in Figure 8.7, for two- or three-layered composites, large stress gaps remain at the interlayers and result in the cracking of bonded or coated materials. In FGM specimens, such stress gaps disappear, and such troubles are avoided.

Powders of Ti, B, and Cu were mixed with stepwise changing of the mixing ratios. The mixed powders were stacked in a vessel with the steps compositionally divided into 2, 4, 6, 11, 16, or 21 units, pressed and ignited in a uniaxial loading system effected by a spring or in an autoclave. After these processes, products containing interlayered materials were formed having controlled compositions that ranged stepwise from Cu to TiB_2.

Another FGM in the TiC-Ni system was fabricated using both SHS and HIPing processes. This method was developed by Y. Miyamoto and colleagues.[20] The procedure used was as follows: The pressed body of reactants was vacuumed, sealed into a glass capsule (just as in the capsule method for HIPing), and put into a crucible filled with a combustion agent. The reaction was initiated by the heat released from the combustion agent under an Ar-gas pressure of <100 MPa.

Be, Be alloys, Be_2C, and C are the ingredients of a class of nanophase $Be/Be_2C/C$ composite materials that can be formulated and functionally graded to suit a variety of applications. In a typical case, such a composite consists of a first layer of either pure Be or a Be alloy, a second layer of Be_2C, and a third layer of nanophase sintered C derived from fullerenes and nanotubes. The three layers are interconnected through interpenetrating sponge-like structures.

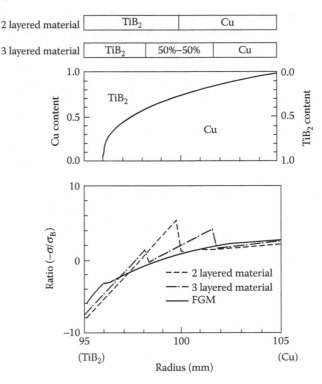

FIGURE 8.7
Relation of composition/thermal stress vs. radius in rocket thrust chamber made of TiB$_2$-Cu FGM.[19]

These Be/Be2C/C composite materials are similar to Co/WC/diamond functionally graded composite materials, except that (1) W and Co are replaced by Be and alloys thereof, and (2) diamond is replaced by sintered C derived from fullerenes and nanotubes. (Optionally, one could form a Be/Be$_2$C/diamond composite.) Because Be is lighter than W and Co, the present Be/Be$_2$C/C composites weigh less than do the corresponding Co/WC/diamond composites. The nanophase C is almost as hard as diamond.

WC/Co is the toughest material. It is widely used for drilling, digging, and machining. However, the fact that W is a heavy element (that is, has high atomic mass and mass density) makes W unattractive for applications in which weight is a severe disadvantage. Be is the lightest tough element, but its toughness is less than that of WC/Co alloy. Be strengthened by nanophase C is much tougher than pure or alloy Be. The nanophase C has an unsurpassed strength-to-weight ratio.

The Be/Be2C/C composite materials are especially attractive for terrestrial and aerospace applications in which there are requirements for light weight along with the high strength and toughness of the denser Co/WC/ diamond materials. These materials could be incorporated into diverse components, including cutting tools, bearings, rocket nozzles, and shields. Moreover, because Be and C are effective as neutron moderators, Be/Be$_2$C/ C composites could be attractive for some nuclear applications.[21]

Nagano and Wakai[22] fabricated an FGM ZrO$_2$ - Al$_2$O$_3$ material by superplastic diffusion bonding. Conditions included a bonding temperature of 1550°C, a time of 30 min, and strains of 17, 33, and 50%. Complete bonding was obtained under all of these bonding conditions. The bonding method is schematically shown in Figure 8.8, while Table 8.2 summarizes some properties of the materials.

The apparent bending strength of the ZrO$_2$–Al$_2$O$_3$ FGM was significantly different according to the direction of applied stress. When the tension side was the Y-TZP, the apparent bending strength of the specimen (strain of 50%) was 1860 MPa; when the tension side was Al$_2$O$_3$, the apparent bending strength (strain of 50%) was 330 MPa.

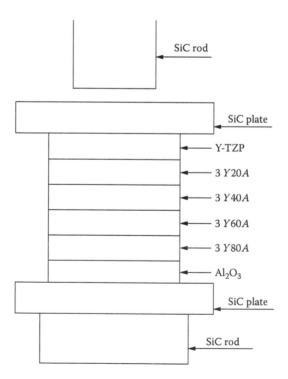

FIGURE 8.8
Illustration of bonding set.

TABLE 8.2

Some Properties of Materials

Material	ZrO$_2$ (wt %)	Al$_2$O$_3$ (wt %)	Density (g cm^{-3})	Bending Strength (MPa)	Grain Size (μm)	Young's Modulus (GPa)	Shear Modulus (GPa)	Thermal Expansion (1/°C($\times$ 10^{-6}))
Y-TZP	99.7		6.05	750	0.59	200	76	12.0
3Y20A	79.9	20.0	5.50	2300	0.77	250	97	11.2
3Y40A	59.9	40.0	5.10	2300	0.79	300	119	10.4
3Y60A	39.9	60.0	4.60	2000	0.80	335	134	9.94
3Y80A	19.9	80.0	4.30	1200	1.06	370	150	9.48
Al$_2$O$_3$		92.0	3.53	360	4.77	280	113	8.71

Fabrication/Processing Methods

A considerable amount of research, development, prototype fabrication and manufacturing, testing, and material evaluation has been conducted in both academia and in industry to produce FGMs for a variety of applications.

A minimum number (14) of approaches that may increase have been identified for producing FGMs. These include (1) dynamic ion mixing, (2) laser alloying and cladding, (3) powder processing with slip casting and pressureless-sintering, (4) tape casting, (5) reactive sputtering, (6) SHS/ HIP, (7) spark plasma sintering, (8) electrophoretic deposition, (9) sedimentation, (10) multi-ingredient powder deposit, (11) powder metallurgy, (12) plasma spray, (13) SPF/DB, and (14) CVD.

Dynamic Ion Mixing. ZrO$_2$-Cu films have been prepared as FGMs that change continuously from Cu to ZrO2. Partially stabilized ZrO2 (PSZ) and the PSZ-Cu FGM films were prepared on a Cu substrate by a dynamic ion mixing process.[23] Oxygen ion irradiation was essential for the FGM formation of the PSZ-Cu films, because films deposited without oxygen ion irradiation were easily peeled from the substrate during the deposition steps. The electron beam deposition rates of Cu and ZrO$_2$ were individually controlled. The FGM structures were analyzed by X-ray photoelectron spectroscopy and TEM. Nakeshima et al.[23] found that the Cu and PSZ crystal grains were finely mixed in the FGM layers. The thermal resistance of the FGMs was examined by a heat test in which the film surface was heated by an Ar+ H$_2$ plasma jet, and no cracks were observed on the PSZ-Cu FGM surfaces after the test. However, for PSZ film (without the FGM layer) on the Cu substrate, cracks were seen.

Laser Alloying and Cladding. The techniques of surface alloying, cladding, and particle injection provide possible routes for producing certain types of FGMs. This resultant FGM has aerospace applications where it is desirable to relax thermal stresses that might otherwise lead to fracture when a large temperature difference occurs across a component. To produce gradients normal to the substrate surface, successive fully overlapping alloyed and clad layers can be deposited using powder feed mixtures with increasing proportions of one or more components to provide the composition range. Control of the dilution from the substrate influences the composition of the first layer, and in succeeding deposits the dilution from each underlying layer is critical. The technique produces a series of layers of essentially discrete composition rather than a gradual composition change.

Abboud and his associates[24] used two approaches in processing FGM materials (1) surface alloying of a near-Ti alloy substrate using a powder feed of Al and (2) cladding unto a commercial purity (CP) Ti substrate using powder mixtures of Al with Ti-6Al-4V alloy. They found that cladding involving low dilutions provided a better approach to producing several discrete layers of increasing Al content than did alloying, which is associated with higher dilutions. This cladding technique used powder mixtures of Al and Ti-6Al-4V (wt%) alloy that involved low dilutions and allowed several layers of discrete composition to be produced having Al contents in the range from 10.5 wt% (approximately 18 at%) to 34.5 wt% (approximately 48 at%).

Jasim, Rawlings, and West[25] applied a technique of injection with a powder feed of a mixture of metal-ceramic that combines the processes of laser alloying, cladding, and injection to study the feasibility of using a continuous wave CO_2 laser to produce an FGM. A 2 kW CO_2 laser was used to produce, on an Ni alloy substrate, single alloy/clad tracks and three totally overlapping clad tracks using powder mixtures of Al-10 wt% SiC, Al-30 wt% SiC, and Al-50 wt% SiC.

They showed that an FGM can be produced by the laser processing of Al-SiC powder mixtures on a Ni alloy substrate (IN 625) and that three successive tracks can be deposited to give a wide range of compositions progressing from the inner to the outer regions. However, the accompanying changes in microstructure through the 3-mm-thick FGM were not fully satisfactory. In particular, it was found to be difficult to retain a significant proportion of the SiC_P other than in the final layer of the FGM. Furthermore, because of the number of alloying additions in the substrate, the microstructures were complex and difficult to interpret, and the results show the need for detailed consideration of the constitution when designing a multicomponent system.

Slip Casting and Pressureless-Sintering. Takebe and Morinaga[26] examined the tetragonal ZrO_2-Ni system as a model to fabricate FGMs by a powder processing technique using slip casting and pressureless sintering. They found that the keys to successful fabrication were (1) well-dispersed ceramic and metal particles in the slip, (2) controlled layered microstructures in the green compacts, and (3) avoidance of fracture of powder compacts by drying and sintering stresses.[27]

They fabricated a multilayer ZrO_2-Ni green compact formed with a stepwise compositional gradient and sintered in Ar+ atmosphere. One final key to successful fabrication was to minimize differences in permeability and pore radius between neighboring layers in the multilayer green compact and thus to avoid fracture during drying in the ambient atmosphere.

Morinaga and his associates[28] also succeeded in fabricating an Al_2O_3/W FGM using a slip cast method. They obtained a gradient material with a continuously changing composition by controlling the fine particle settling rate, which is dependent upon specific gravity and grain size, in the slip cast method thickening process. When they use a thin 5% or 10% slip density, the Al_2O_3 layer and W layer separated, confirming clear continuous inclined layers. In the past, vapor deposition, ion implantation, and diffusion were used to fabricate gradient materials, but this technique has attracted attention and could make it possible to fabricate a variety of FGMs using thin slip densities and fine-particle raw materials with differing specific densities and grain sizes.

The addition of a pressure during slip casting increases the rate of material deposition and enables larger pieces to be fabricated; this was found in a study by Grazzini and Wilkinson.[29] Their work covered processing of monolithic Al_2O_3 and Al_2O_3-Si composites with slip casting formulations based on both dispersed and coagulated slurries that were analyzed, and excellent results were achieved using coagulated slurries, with no cracking present after drying. The effect of pressure was found to increase the green density with a consequent increase in the fired density. In addition, homogeneous microstructures were achieved in the composite system, despite the large difference in particle size used.

Tape Casting. O'Day et al.[30] examined tape casting of ceramic materials, which offers the flexibility of gradually altering the electronic or structural properties of two dissimilar systems in order to improve their compatibility. Their research outlined the processing and fabrication of two systems of FGMs. The systems are both electronic ceramic composites consisting $Ba(1-x)Sr(x)TiO_3$ and Al_2O_3 or a second oxide additive. These composites would be used in phased array antenna systems; therefore, the electronic properties of the material have specific requirements in the microwave frequency regions. The composition of the tapes were varied to provide a graded dielectric constant, which gradually increased from that of air (dielectric constant = 1) to that of the ceramic (dielectric constant = 1500). This allowed maximum penetration of incident microwave radiation as well as minimum energy dissipation and insertion loss into the entire phase shifting device.

Sabljic and Wilkinson[31] developed a process that uses tape casting and lamination to form a metal/ceramic composite. It is well known that in the processing of metal/ceramic composites, severe problems are often associated with the significant differences in coefficients of thermal expansion. Stresses that develop between the ceramic and metallic layers can often cause delamination failures. FGM systems have been developed to minimize this differential at the ceramic/metal interface. They used a series of thin tape-cast layers, each with a slightly different ceramic/metal composition. These layers were then laminated together, and after sintering, they formed a gradual transition from a "pure" ceramic surface layer to a "pure" metal surface layer. Tape casting was the layer forming technique of choice because the same organics can be used throughout the composite structure. This, in turn, allowed the same lamination and organic burnout schedules to be used during the final fabrication process. The composition and thickness of the layers were adjusted to yield the desired compositional gradient.

Atarashiya, Kurokowa, Nagai, and Uda[32] evaluated the FGM system Ni-MgO using Ni-NiO as joining fillers by pressureless sintering. Joining of Ni to MgO was completely achieved at 1300°C using a FGM of Ni-NiO. The fracture strength of these joints ranged from 72 to 128 MPa, compared to 30 to 60 MPa for a conventionally joined part. An FGM of Ni-MgO was used to join the same system, achieving a fracture strength of 110 MPa. Good adhesion was also obtained when joining Si_3N_4 to Ni-Si_3N_4. In addition, Fe was successfully joined to an FGM of Al-AlN at N at 650°C for 48 h. The idea of using FGM blocks as filler metal and as an aid in joining is useful in reducing residual thermal stress. (See Figure 8.9.)

The technique above employs green compacts, which are heated at temperatures ranging from 627° to 1300°C (depending on composition) under null pressure. These materials are then subsequently used to join metals to ceramics of the various compositions.

Reactive Sputtering. Inoue and Masumoto[33, 34] created Al-based amorphous alloys with a gradually changing composition in the depth direction. The alloys were made by the reactive sputtering method with the N_2 concentration controlled by the voltage. Comparing these newly developed amorphous FGMs (Al-Ti, Al-Zr, and Al-Cr alloys) with single-phase compositions shows the advantages of the functional gradient: the temperature resistance was nearly doubled to 200–300°C. The abrasion resistance increased several times, the hardness multiplied by a factor of 10 to 11, and the impact resistance was also much enhanced.

The above amorphous alloys were deposited on a substrate in Ar+ gas with N_2 added. The N_2 concentration was controlled to make the alloy composition change gradually.

With the gradient in the composition, the new amorphous alloys demonstrate the favorable qualities of the individual single-phase alloys without the disadvantages. The Al-Ti alloy begins with an amorphous metal and ends with crystalline ceramic material.

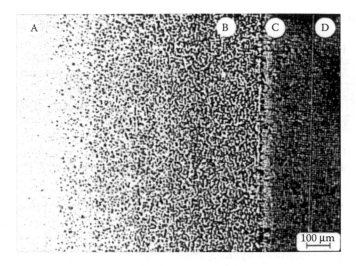

FIGURE 8.9
Optical micrograph of a cross section of joined Ni/Ni-NiO(FGM)/MgO, showing a composition gradient in the region of the joint between the nickel and the solid solution. The regions marked on the micrograph represent A: nickel; B: Ni-NiO(FGM); C: the solid solution; and D: MgO.

The Al-Zr alloy is a metal whose composition changes gradually. The Al-Cr alloy begins with an amorphous metal and gradually changes into an amorphous ceramic material.

Inoue et al.[35] used the R.F. reactive sputtering of a Ti target in a mixed gas of Ar+ and N_2 to fabricate compositionally gradient films consisting of Ti and TiN phases. The compositionally gradient films Ti-TiN films were deposited onto microscope glass slides. The substrate was at room temperature. The crystallographic structure, the composition and the morphology of the deposited films were characterized by x-ray diffractometry, Auger electron spectroscopy and SEM, respectively. The scientists found that R.F. power control method can be used to grow compositionally gradient films. They demonstrated that the deposited film had a structure with a nonoriented TiN layer on a c-axis perpendicular Ti layer, and that the morphology of the films is independent of the deposition method. It was also shown that optical emission spectroscopy and mass spectroscopy were useful tools to monitor the growth of these films.

Dr. Greg Carman[36] reported a significant breakthrough in the manufacture of functionally graded, thin film microactuators.

Significantly, the large deformations generated during the original memory spring back of NiTi produce the largest energy density of any smart material available — an order of magnitude larger than piezoelectric materials. However, slow cycling speeds (1 Hz) and one-way actuation

Sputtering system   Heated target   Deposited film

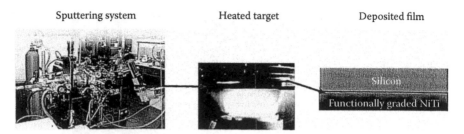

FIGURE 8.10
Sputtering system, heated target, and illustration of functionally graded shape memory
materials.[36]

limit the usefulness of SMAs in many engineering applications. To overcome this problem, Active Materials Laboratory (AML) researchers developed micron-sized (one millionth of an inch), thin film actuators, thus
increasing the possible cycling speed from 1 Hz to 100 Hz. However,
researchers found that producing the thin films necessary for such a large
improvement was problematic due to the loss of Ti during the sputtering
deposition process, where the particles are transferred to a substrate. To
address this issue, AML researchers pioneered a novel approach that
limits the Ti loss during sputter deposition by using a heated target (see
Figure 8.10). They found that target temperature influences the Ti dispersion during sputtering and could be used to control the composition of
the film during deposition. This advancement led to the development
of functionally graded, thin film material.

By varying the target temperature during sputtering, AML researchers
produced thin film with a gradation in Ti content through the thickness,
resulting in functionally graded thin film. As the Ti content increases in
the micron-thick film, the material properties change from pseudoelastic
(similar to rubber) to shape memory The seamless integration of pseudoelastic with shape memory characteristics produces a two-way actuator:
one of the first materials to exhibit this behavior. When researchers heat
the actuator with a small electrical current, the actuator deforms into a
memorized shape. Upon cooling, the material returns to its original shape.
Due to its small dimensions, the cyclic response of the device can approach
100 Hz, over an order of magnitude improvement when compared to
commercially available shape memory material. Given that power output
is the product of energy times frequency, the power output of these small-
scale actuators is considerable.

Using microelectromechanical systems (MEMS) manufacturing techniques, AML researchers have manufactured microactuators. One of the
first actuators they developed was a bubble actuator for flow control on
aircraft systems (see Figure 8.11). The bubble actuator's size is comparable

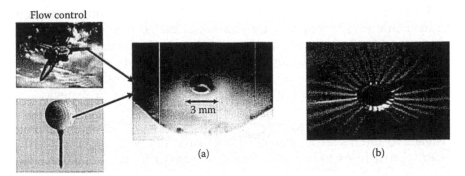

FIGURE 8.11
Flow control applications using (a) bubble actuator concept and (b) more complex geometry.[36]

to the geometry of an inverted golf ball dimple. When a small current passes through the bubble, it dimples outward into the flow (Figure 8.11). When the current is off, the film quickly cools and the dimple returns to the original shape. Correct placement on a missile or an aircraft may decrease drag and provide improved aerodynamic maneuverability. AML researchers also investigated utilizing more sophisticated actuation geometries (Figure 8.11) to generate complex flow patterns for enhanced control.

In addition to flow control actuators, AML researchers fabricated a microwrapper using the functionally graded, thin film shape memory material. The overall dimension of the microwrapper arms was 100 μm, approximately the diameter of a human hair. Doctors might use this microwrapper to manipulate microorganisms or possibly in minimally invasive surgery to remove anomalies such as tumors (see Figure 8.12).

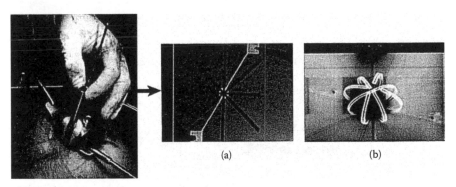

FIGURE 8.12
Medical device applications using the micro-wrapper (a) open due to applied current and (b) closed around a 50 μm particle.[36]

In Figure 8.12, the microwrapper has a small current passing through it to maintain the flat shape. Upon removal of the current, the small arms close (Figure 8.12) to form a cage approximately 50 μm in diameter. Current biomedical applications for NiTi SMAs include vena cava filters and surgical stents. With the advent of these developed thin film MEMS shape memory devices, a much larger array of biomedical devices is possible.

Self-Propagating High-Temperature Synthesis/Hot Isostatic Pressing (SHS/HIP). Miyamoto and his associates[34] at Osaka University have prepared a symmetrical structure $Cr_3C_2/Ni/Cr_3C_2$ (FGM) from Cr_3C_2 and Ni by the SHS/HIP method using Si fuel (see Figure 8.13).

The Si fuel burns in N_2 at pressures >3 MPa and forms Si_3N_4. The Si fuel makes the SHS/HIP process safer and more economical. Using thermite pellets, which are placed at desired points in a chemical oven and reacted at temperatures <1000°C, eliminates the need for an electrical ignition system.

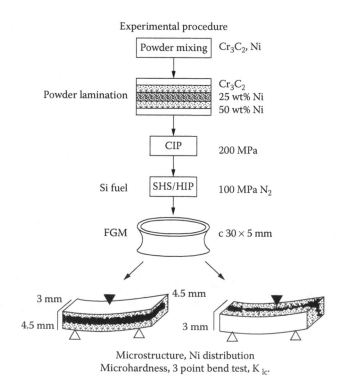

FIGURE 8.13
A flow diagram of the experimental procedure for the fabrication of the symmetrical FGM, $Cr_3C_2/Ni/Cr_3C_2$.

The SHS/HIP method using Si fuel can produce dense materials instantaneously due to the high heat generation of Si nitridation in a pressurized N_2 atmosphere. This new high-temperature rapid process has advantages for energy saving and potential for creation of new materials. The Cr_3C_2/Ni/Cr_3C_2 with a symmetrical gradient composition demonstrated surface reinforcement due to the residual compressive stress at the surface. As a result, the material has a bending strength of 85 kg/mm^2, 50–60% higher than present FGMs.

The material was also found to have improved fracture toughness (10 MPa m$^{1/2}$) at the Cr_3C_2 rich surface compared with monolithic Cr_3C_2 (5.3 MPa m$^{1/2}$). This is believed to be due to the strong residual compressive stress (580 MPa) induced at the surface by contraction of the Ni rich center layer (which has a higher rate of thermal expansion than the surface), during the rapid heating and cooling in the SHS/HIP process. A slightly diffused Ni metal dispersion also contributes to the improvement in toughness. The researchers suggest that this toughening mechanism, which they call gradient toughening, is applicable to other ceramics and metals.

The SHS/HIP method is useful for fabrication of FGMs because graded compositions with differing sintering temperatures can be consolidated due to the rapid densification. For example, FGMs such as titanium carbide-nickel (TiC/Ni), titanium diboride-nickel (TiB_2/Ni), chromium carbide-nickel (Cr_3C_2/Ni), and molybdenum disilicide-silicon carbide/ titanium aluminide ($MoSi_2$-SiC/TiAl) have been fully densified.

The Si fuel used is a byproduct of Zn smelting. During the sintering part of the process, the Si is nitrided by N_2 gas, which is also the source of pressure for the HIP. The combustion product, a brittle solid consisting of β-Si_3N_4 and a small amount of Si, is potentially usable for refractories, abrasives, and other applications. A flow diagram for the entire process is shown in Figure 8.14.

These Osaka researchers[34] used a chemical oven plus a hot isostatic press, which essentially combined self-propagating high-temperature synthesis with hot isostatic pressing. The chemical oven consisted of a graphite crucible containing the Si fuel and a glass container in which the green compact was sealed by an automatic glass encapsulation method. Ignition was initiated either electrically at low temperature by passing 60 A of current for 3 sec through an ignition heater or chemically using thermite agents. For example, a mixture of iron oxide (Fe_2O_3) and Al produced a strong exothermic reaction at about 927°C and 100 MPa N_2 pressure.

The developers[34] feel that the chemical approach is advantageous over electric ignition because heat energy can be supplied uniformly to large samples by evenly distributing the thermite pellets, and because several

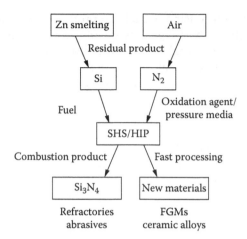

FIGURE 8.14
A flow diagram of the gas pressure combustion sintering process (SHS/HIP) using silicon as the fuel.

chemical ovens can be installed in a hot isostatic press and ignited simultaneously. Also, by changing the volume of the fuel used in each of the ovens, the heating conditions can be adapted for different materials. Compared with other powerful sources of heat energy, such as lasers and plasma, which can only heat a small area or volume, a chemical oven can deliver a large quantity of heat instantaneously to a large volume of material.

Additional work at Osaka University has combined SHS/HIP to prepare TiC-Ni. The pressed body of reactant powders is vacuum-sealed into a glass capsule and placed into a crucible filled with a combustion agent. The reaction is initiated by the heat released from the combustion agent under an Ar+ pressure below 100 MPa. Flexural strengths at room temperature and 500°C ranged from 1100 to 1340 MPa and 630 to 920 MPa, respectively, depending on composition.

X. Ma et al.[37] used the SHS/HIP process, which they called Gas-Pressure Combustion Sintering (GPCS), to synthesize and densify the refractory ceramics or composites simultaneously and rapidly. The GPCS was suited to fabricate the FGMs because the diffusion of elements was limited due to the rapid process, so that the graded arrangement of compositions could be saved.

Their study involved the combination of TiC and Ni also. It was easy to fabricate the dense TiC and composites with Ni by combustion sintering because the TiC has a high formation energy (180 kJ/mole).

They found that the TiC/Ni FGMs with different graded compositions had a unique microstructure, with TiC grains becoming finer from the TiC ceramic to the Ni metal end. Thermal and mechanical properties of non-graded TiC/Ni composites were measured and used for FGM design. By cyclic loading, the temperature difference was about 327°C between the top and bottom of the FGM, at which the subcritical cracks appeared, but the graded composition of n = 0.5–0.7 showed relatively good thermal durability.

Other research[34] has synthesized a symmetrical structure Al_2O_3/TiC/Ni/TiC/Al_2O_3 FGM which is expected to have a self-repair function for passivation due to TiC oxidation at high temperatures.

Dongliang et al.[38] sintered α-SiC fine powder clad with special glass by HIP processing under 200 MPa at 1800°C. The density of the material obtained was close to 97.5% of theoretical density; bending strength reached 582 MPa, which is about 50% higher than that of common pressureless sintered SiC. After post-HIP treatment, density was further increased to above 98.5% of theoretical density. Experimental work was successful in fabricating SiC-Si_3N_4 gradient composite material. This kind of new SiC-Si_3N_4 gradient composite material resulted in a strength and fracture toughness of 900 ± 100 MPa and 8.4 MPa m$^{1/2}$ respectively, which were about 1 time higher than those of hot-pressed SiC material.

Feng and Moore[39] investigated the FGM system TiB_2/Al_2O_3/Al by utilizing SHS as an efficient processing route for the synthesis of the ceramic-metal composite system. Thermal explosion of the green FGM pellet coupled with a light squeezing compaction, simultaneously applied while the reaction proceeds, offered a means by which the FGM material was consolidated. Squeezing at pressures as low as 1.4 MPa provided substantial improvements in product density, and this combination of combustion synthesis and squeezing was needed to produce dense FGMs even when the excess liquid metal infiltrant concept was used.

Using two specially developed techniques, an automated powder spraying and stacking device, shown in Figure 8.15, and hydrostatic pressing of the ignited sample to simultaneously synthesize and form the FGM, a TiB_2-Cu FGM (30 mm in diameter and 1 mm thick) was deposited by SHS on a Cu substrate, according to N. Sata.[40]

The process steps involved first creating the desired gradient composition by computer control of both the rate that the powder suspensions are transported to the substrate surface and the x-y stage moving pattern of the stacking area. A spraying program was developed as seen in Figure 8.16. The stacked samples were then cold isostatically pressed (CIP) and degassed before simultaneous SHS and consolidation. By using the appropriate composition path in which Cu is only present as a liquid phase (i.e., is not gasified), the resulting TiB_2-Cu FGM was crack-free and contained no large pores. From the x-ray intensity patterns of each element in the sample, it was seen that the concentration changed gradually from

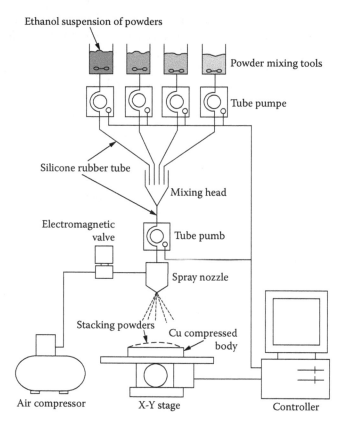

FIGURE 8.15
Schematic of automated powder spraying and stacking system for fabricating FGMs by self-propagating high-temperature synthesis.

the Cu substrate to the surface. Because of the use of CIP, the developers believe that large, complex shapes and sheets could be fabricated.

Spark Plasma Sintering (SPS). Using the SPS process, Sumitomo Mining Co., Ltd.[41] succeeded in synthesizing FGM made of glass and metal.

Glass and metals are entirely different substances, so producing FGMs with these materials by the conventional method was quite difficult. The company established an SPS process sintering the powders of these materials by applying a high pressure and an electric current at the same time. The material could be used for producing sensors for use in high-temperature, high-pressure environments and in the sector of optoelectronics.

The SPS process is a unique synthesis process in which electric energy is introduced between the gaps of green compact particles, and the high energy of the discharge plasma generated instantaneously is applied effectively to enable sintering or sintered bonding at a lower temperature and

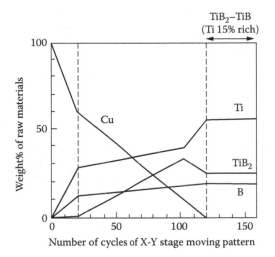

FIGURE 8.16

A spraying program for the component powders for a TiB$_2$-Cu FGM.

in a shorter time compared with conventional processes. It may be regarded as one of a new, next-generation type of sintering process that most effectively utilizes the spontaneous self-heating effect arising inside the particle specimen, as with the SHS process.

With the electrical method, however, the powders were heated so quickly that they were sintered into an integrated whole before thermal expansion had a chance to ruin things.

The synthesized FGM was produced by using glass and Si or glass and stainless steel in combination, and was produced in disk form with a diameter of 2 cm and height of 1 cm. First, a pressure of 30 atm was applied to mix and solidify the powdered glass and metal without any voids. A large current of 500 A was applied intermittently at a fixed interval, after which the mixed powder was heated to 800°C. Synthesis was accomplished as quickly as 15 min. When the distribution of composition was investigated, it was confirmed that the material changes gradually from the surface to the center, indicating that the material was completely synthesized as an FGM.

Additionally researchers[42] in Japan's National Defense Academy have used the new electrical method to construct FGMs from a Ti-Al alloy and a Zr alloy — two materials with vastly different melting points.

Although the Ti-Al alloy melts at 1200°C, and the Zr alloy melts at 1600°C, they were able to bridge the 400°C temperature differential by putting the initial mixture of powders into a convex graphite mold.

When a strong current was applied to the convex mold, the current was densest at the mold's apex, producing a higher temperature in that area.

A 1200 A current can create a temperature differential of at least 900°C between powder in the bottom and powder in the top of the mold.

The electrical method makes it possible to sinter the material in about 6 min, while the standard technique would take more than 6 h.

Speed and low cost are two huge advantages of the electrical sintering method.

Electrophoretic Deposition (EPD). An electrophoretic deposition and sintering route was used by Sarkar, Huang, and Nicholson[43] to prepare successfully YSZ/Al_2O_3 composites with a compositional gradient. The YSZ content was continuously decreased from the YSZ-rich surface to the Al_2O_3-rich surface. Microstructural and Vickers hardness (16–24 GPa) evidence tracked the compositional development, and the indentation fracture toughness was found to vary across the section (10 to 3 MPa $m^{1/2}$).

Sedimentation. A modified sedimentation process was used in the production of an FGM, NiAl/Al_2O_3, by Miller and Lannutti.[44] Four-point bend tests were conducted to establish the mechanical load-displacement behavior of a single interlayer FGM.

The bend test values at room temperature for these composites were about 3–4 times that of both unreinforced NiAl and sapphire-fiber-reinforced NiAl (570 MPa vs. approximately 150 MPa).[45] At 572°C the final bend test values were approximately 50% higher (150 MPa vs. 100 MPa),[45] and at 727°C they were about twice as large (100 MPa vs. 50 MPa).[45] At elevated temperatures, composite fracture occurred in a gradual, non-catastrophic mode involving NiAl retardation of a succession of cracks originating in the Al_2O_3 face.

Multi-Ingredient Powder Deposit. The Miyagi Institute of Technology and Makabe Co.[46] have jointly developed a process for making bulk FGMs. Under computer control, powders of materials were made into a slurry mixture, which was sprayed by a supersonic nozzle to form an even, continuous deposition. The deposit was simultaneously shaped and sintered to form disks up to 90 mm in diameter. The process provided inexpensive, reproducible large samples of FGM, free of defects such as the tendency to bend or delaminate.

This new process, called the "Multi-Ingredient Powder Deposit Process," is concerned with two materials: ZrO_2 and stainless steel. The supersonic nozzle and the movement of the deposition stage allow the thick deposit to be even in thickness. The design allows no part of the sprayed matter to bounce back from the target, so that source material is saved. The process is applicable to fast, easy production of large cylindrical deposits and more intricate shapes such as a film insert for junctions as well as planar deposits.

Powder Metallurgy. Bishop et al.[47] investigated a bulk production system to fabricate FGMs. The metal-ceramic system selected for investigation has potential applications in orthopedics and dentistry. The metallic component was Ti, which is currently widely used in surgery, and the ceramic was hydroxyapatite (HA). HA is bioactive; that is, it reacts chemically with living bone to give a strong bond that can resist high stresses. Thus, an implant fabricated from a bioactive material becomes quickly and firmly fixed without the need for any mechanical means of fixation such as using screws or bone cement. It is hoped that the incorporation of HA into Ti will enhance the bonding to bone, and that by producing an FGM this will be achieved without a significant degradation in mechanical performance.

The work by Bishop et al.[47] demonstrated that an FGM may be produced by a working process that involves low temperatures compared with those required for sintering, and that has the potential for rapid fabrication. Besides an economic benefit, the low temperatures reduce problems that can arise due to the difference in properties, such as melting point and CTE, of the constituents.

The above researchers used mixtures of dry Ti and HA powders containing 10, 20, and 30 wt% HA to give the following layered arrangement: Ti, Ti-10% HA, Ti-20% HA, Ti-30% HA, Ti. The layered powders were cold compacted at a pressure of 500 MPa for 1 min to give a "green billet" approximately 60 mm high. The cold-compacted green billets were placed in a furnace for preheating to the consolidation temperature. They were held at 500°C for 30 to 60 min. Each billet was transferred to a 60-mm-diameter die and hot-pressed to a maximum pressure of about 1630 MPa.

As far as the specific system studied is concerned, the results are most encouraging. Not only may the processing route be used to produce FGMs, it could also be employed to fabricate homogeneous Ti-HA composites. Thus, a Ti-HA FGM with a strong, tough core and an outer layer of high HA content is feasible to produce.

FGMs via Powder Spray Forming. Powder metallurgy has been used at Tohoku University[48] to fabricate an FGM with a gradually controlled composition gradient of partially stabilized Zr (PSZ) powder (average particle size 0.07 µm) and type 304 stainless steel powder (average particle size 3 µm). Figure 8.17 shows the automated powder spray forming process used to deposit the powders, which are suspended in ethanol, onto a preheated green substrate made of stainless steel powder (14 mm in diameter and 2 mm thick). The ratio of the PSZ to stainless steel deposited was gradually varied according to a predesigned composition profile that could be controlled by a microcomputer within 0.01 mm. To complete the process, the FGM-coated substrate was isostatically pressed, encapsulated, and hot isostatically pressed. Examining the microstructure, one finds that at the low concentrations of PSZ, the PSZ phase is dispersed in the stainless steel. As the mixing ratio is increased, a PSZ network structure

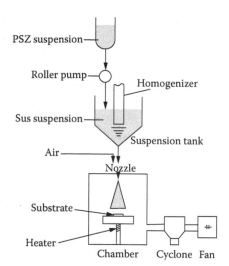

FIGURE 8.17
Schematic of the automated powder stacking apparatus. To vary the mixing ratio of the metal (SUS) and ceramic powders (PSZ), the flow rate of the ceramic powder suspension is changed by controlling the rotating speed of the roller pump with a microcomputer.

forms in which the stainless steel particles are dispersed. The thermal shock resistance of this FGM is about 2.5 times that of monolithic PSZ. Thermal shock-resistant FGM coatings of PSZ/Mo and PSZ/W have been similarly made by the Tohoku researchers.

Other P/M methods have evolved in preparing FGMs of Ni-MgO, Ni-NiO, Ni-Si$_3$N$_4$, and Al-AlN. A mixture of metal and ceramic powders with compositional gradient were cold-pressed at 20 to 32 MPa.

Plasma Spray. Erolu et al.[49] chose an NiCr-Al /MgO-ZrO$_2$ system for a study as a thermal barrier coating used in the aerospace and automotive industries. The coatings of this system usually consist of two layers of NiCr-Al and ZrO$_2$ stabilized by MgO. The first layer of NiCr-Al serves as a bond coat for a ceramic MgO-ZrO$_2$ topcoat that has the desired surface properties, such as low thermal conductivity. The bond coat/ceramic interface is the area where coating failure most frequently occurs when the coating is subjected to high temperature gradients. The graded zone between the metallic and ceramic layers leads to gradual changes in properties, such as hardness or thermal expansion coefficient, which can improve the coating resistance to failure.[50]

Erolu and associates developed a methodology for the synthesis of graded coatings in a single torch plasma spray reactor, whereby a graded coating was synthesized in discrete layers with ZrO$_2$ content increasing with the coating thickness.

A new type of plasma spray gun with four powder carrier ports has been developed at Nippon Steel by Shimada[51] for low-pressure plasma spray deposition of FGM coatings. With this gun, the ceramic and metal powders can be injected at the same time. A 1-mm-thick coating with a gradient composition of 8% Y_2O_3 stabilized ZrO_2 and Ni-20% Cr on stainless steel or Cu (30 mm wide and 5 mm thick) was able to withstand a temperature difference of 377°C between the top and bottom surfaces for 80 cycles without damage. Thin sheet lamination using this low-pressure plasma spray technique is considered promising for the mass production of large complex structures.

Researchers at Plasma Processes Inc.[52] have produced an FGM through advanced vacuum plasma spray processing for high heat flux applications. They developed manufacturing methods for a four-component FGM of Cu, W, B, and BN. The FGM was formed with continuous gradients and integral cooling channels eliminating bondlines and providing direct heat transfer from the high-temperature exposed surface to a cooling medium. Metallurgical and x-ray diffraction analyses of the materials formed through innovative vacuum plasma spray (VPS) processing was also developed. Applications for this functional gradient structural material range from fusion reactor plasma facing components to missile nose cones to boilers.

Superplastic Forming/Diffusion Bonding (SPF/DB). Nagano and Wakai[53] successfully fabricated FGM (ZrO_2-Al_2O_3) by SPF/DB at a bonding temperature of 1550°C and at strains of 17, 33, and 50%.

The apparent bending strength of the ZrO_2-Al_2O_3 FGM was significantly different according to the direction of applied stress. When the tension side was the Y-TZP, the apparent bending strength of the specimen (strain of 50%) was 1860 MPa; when the tension side was Al_2O_3, the apparent bending strength (strain of 50%) was 330 MPa.

The authors determined that the phenomenon was caused by increasing residual compressive stress in the Y-TZP surface with increasing warp of the FGM and increasing strength of the base material with increasing strain in compressive deformation.[54]

Chemical Vapor Deposition. Kude[55] demonstrated the ability to deposit an FGM on a complex shape by CVD of a six-step SiC FGM on C fibers, on a cylindrical graphite tube, and on a graphite plate. The source gases for the coatings were methane (CH_4) for the C and silicon tetrachloride ($SiCl_4$) and methane for the SiC. The C content in the deposits was changed by changing the molar ratio of the two gases; that is, changing the ratio of Si/C in the gas phase changes the ratio of C/(SiC + C) in the deposit.

By controlling the CVD conditions, the microstructure of the FGM coatings can be changed from dense to porous. The porous structures were found to resist both delamination and crack propagation, and the researchers concluded that by controlling the microstructure of these FGMs, both their thermal barrier and thermal fatigue properties can be controlled.

In the application of CVD to create an FGM is a C/C composite in which the FGM structure has been created internally by penetrative CVD. The net effect is that the Si content is changed gradually within the C/C composite. This FGM composite with an FGM SiC/C CVD coating has been tested for thermal fatigue in a rocket combustor environment with excellent results.[55]

Joining FGMs

Joined ceramic-intermetallic composites consisting of TiC and NiAl have been successfully fabricated by Kudesia et al.[1, 56] Self-propagating high-temperature synthesis was used to prepare both the FGM joint layer and the two materials to be joined. Successful densification and retention of the composition gradient following fabrication were confirmed by SEM and EDS characterization, respectively. Design calculations were performed for thermal shock of TiC/NiAl composites joined by an FGM layer and illustrated the importance of sample geometry parameters on the development of transient thermal stresses. Subsequent design calculations were performed for fixed sample geometry and varied material property parameters, to illustrate the importance of material property interactions. This design-fabrication iterative process — along with planned thermal shock and fatigue experiments — is expected to yield joined materials that exhibit superior performance under expected-use conditions.

A very active joining program aimed at establishing FGM cooling panel fabrication technology was reported by Fujiwara et al.[57] using the thin film lamination method.[57]

1. In order to ensure a brazing clearance in brazing an FGM plate to a metallic plate, the FGM plate must be formed into a shape that will produce a flat surface when it is heated to brazing temperature.

2. In brazing an FGM plate to a metallic plate via an Ni brazing filler material, pore-like defects tend to appear within the eutectic phase formed between the plates. An effective way to prevent such defects from forming is to use liquid diffusion brazing for isothermal solidification of the filler material at brazing temperature.

Two-stage steps were taken in the fabrication of an FGM cooling panel. In the first brazing stage Nicrobraze #150 (Ni-15 Cr-3.5 B) was used as a brazing filler metal (BFM). Brazing conditions were 1075°C temperature heating for 5 h. In step two, a 304 stainless steel cooling structure panel, 304 manifolds, an Ni sheet, and an FGM plate were then brazed using BNi-2 as a BFM. Brazing conditions were 1030°C for 5 h.

Properties

FGMs can be used in aggressive environments at elevated temperatures because they provide the possibility of minimizing wastage of materials. Gradation of the volume fraction of hard particles through the layers means that thermal cycling effects are less severe than for many conventional metal-substrate systems. Because such materials may provide resistance to wear and corrosion (by using a corrosion-resistant matrix), it is thought that they may have applications to environments at elevated temperatures, in which materials selection involves a compromise between corrosion resistance and high yield strength. Stark et al.[58] conducted a study to investigate the erosion resistance of the various layers of a candidate FGM that consisted of WC particles in an Ni-Cr matrix. The performances of the various composite layers were considered separately in order to establish the variation of erosion rates through the graded structure. The effects of temperature, volume fraction of hard particles, and erodent size were investigated in a laboratory-simulated fluidized bed erosion rig. SEM and thickness loss measurements were used to characterize the surfaces following exposure. The results showed that the erosion rate at room temperature was at a minimum at intermediate volume fractions of WC particles. However, this behavior reversed for erosion with larger particle sizes. Although the thickness losses increased with increasing temperature for all volume fractions of reinforcement particles, a reduction in the thickness loss at the highest temperature studied was observed for exposure to both large and small erodents (600 and 200 μm Al_2O_3.)

Tohgo, Sakaguchi, and Ishii[59] investigated FGMs in which the ceramic particles were dispersed in the ductile metal matrix and a particle volume fraction was linearly changed. The elastic and elastic-plastic analyses of a crack were carried out based on a newly developed finite element method.[60–62]

Wetherhold, Seelman, and Wang[63] investigated the use of FGMs to eliminate or control thermal deformation.

If we wish to minimize the thermal deformation of a component or to match it to the thermal deformation of another component, we can accomplish this by using a composite whose fibers have a negative axial thermal expansion coefficient. By varying the fiber volume fraction within a symmetric laminated beam to create an FGM, certain thermal deformations can be controlled or tailored. Specifically, a beam can be designed that does not curve under a steady-state through-thickness temperature variation. This result is independent of the actual temperature values, within the limitation of constant material properties of the constituents. The beam can also be designed to match or eliminate an in-plane expansion coefficient or to match a desired axial stiffness. Combining two fiber types to create a hybrid FGM can offer desirable increases in axial and bending stiffness while still retaining the useful thermal deformation behavior.

Gupta and Jena[64] conducted a series of programs on functionally graded ceramic materials with gradually changing pore structures, which are being increasingly used for high-tech applications. The effectiveness of such materials depends on the pore structure of individual layers. Pore structures of ceramic materials are normally determined by porosimetry or flow porometry, in which the flow occurs parallel to the thickness of the sample.

Using a new technique based on flow porometry, the largest pore diameter, the mean flow pore diameter, and the pore size distribution of each layer of a two-layered ceramic composite were determined. In comparison, porosimetry could not measure any property of Layer 1.

This new technique can be used to accurately determine the pore structures of the individual layers of these composites, thus ensuring the quality of the final product.

Applications

Dental Implant

Researchers at Tokyo Medical and Dental University[65] have developed a dental implant (artificial tooth root) using an FGM. The material consists primarily of apatite that has a high biocompatibility with bone and titanium that provides the necessary strength.

Implants are produced today using a ceramic (apatite) and metals such as Ti. Apatite is biocompatible with the bone but lacks strength and is brittle, whereas Ti has a greater strength than natural tooth but less affinity with the bone. The research team attempted to use the advantages of both these materials by making a implant made of an FGM consisting primarily of apatite and Ti.

Mixed powders with the different ratios of Ti and composite materials (apatite, ZrO_2, fused quartz, and Pd) were packed into the Si rubber mold, changing the concentration gradually from one end to the other. The specimens were compressed at 400–500 MPa by CIP and sintered in Ar+ by high-frequency induction heating at 900–1300°C.

FGM specimens were formed with the content changing gradually from pure Ti at one end to 9% ceramic at the other end. The strength of the FGM specimens was approximately same as the 9% ceramic specimens.

The prototype implant was used in various strength tests. When apatite and Ti were mixed at a ratio of 1:9, the compressive strength was about 500 MPa and the flexural strength was about 100 MPa. When tests were conducted on implants made only of apatite, the compressive and flexural strengths were only a few dozen MPa.

Building Materials

Takahashi et al.[66] has successfully developed an FGM with moisture absorption and release functions made by stepwise or continuously changing the composition of two components, a zeolite or ligneous material-based humidity-conditioning material, capable of absorbing and releasing moisture, and calcium silicate-based concrete. Compared to lumber and concrete, this FGM can absorb a maximum of approximately 2.5 times more moisture and is particularly good in its initial response to a humidity increase or decrease.

In addition, the new material hardly changes its dimensions by expansion or contraction due to humidity, and it is strong, fireproof, frost damage-resistant, and quite amenable to shape forming from flat board to curved corner material. As a building material, this material with a gradient moisture-conditioning property from its outer wall to inner wall could eliminate extra construction steps for faster completion.

This material can be made in a 300 mm^2 piece but future plans are to continue the joint research project to produce larger sizes of up to 600 × 900 mm, produce porous concrete compositions for weight reduction and better insulation, and develop a mass-production facility.

Ballistic Protection

Professor Ma Jan[67] investigated a ballistic protection application of FGMs that involved the processing of a Ti-TiB_2 FGM system using the powder metallurgy method. The effect of starting particle size and sintering aid on the sinterability and mechanical properties of the FGM system were studied.

It was found that smaller starting particle size will improve the densification of the ceramics, and hence the overall FGM system. SiC as sintering aid showed an enhancement in the sinterability of the ceramic layer and resulted in tremendous improvement on both flexural strength and toughness of the FGM. Needle-like TiB phase was also confirmed to be present after the sintering process and was believed to provide considerable contribution to the excellent properties of the system.

Temperature Sensors

Peters et al.[68] investigated $MoSiO_2$ and Al_2O_3 FGM tubes because they are thermodynamically stable high-temperature materials whose thermal expansion coefficients match closely. Composites of these materials have potential for use in applications such as protective sheaths for high-temperature sensors.

Pt coatings are widely used on Al_2O_3 sheaths for thermocouples in the U.S. glass industry to protect the thermocouple wires and Al_2O_3 sheath from corrosion and dissolution of the temperature-sensing unit. The cost associated with Pt coatings can be prohibitively high, considering the steps necessary at glass plants to maintain and secure an inventory of Pt. There also is a need to improve the performance of the Pt-coated Al_2O_3 sheaths, because the failure rate of the thermocouples can be as high as 50%. These issues are driving the glass industry to search for alternative materials that can replace Pt and still provide the durability and performance needed to survive in an extremely corrosive glass environment.

Plasma spraying is a very affective method to produce $MoSi_2$ and $MoSi_2$ composite coatings and spray-formed components.[69] For example, studies of plasma spray-formed $MoSi_2$-Al_2O_3 composite gas-injection tubes show they have enhanced high-temperature thermal shock resistance when immersed in molten Cu and Al.[70] The composite tubes outperform high-grade graphite and SiC tubes when immersed in molten Cu and had similar performance to high-density graphite and mullite when immersed molten Al.

Energy-absorbing mechanisms such as debonding (between the $MoSi_2$ and Al_2O_3 layers) and microcracking in the Al_2O_3 layer contribute to the ability of the composite to absorb thermal stresses and strain energy during the performance test. $MoSi_2$ and Al_2O_3 are chemically compatible and have similar thermal expansion coefficients.[71, 72]

Layered and continuous FGM strength distribution plots are shown in Figure 8.18. Both continuous and layered FGM microstructures have similar mean Weibull strengths (σ_f -70 MPa), which were calculated using the following equation:

$$\sigma_f = a^{-1/\beta} \, \Gamma \, [1 + 1/\beta]$$

where beta (β) is the Weibull modulus and gamma (Γ) is a function of the sample size.[73]

However, the spread of the data for the continuously graded material was smaller, thus, a larger Weibull slope (13.38 for the continuously graded samples vs. 7.635 for the layered graded samples). FGM fracture energy (qualitatively determined from the area under the load-displacement

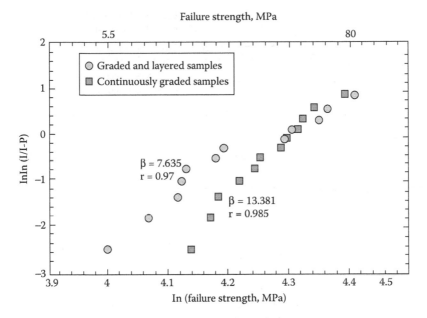

FIGURE 8.18
Results of C-ring tests performed on continuously graded and layered-graded Al_2O_3-$MoSi_2$ coatings.

plot of the C-ring tests) is significantly higher (~3 tunes) than that of monolithic Al_2O_3 and $MoSi_2$.

Continuously graded and layered FGM fracture surfaces contain extensive microcracking and roughening. Increased toughening of the composite is believed to be a direct result of microcracking. This work indicates that the strength and toughness of FGM tubes meet the performance requirements of this application.

An actuator was prepared by diffusion bonding of two plates having different piezoelectric constants in the system of PZT-NiNb. The compositionally gradient intermediate layer, which was bonded by firing to reduce stress that remained between the two plates, resulted in a higher reliability of the material as an actuator. The structure of the actuator prepared in this manner is schematically shown in Figure 8.19.

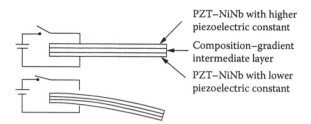

PZT–NiNb with higher
piezoelectric constant

Composition–gradient
intermediate layer

PZT–NiNb with lower
piezoelectric constant

FIGURE 8.19
Structure of FGM piezo-electronic actuator.

References

1. L. M. Sheppard, Enhancing Performance of Ceramic Composites, *Am. Ceram. Soc. Bull.*, 71(4), 617–631, 1992.
2. M. Koizumi, Recent Progress of Functionally Gradient Materials in Japan, *Proc. 16th Ann. Conf. Composites Adv. Ceram. Mater.*, M. Mendelson, Am. Ceram. Soc., pp. 333–347, January 7–10, 1992, Cocoa Beach, FL.
3. M. M. Schwartz, *Emerging Engineering Materials*, Lancaster, PA: Technomic Publishing Company, 1996, p. 292
4. Report from Science and Technology Agency of Japanese Government of "Research on the Basic Technology for the Development of Functionally Gradient Materials for Relaxation of Thermal Stress," 1987.
5. M. A. Steinberg, *New Materials in the Aerospace Field*, Science Japanese Edition, Nikkei Sci. Inc., Tokyo, 1986, pp. 29–35.
6. Y. Fukui and Y. Watanabe, Analysis of Thermal Residual Stress in a Thick-Walled Ring of Duralcan-Base Al-SIC Functionally Graded Material, *Metall. Mater. Trans. A — Phys. Metall. Mater. Sci.*, 27(12), 4145–4151, 1996.
7. G. Sacks, Z. *Metallkd.*, 19, 352–357, 1927.
8. H. Kimura and K. Toda, Design and Development of Functionally Graded Material by Pulse Discharge Resistance Consolidation with Temperature-Gradient Control, *Powder Metall.*, 39(1), 59–62, 1996.
9. H. Kimura and S. Kobayashi, in O. Izumi (Ed.), *Intermediate Compounds*, Sendai: Japan Institute of Metals, 1991, p. 985.
10. H. Kimura and S. Kobayashi, *J. Jpn. Inst. Met.*, 58, 291, 1994.
11. H. Kimura and S. Kobayashi, *Mater. Trans. JIM*, 36, 982, 1995.
12. T. Hirano, T. Yamada, et al., Fundamental Design, Multiobjective Optimization for FGM, FGM '91 Sym., October 8–9, 1991, pp. 15–29.
13. Y. Obata, N. Noda, et al., Steady State Thermal Stress in FGM Plates, FGM '91 Sym., October 8–9, 1991, pp. 36–44.
14. Y. Matsuzaki, M. Kawamura, et al., Fundamental Studies of TiAl-Based FGMs, FGM '91 Sym., October 8–9, 1991, pp. 45–59.
15. A. Kawasaki and R. Watanabe, Powder Metallurgical Fabrication of the Thermal-Stress Relief Type of Functionally Gradient Material, "Sintering '87 Tokyo," Somiya et al. (Eds.), London: Elsevier, 2, 1197–1202, 1988.

16. K. Watanabe, A. Kawasaki, and N. Murahashi, Fabrication of Thermal Stress Relief Type of Functionally Gradient Material in Molybdenum/Zirconia System, *J. Assoc. Mater. Eng. Res.*, 1, 36–44, 1988.

17. T. Fukushima, S. Kuroda, and S. Kitahara, Gradient Coatings Formed by Plasma Twin Torches and Those Properties, *Proc. First Intl. Sym., FGM*, Sendai, pp. 145–150, 1990.

18. N. Shimoda, S. Kitaguchi, T. Saito, H. Takigawa, and M. Koga, Production of Functionally Gradient Materials by Applying Low-Pressure Plasma Spray, *Proc. First Intl. Symp., FGM*, Sendai, pp. 151–156, 1990.

19. N. Sata, N. Sanada, T. Hirano, and M. Niino, Fabrication of a Functionally Gradient Material by Using a Self-Propagating Reaction Process, *Proc. First Intl. Symp. on Combustion and Plasma Synthesis of High-Temperature Materials*, VCH Publishers, pp. 195–203, 1990.

20. Y. Miyamoto, H. Nakanishi, I. Tanaka, T. Okamoto, and O. Yamada, Gas Pressure Combustion Sintering of TiC-Ni FGM, *Proc. First Intl. Symp., FGM*, Sendai, pp. 257–262, 1990.

21. O. A. Voronov and G. S. Tompa, Functionally Graded Nanophase Be/Cu Composites, L.B. Johnson Space Center, *NASA Tech Briefs*, November 2003, p. 52.

22. T. Nagano and F. Wakai, Fabrication of Zirconia-Alumina Functionally Gradient Material by Superplastic Diffusion Bonding, *J. Mater. Sci.*, 28(21), 5793–5799, 1993.

23. S. Nakashima, H. Arikawa, M. Chigasaki et al., ZrO_2 and Cu Functionally Gradient Materials Prepared by a Dynamic Ion Mixing Process, *Surface and Coatings Technol.*, 66(1–3), 330–333, 1994.

24. J. H. Abboud, R. D. Rawlings, and D. R. F. West, Functionally Gradient Layers of Ti-Al-Based Alloys Produced by Laser Alloying and Cladding, *Mater. Sci. Technol.*, 10(5), 414–419, 1994.

25. K. M. Jasim, R. D. Rawlings, and D. R. F. West, Metal-Ceramic Functionally Gradient Material Produced by Laser Processing, *J. Mater. Sci.*, 28, 2820–2826, 1993.

26. H. Takebe and K. Morinaga, Fabrication of Zirconia-Nickel Functionally Gradient Materials by Slip Casting and Pressureless Sintering, *Mater. Manufact. Proc.*, 9(4), 721–33, 1994.

27. K. Takebe and K. Morinaga, *J. Ceram. Soc. Jpn.*, 98, 1250, 1990.

28. K. Morinaga, K. U. Toto, Fabricate Thick Alumina-Tungsten FGM Using Slip Cast Method, *Sci. Technol.*, p. 4, December 1993..

29. H. H. Grazzini and D. S. Wilkinson, Slip Casting Under Pressure, Dept. of Mater. Sci. and Engrg., McMaster University, Hamilton, Ontario L8S 4L7, Canada, *Proc. 16th Ann. Conf. on Composites & Advanced Ceram. Mater.*, M. Mendelson, Am. Ceram. Soc., pp. 408–418, January 7–10, 1992, Cocoa Beach, FL.

30. M. E. O'Day, L. C. Sengupta, E. Ngo et al., Processing and Characterization of Functionally Gradient Ceramic Materials, ARL-TR-337, Army Research Lab., Adelphi, MD, February 1994, p. 19.

31. R. E. Mistler, Tape Casting: An Enabling Fabrication Technology, *Ceram Ind.*, 27–30, 2000.

32. K. Atarashiya, K. Kurokawa, and T. Nagai, Functionally Gradient Material of the System Ni-MgO, Ni-NiO, Ni-Si$_3$N$_4$ or Al-AlN by Pressureless Sintering, Hokkaido University, Japan, and M. Uda, Nisshin Steel Co., Ltd., Japan, *Proc. 16th Ann. Conf. on Composites & Advanced Ceram. Mater.*, M. Mendelson, Am. Ceram. Soc., pp. 400–407, January 7–10, 1992, Cocoa Beach, FL.

33. Functionally Gradient Al-Based Amorphous Alloys, *JETRO*, 22(2), 17, 1994.

34. Y. Miyamoto, Economic Process for Rapid Densification of Ceramics, Metals, and Functionally Gradient Materials, Processing Research Center for High Performance Materials, Institute of Scientific and Industrial Research, Osaka University, Ibaraki, Osaka 567, Japan, 8(5/6), May/June 1993, and Functionally Gradient Materials by SHS/HIP, JETRO, 21(4), 30, 1993.

35. S. Inoue, H. Uchida, K. Takeshita, K. Koterazawa, and R. P. Howson, Preparation of Compositionally Gradient Ti-TiN Films by R.F. Reactive Sputtering, *Thin Solid Films*, 261(1–2), 115–119, 1995.

36. G. Carman, Functionally Graded Thin Film Shape Memory Alloy Micro-Actuators, *AFRL Technol. Horizons*, September 2002, pp. 38–39.

37. X. Ma, K. Tanihata, Y. Miyamoto, A. Kumakawa, S. Nagata, T. Yamada, and T. Hirano, Fabrication of TiC/Ni Functionally Gradient Materials and Their Mechanical and Thermal Properties, *Proc. 16th Ann. Conf. on Composites & Advanced Ceram. Mater.*, M. Mendelson, Am. Ceram. Soc., pp. 356–364, January 7–10, 1992, Cocoa Beach, FL.

38. J. Dongliang, S. Jihong, T. Shouhong et al., SiC-Si$_3$N$_4$ Gradient Composite Ceramics by Special HIP Processing, Shanghai Institute of Ceramics, Academia Sinica, Developments in Science and Technology of Composite Materials, *Proc. Fourth European Conf. on Composite Materials* (Stuttgart), September 25–28, 1990, European Association for Composite Materials, Elsevier, ECCM-4, pp. 416–423.

39. H. J. Feng and J. J. Moore, The Effect of Pressure on the Combustion Synthesis of a Functionally Graded Material: TiB$_2$-Al$_2$O$_3$-Al Ceramic Metal Composite System, ASM Int., JMEPEG, *J. Mater. Eng. Performance*, 2(5), 645–650, 1993.

40. N. Sata and Y. Miyamoto, FGMs Via Self-Propagating High-Temperature Synthesis, *Mater. Proc. Report*, March/April 1992, pp. 4–5.

41. Functionally Gradient Material Made of Glass and Metal, JETRO, *New Tech. JAPAN*, 21(2), May 1993.

42. *Science & Technology*, May 28, 1993, p. 74.

43. P. Sarkar, X. Huang, and P. S. Nicholson, Zirconia/Alumina Functionally Gradiented Composites by Electrophoretic Techniques, *J. Am. Ceram. Soc.*, 76(4), 1055–1056, 1993.

44. D. P. Miller, J. J. Lannutti, and R. D. Noebe, Fabrication and Properties of Functionally Graded NiAl/Al$_2$O$_3$ Composites, *J. Mater. Res.*, 8(8), 2004–2013, 1993.

45. R. R. Bowman, in D.B. Miracle, D.L. Anton, and J. A. Graves (Eds.), Intermetallic Matrix Composites II, *Mater. Res. Soc. Symp. Proc.*, 273, Pittsburgh, PA, 1992.

46. Supersonic Particle Process for Making Bulk Functionally Gradient Material, JETRO, *New Tech. JAPAN*, 21(7), October 1993.

47. A. Bishop, C.-Y. Lin, M. Navaratnam et al., A Functionally Gradient Material Produced by a Powder Metallurgical Process, *J. Mater. Sci. Lett.*, 12(19), 1516–1518, 1993.
48. R. Watanabe, FGMs Via Powder Spray Forming, *Mater. Proc. Report*, March/April 1992, pp. 3–4.
49. S. Erglu, N. C. Birla, M. Demirci et al., Synthesis of Functionally Gradient NiCr-Al/MgO-ZrO$_2$ Coatings by Plasma Spray Technique, *J. Mater. Sci. Lett.*, 12(14), 1099–1102, 1993.
50. A. Bennett, *Mater. Sci. Technol.*, 2, 257, 1986.
51. N. Shimoda, FGMs Via Low Pressure Plasma Spraying, *Mater. Proc. Report*, March/April 1992, pp. 3–4.
52. T. N. McKechnie and E. H. Richardson, Continuous Spray Forming of Functionally Gradient Materials, Sandia National Labs, Albuquerque, NM, 1995, 6 p, SAND-95-2649C, C, CONF-9509182-10, 1995 National Thermal Spray Conference, Houston, TX, 11–15 Sep 1995; *NTIS Alert*, 96(12), 24, 1996.
53. T. Nagano and F. Wakai, Fabrication of Zirconia-Alumina Functionally Gradient Material by Superplastic Diffusion Bonding, Research and Development Corporation, Takatsuka-cho, Japan, *J. Mater. Sci.*, 28, 5793–5799, 1993.
54. T. Nagano, H. Kato, and F. Wakai, *J. Mater. Sci.*, 26, 4985, 1991.
55. Y. Kude, FGMs Via Chemical Vapor Deposition, *Mater. Proc. Rept.*, Nippon Oil Co., Ltd., Elsevier, March/April 1992, pp. 2–3.
56. R. Kudesia, S. E. Niedzialek, G. C. Stangle, J. W. McCauley, R. M. Spriggs, and Y. Kaieda, Design and Fabrication of TiC/NiAl Functionally Gradient Materials for Joining Applications, *Proc. 16th Ann. Conf. on Composites & Advanced Ceram. Materials*, M. Mendelson, Am. Ceram. Soc., pp. 374–383, January 7–10, 1992, Cocoa Beach, FL.
57. C. Fujiwara, S. Nagata, S. Kiyotoh et al., Functionally Gradient Materials Symp., Selected papers; *Science and Technology*, pp. 60–69, October 2, 1992, p. 163, and Fabrication Process for Actively Cooled FGM Plate, Tokyo FGM '91, in Japanese, October 8–9, 1991, pp. 235–243.
58. M. M. Stack, J. Chacon-Nava, and M. P. Jordan, Elevated-Temperature Erosion of Range of Composite Layers of Ni-Cr-based Functionally Graded Material, *Mater. Sci. Technol.*, 12(2), 171–177, 1996.
59. K. Tohgo, M. Sakaguchi, and H. Ishii, Applicability of Fracture-Mechanics in Strength Evaluation of Functionally Graded Materials, *JSME Intl. J. Ser. A — Mech. Mater. Eng.*, 39(4), 479–488, 1996.
60. K. Tohgo. and T. W. Chou, Incremental Theory of Particulate-Reinforced Composites Including Debonding Damage, *JSME Intl. J. Ser — A.*, 39(3), 389, 1996.
61. K. Tohgo and G. J. Weng, A Progressive Damage Mechanics in Particle-Reinforced Metal Matrix Composites Under High Triaxial Tension, *Trans. ASME, J. Eng. Mater. Technol.*, 116, 414, 1994.
62. K. Tohgo, N. Suzuki, and H. Ishii, Influence of Debonding Damage on a Crack Tip Field in Particulate-Reinforced Ductile-Matrix Composite, *Intl. J. Damage Mech.*, 5, 150, 1996.

63. R. C. Wetherhold, S. Seelman, and J. Z. Wang, The Use of Functionally Graded Materials to Eliminate or Control Thermal Deformation, *Comp. Sci. Technol.*, 56(9), 1099–1104, 1996.

64. N. Gupta and A. Jena, Measuring in Layers, *Ceram. Ind.*, 24–29, 2001.

65. Dental Implant Using Functionally Gradient Material, *New Tech. Jpn.*, JETRO, 20(11), 17, 1993.

66. H. Takahashi and Tohoku U., Group Develops Moisture Absorbent/Proof Functionally Gradient Material, *Sci. Technol.*, December 21, 1993, p. 3.

67. M. Jan, Ballistic Protection Application of FGMs, p. 1, May 6, 2003, www.ntu.edu.sg/sme/research/res_cer_6.htm.

68. M. I. Peters, R. U. Vaidya, R. G. Castro, J. J. Petrovic, K. J. Hollis, and D. E. Gallegos, Functionally Graded $MoSi_2$-Al_2O_3 Tubes for Temperature-Sensor Applications, *Industrial Heating*, October 2001, pp. 105–110.

69. R. G. Castro, H. Kung, K. J. Hollis, and A. H. Bartlett, 15th Intl. Thermal Spray Conf., Nice, France, May 25–29, 1998, pp. 1199–1204.

70. A. H. Bartlett, R. G. Castro, D. P. Butt, H. Kung, Z. Zurecki, and J. J. Petrovic, *Industrial Heating*, January 1996, pp. 33–36.

71. R. U. Vaidya, P. Rangaswamy, M. A. M. Bourke, and D. P. Butt, *Acta Metallurgica*, 46, 2047–2061, 1998.

72. W. D. Kingery, H.K. Bowen, D.R. Uhlmann, *Introduction to Ceramics*, Second Ed., New York: John Wiley & Sons, 1976.

73. W. A. Weibull, *J. Appl. Mech.*, 18(3), 293, 1951.

Bibliography

S. Amada, T. Munekata, and Y. Nagase, The Mechanical Structures of Bamboos in Viewpoint of Functionally Gradient and Composite-Materials, *J. Comp. Mater.*, 30(7), 800–819, 1996.

K. Atarashiya, Joining Metals to Ceramics Using FGMs, *Mater. Proc. Rept.*, Hokkaido University, March/April 1992, Elsevier, pp. 5–6.

S. B. Bhaduri and R. Radhakrishnan, Characterization of Functionally Gradient Materials in the Ti-B-Cu System, *Proc. 16th Ann. Conf. on Composites & Advanced Ceram. Materials*, M. Mendelson, Am. Ceram. Soc., pp. 392–399, January 7–10, 1992, Cocoa Beach, FL.

K. Fujii, J. Nakano, and M. Shindo, Development of Oxidation Resistant SiC/C Compositionally Gradient Materials, JAERI-M-94-001, Japan Atomic Energy Research Inst., Tokyo. *NTIS Alert*, 95(12), 12, 1995.

V. Gupta and A. K. Jena, *Advances in Filtration and Separation Technology*, 13b, 833–844, 1999.

T. Hirano, T. Yamada, and J. Teraki, *Functionally Gradient Materials Symp., Selected Papers; Sci. Technol.*, Japan, JPRS-JST-92-025, pp. 15–27, October 2, 1992, p. 163, and Fundamental Design, Multiobjective Optimization for FGM, Tokyo FGM '91, in Japanese, October 8–9, 1991, pp. 199–208.

Improved Shuttle Tile, *Aerospace Engr.*, June 1994, p. 28.

A. K. Jena and K. M. Gupta, *J. Power Sources*, 80, 46–52, 1999.

D. P. Miller, J. J. Lanutti, and R. N. Yancey, Functionally Gradient NiAl/Al$_2$O$_3$ Structures, *Proc. 16th Ann. Conf. on Composites & Advanced Ceram. Materials,* M. Mendelson, Am. Ceram. Soc., pp. 365–373, January 7–10, 1992, Cocoa Beach, FL.

X. Wang, E. A. Olevsky, and M. A. Meyers, Synthesis of Functionally Graded Nanomaterials by Electrophoretic Deposition and Microwave Sintering, SAMPE 2004 National Meeting, Session 6B, San Diego, CA, November 18, 2004.

9

Microelectromechanical Systems

Introduction

When a car collides with another car, a tiny device called an accelerometer detects the change in motion and sets off an airbag, an innovation that has saved many lives.

The accelerometer is one of the most common uses of microelectromechanical systems (MEMS), but scientists and engineers also are starting to use them in devices ranging from angioplasty pressure sensors and pacemakers to optical disk drives.

MEMS, also known as micromachines, are a relatively new technology that uses existing microelectronics manufacturing methods to create complex machines with micrometer feature sizes. MEMS devices represent a rapidly growing component of the semiconductor industry.

Many micromachines contain moving parts that are combined with integrated circuits. Like most high-tech devices, they must be made with precise dimensions and materials properties to operate properly. To help manufacturers ensure that their devices meet these exacting specifications, National Institute of Standards and Technology (NIST) scientists and engineers helped develop three ASTM International standard test methods for the thin films used to make micromachines.

The test procedures, which are the first such standards in the world, were published in *The Annual Book of ASTM International Standards* in 2003. The standards are expected to facilitate global commerce in MEMS technologies by enabling measurements that will lead to the development of more reliable and reproducible MEMS devices. The three standards provide detailed instructions for measuring thin-film dimensions and "strain," a property related to the stress in the thin film. A Web site[1] has also been created to help semiconductor manufacturers perform the complex mathematical calculations required by the new standard test methods.

What Are MEMS?

If Jonathan Swift's Lilliputians had mastered electronics, the innards of their gadgets would probably look something like this. Microelectromechanical systems squeeze by the score entire assemblies of mobile machinery-like gears, springs, and mirrors onto chips measuring just millimeters across. MEMS are key components of a growing list of sensors, and because they are built employing the same methods used to make computer chips, they are, like those chips, cheap and easy to mass-produce.

MEMS is an acronym for microelectromechanical systems: microscopic mechanical devices fabricated from semiconductors and compatible materials using photolithographic techniques. Mechanical structures small enough to be flexed over a limited range of angles are chemically etched from layered structures, where they remain suspended above a substrate. Electronic circuits on the substrate control their motion by applying voltages or currents, generating electrostatic or magnetic forces that attract part of the flexible component (see Figure 9.1). In the best-known optical MEMS devices, the moving components are mirrors that are tilted or moved vertically. Other moving optical MEMS components include microlenses and optical waveguides.

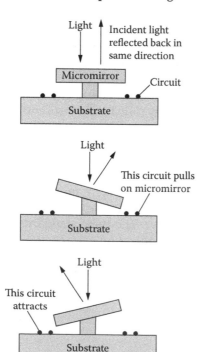

FIGURE 9.1
In a simple tilting-mirrors optical MEMS, current passing through a circuit on the substrate, or a charge accumulated on the substrate, pulls on an elevated mirror, tilting the mirror and bending the pillar that holds it.

A tad too large to qualify as nanotechnology, but still as tiny as a speck of dust, MEMS are making a big splash. Once considered a laboratory novelty, MEMS are functional micromachines that use mechanical pumps, valves, cantilevers, gears, or switches to perform physical tasks. Now they are beginning to appear in commercial applications such as consumer electronics and automotive systems. The breakthrough? Inexpensive manufacturing techniques that stamp out micron-sized mechanical assemblies in much the same way that circuits are etched on Si to create processor chips. MEMS nozzles are already used in ink-jet printers, and MEMS velocity sensors tell most new cars when it is time to trigger airbags. Untapped opportunities lie in the fields of medicine and telecommunications, where MEMS pumps could provide precision-controlled drug-delivery systems, while MEMS optical switches could reduce the cost and complexity of fiber-optic networks.

MEMS, an enabling technology for a number of systems important to the Air Force and national security, are particularly crucial for miniaturization and integration, and enhancing capabilities in spaceborne military systems. They are also very important to commercial industry. Examples include miniature satellites, actuators, gyroscopes, microwave switches, accelerometers, sensors, airflow control, aircraft turbine engines, and unattended air vehicles. MEMS are extremely small, batch-fabricated systems comprised of electrical and mechanical components on the millimeter to nanometer scale. Their principal advantage over conventional systems is an ability to significantly reduce the size, weight, power consumption, and cost of aerospace and space systems. MEMS technology and processes include lithography, bulk micromachining, surface micromachining, etching, film deposition, and packaging. Examples of structural materials used to build them include polysilicon, SiC, Si_3N_4, and diamond.

MEMS can enhance capabilities affecting flight safety and mission success. This is particularly true for onboard diagnostics and reconfiguration of microsatellites. Researchers understand that protecting them against friction, wear, suction, adhesion, and other phenomena that hinder their performance and shorten operating life is a formidable challenge.

MEMS Design

The question now is that when the floodgates open and more MEMS devices pour out into commercial industries, who will make them, and how small will they get? MEMS-based designs can produce systems on a chip, in which a transceiver, batteries, sensors, and microprocessor are all on a single component not much larger than a postage stamp. The trick is designing each miniature piece of the system and making sure they are manufacturable and durable enough to operate.

MEMS devices are surprisingly rugged and can operate for long periods of time on little power. They must be able to endure damaging internal heat buildup and withstand excessive structural loads, ambient temperature swings, and severe shock and vibration.

MEMS is a unique technology, so there will always be unique design challenges. For every commercial MEMS device that is developed, both the device itself and the process to make it need to be developed. Another challenge is that MEMS requires a number of engineering disciplines. In order to create a new MEMS device, a company needs chemical engineers, process engineers, mechanical engineers, electrical engineers, potentially fluidic engineers: a great deal of diverse expertise.

That expertise includes the ability to simulate and analyze how these devices will perform. Software vendors have even begun to specifically address MEMS development as a capability of their products.

For example, a supplier of finite-element analysis (FEA)-based simulation software has incorporated tools for simulating MEMS. The company's solution links electrostatic analysis to structural analysis with a graphical user interface that works within many CAD systems and includes an FEA model-building tool. This MEMS solution will enable engineers to design the devices that promise to make the next generation of electronic products smarter and cheaper.

With all of this engineering expertise, how small can MEMS devices get? Theoretically, there is not a size limit, but there are practical limitations. The inertial mass is essentially what gravity or acceleration is acting upon when it comes in contact with an accelerometer. If you are concerned about size, you can always make them smaller. But the larger the mass, the more sensitive the device. On the other hand, the industry is being pushed to make these things smaller.

Physics is ultimately what will define the smallest size possible and the capabilities of the materials used. On a more practical level, the applications will define how small devices need to be for each application.

Types of MEMS

One type of MEMS are sensors, called microcantilevers, which resemble rows of diving boards or spatulas, each vibrating spontaneously. They can be made so thin that 100 would fit snugly inside a human hair. The cantilever is coated on one side with a chemical that specifically binds a target molecule; say, a cancer-related protein or a plastic explosive. When that molecule sticks to it, the cantilever bends and the frequency of its vibration changes, which can be measured by bouncing a laser beam off its surface. In fact, researchers at ORNL[2] completed a sensitive handheld detector that will suck in air and search for a variety of explosives. Because the cantilever sensors are carved out with the same technology used to build computer chips, the detector should ultimately cost only tens of dollars. In airports today, the only machine as sensitive — the mass spectrometer — is too large to carry around, costs $100,000, and can't "sniff out" explosives molecules in the air.

Sensors can detect proteins associated with prostate cancer. Arrays have been built to detect markers for other cancers, heart disease, and even mutant genes.

MEMS devices known as accelerometers have been built. Although most people may not be familiar with the term *accelerometer*, this tiny machine plays a critical safety role each time someone gets behind the wheel of an automobile. Accelerometers detect movement; in their most important job, they quickly sense the sudden stop of a car during an accident and trigger inflation of protective airbags.

The accelerometers combine new and existing technologies. Integrated circuit technology widely used in the semiconductor industry is combined with a microscopic heater. Acceleration affects the convection process of gases, creating minuscule differences in temperature on both sides of the microheater. Sensors built into the chip then detect the temperature differences.

Accelerometers can be employed in a variety of ways besides their common use in airbags. For example, the global positioning system network of satellites to determine precise positions on Earth can use accelerometers to track acceleration and direction of cars or other vehicles.[3]

Smart dust is the name for wireless networks of sensors, called motes. Each mote has a chip about the size of a grain of rice that detects and records things like temperature and motion at its location. Attach it to a battery the size of an aspirin, and a mote will keep doing this for longer than a year; add a power source the size of a bottle cap, and your mote is good for a decade. Most important, the motes have minuscule radio transmitters that talk to other motes (or to a base station connected to a PC) within 100 feet or so. With a single network of 10,000 motes, the upper limit, you could cover 9 square miles — and get information about each point along the way.

Tiny motes will have a transforming effect on how we monitor the world. An experiment has been performed for the U.S. Army in which a mere eight motes were dropped from a plane and used to detect a fleet of vehicles on the ground. Homeland Security plans to start using smart dust in a pilot project to protect ports in Florida. And firms have started using motes in supermarkets to make giant refrigerators more energy-efficient.

Fabrication

UV Lasers

Ultraviolet (UV) laser micromachining of MEMS and micro-optoelectro-mechanical systems (MOEMS) is emerging as a viable, and often preferable, alternative to wet processing and other methods, thanks to its speed, accuracy, and simplicity. It can be used in applications such as die separation, where wet processing cannot be used, and is superior to mechanical cutting because of its lack of vibration. One of the key concerns of potential users of this technique is the type of byproducts or debris created by UV laser micromachining: its nature, effect, and removal procedures. Although the amount and type of debris created by UV laser micromachining (photo ablation) varies with the material being machined, there are methods of removal and remediation that are effective and support the successful application of UV lasers to MEMS manufacturing.

MEMS and MOEMS are the integration of mechanical and optical elements, sensors, actuators, and electronics on a common substrate (predominantly Si) through microfabrication technology. Although the electronics are fabricated using IC process sequences, the micromechanical components are fabricated using compatible micromachining processes that selectively remove parts of a wafer to form the mechanical and electromechanical devices. Chemical etching is one commonly used method. Others include ion beam and plasma etching, molding devices, and mechanical sawing. Increasingly, however, UV lasers are proving to be a better choice for MEMS micromachining processes (see Figure 9.2).

The emerging demands of MEMS micromachining require extremely precise, tight tolerances; high repeatability; and cost-effective processing. Short-wavelength (157 to 248 nm) excimer and UV diode-pumped solid-state (DPSS) lasers are ideal for such applications, particularly with regard to processing difficult materials such as borosilicate glass, quartz, fused silica, and sapphire. These lasers exhibit the ability to execute complex features with large-area and ganged processing capabilities and characteristic smooth cuts in applications such as precise drilling of microscopic apertures.

MEMS micromachining may require complex features, holes, cones, channels, and sample chambers of microscopic size, of uniform and consistent size, with certain essential characteristics. These may include

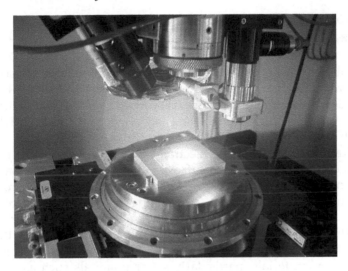

FIGURE 9.2
(Above) The shorter-wavelength resolution of UV excimer lasers, such as the one pictured here, enable smaller spot sizes for MEMS micro-machining.[4]

sharply defined features, smooth walls, and optically clear surfaces, and they must be produced with high repeatability and at production speeds sufficient to make their production economically feasible. Chemical etching, compared to laser micromachining, requires more process steps, poses environmental concerns because of chemicals used, and has materials compatibility issues.

In laser micromachining, different materials absorb laser energy differently; the greater the absorption of the material, the easier it is to machine it cleanly and consistently. Many materials can be effectively micromachined with longer-wavelength lasers (e.g., Nd:YAG); however, many materials such as certain types of glass and sapphire cannot tolerate longer wavelengths without cracking, melting, or shattering. Other materials will exhibit rough holes (caused in part by low resolution) and edges that do not meet the strict requirements of the application. Difficult materials, such as quartz and fused silica, can be effectively processed using short-wavelength (157 nm) UV lasers. Due to the high absorptivity of short-wavelength UV by the material, micromachining is crisp, precise, and repeatable. UV lasers have high to average power sources that operate at a variety of user-selectable wavelengths from 157 to 351 um. This allows processes to be optimized based on absorption. Submicron layers of materials can be removed with each laser pulse. Short UV laser wavelengths can be lithographically projected onto material with very high resolution. Even with the use of simple lenses to shape and direct the beam, micron resolution is easily achieved.

Photo Ablation

The method of materials removal with UV lasers is unique and a direct function of the laser's characteristic form and energy type. Known as photo ablation, this occurs when small volumes of materials absorb high peak power laser energy. When matter is exposed to focused UV laser light pulses, the energy of the pulse is absorbed in a thin layer of materials, typically <0.1 µm thick, due to the short wavelength of deep UV light. The high peak power of a UV laser light pulse, when absorbed into this tiny volume, results in strong electronic bond breaking in the material. The molecular fragments that result expand in a plasma plume that carries any thermal energy away from the workpiece. As a result, there is little or no damage to the material surrounding the feature produced. Each laser pulse etches a fine submicron layer of material. The ejecting material carries the heat away with it. Depth is obtained by pulsing the laser repeatedly; depth control is achieved through overall dosage control.

With many materials, particularly softer materials such as polymers, the byproducts of photo ablation are nearly vaporized and carried off in the plasma plume, with only a small fraction of the ablated material (such as C) remaining to redeposit on the workpiece. This is especially true of light micromachining operations. With drilling and complex patterns where typically a greater volume of material is removed, and if the material is particularly hard (such as fused silica and sapphire), there will be a greater volume of residue remaining. It is much finer and decreased in volume than, for example, the rough byproducts of diamond saw cutting and mechanical scribing.

Due to the small size of their moving parts and close tolerances between them, MEMS are particularly sensitive to any type of particulate contamination that could result from a micromachining process other than chemical etching. Although the ejected submicron fragment plume resulting from laser photo ablation is often of minimal concern, it is prudent to remove the material so that it will not affect the operation and reliability of the MEMS. Removal of the dust is sometimes problematic. Some MEMS cannot see a wet process, so rinsing or water removal is not possible. Vacuum and purge removal is effective when carefully combined with an airbrush or airknife, but must be balanced so as not to merely blow the dust onto other MEMS or into areas where it cannot be removed and can create reliability issues.

Protective Film Technique

For MEMS applications where the use of water can be tolerated, a water-soluble film protective technique has been developed that is applied to the device before micromachining begins. This film traps the small remnant debris and allows it to be removed with the film after micromachining is finished. Methods have been investigated to effectively apply this protective cover in a mass-production environment. Difficulties with the process include the small size of the MEMS parts, as well as their large numbers.

UV laser micromachining is typically accomplished using short-pulse and short-wavelength DPSS lasers with shorter pulse widths than are typically available. All UV laser wavelengths can be used, including 157, 193, and 248 nm.

This method has a number of advantages over wet processing. There are no chemical materials involved and fewer process steps. Material is removed in a one-step process, faster than plasma etching. Laser micromachining allows 3-D features to be created in a single step by controlling the laser exposure. Additionally, lasers produce a limited taper angle that eliminates the problem of undercutting associated with wet processing.

In UV laser processing, the shorter the UV wavelength used, the better the material will absorb energy (see Figure 9.3). This allows the operator to strip very fine and controlled layers of material off with each laser pulse. The UV laser is a very high-powered laser that can run at hundreds of Hz, up to kHz repetition rates. This gives the user precise depth control based on the absorption of the material, and most materials absorb strongly in the UV wavelength range. Additionally, the shorter the wavelength, the finer the resolution one can achieve. The ability to focus to smaller spot sizes is a key factor in MEMS micromachining capability.

FIGURE 9.3
Controlling laser exposure can create 3D features for MEMS and MOEMS applications. The features of this ink-jet nozzle, shown in a SEM photo, were micro-machined using short-wavelength UV excimer laser technology.[4]

Virtually all of the semiconductor materials — Si, GaAs, gallium nitride, sapphire, glass, the full range of ceramics, polymers (used in microfluidics) — are readily etched by UV lasers. When substrate material is photoablated by a UV laser pulse, the material, whether vaporized or reduced to a submicron-sized particulate, does not necessarily go away. If it is a finely divided powder, for example (particularly with semiconductor materials or ceramics), it may be redeposited on the surface of the substrate.

What Is Left?

With a longer-pulsed laser, the materials removed at the beginning of the pulse and at the tail end of the pulse tend to be absorbed by the plasma plume that is formed. Superheating scatters the material. With shorter laser pulses, there is less scattering of the debris field. That material is ejected out in the ablation process, supersonically, and carries away the heat of the laser pulse and ablation process with it. This allows the micromachine process to operate virtually heat-free, so that there is little or no thermal effect on surrounding material or microstructures.

The size of the debris particles depends on the material that is being ablated. With polymers, the composition of the debris can go down to the molecular level, consisting of cracked polymers, various compounds, and gases. With solids such as SiO_2, Si, or ceramics, the debris cloud can consist of constituent metals and oxides, very fine submicron particles. These types of particles are a concern for redeposition. The range of the debris field can range from 10s to 100s of microns, depending on how much material is being removed. Generally, adjacent devices are within that range, particularly when conducting a dicing operation where one is actually creating the device itself.

Debris Removal Methods

Removal of the debris created by the laser micromachining process can involve a variety of methods and techniques. One method (where wet processing or contact with water or liquids is allowed) is the use of a liquid assist with the laser, where the waste materials are actually carried or rinsed away with the liquid. With UV lasers, a liquid stream or a fine mist aimed near the target area will carry away the debris, but this has the disadvantage of requiring an increase in the number of laser pulses needed in order to etch the part. The use of CO_2 "snow" to collect debris particles is another method.

A unique approach is to use a "sacrificial" soluble coating or layer that the laser will burn through to etch the part; the ejected debris collects or settles out on this coating, which is then washed away after processing, taking the debris with it. There are a wide variety of coatings, but of course they must be compatible with the end result that one is trying to achieve.

Still another method is to process in a vacuum, where the mean free path for ejected plasma is increased. This allows the debris material easier egress from the ablation area (fewer air or gas molecules to collide with). But processing in a vacuum is complicated and impractical for most facilities and production situations with technical and equipment capitalization issues.

Where liquid assist is not or cannot be used, the most common debris removal method is to use a jet of purge gas in conjunction with a vacuum nozzle, whereby the vacuum nozzle creates a Venturi effect to draw out matter from the debris field. The purge gas is introduced into the ablation zone, with the vacuum creating a low-pressure zone that serves as an exhaust that literally scoops the debris up off the surface and carries it away through the vacuum nozzle. Many different types of purge gas can be used. However, in UV laser processing, it is more commonplace to use an inert gas such as N_2, Ar, or He rather than a reactive gas such as air or O_2. He offers the advantage of a very high ionization potential, resulting in a very long mean free path for debris ejection due to the small, light nature of He atoms and He's plasma suppression qualities. A high-pressure, aggressive flow of He, therefore, reduces the amount of debris generated to the surface of the immediate area adjacent to an ablation. The disadvantage to this flow is that it can negatively affect the laser beam in high-resolution imaging. The flow must be strictly controlled or beam distortion can result. He increases the etch rate in UV laser ablation, possibly due to a reduction in the volume of plasma in the ejection plume, because plasma will absorb the energy of the laser beam.

The larger the cut that is being ablated, the more debris is produced. Material removal must be managed so that it does not interfere with the ablation process or leave residue behind. The removal process must match the rate at which debris is being generated.

UV laser photo ablation is a viable method offering many advantages over other techniques for micromachining MEMS. The process creates debris that, depending on the material being removed and its volume, may require removal to prevent redeposition on the substrate or on adjacent parts. A number of removal technologies have proven effective, offering the user the ability to capitalize on UV laser processing's strengths to meet the ever-evolving challenges and promise of advancing MEMS technology.

Plasma-Enhanced Chemical Vapor Deposition

Many techniques are suitable for depositing tribological coatings on conventional machine elements, but the miniature nature of MEMS elements presents a tough challenge to the coatings community. Handling of individual parts is impractical, and applying coatings by such techniques as resin bonding and burnishing is not possible. Physical vapor deposition is a line-of-sight process, making it difficult to deposit material on the sidewalls where it is needed. Chemical vapor deposition of tribological coatings such as diamond, though conformal in nature, requires a high-temperature process that could potentially alter the microstructure of the base Ni alloy.

To circumvent these problems, a novel coating strategy has recently been developed at Sandi National Laboratories.[5] Coatings have been applied by the plasma-enhanced chemical vapor deposition (PECVD) technique, in which ionized species are deposited onto biased substrates, thereby limiting the substrate temperatures to below 150°C. Furthermore, instead of coating the individual parts, the entire wafer is coated before releasing the parts, but after electroplating, planarizing, and dissolving the mold material.

LIGA is a microfabrication process in which individual mechanical elements are created by deposition of material into lithographically formed molds, followed by assembling the micromachine elements into a microelectromechanical device. LIGA is an acronym for the German phrase "Lithographie, Galvanoformung, und Abformung" (lithography, electroplating, and molding).

The various steps involved in a typical LIGA process are

a) Expose

b) Develop

c) Electroform

d) Planarize

e) Release

The first step is the creation of a micromold by deep x-ray lithography via synchrotron radiation. An x-ray mask with the two-dimensional lay-out of parts is prepared. Typical masks consist of Be, Si, C, thin membranes of diamond, SiC, or Si_3N_4 (depending upon the x-ray source and the patterns), onto which layers of W or Au are patterned to act as x-ray absorbers. Currently, polymethylmethacrylate (PMMA) is the material of choice for the x-ray photoresist. It is attached to a metallized substrate (Si, glass, etc.) either by direct polymerization of methylmethacrylate or by bonding a sheet of PMMA with the MMA monomer. The exposed regions of PMMA are removed by suitable developers to create a precision micromold.

The base of the mold must contain a plating base (conduction layer) suitable for electroplating. A typical example of a plating base is a throes layer film of Ti/Cu/Ti or Ti/Au/Ti, in which the lower Ti layer provides adhesion to the substrate, and the top Ti layer provides photoresist adhesion while also protecting the electroplating seed layer of Au or Cu. The top Ti layer is removed after PMMA development and before electroplating.

Synchrotron radiation, which is highly collimated, is used to expose thick PMMA layers. This creates high-aspect-ratio straight mold walls that are all perpendicular to the plating base (for normal-incidence exposure). Several other techniques are being explored to produce micromolds without synchrotron radiation, for example, ultraviolet radiation, deep reactive ion etching, and laser ablation.

The mold is filled by conventional electrodeposition. Structures can be made from virtually any material that can be electroformed at low stress from solution. Ni and Ni alloys (Ni-Co, Ni-Fe) are the most common materials, but a much wider set of materials including Cu, Au, and Ag can be electroplated to produce LIGA parts.

By incorporating particles of different materials, such as solid lubricants or hard particles, into the plating bath, self-lubricating or dispersion-strengthened composites can be fabricated. Alternatively, a metal micromold can be fabricated as a negative of the final part. This metal mold can then be filled with ceramics, polymers, or polymer composites, further expanding the materials set for design of microsystems.

The current LIGA technology produces individual machine elements that need to be assembled into larger functional microsystems. However, vacuum diffusion bonding is being developed to produce more complex multilayer LIGA structures. Techniques for assembling the parts into LIGA MEMS utilizing robots are currently under development at Sandia.[5] It is possible that future LIGA microsystems may be fabricated by means of sacrificial layers and diffusion bonding to eliminate the need for assembly altogether, producing a fully integrated device.

One of the most effective coating application technologies is plasma-enhanced chemical vapor deposition, in which a plasma is formed from a siloxane precursor. This technique has been applied to deposit a 100-nm-thick diamond-like nanocomposite (DLN) on Ni LIGA parts. The films contain a diamond-like network of a C:H and a second network of Si:O with minimal bonding between the two networks. This structure is claimed to have significantly lower internal stresses than conventional diamond-like C, resulting in good adhesion to most substrates, with no interlayer.

In principle, the procedure described above can be applied to deposit any tribological coating. Although fabrication of systems in the LIGA technology currently requires some assembly, the advantage is that a wider choice of materials is available and that the surfaces are more accessible to coatings. By suitably modifying the conventional coatings processes, more traditional tribological coatings (hard coatings and solid lubricants) or surface treatments (carburizing, nitriding, and ion implantation) can be explored to mitigate friction and wear.

Wafer-Level Membrane Transfer Process

A process for transferring an entire wafer-level micromachined Si structure for mating with and bonding to another such structure has been devised. This process is intended especially for use in wafer-level integration of MEMS that have been fabricated on dissimilar substrates.

Unlike in some older membrane-transfer processes, there is no use of wax or epoxy during transfer. In this process, the substrate of a wafer-level structure to be transferred serves as a carrier and is etched away once the transfer has been completed. Another important feature of this process is that two wafer-level structures to be integrated with each other are In-bump-bonded together; this is advantageous in that it produces less (in comparison with other bonding techniques) stress during bonding of structures formed on two dissimilar wafers. Moreover, unlike in some older membrane-transfer processes, there is no incidental release of HF from the final structure: an advantage when In, Al, or another soft metal is used for bonding.

This process was demonstrated[6] by applying it to the joining of (1) a corrugated polycrystalline Si (polysilicon) membrane that had been fabricated by patterning and etching on an Si-on-insulator (SOI) wafer with (2) an Si substrate. The transferred membrane with underlying electrodes constitutes an electrostatic actuator array. An SOI wafer and an Si wafer (see Figure 9.4) were used as the carrier and electrode wafers, respectively. After oxidation, both wafers are patterned and etched to define a corrugation profile and electrode array, respectively. The polysilicon layer is deposited on the SOI wafer. The carrier wafer is bonded to the electrode wafer by using evaporated In bumps. Electrostatic actuators with various electrode gaps have been fabricated by this transfer technique. The gap between the transferred membrane and electrode substrate is very uniform, and Figure 9.5 depicts the finished product.

Micromachines

Researchers at Sandia National Laboratory[5] are building micromachines from amorphous diamond, second in hardness only to crystalline diamond. The material should let micromachines better resist wear and reduce stiction, a combination of friction and stickiness. Amorphous diamond micromachines are also biocompatible and can be implanted in the human body. Diamond-coated devices should last 10,000 times longer than those of polysilicon, the current material of choice for MEMS. Sandia

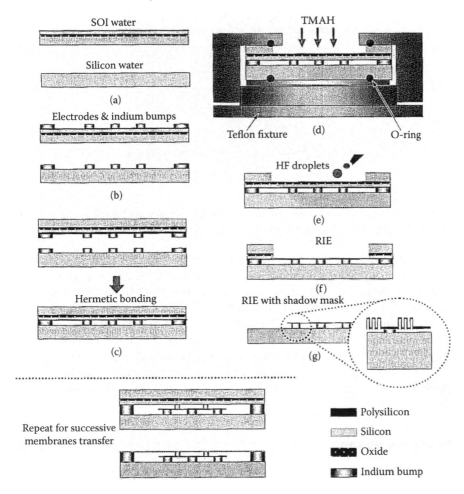

FIGURE 9.4
An outline of the process shows the key steps.

can make diamond MEMS using laser deposition in about 3 hours; annealing to prevent internal stresses takes only a few minutes.

Manufacturing MEMS

Today, the capacity for manufacturing MEMS devices is outpacing the number of devices being manufactured. Companies new to the MEMS manufacturing business are taking their cue from the semiconductor industry and looking to "fab-less" manufacturing. When companies start

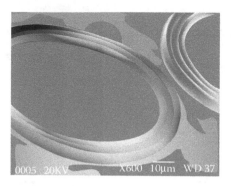

FIGURE 9.5

A corrugated polysilicon membrane, only 1 μm thick, was transferred onto a silicon substrate to form an array of electrostatic actuators. The actuators were found to function as intended.

up in the MEMS business, they are no longer building a facility and buying fabrication equipment because there is so much equipment that is not being utilized. There is so much capacity to make MEMS, and less than 10% of that capacity is being utilized.

MEMS manufacturing facilities often are categorized by the process they use. Some specialize in surface micromachining, which uses a lithography process similar to chip manufacturing, and others use bulk micromachining, which involves etching onto wafers. Other companies produce metal parts.[7]

Applications

Perhaps the most innovative and exciting applications for MEMS technology are coming not from commercial industry, but from university, government, and independent research labs. NASA has been a leader in the development of MEMS technology for missions requiring lower cost, smaller size, lower weight, and less power consumption. Nanotechnology development has been under way at a number of NASA centers, including the Center for Space Microelectronics Technology's Microdevices Laboratory at NASA's Jet Propulsion Lab in Pasadena, CA. The Center focuses on sensors, electronics, environmental and biomedical technologies, and high-performance computing devices that use micro- and nano-sized devices. Other NASA Research Centers have been developing and commercializing microsystems. These include high-temperature pressure sensors, chemical sensors, bio-MEMS, and SiC microdevices.

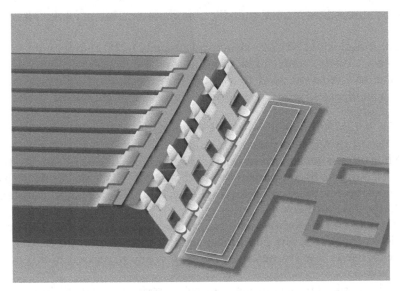

FIGURE 9.6
MEMS electrostatic lateral output motor used to test bound and mobile phase lubricants.

The future of MEMS is in the applications, according to Albert Swiecki of IntelliSense, a subsidiary of Corning.[8]

Researchers Jeffrey Zabinski and Joseph Meltzer, Jr. of the Air Force Research Laboratory,[9] working with the University of Dayton Research Institute (UDRI), successfully demonstrated Fomblin[R] Z-DOL 2000, a lubricant system with bound and mobile phases, on a miniature electrostatic lateral output motor (see Figure 9.6). They chose the motor as the test case because of its large electrode area, which provides a sufficient level of force, and because it has a number of different contact interfaces for tribological studies and ensures the relevancy of any lubrication scheme. The output motor also provides a platform for several interesting experiments. They chose Fomblin Z-DOL 2000 because it provides the proper bound and mobile phase mix after being heat-treated, and it has already proven successful as a computer hard disk lubricant.

Bound monolayer lubricants wear away quickly. The research team demonstrated that the addition of a mobile phase creates a lubricant system with a self-healing capability. This new system may hold the key to building MEMS that perform better and last longer.

Industry observers predict that MEMS will do for mechanical components what microchips did for electronics.

Some MEMS, including those applicable to biomedicine (such as prostheses), defense systems (such as sensors to detect chemical and biological weapons), and portable consumer products (power supplies, for example), require three-dimensional metallic components. This is where Ni comes in.

Electrodeposited Ni gives structural integrity to 3D components. In fact, microelectrodeposition can be thought of as the small-scale equivalent of casting and welding used to manufacture mechanical structures at the macro level, says George Whitesides, professor of chemistry at Harvard University.[10]

"The actual amount of Ni that's involved is minuscule, but the value that the Ni provides is very high," says Whitesides.

Whitesides is testing Ni for use in 3D metallic structures ranging from heat exchangers to components of small aircraft. His team combines electrodeposition with lithography, a set of techniques for pattern transfer, to build the microtrusses.

"The reason for working with Ni," Whitesides says, "is that it responds to electrochemistry, has good mechanical and corrosion properties, and is inexpensive. It is strong and cheap and easily processed in this particular style."

For example, one technique routinely produces metal features at the 1–100 μm scale. Electrodeposition then transforms the planar metallic structures into miniature 3D devices by joining the separate 2D components together (see Figure 9.7).

Ni could even be used in biomedical applications, such as implants that dispense drugs, though the metal would likely be coated to prevent possible allergic reactions. The metal's magnetic properties make it a natural choice for magnetic applications.

Alternative materials for 3D applications include welded Cu, material that has been machined out of Si, and, for biomedical purposes, stainless steel, Ti, and Au-plated materials.

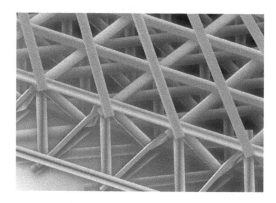

FIGURE 9.7
Pure nickel trusses on the 1-to-100-micrometer scale could be used in a wide range of miniature devices.

This development covers a range of microstructures that cannot be produced economically by conventional means.

MEMS is aiding the resulting new generation of smart sensors to not only communicate with one another to share data and configuration information, but also talk with controllers far away. These sensors include an embedded chip containing transducer electronic data sheets, or Teds, that store parameters for self-identification and self-description. Teds promise to simplify sensor setup, use, and maintenance by automatically gathering calibration data, eliminating time-consuming manual methods and potential errors. It should make the process of integrating sensors into systems as easy as plugging a mouse into a computer.

The ability to put more computing power on the sensor is eliminating the need for some ancillary electronics and software. In the past, sensors just gathered data. Now, smart sensing systems can analyze and manipulate captured data before sending it back to a computer, eliminating A/D converters and lightening the computational load on system controllers.

The steady growth of MEMS is also having more of an impact. MEMS technology has become commonplace in sensing pressure, acceleration, and altitude. But the next generation of MEMS devices will look more like microsystems than sensors.

The new paradigm is to integrate MEMS with microcontrollers and communication links into a single package. The resulting sensors can gather data, process and analyze it, and send it over a wireless link.

MEMS offer the advantage of high-volume production, lowering cost. Because they are manufactured in a clean room, mass production offsets the high capital investment. MEMS sensors are also relatively small and can squeeze into tight places where ordinary sensors cannot. This is especially important in biomedical applications where sensors may need to be implanted in the body.

MEMS-based sensors are expected to grow in areas that include medical infusion pumps, oxygen concentrators, hemodyalisis machines, and blood glucose monitors. Automotive applications will also be big for inertial sensors and tire-monitoring systems. Other areas are biometric ID systems, photonics, electronic warfare systems, and chemical biosensing, all of which are critical for homeland security and environmental monitoring.

To study the potential of MEMS, the Southwest Research Institute (SWRI) has built what it says is the first facility for developing and testing MEMS in vacuum conditions, and recent tests have yielded significant findings about how MEMS devices can be used in space applications.[11]

The cost of launching payloads into space, thousands of dollars per pound depending on the launch vehicle, makes MEMS desirable for that venue. The miniaturization of electronics, power supplies, and structures has been done for years, but miniaturization of science instruments is difficult because many require large aperture sizes to collect samples.

Though possible, miniaturizing many space instruments overall is not practical because the smaller size gives the sensor less signal, such that it receives fewer particles or photons and cannot measure the highly tenuous particle distributions or dim emissions they were intended to measure.

MEMS will enable space instruments to have large aperture sizes in a flat-panel shape that can be much thinner than current sensors. Space instruments will use arrays of many thousands of identical MEMS, which will permit an instrument that suffers failure of a small number of devices to continue to operate at nearly full sensitivity.

"With MEMS, the laws of physics are of course the same, but how they work on that scale is quite different," says David McComas, head of the SWRI MEMS facility and executive director of its Space Science and Engineering Division. "Effects that you're used to seeing in normal life — gravity and inertia — mean very little, while small electrical forces and the damping of motions in air are incredibly important."

Researchers at SWRI found that MEMS operate in a vacuum differently than they operate in normal atmosphere in two ways: The voltages required for resonant operation are much lower, and the energetic amplifications are much larger. The researchers found during testing that oscillators needed only one tenth of the voltage normally required in air.

That is significant for space applications because space MEMS might be able to run on standard low voltages of only 10–15 V, instead of units operating at much higher voltages that are heavier and more expensive to launch.

Researchers had worried that "stiction" and vacuum welding, the tendency for metal parts to bond together in vacuum conditions, could be major factors in space MEMS. Yet that has not been the case thus far.

Water vapor and air act as lubricants for MEMS surfaces that slide on or touch each other. In a vacuum, however, parts that touch lack that layer of gas between the surfaces, leading to the possibility that surfaces could exchange atoms and eventually bond.

Among the concepts targeting space applications, most, such as magnetometers and gravity gradient monitors (nano-g accelerometers), can work in sealed and potentially shielded packaging, which mitigates the concern of open exposure to space.

MEMS employ microfabrication technologies developed in recent years to build microscale mechanical structures. Such structures can have many advantages over their macroscale counterparts, including small mass and size, lower power requirements, tight dimensional control, repeatability, potential for low cost with large production volumes, and high reliability.

Researchers hope to partially restore sight in macular degeneration patients by replacing the faulty light receptors with a MEMS array. A module containing 1000 tiny MEMS electrodes would locate on each retina within the vitreous humor. The chips would connect to the micron-sized retinal nerves with spring-loaded electrodes. The electrodes make good electrical contact with minimal force, important because the retina cannot take much pressure, explains one researcher. Oddly enough, rods and cones lie beneath nerves, not above them, which makes it slightly easier to connect the arrays directly to the nerves. The tissue housing the nerves is relatively clear.

A tiny camera and RF transmitter affixed to a patient's glasses will transmit information and power to the MEMS array through a receiving antenna and interface module implanted in the eyeball. The idea is to directly stimulate some of the nerve endings to produce images good enough to read large print and distinguish between objects in a room. "We'll initially use a crude, shotgun approach that fires groups of nerves, though stimulating individual nerves is the ultimate goal," says Sandia National Labs manager Mike Daily.[12]

MEMS chips have been used in airbags since the 1990s. Today they are being built into toys, pacemakers, and projectors. Engineers are even inserting them into the earpieces of Indy racers to record the g-forces their heads endure, and they are just getting started.

Building small allows decentralized sensing and control, fast local response, and low-power requirements. Semiconductor fabrication techniques permit batch manufacturing, in turn making many new products — from the automotive airbags to ink-jet printers to disposable blood pressure sensors — cost-effective. The following examples represent just a few of the many new MEMS designs looking for a home.[13] The applications in addition to those previously mentioned or discussed include motor-speed and stop control, micromotors, trackballs and joy sticks, conveyor belts, robotics, positioning controls for video and motorized toys, and program switches for household appliances.

A nanoguitar from Cornell University has been shown as a demonstration vehicle for ways of making submicroscopic mechanical devices using techniques designed for building microelectronics.

The strings of the nanoguitar are Si bars, 150×200 nm in cross-section and ranging from 6 to 12 mm in length. The strings vibrate at frequencies 17 octaves higher than those of a real guitar. The device is "played" by hitting the strings with a focused laser beam. Vibrating strings create interference patterns in the light reflected back. The detected patterns are electronically converted to audible notes.

The ability to make tiny things vibrate at high frequencies opens the door for many potential electronics applications. For example, cell phones and other wireless devices use the oscillations of a quartz crystal to generate the carrier wave on which they transmit or to tune in incoming signals. A tiny vibrating nanorod could do the same job in less space while drawing only milliwatts of power. Supersharp filtering is another possibility. They may also detect vibrations to help locate objects or detect faint sounds to predict machinery or structure failure.

Because nanoelectromechanical systems (NEMS) can modulate light, they might make fiber-optic communications less expensive. Instead of using a laser at each end of a fiber-optic cable, a powerful laser at one end could send a beam that would be modulated and reflected back by a less-expensive NEMS device.

Researchers at Ohio State University, in collaboration with iMEDD Inc., Mountain View, CA, are developing several MEMS devices that can be implanted in humans to control the release of medication. The devices use micropore filters that are highly uniform and serve as rate-limiting membranes that supply a constant level of drug from a drug reservoir. This lets a single implant replace multiple injections. Researchers are also exploring the idea of implanting several different drugs at once and using the membrane to dispense them with variable-release rates and at different lag times. This will make the tiny MEMS implant suitable for anticancer chemo- and biological therapies.[14]

Researchers are also looking at biocapsules containing these filters. One type is specially suited for implanting clusters of cells in a way that heads off rejection by the immune system and is aimed at the treatment of type I diabetes mellitus. The transplantation of these islet cells will provide a natural means of regulating glucose levels without external insulin injections. A similar system will help patients suffering from hemophilia or neurodegenerative diseases.

Another MEMS biocapsule under development offers a way to implant enzymes in patients suffering from enzyme deficiency disorders like Canavan's disease. Attempts to replace enzymes in the body are often hampered by the rapid removal of the enzyme, by the body, following injection. The implantable biocapsule, called an enzymatic reactor, keeps the enzyme biologically available longer.

Another MEMS device precisely dispenses coatings, molten solder, and inks using sound waves. Other dispensing devices such as ink-jet printers use a piezoelectric device to generate internal pressure behind a nozzle. The acoustic microdispenser, however, is completely nozzleless. Instead, it uses acoustic radiation to form a pressure zone that can either eject nanodroplets or form low-velocity fountains on the surface of a free pool of liquid.

The system contains a transducer made from a quartz or sapphire buffer rod with a diffraction-limited lens on the ejection end and a disc of piezoelectric material bonded to the opposite end. The transducer emits bursts of ultrasonic waves focused by the lens at the liquid's surface. The pressure from the focused waves makes the liquid rise and form either a fountain or droplets. The fountain height or drop size is proportional to the acoustic wavelength emitted by the transducer. Frequencies above 100 MHz are the norm, as compared to ink jets that operate at 10 to 20 kHz. For this reason, MEMS dispensers are more akin to acoustic microscopes than ink-jet printers.

One obvious advantage of the system is that it has no nozzle to clog, so it can easily dispense molten metals. Furthermore, the device easily ejects droplets at low velocities, a feat that has proven nearly impossible for nozzle-based technologies. A near-zero velocity dispensing system makes acoustic MEMS suitable for space applications.

The system works under vacuum and at high temperatures and is capable of controlling droplet size from between 25 and 250 µm. Developed at the NASA Glenn Research Center, Cleveland, OH, the microdispensing system in fountain mode etches or plates without the need of masks and is said to precisely place solder bumps and balls with the accuracy of an ink-jet printer.[14]

MEMS serve as a miniature broadband light source providing up to 250 MW of optical power over a wavelength range of 500 to 900 nm. NASA Glenn, the Jet Propulsion Laboratory, and Lighting Innovations Institute in Cleveland are developing devices for use in optical sensors, spectrometer calibration sources, and display lighting.

The MEMS light source has a planar geometry and connects easily with fiber optics and drive electronics. It requires less electrical input power than commercial light sources and is small, rugged, and lightweight.[14]

A new generation of micromachines is coming to market. So far, commercial versions of tiny sensors and actuators built on silicon chips have performed simple tasks such as switching telecommunication signals on and off or triggering airbag deployment. The next wave of MEMS, however, should bring devices that perform a broader range of tasks cheaply and efficiently. These include advanced versions of the technology in products that could transform everything from eye surgery to cell-phone reception.

The key to these new applications is a sophisticated yet affordable fabrication process that stacks five layers of 2.5-µm-thick films onto chips with exact precision. The extra layers of tiny sensors, gears, and electronics — most commercial MEMS devices have only two or three layers — enable not only more complex machines but also more flexibility in product design.

Sophistication could soon go to work in your mobile phone. Developments have produced tunable cell phone components that sense signal fluctuations caused by changing weather conditions and distances to cellular towers and can automatically adjust a phone's circuits to compensate. That will mean fewer dropped calls, better sound quality, and longer battery life. Also being built are movable arrays of mirrors that will let patients preview the effects of certain eye surgeries by looking through an eyepiece. The arrays work by precisely filtering the light that interacts with the cornea on the basis of computer models of the surgery.

Expect to see cell phones with the tunable components in 2005 and the first eye surgery simulators in 2007. With the support of large investors and recent grants from the U.S. government, farther-out products like microsurgical tools and implantable biodevices for automatic drug delivery have been created.

Future Potential

The key to broader adoption of MEMS technology is packaging.

A package, by this definition, is a device that protects MEMS and ICs from damage and contamination. There are two basic packaging approaches. The first installs a separate package on each MEMS unit after die singulation. Die singulation means to slice individual devices (die) from the wafer from which they were deposited or etched. Packaging after singulation is mostly used for prototyping and low-volume production. Another approach better suited for high production adds the package before singulation. There are many variations on this theme.[15]

Another problem is that one thing the MEMS industry lacks today is standards. Because there are so many diverse applications for MEMS devices and technologies, composing one set of standards is a daunting, if not impossible, task. For example, how does one write standards that govern MEMS devices used in both medical and automotive applications?

The solution seems to be common manufacturing standards and separate application standards. The Semiconductor Equipment and Materials International (SEMI), the international standards group for the semiconductor manufacturing industry, has established a committee to draft MEMS manufacturing standards.

The solution might be to look at standards that have already been developed for the semiconductor industry that have a high degree of applicability to MEMS. Many of the processes in MEMS are similar to those in semiconductor manufacturing; therefore, the people designing MEMS need to know what those standards are.

The U.S. market for MEMS devices and products will increase better than 20% per year through 2006.

Fueling gains will be technological innovations resulting in lower costs and improved performance, allowing the scope of applications for MEMS to widen considerably, into such areas as telecommunications, biotechnology, and consumer electronics, among others. Particularly important are the development of standardized designs and fabrication methods — as well as packaging and testing techniques — that will promote high-volume production and the creation of economies of scale.

The best growth prospects for MEMS products through the mid/latter part of the decade will be found in the telecommunications sector, reflecting the advent of all-optical switching in fiber-optic transmission networks, and the initial penetration of radio frequency (RF) switches and other MEMS-based devices in next-generation wireless handsets. Biomedical-related markets will also grow briskly, as MEMS use is extended into such areas as implantables, nerve stimulation, glucose testing, and biotechnology (biochips, lab-on-a-chip, etc.). The better established automotive MEMS market will grow less rapidly, although significant opportunities will be found in emerging applications such as tire pressure monitoring, occupant position detection for safety systems, and telematics (high-end wireless automotive electronics). In addition, MEMS technology holds good prospects in consumer electronics applications, especially for visual/optics-based items including home theater systems, digital televisions, and high-end video games. Finally, the military/aerospace and industrial sectors will register rapid growth in MEMS demand, but remain comparatively small markets as applications tend to be more specialized.

Packaging Advancements

There are design considerations required to provide a better understanding of the unique challenges facing MEMS packaging engineers, when compared to traditional IC packaging. Automatic MEMS assembly requires optimized equipment and processes.

MEMS technology-based products continue at relatively low volumes compared to large markets such as memory or microprocessors. However, many market forecasts predict that MEMS will grow at a compound annual growth rate of 17 to 20% over the next 4 years.[16]

MEMS Component Challenges

The particular MEMS application dictates the functional performance and reliability requirements of the packaged device. Packaging technical challenges include cost, size, package stresses, electrical shielding, particles, and hermeticity.

IC packaging is mature, and approximately 30 to 95% of the whole manufacturing cost. MEMS package costs account for 70 to 90% of the device. Package stress, particle protection during manufacturing, hermeticity requirements, and lower production volumes are the primary drivers for increased cost in MEMS packaging. Material selection is critical to minimize package stress and out-gas contamination of tiny MEMS structures.

Design modeling will help designers remove redundant features in MEMS component packages to help drive cost down. However, modeling total system performance is difficult when combining the package, components, adhesives, interconnections, and possible effects on the package from board mounting, thermal conditions, etc.

Assembly Automation Design Considerations

Automated assembly using pick-and-place equipment is considered by many to be the most flexible technique for creating a total MEMS package. Pick-and-place machines allow designers to choose components from multiple sources and attach them together in a single package. Although capping can be accomplished with wafer-to-wafer bonding and semiconductor processing techniques, these components are still relatively delicate and need to be handled with care.

Design areas to consider when assembling a MEMS package include component presentation to the equipment; fiducial marks for the package and MEMS device; MEMS no-touch zones and special pick tool requirements; material compatibility with the component and application; cleanliness of the process; placement accuracy and alignment algorithms (global or relative placement); attachment methods and materials; single level or die stacking; flip chip or circuit up; and wire bond.

Component Presentation to the Equipment. MEMS components can be presented to the equipment in a variety of formats such as Waffle Pack, Gel Pak, Tape and Reel, and Wafer. The package may be lead-frame, strip, panel, or individual packages. Depending on the production volume and manufacturing strategy, assembly cells may stand alone with operators feeding material by hand or be fully automated material handling systems with inline integration of upstream and downstream equipment processes.

Fiducial Marks for the Package and MEMS Device. Physical features or marks that are recognizable by the machine's vision processing systems is critical for robust automated assembly. The location, size, shape, and materials used for fiducial marks should be reviewed early in the design process. Fiducial marks are ideally created during the same process step as the features that are being aligned inside the package. Package fiducial marks are typical larger and less repeatable when compared to MEMS component fiducial marks. Alignment accuracy of the MEMS to package will depend on both the package and MEMS fiducial marks.

MEMS No-Touch Zones and Special Pick Tool Requirements. Many MEMS have sensitive areas that are damaged if touched by a pick tool. Custom pick tools can be designed to avoid the sensitive areas of the MEMS. In some cases, two- or four-sided collets are used to pick from the very edges of the MEMS device. The pick-and-place equipment must have the ability to align these custom pick tools over the die. Flat-bottom tools may also be used if the tool can be positioned on a durable area of the MEMS.

Material Compatibility with the Component and Application. The materials are chosen for chemical, mechanical stress, electrical, and optical compatibility. The package, MEMS, and attach materials form a complex system that can affect the performance of the device adversely when exposed to the manufacturing processing and environmental conditions. Optical MEMS have created some of the most difficult challenges for the packaging community.

Cleanliness of the Process. Assembly, joining, and aging should not generate contamination that will adversely affect the device. Attach materials and joining or curing processes must be chosen carefully to consider the affect on the device during manufacturing processes.

Placement Accuracy and Alignment Algorithms. Global placement relies on one set of package fiducial marks to which all components are placed. Relative placement algorithms may place some of the components relative to previously placed components. Relative placement algorithms will visually reference a component in the package and then place the next component relative to the found component location.

Single Level or Die Stacking. There is a growing trend to deliver more functional performance per square centimeter of package area. Manufacturers are now stacking dies on top of each other where possible to save area. Die stacking requires relative placement and possibly adhesive dispensing on the same platform if stacking in a single pass through the assembly machine.

Flip Chip or Circuit Up/Wire Bond. The choice between flip chip or circuit up/wire bond affects many of the areas discussed above and must considered carefully for cost and reliability.

Future Work

MEMS-based products cover a broad range of applications (see Table 9.1). MEMS-based packages require a variety of different form factors and package styles. As such, equipment that serves MEMS assembly will also need flexibility to handle multiple assembly criteria.

TABLE 9.1

Uses in Today's Market

Segment	Application
Automotive	Pressure Sensors (HVAC, fuel injection, MAP/BAG, gearbox, tire)
	Flow sensor
	Accelerometers (air bag, suspension)
	Gyroscopes (anti-rollover, navigation, ABS dynamic control)
Aeronautics, space, and military	Temperature sensors
	Chemical sensors (gas, analysis nose)
	Active flight control surfaces
Medical and biomedical	Pressure (blood, intravascular)
	Micro-nozzle injection systems
	Micro-fluidic sensors
	DNA testing (gene probes)
	Implantable (pacemakers)
	Hearing aids
Information technology	Data storage (read/write heads)
	Displays
	Video projectors (mirrors)
	Ink-jet print heads
Telecom	Switches (RF/optical)
	Variable optical amplifiers
	Optical add/drop multiplexes
	Tunable lasers
	Inductors
	Resonators
	Millimetric wave sensors
Consumer	Accelerometers (toys)
	Gyroscopes (image stabilizer)

References

1. www.ceel.nist.gov/812/test-structures.
2. P. J. O'Rourke, Beyond the Sixth Sense, *Time*, January 12, 2004, pp. 40–45.
3. *NIST Tech Beat*, March/April 2001, p. 3.
4. J. P. Sercel, UV Lasers for MEMS & MOEMS Micro-Machining, *Photonics Tech Briefs*, January 2004, p. 2a.
5. S. Prasad, T. Christenson, and M. Rugger, Tribological Coatings for LIG-AMEMS, *AM&P*, December 2002, pp. 30–33.
6. E.-H. Yang and D. Wiberg, Wafer-Level Membrane — Transfer process for Fabricating MEMS, *NASA Tech Briefs*, January 2003, pp. 58–60.
7. C. R. Forest, The Future of MEMS: Big Expectations for Small Products, *NASA Tech Briefs*, August 2002, pp. 25–28.
8. J. M. Jackson, MEMS: Smaller Is the Next BIG Thing, *NASA Tech Briefs*, August 2001, pp. 16–20.

9. J. S. Zabinski and P. Meltzer, Jr., Monolayer Protective Coatings, *AFRL Technol. Horizons*, June 2003, pp. 34–35.
10. V. H. Heffernan, Nickel in Nanotechnology, *Nickel*, 15(4), 11, 2000.
11. B. Rosenberg, MEMS the Word in Space Instruments, *Av. Wk. Sp. Technol.*, March 24, 2003, p. 71.
12. L. Kren, Helping Blind People See, *Machine Design*, October 24, 2002, p. 50.
13. D. Bak, J. Ogando, and D. Normile, Small Parts, Big Potential, *Design News*, January 21, 2002, pp. 60–64.
14. R. Khol, No More Needles: Implantable Microdevice Dispenses Medication, *Machine Design*, April 6, 2000, pp. 128–130.
15. T. Glenn and S. Webster, Packaging Microscopic Machines, *Machine Design*, December 7, 2000, pp. 126–130.
16. D. D. Evans, Jr., Advances in MEMS Packaging, *Advanced Packaging*, April 2004, pp. 19–21.

Bibliography

J. Hecht, Optical MEMS Are More than Just Switches, *Laser Focus World*, September 2003, pp. 95–98.

J. P. Sercel, Ultraviolet Laser-Based MOEMS and MEMS Micromachining: An Alternative to Wet Processing, *Advanced Packaging*, April 2004, pp. 29–31.

Ultrahigh Density Scalable Digital Control of Microelectromechanical Systems, p. 1, http://jazz.nist.gov/atpcf/prjbriefs/prjbrief.cfm?ProjectNumber=00-00-5352.

P. Vettiger and G. Binnig, The Nanodrive Project, *Scientific American*, January 2003, pp. 47–53.

D. Wiberg, S. Vargo, V. White, et al., Miniature Gas-Turbine Power Generator, *NASA Tech Briefs*, January 2003, p. 56.

10

Fuel Cells

Introduction

The future of power generation has already begun. One hundred years ago, energy meant steam engines and so on. There was plenty to go around. Electricity was still largely a novelty. Things changed, and by midcentury most of the industrialized world were primarily running on electrical energy produced by hydroelectric power. Hydropower is now essentially working at its limit, and Mother Nature can cripple it with periods of drought. Crude oil is not an infinite resource. Then there are issues of pollution from coal and crude coal and oil. There is tremendous interest in distributed power generation, which conserves resources from climate and is environmental friendly.

Today, we are still producing energy pretty much as we did 50 years ago. The picture has changed. Increasing populations and industrialization increase the demand for energy every year, but the ability to meet demand is increasingly questionable. The age of fossil-fueled power plants is drawing to a close with dwindling oil reserves and environmental concerns. In the United States, for example, the 2001 energy crisis mingled flawed energy policies with energy distribution glitches. The Enron scandal revealed a questionable energy policy manipulation as it did corporate mismanagement. It would appear that with every type of traditional source of energy, humans have found ways to complicate the problem through mismanagement.

Humans are not likely to change, but it appears that the methods and patterns of energy production are going to be required and we shall need more and more power; power that is nonpolluting, affordable, plentiful, and adaptable to our ever-increasing and mobile applications. Hydrocarbons are not the future; they are being exhausted and their byproduct pollutes the planet. The handwriting is on the wall. Now what? Fuel-cell (FC) commercialization opportunities are focused in several large-scale areas: repowering, central power plants, industrial generators, and commercial/residential generators, to name a few.

We are entering the age of renewable natural power sources — and the fuel cell. Fuel-cell technologies are found in some stationary power, residential power, and portable power applications. Future applications appear to be in bus engines and car engines.

Which fuel-cell technology will benefit first and where?

Fuel cells offer outstanding efficiency, negligible emissions, and inherent simplicity. These characteristics, together with massive market potential in the power generation, automotive, and portable power sectors, place FCs amongst the most compelling of the emerging energy conversion technologies, They are now seen as front-runners in key business areas, such as zero- and low-emission vehicular powertrains and distributed generation markets. New energy sources will appear over the next decade as advances are made in new generations of alternative energy and batteries. Fuel cells will likely be available in large sizes for tanks and small sizes for soldiers.

What Is a Fuel Cell?

A fuel cell is an electrochemical device that converts H_2 fuel into electricity without using combustion. The only significant byproducts are heat and water vapor, making this an environmentally ideal means of generating electricity.

There are myriad uses for the technology, from providing primary or backup power for buildings and plants, to powering mobile phones, laptop computers, and other portable electronics. But the most publicized use of fuel cells is in automobiles, driven in part by strong government endorsement in the United States, the EU, and Japan.

Although development of this industry is moving slowly, the growth potential is significant: The fuel-cell market could reach $35 billion by 2013, according to a report by market researcher Allied Business Intelligence (ABI, Oyster Bay, NY). Growth is dependent on technology improvements and the increased availability of methanol and H_2, fuels used by fuel cells in the automotive and portable electronics fields, respectively. By 2006, new products using fuel cells are expected to deliver performance gains and sell in larger numbers[1] (see Figure 10.1).

There are, however, considerable performance and cost issues to overcome. ABI predicts automotive fuel-cell penetration will be limited by strict technological challenges, in addition to cost concerns, infrastructure challenges, and lack of certainty over fuel choice in the near term.

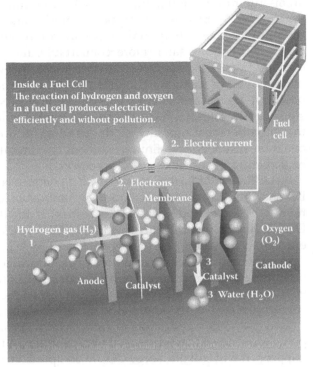

Inside a Fuel Cell
The reaction of hydrogen and oxygen in a fuel cell produces electricity efficiently and without pollution.

Fuel cell

2. Electric current

1. **Hydrogen** enters the anode of the fuel cell, where a catalyst separates the molecule into hydrogen ions and electrons.

2. Electrons

Membrane

Hydrogen gas (H$_2$)

1

Oxygen (O$_2$)

3

Cathode

Catalyst

Anode   Catalyst

3 Water (H$_2$O)

2. **Electrons** cannot pass through the special membrane of the cell and must take an external route, creating electricity.

3. **The hydrogen ions** continue through the membrane, combining with oxygen ions on the other side, creating only small amounts of heat and water.

FIGURE 10.1
Fuel cell.

Major Types and Operations

The advent of the automobile led to dramatic alterations in people's way of life as well as the global economy — transformations that no one expected at the time. The ever increasing availability of economical personal transportation remade the world into a more accessible place while spawning a complex industrial infrastructure that shaped modern society.

Now another revolution could be sparked by automotive technology: one fueled by H$_2$ rather than petroleum. Fuel cells — which cleave H$_2$

atoms into protons and electrons that drive electric motors while emitting nothing worse than water vapor — could make the automobile much more environmentally friendly. Not only could cars become cleaner, they could also become safer, more comfortable, more personalized — and even perhaps less expensive. Further, these fuel-cell vehicles could be instrumental in motivating a shift toward a "greener" energy economy based on H_2. As that occurs, energy use and production could change significantly. Thus, H_2 fuel-cell cars and trucks could help ensure a future in which personal mobility — the freedom to travel independently — could be sustained indefinitely, without compromising the environment or depleting the earth's natural resources.

A confluence of factors makes the big change seem increasingly likely. For one, the petroleum-fueled internal-combustion engine (ICE), as highly refined, reliable, and economical as it is, is finally reaching its limits. Despite steady improvements, today's ICE vehicles are only 20 to 25% efficient in converting the energy content of fuels into drive-wheel power. And although the U.S. auto industry has cut exhaust emissions substantially since the unregulated 1960s (hydrocarbons dropped by 99%, CO by 96%, and nitrogen oxides by 95%), the continued production of CO_2 causes concern because of its potential to change the planet's climate.

Even with the application of new technologies, the efficiency of the petroleum-fueled ICE is expected to plateau around 30% — and whatever happens, it will still discharge CO_2. In comparison, the H_2 fuel-cell vehicle is nearly twice as efficient, so it will require just half the fuel energy. Of even more significance, fuel cells emit only water and heat as byproducts. Finally, H_2 gas can be extracted from various fuels and energy sources, such as natural gas, ethanol, water (via electrolysis using electricity), and eventually renewable energy systems. Realizing this potential, an impressive roster of automotive companies are making a sustained effort to develop fuel-cell vehicles, including Daimler Chrysler, Ford, General Motors, Honda, PSA Peugeot-Citroën, Renault-Nissan, and Toyota.

There are several key fuel-cell technologies, including proton exchange membrane (PEM), solid oxide (SOFC), molten carbonate, alkaline, phosphoric acid, and metal-air power cells (Zn-Air and Al-Air). The key difference in the various types is temperature, which ranges from less than 100°C in PEM, to 190°C in phosphoric acid, to 650°C in molten carbonate, to 1000°C in SOFC.

Proton Exchange Membrane

The fuel-cell concept most associated with the automotive industry is called PEI\d, for proton exchange membrane, shown schematically in Figure 10.2. The PEM fuel cell is favored for its low-temperature (<100°C)

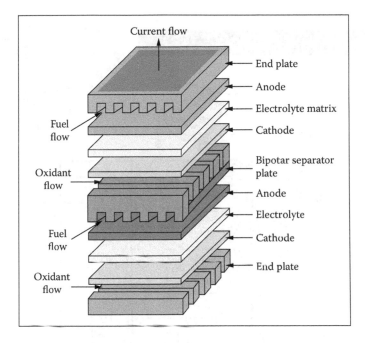

FIGURE 10.2
A schematic of a proton exchange membrane. The complexity of the design and high number of plates present unique joining challenges.

and low-pressure (<100 psi) operating conditions. Alternating layers of "flow field plates" or "bipolar plates" and membranes are usually bolted into a stack, which then comprises the fuel cell. The plates that surround the active membrane area are comprised of a pair of thin, internally cooled metal sheets, and dozens or even hundreds of them may be required in a fuel-cell stack (see Figure 10.3).

These metal sheets must be sealed along their edges to provide a leak-tight gas manifold. Anode and cathode gases cannot cross over, and leak rates must be maintained at very low levels. This requirement means there are hundreds of inches of seal area in the average fuel-cell stack. The seal issue is exacerbated by the use of metals chosen for low coefficient of expansion or corrosion properties, not for weldability. Hence, manufacturing costs can be very high due to rework and repair for leaks.

If fuel cells of this sort come to fruition in the automotive industry, someone will need to be able to produce hundreds of millions of them per year. Now, that's a manufacturing challenge worthy of an early start. Joining processes now must be evaluated and selected to process the components.

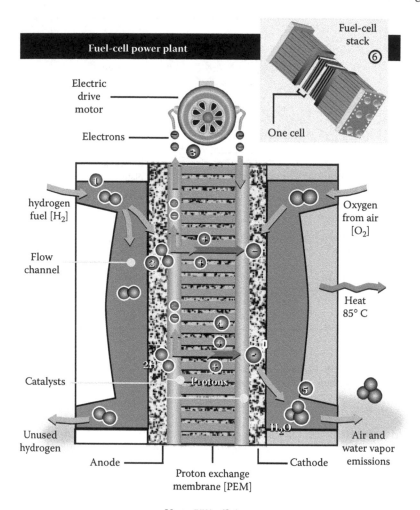

Up to 55% efficiency

FIGURE 10.3

Electrochemistry vs. combustion: A proton exchange membrane [PEM] fuel cell comprises two thin, porous electrodes, an anode and a cathode, separated by a polymer membrane electrolyte that passes only protons. Catalysis coat one side of each electrode. After hydrogen enters [1], the anode catalyst splits it into electrons and protons [2]. The electrons travel off to power a drive motor [3], while the protons migrate through the membrane [4] to the cathode. Its catalyst combines the protons with returning electrons and oxygen from the air to form water [5]. Cells can be stacked to provide higher voltages [6].[4]

The components holding the most promise for plastics, experts say, are bipolar plates, end plates, plate assemblies, manifolds, and peripheral system components used in fuel-cell stacks. Bipolar plates can be fabricated from thermoplastic or thermoset resins with electrically conductive fillers and have elastomeric seals pressed in place, overmolded, or dispensed onto them for plate/seal assemblies.[2,3]

Other components include crossovers, collectors, water balancers, plastic plates with needles, and various types of membranes:

- Ceramic membranes
- Glass membranes
- Hetropolyacid membranes
- High-conductivity membranes
- Metallized biocellulosic membranes

In order to advance the technology of PEMs, various processes are being developed and utilized to fabricate parts. Development and work suggests integrated injection-molded plates made of glass-reinforced Fortron PPS, which can provide significant cost and weight savings in PEM fuel cells. By integrating the end plate and adjacent insulating plate into one functional unit, Fortron PPS may reduce cost and weight by as much as 90% compared to current fuel-cell prototypes.

Solid Oxide Fuel Cell

Basic Science behind SOFCs

SOFCs are based on Nernst's century-old observation that solid-state electrolytes will produce an open-circuit voltage:

$$E = \frac{RT}{4F} \ln \left\{ \frac{PO_2 \text{ (oxidant)}}{PO_2 \text{ (fuel)}} \right\}$$

where R is the gas constant, T is the cell temperature, F is the Faraday constant, and PO_2 is the O_2 partial pressure. The most commonly used SOFC electrolyte is ZrO_2. The electrolyte must be fully dense so that the fuel and oxidant gases do not mix, which by the above equation would reduce or eliminate the electrical potential. The anode, generally a Ni-yttria zirconia cermet, and the cathode, generally a doped $LaMnO_3$, are both porous so that fuel and oxidant can rapidly ionize, react and diffuse into or out of the surface of the electrolyte, and react as required. Conduction in the electrolyte will depend upon the concentration of O_2 vacancies, which will depend upon the level of doping and the temperature.

ZzO_2 SOFCs generally operate at 1000°C. ZrO_2 is typically doped with ~10% Y_2O_3, which provides stabilization of the cubic phase and a good

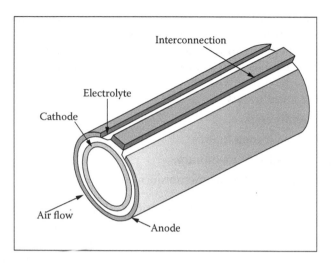

FIGURE 10.4

A schematic of a solid oxide fuel cell. EWI is aligning staff, equipment, and lab resources to meet the materials joining needs of fuel cell manufacturers.

trade-off between cost and conductivity. To minimize internal cell resistances, the electrolyte is generally made as thin as possible (~40–50 µ).

Individual fuel cells are connected into stacks to yield systems with desired voltages and power ratings. Individual cells may be connected in series, parallel, or a combination. Interconnects provide electrical contact from the anode of one cell to the cathode of the next cell in the stack.[5] The most common interconnect material for ZrO_2 SOFCs is doped $LaCrO_3$.

This type of cell is based on SOFC, shown schematically in Figure 10.4. Other SOFC configurations include

- Tubular
- Closed end ceramic tubes
- Flat plate design
- Monolithic

They are very different from PEM fuel cells. The SOFC cells operate at very high temperatures (>800°C) and are therefore composed of ceramics and high-temperature superalloys. This cell produces electricity by electrochemically combining the fuel and oxidant gases across an ionically conducting oxide membrane (Figure 10.5). To build up a useful voltage, a number of cells or PENs (positive cathode/electrolyte/negative anode) are electrically connected in series in a stack through bipolar plates that are also known as "interconnects." A key challenge is the search for the right material for interconnects.[6]

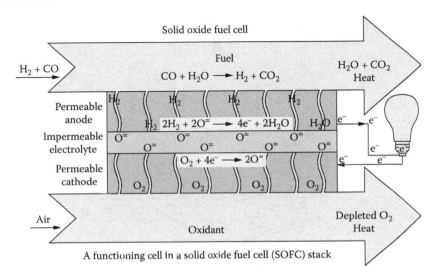

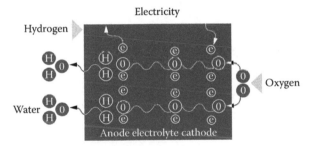

FIGURE 10.5
A solid oxide fuel cell is a solid state energy-conversion device in which conventional fuels produce electricity. Hydrogen from the fuel passes over the anode. Oxygen from the air passes over the cathode, gaining electrons at the interface with the electrolyte. These oxygen ions pass through the electrolyte and react with the hydrogen from the anode to produce electricity, water, and heat.[6]

Figure 10.6 shows a schematic of the repeat unit for a planar stack, which is expected to be a mechanically robust, high power-density, and cost-effective design. In the stack (Figure 10.7), the interconnect is simultaneously exposed to both an oxidizing environment on the cathode side and a reducing environment on the anode side for thousands of hours at temperatures of 700 to ~1000°C. The oxidizing environment is air, and the reducing environment consists of fuels such as H_2 or natural gas. Other anticipated conditions include water vapor in both environments and sulfide impurities in the fuel.

The interconnect must be stable toward any sealing materials with which it is in contact, under hundreds of thermal cycles. It must also be

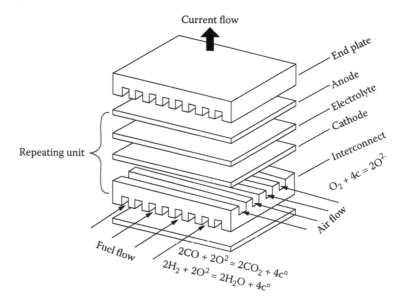

FIGURE 10.6
This schematic shows the layers in a solid oxide fuel cell.[6]

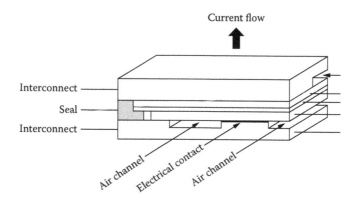

FIGURE 10.7
The interconnect and the repeat unit for a planar stack, which is expected to be a mechanically robust, high power-density, and cost-effective design.[6]

chemically compatible with electrical contact materials for minimizing interfacial contact resistance or the electrode materials. Therefore, the interconnect materials should possess the following properties:

- Good surface stability: Interconnects must have resistance to oxidation and corrosion in both cathodic (oxidizing) and anodic (reducing) atmospheres.

- Thermal expansion: The coefficient of thermal expansion (CTE) must match the ceramic PEN and other adjacent components, all of which typically have a CTE in the range of 10.5 to ~12.0 × 10^{-6} K^{-1}.

- High electrical conductivity: Both the bulk material and in-situ formed oxide scales must provide high electrical conductivity.

- Reliability and durability: Satisfactory bulk and interfacial mechanical/thermomechanical reliability and durability must be provided at the SOFC operating temperatures.

- Compatibility: Good compatibility with materials in seals and electrical contacts is required.

Until recently, the leading candidate material for the interconnect was doped $LaCrO_3$, a ceramic material that can easily withstand the traditional 1000°C operating temperature. However, several issues have prevented its acceptance, including the high cost of raw materials and fabrication, difficulties in producing high-density chromite parts at reasonable sintering temperatures, and the tendency of the chromite interconnect to partially reduce at the interface between the fuel gas and the interconnect, causing the component to warp and the peripheral seal to break.

Fortunately, these challenges may have been rendered moot by the recent trend in developing lower-temperature designs. These are also more cost-effective cells, and they feature electrolytes that are anode-supported and several microns thin, or new electrolytes with improved conductivity. They make it feasible for $LaCrO_3$ to be replaced by pure metals or alloys as the interconnect materials. Compared with doped $LaCrO_3$, metals or alloys offer significantly lower raw material and fabrication costs.

Considering the property requirements for the SOFC interconnect, potential candidates are oxidation-/heat-resistant alloys that demonstrate overall oxidation resistance at elevated temperatures. The oxidation-/heat-resistant alloys of interest may include Ni, Fe, and Co-base superalloys; Cr-base alloys; and the stainless steels. All typically contain Cr or Al, which can provide oxidation resistance by forming oxide scales of Cr_2O_3 and Al_2O_3, respectively.

If Cr_2O_3 is the continuous, dominant phase in the oxide scale, the alloy is designated a chromia former, while an alloy with Al_2O_3 as the continuous, dominant phase in the scale is classified as an Al_2O_3 former. For the chromia formers, enough Cr must be present in the alloys to form a continuous oxide scale and to effectively provide oxidation resistance. The

Al content ought to be controlled at a minimum to avoid formation of a continuous Al_2O_3 layer, because it would act as an insulator and produce an unacceptable electrical-resistance scale in the semiconducting chromia scales.

For Ni- and Fe-based superalloys and stainless steels, a minimum Cr content of about 18 wt% is recommended, with optimum content over 20% for long-term exposures to combat corrosion and prevent possible depletion of Cr in the matrix. For Co-base superalloys, 22% should be considered the minimum.

To maximize scale electrical conductivity, it is beneficial to have the Al (or Si) content as low as possible, but a small amount (less than 1%) of Al or Si is added in many traditional alloy compositions to enhance resistance to oxidation and corrosion. For Al_2O_3-forming alloys, the Al content should be at least 4 wt% in order to form a continuous Al_2O_3 layer for optimum protection.

Common Al_2O_3-forming alloys, such as Fecralloys, have an Al content in the range of 4.5 to 5.5 wt% and also contain substantial Cr (over 15%) to prevent internal oxidation and depletion. Traditional alloy compositions that satisfy the previously described composition criteria (Co-base alloys not included) are classified into several groups, which are schematically shown in Figure 10.8. Properties that are considered relevant to SOFC application are listed in Table 10.1.

The selection and suitability of a heat-resistant alloy is closely correlated to the specific stack design under consideration. Among different groups of alloys, the ferritic stainless and Cr-base compositions, which are characterized by a body-centered cubic (BCC) matrix structure, offer better

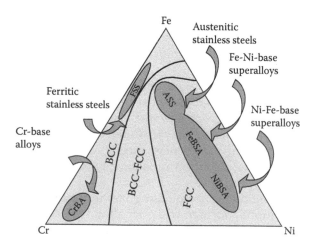

FIGURE 10.8
These alloys are being considered for SOFC interconnect materials.[6]

TABLE 10.1

Comparison of Key Properties of Different Alloy Groups for SOFC Applications[6]

Alloys	Matrix Structure	CTE $\times 10^{-6} \cdot K^{-1}$ (RT-800°C)	Oxidation Resistance	Mechanical Strengths	Manufacturability	Cost
CrBA	BCC	11.0–12.5	Good	High	Difficult	Very expensive
Ferritic stainless steel	BCC	11.5–14.0	Good	Low	Fairly readily	Inexpensive
Austenitic stainless steel	FCC	18.0–20.0	Good	Fairly high	Readily	Inexpensive
FeBSA	FCC	15.0–20.0	Good	High	Readily	Fairly expensive
NiBSA	FCC	14.0–19.0	Good	High	Readily	Expensive

thermal expansion match to the PEN structure than the Ni-, Fe-, or Co-based superalloys or austenitic stainless steels, all of which have a face-centered cubic (FCC) matrix structure.

Al_2O_3-forming alloys may be suitable for stack designs in which interconnect components do not carry electricity, or in which the electrically insulating Al_2O_3 scale can be excluded from the electrical path. Al_2O_3-forming alloys possess a much better surface stability against oxidation and corrosion than Cr_2O_3-forming compositions and also offer higher mechanical strength than other ferritic structures. For example, Fecralloys typically demonstrate a scale growth rate several orders of magnitude lower than the Cr_2O_3-forming alloys and have a yield strength over 550 MPa at room temperature, compared with around 301 MPa for Fe-Cr ferritic alloys.

Considering these advantages, some SOFC developers have tried to find applications in SOFC stacks for the Al_2O_3-forming alloys, especially those with a ferritic structure, such as Fecralloys, that have a good CTE match to the PEN. For instance, the interconnect can be constructed with Fecralloy as the matrix, and then it can be cladded with a Cr_2O_3 former at the contact with the cathode, thus reducing the interfacial resistance between interconnect and electrodes. Fecralloys also are suitable for the frame, which functions as a support to the PEN in some SOFC stacks.

In a NASA-sponsored program at the Texas Center for Superconductivity and Advanced Materials (TcSAM) at the University of Houston,[7] researchers have found a way to build SOFCs that operate at half the temperature of current designs.

The devices operate at 500°C, which the researchers hope could make this kind of fuel cell less expensive to manufacture and easier to fuel. This comes about partly because exotic materials and elaborate heat-dissipation systems can be eliminated.

The key to the advance was making the heart of the fuel cell: the sheet of electrolyte that controls the flow of electrically charged ions out of a thin film only 1 µm thick. The thinness cuts down internal resistance to electric current, and the electrolyte is created by epitaxy, the deposition of one layer of atoms at a time.

The thin-film variety being developed at TcSAM can run on methanol or gasoline now and switch to pure H_2 as it becomes available.

Contrasting the previously described off-the-shelf SOFCs of today have electrolyte layers 100 µm thick or more (a micron is one thousandth of a millimeter). The thinness cuts down internal resistance to electric current, so it is possible to get comparable power output at much lower operating temperatures.

To make this ultrathin layer, one doesn't simply shave down a chunk of bulk material until it is thin enough. Instead, you grow the electrolyte atom by atom by epitaxy. The thin films in these fuel cells are about 1000 atoms thick.

Squeezing out the same power at half the temperature creates a domino effect of cost savings. For one, cheaper materials can be used to build them, rather than the expensive heat-tolerant ceramics and high-strength steels demanded by 1000° fuel cells. And the automobiles and personal electronics that could use these fuel cells can also forgo exotic materials and elaborate heat-dissipation systems, lowering manufacturing costs. All of this tips the scales of economic feasibility in the right direction.

The National Institute of Standards and Technology (NIST) has funded a 3-year project to design and demonstrate an affordable SOFC system using a multistage concept to achieve ultrahigh (70%) fuel-to-electricity efficiency in system sizes starting at 1 kW with Technology Management Inc.[8] Multiple systems could be combined for higher power outputs with the same performance advantages. The key innovation will be the demonstration of "progressive oxidation" stacks, which maximizes efficiency by approaching 100% electrochemical fuel utilization (compared to about 75% in conventional systems). The overall challenge is to optimize high fuel use and high cell voltages with affordable costs and reliability, and to develop large, high-temperature seals to prevent efficiency-sapping stack leakage. Once developed, the stacks will be incorporated into a relatively simple, low-cost fuel-cell system design that could reduce electricity costs 20% below targets for current SOFC designs. In addition, the very low system emissions make it among the "greenest" of all fossil-fuel-powered technologies for making electricity.

A 250 kW SOFC demonstration power plant using more than 600 kg of Ni went into operation recently (see Figure 10.9). This single SOFC consists of a vertical ceramic tube, closed on the bottom end, composed of concentric layers: an inner cathode, an outer anode, and an electrolyte between. The three layers include the following materials:

- A doped lanthanum manganese oxide cathode, $LaMnO_3$, on the inside surface
- An yttrium-stabilized zirconia electrolyte, YSZ, in the middle
- A cermet of Ni metal and YSZ, as an anode, on the tube's outside

The cell is 1.7 m long with an inside diameter of 2.2 cm. Process air is injected through an Al_2O_3 tube concentric with the cell and it flows down to the bottom closed end and back up the annular space between the two tubes in Figure 10.9. O_2 in the air is ionized as it flows along the inner cathodic surface of the cell, the O_2 ions pass through the intermediate electrolyte layer and react, at the outer anodic surface, with H_2 and CO, to produce H_2O, CO_2 and electricity. The H_2 and CO are generated from "reformed" natural gas, supplied as the fuel to the plant, and are directed along the outer surface of the tube.

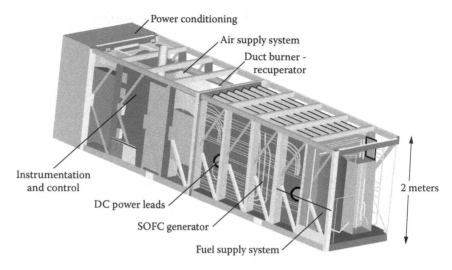

FIGURE 10.9
Isometric cutaway of 250 KW SOFC.

A single SOFC generates about 150 W of power at about 0.65 Vs. For a power plant producing kilowatts of power, many individual cells must be connected in an array of cell bundles, with 24 cells in each bundle. For the 250 kW plant, for example, the complete cell "stack" consists of 2300 cells, arranged in 96 bundles. All these cells are interconnected electrically by Ni "felts" of electrolytic-grade Ni. The stack, in turn, is connected to the output of the generator by DC bus bars of N02200 Ni and to electrical contact plates, also of N02200 Ni.

Tests have shown that modification of chemical compositions can increase the strengths and fracture toughness of SOFC electrolytes. Heretofore, these solid electrolytes have been made of YSZ, which is highly conductive for O_2 ions at high temperatures, as needed for operation of fuel cells. Unfortunately, YSZ has a high CTE, low resistance to thermal shock, low fracture toughness, and low mechanical strength. The lack of strength and toughness are especially problematic for fabrication of thin SOFC electrolyte membranes needed for contemplated aeronautical, automotive, and stationary power-generation applications.

The modifications of chemical composition that lead to increased strength and fracture toughness consist of the addition of Al_2O_3 to the basic YSZ formulations. Techniques for processing of PSZ/Al_2O_3 composites containing as much as 30 mole% of Al_2O_3 have been developed. The composite panels fabricated by these techniques have been found to be dense and free of cracks. The only material phases detected in these composites has been cubic ZrO_2 and Al_2O_3: This finding signifies that no undesired chemical reactions between the constituents occurred during processing at elevated temperatures.

The flexural strengths and fracture toughness of the various ZrO_2-Al_2O_3 composites were measured in air at room temperature as well as at a temperature of 1000°C (a typical SOFC operating temperature). The measurements showed that both flexural strength and fracture toughness increased with increasing Al_2O_3 content at both temperatures. In addition, the modulus of elasticity and the thermal conductivity were found to increase and the density to decrease with increasing Al_2O_3 content. The O_2-ion conductivity at 1000°C was found to be unchanged by the addition of Al_2O_3.[9]

Molten Carbonate Fuel Cell (MCFC)

A fixed-station fuel cell consists of a sandwich of an anode, made of porous Ni strip and a cathode made of NiO strip, separated by a ceramic-based matrix layer. A carbonate electrolyte, soaked into the matrix layer, facilitates the electrochemical reaction between the anode and the cathode.

The Ni strip used for both the anode and cathode is made using a powder metallurgy technique. A slurry of micron-sized Ni powder is evenly distributed on a moving belt that then passes through a sintering furnace and through rolls that compact and increase the density of the powder strip. This process is called tape casting.

Ni is the ideal material for this, as it is a good conductor of heat and is resistant to corrosion. It is not consumed in the process.

A hydrocarbon, such as natural gas, enters the fuel cell and is chemically dissociated, or reformed, by the catalyst so that H_2 feeds the anode and air (O_2) enters the cathode. The reaction splits the fuel into ions and electrons. The electrons move from the anode by means of a conventional Cu bus bar external circuit, while the ions move through the electrolyte to produce CO_2 and H_2O byproducts. The heat generated (in the 370°C range) can be captured and used.

A single fuel cell generates only a small amount of electricity and measures 1.2 m by 0.7 m by 0.63 cm thick. However, by stacking 350 to 400 fuel cells in a module, that module will produce 250 kW of electricity. Multiple modules can be packaged into larger units for 1 to 2 MW of power.

Efficiency of electricity generated by this type of fuel cell is about 50% and can be as high as 80% with cogeneration use of the byproduct heat.

Phosphoric Acid Fuel Cell (PAFC)

Besides PEMs, SOFCs, and MCFCs, there are PAFCs as denoted by their electrolytes; Table 10.2 provides the characteristics of each family of fuel

TABLE 10.2

Fuel Cell Families and Characteristics[5]

	PAFC	MCFC	PEM	SOFC
Operating temperature (°C)	190	650	80	1000
Fuels	H2, R*	H2, R*, CO	H2, R*	H2, R*, CH4, CO
Internal reforming	No	Yes	No	Yes
Efficiency (HHV)	~40%–50%	~50%–60%	~40%–50%	~45%–55%

* Reformate.

cells. Because the PEM fuel cells operate at low temperature, are compact, and are believed to be manufacturable at an acceptable cost, they are the fuel cell of choice for automotive use.

At first glance, the table might appear to suggest that SOFCs have little to differentiate themselves from the competing technologies. But they are extremely durable, have the widest range of fuel capability, and, perhaps most importantly, run at high temperature. The high temperature of operation means that all gases passing through the system will exit at a higher temperature than for any competing fuel-cell technology. Thus, as the high-temperature component for a combined cycle stationary power system, SOFCs will provide the highest efficiency to a turbine-based bottoming cycle.

A graphitized phenolic thermoset bipolar plate compound for use in high-temperature (160°C) PEM fuel-cell environments that operate using phosphoric acid has been developed recently.

The phenolic material can stand up to this operating temperature and a phosphoric acid environment, while the standard vinyl ester cannot.

Other Fuel Cells

Metal-air power cells have been experimentally developed using Zn-air and Al-air.

Researchers at the University of Massachusetts, Amherst,[10] have discovered a microorganism — Rhodoferax ferrireducens — capable of stable, long-term electricity production by oxidizing carbohydrates. The organism transfers electrons directly onto an electrode as it metabolizes sugar into electricity, producing CO_2 as a byproduct.

Professor Derek Lovley said, "In the past, [microbial fuel cells] have converted 10% or less of the available electrons, and we're up over 80%. And previous attempts to convert carbohydrates to electricity have required an electron shuttle, or mediator, which is typically toxic to humans. This organism doesn't require a mediator because it attaches directly to the surface of the electrode."

"That's one of the big advances," he said. "People have done it without a mediator before, but their recovery of energy was less than 1%. And not having to use toxic elements is an obvious advantage in creating electricity. In the end, the electrons in the fuel cell are transferred to O_2, so what we are really doing is putting a wire in between the microbe and the O_2 and harvesting this electron flow that otherwise would just go directly to O_2."

Because sugars are a substantial component of many types of waste and carbohydrate-rich crops, they could become economical alternatives to fossil fuels in the production of electricity. In theory, this method would allow a cup of sugar to power a 60 W light bulb for 17 hours; however, the device does need improvement before it can be used commercially.

In another investigation,[11] a fuel cell based on catalysts sourced from enzymes found naturally in the environment has been reported by researchers from Oxford University's chemistry department. The enzyme catalysts compare favorably with Pt catalysts currently used in fuel cells that convert H_2 and O_2 into electricity. It is hoped that the biological fuel-cell invention would allow cheap, robust, and clean energy production in medium-scale domestic energy applications.

According to Professor Fraser Armstrong, "The technology is immensely developable. We are at the tip of a very big iceberg, but there is still much to do before this generation of enzyme-based fuel cells becomes commercially viable."

A class of developmental membrane electrolyte materials for methanol/air and H_2/air fuel cells is exemplified by a composite of (1) a melt-processable polymer [in particular, poly(vinylidene fluoride) (PVDF)] and (2) a solid proton conductor (in particular, cesium hydrogen sulfate). In comparison with previously tested membrane electrolyte materials, these developmental materials offer potential advantages of improved performance, lower cost, and greater amenability to manufacturing of fuel cells.

A principal limitation on the utility of the other tested membrane electrolyte materials is that they must be hydrated to be able to conduct protons. This requirement translates to a maximum allowable operating temperature of about 90°C, and the presence of H_2O in the polymer matrices undesirably gives rise to high permeability by methanol. It would be desirable to reduce permeability by methanol to increase cell performance and fuel-utilization efficiency, and it would be desirable to operate fuel cells at temperatures as high as 140°C to increase their tolerance to CO from reformate streams. Therefore, what are needed are membrane materials that conduct protons in the absence of H_2O.

In a composite material of the type undergoing development, the polymer serves as a matrix to support the solid proton conductor. In cesium hydrogen sulfate, proton conduction occurs by a mechanism that does not depend on H_2O. At room temperature, the protons are in a bound state, and so there is little or no proton conduction. However, as the temperature rises past 130°C and toward a value between 135 and 145°C, the cesium hydrogen sulfate undergoes a phase transition to a state in which the H_2 ions have a significant amount of mobility; that is, the material becomes a proton conductor. The conductivity can be as high as $0.1^{-1}\Omega cm^{-1}$ — of the order of the conductivities of previously tested membrane electrolyte materials. These membranes would be used to fabricate membrane/electrolyte assemblies for testing in future fuel cells.[12]

Finally, polymer electrolyte fuel cells (PEFCs) are a promising power source for commercial application — except for a variety of fuel-related issues. They run best on H_2, but H_2 is expensive and difficult to transport and store. More readily available fuels such as gasoline or methane (natural gas) can be used, but only by using a fuel processor to "reform" the fuel into an H_2-enriched gas stream. The reformer is costly and must incorporate a cleanup step to eliminate reaction byproducts (CO in particular) that poison the PEFCs Pt catalyst. One solution is to design a catalyst-coated membrane (CCM) for the PEFC that is much more tolerant of CO. The reformer would not need CO-scrubbing steps and could be made smaller and at a much lower cost. Several companies have proposed to develop a novel high-temperature material for a CCM and a new continuous manufacturing process to produce the membranes and electrode assemblies in high volumes at low cost. One proposed CCM will operate at much higher temperatures (above 100°C) than current systems, use less Pt, and tolerate CO levels as much as 1000 times higher than conventional PEFCs. In addition, unlike conventional low-temperature fuel-cell membranes, they would require much simpler water-management systems. Novel methods are being developed to make CCMs for portable power applications and for stationary sources, including an aerosol-assisted chemical vapor deposition (CVD) process to deposit very thin, highly porous Pt films on the electrodes.

Recently, a prototype of a direct methanol fuel cell (DMFC) was unveiled to power digital audio players and wireless headsets for mobile phones. The new fuel cell outputs 100 mW of power, weighs 8.5 grams, and can power an MP3 music player for as long as 20 h on a single 2 cc charge of highly concentrated methanol.

Applications in Transportation

Fuel cells are not new. Astronauts have been using them for power aboard spacecraft since the 1960s. Soon, perhaps, they'll be just as common on Earth: powering cars, trucks, laptop computers, cell phones, etc.

Because fuel cells can produce plenty of electric power while emitting only pure H_2O as exhaust, they are also so clean that astronauts actually drink the H_2O produced by fuel cells on the space shuttle.

In recent years the interest in bringing this environmentally friendly technology to market has become intense. But there are problems: You cannot "fill 'er up" with H_2 at most corner gas stations. And fuel cell-based cars and computers arc still relatively expensive. These obstacles have relegated fuel cells to a small number of demo vehicles and some specialty uses, such as power aboard the space shuttle and backup power for hospitals and airports.

Government-sponsored research monies have helped tackle and overcome these obstacles with the hope to make the various kinds of fuel cells both cheaper to manufacture and easier to fuel.

What is lacking is a fueling infrastructure for H_2, and regulations that will allow its use across the nation are also being addressed.

Automotive

All the large and major automotive motor companies in the world have either experimental, demonstration, or production-type vehicles in some stages of fuel-cell development programs.

H_2-powered Ford Focus vehicles will soon be cruising the streets of Vancouver. They are part of a 3-year collaborative development program to test fuel-cell vehicles in the real world.

The vehicles use a Canadian-made Ballard fuel-cell engine and a Dynetek compressed H_2 storage tank (see Figure 10.10). A regenerative

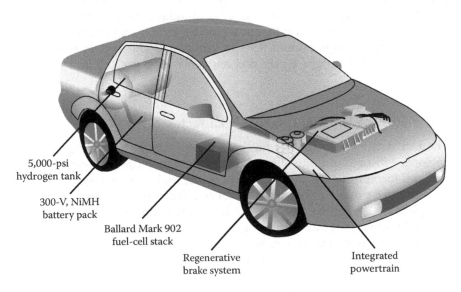

5,000-psi hydrogen tank

300-V, NiMH battery pack

Ballard Mark 902 fuel-cell stack

Regenerative brake system

Integrated powertrain

FIGURE 10.10
Test fuel cell vehicle.

system captures energy spent braking, and an advanced Ni-metal hydride battery handles energy storage.

One manufacturer has produced a fuel-cell system whose H_2 storage tank is located in the center of the vehicle, enabling the body height to be lowered to 1.25 m, while another has produced a city commuter car that combines a Nissan newly developed Li-ion battery with a fuel-cell stack; a third (Jeep) has designed and built a three-passenger fuel-cell car that utilizes drive-by-wire technology; and another, Suzuki, a has six-passenger van that also uses drive-by-wire technology.

A Japanese manufacturer has developed a compact next-generation fuel cell stack that delivers higher performance with increased range and fuel efficiency, and is designed to operate at temperatures as low as $-20°C$. Cold weather operation is one of the most significant technical barriers to the mass-market application of fuel-cell technology.

General Motors Corp. (GM) aims to sell 1 million units of its fuel-cell vehicles by 2020, according to their management personnel.

Motorcycles, Scooters, and Water Taxis

Yamaha Motor Company[13] has developed a fuel cell for motorcycles, becoming the world's first major motorcycle maker to create such a device. The fuel cell is powered by methanol and measures about 40 cm^2.

Yamaha[13] is also developing fuel cells for 50-cc-class motor scooters. A prototype reached 40 kph (about 25 mph), with an output of 500 W, which is generally equal to the performance of standard 50-cc scooters.

An electric fuel-cell water taxi powered by an H_2 fuel system is successfully undergoing operational tests in California. The system is installed in a Duffy 22-passenger water taxi and will serve the public 10 to 12 h daily.

Locomotives, Submarines, Ferries, and Yachts

The U.S. Army National Automotive Center (NAC)[14] is funding an international consortium to develop and demonstrate a 109-metric-ton, 1 MW fuel cell locomotive for military and commercial railway applications.

In another project in Nevada, a fuel cell mine locomotive successfully completed a testing program and will continue service for further evaluation. Another program[15] running concurrently was the operation of a coal mine with methane gas, where the purpose of the project was to demonstrate the feasibility and advantages of methane from coal mines to generate electricity cleanly and efficiently.

Germany's largest shipbuilder, Howaldtswerke Deutsche Werft,[16] has unveiled what it described as the world's first submarine to be powered by fuel-cell technology. The submarine, the first of four in the company's new 212A class being built for the German navy, was launched from the company's shipyard for testing in the Baltic Sea. The H_2-powered fuel cell vessel is expected to be evaluated for deep-water testing. The technology is designed to cut out noise and emissions.

The Greek navy has taken PEM fuel cell modules for integration into the existing propulsion systems of three class 209 submarines, with the intention of helping increase the vessels' submersed range to that of new ships.

A series of tests and observations were demonstrated to determine a fuel cell system's response to simulated marine power loads. The tests demonstrated that the system met all regulations as they apply to rotating, alternating, and direct current generators. This work was done for the San Francisco Bay Area Water Transit Authority,[17] which is actively developing a fuel cell-powered ferry that will serve Treasure Island, CA.

Finally, a 12-m-long yacht is operating in Germany with a PEM fuel cell propulsion system.

Trucks and Buses

A Japanese truck manufacturer is selling a zero-emission H_2-powered pickup truck running on a fuel cell/battery/hybrid engine. The Clean Urban pickup truck was modified by Nissan and is running with 6 kW PEM fuel cells. The truck accelerates from 0 to 60 mph in 10 seconds, can achieve speeds of up to 75 mph, and can travel 250 miles within city environments. Although this range is similar to conventional vehicles, this initial version of the Clean Urban pickup will be limited to 60 miles on highways in order to keep the price under $100,000.[18]

The U.S. Army Tank-automotive and Armaments Command (TACOM) National Automotive Center (NAC) recently introduced a class-eight Freightliner truck fitted with a methanol-fueled fuel cell auxiliary power unit (APU).[19]

The 5 kW APU, which includes a fuel cell stack, will provide electricity for on-board demands and external devices, including computers, satellite dishes, and three-dimensional mapping systems in military trucks.

General Motors and the U.S. Army unveiled a diesel hybrid military pickup truck equipped with a 5 kW PEM regenerative fuel-cell APU, designed and built by Hydrogenics.[20] When the vehicle is driven, the PEM electrolyzer uses electricity provided by the diesel engine to separate H_2O into H_2 and O_2, with the H_2 stored for future use. The H_2O byproduct of the fuel cell's use for producing electricity is stored and used to repeat the cycle.

Daimler Chrysler[21,22] has spearheaded a fuel cell project to field-test 30 buses in a fleet of electric vehicles powered by the technology, which is essentially hydrolysis in reverse. In fuel cells, ambient O_2 reacts with H_2 (stored as compressed gas), producing mere H_2O, thus achieving the California-mandated goal of zero smog and zero greenhouse emissions.

A bus powered by a UTC Fuel Cells (UTCFC) 75 kW power plant has become the first fuel cell hybrid bus to enter passenger service in California. The 30-ft ThunderPower bus is operated in the greater Palm Springs desert resort area. The bus travels 100 miles a day on its route.[23,24]

Heavy-duty fuel cell engines have been provided in Mercedes-Benz buses for the public transport system in Perth, Western Australia.[25]

The transit authority of Amsterdam has begun operating H_2-fueled buses on two routes in the city. The buses extract H_2 from nine roof-mounted H_2O tanks, each capable of supplying enough fuel to power the bus for 250 km (155 mi.).[25]

An electric fuel Zn-air fuel cell zero-emission bus is being tested in New York State with new ultracapacitor and controllers. Metro of Washington, D.C., is testing a fuel-cell-powered bus with methanol to fuel the bus and then put into revenue service.

Auxiliary Power Units

Fuel cell-based APUs[26] have been combined with reformer technology and H_2 purification systems to allow the use of propane or methanol fuel storage systems. Applications include auxiliary power for recreational vehicles, marine, and highway trucks.

The U.S. Army TACOM[27] has taken a regenerative fuel cell APU for deployment and testing on Army vehicle platforms.

The PEM electrolyzer module will recharge the H_2 supply while the vehicle engine is operating. This will supply the H_2 storage subsystem with sufficient fuel to operate the fuel cell auxiliary power system for up to 5 h with a load of 3 kW average, and peak demand of 5 kW.

Applications in Commercial Sector

Cell Phones/Computers

Japan's largest wireless phone carrier plans to introduce cell phones powered by miniature fuel cells, which run on H_2 or methanol. Additionally, they will also show up as expensive add-ons for high-end laptops.

An Si chip maker in Geneva, Switzerland,[28] is developing tiny fuel cells for cell phones that can be refilled with fuel as needed. They are grappling with the fact that the phone consumes 300 mA of current at 3.6 V, and its power source cannot occupy more than 12 cm^3. The output current of a fuel cell is directly related to the common surface area between the electrodes and membrane and, to obtain 300 mA using conventional fuel cells, would require a surface area of around 60 cm^2, too large for a cell phone (see Figures 10.11 and 10.12).

To overcome this issue, one approach is to implement the fuel cell as a 3D structure containing thousands of microchannels that maximize the contact area between the gases, catalysts, and electrodes. In the same vein, a research team has fabricated a nanoporous layer of Si containing millions

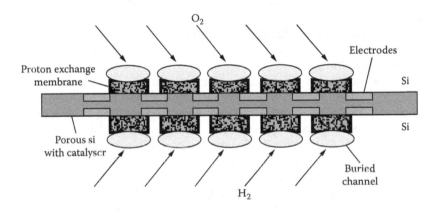

FIGURE 10.11
Mini fuel cells for power cell phones.

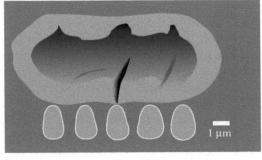

Small pores

FIGURE 10.12
Micro fuel cell with small pores.[28]

of pores, each measuring a few nanometers in diameter. The small pores give the layer a large effective surface area, boosting the catalysts' efficiency.

Prototypes of fuel cells will find their way into notebook computers in the next year.

Direct-methanol fuel cells weigh about 900 g and can power a system for 5 h with one methanol fuel fill-up. This fuel cell has an output density of 50 mW/cm^2 and consists of a polymer electrolyte sandwiched by C electrodes with catalyst particles. Engineers use single-walled C nanohorn aggregates, a type of C nanotube, to adsorb the Pt-catalyst particles. The Pt-catalyst particles are supported on the C nanohorns at about 2 nm, which is claimed to be much finer than possible with conventional C-black materials. The prototype cell has an average output of 14 W (24 W maximum) at 12 V and a fuel concentration of about 10% methanol.

The prototype cell by Toshiba generates 12 W average and 20 W maximum. It powers a notebook computer for nearly 5 h on a single, 50 cc cartridge of methanol.

Methanol generates power most efficiently in a fuel cell when its concentration in H_2O is 3 to 6%. However, thin methanol liquid has little energy, so a bulky fuel tank has been necessary for long electricity generation. Engineers solved this issue by making use of H_2O, a byproduct of the cell itself. A dilution device makes it possible to store highly concentrated methanol and reduce the tank size and weight to 33 × 65 × 35 mm and 72 g in the case of a 50 cc methanol cartridge. There is also a 100 cc cartridge, which is claimed to power a PC for 10 h. Engineers are now working down the size of the fuel cells and cartridges.

PEFCs discussed previously had been initially aimed for use in recharging two-way radios and cell phones, and fuel cells for stationary power units for homes. PEFCs could supplant battery technologies made overseas for portable applications and reduce dependence on less efficient and more polluting fossil fuel plants for distributed power generation. The new technology also could be used in automotive applications. In addition, the new deposition processes could be used for a wide variety of applications in the electronics, display, and communications industries.[29]

The leading rechargeable, portable power source is currently the Li-ion battery, a market that is dominated by Asian firms. An alternate power source that could offer greater energy density as well as "instant recharging" — if significant technical and cost barriers can be overcome — is liquid-feed direct methanol fuel cells. This type of fuel cell converts a simple alcohol, methanol, into electricity and H_2O by reacting it with the O_2 from air. Liquid-feed fuel cells are well suited to low-power consumer electronic devices. In a 3-year project, practical DMFCs are being developed as low-cost membrane electrode assemblies (MEAs). The MEA is the most critical component of a fuel cell, and the source of 40% of the manufacturing cost. The proposed MEAs will reduce the amount of expensive precious metals (such as Pt) used as catalysts. They will be deposited with a high-volume, digital ink-jet printing process in a three-dimensional pattern to minimize fuel loss and other inefficiencies.

The key innovation is the formulation of, and developing printing technologies for, ink-jet printing of functional inorganic materials such as electrocatalysts. A spray-based process is used to make electrocatalyst powders that can be deposited by ink-jet printing. These powders will have increased surface area compared to catalysts prepared by traditional methods, increasing catalytic activity (and reducing the precious metal content) by more than an order of magnitude. If successfully developed,[29] the new technology would displace or augment Li-ion batteries and give the United States a share of the portable power source market. It also would reduce dependence on foreign sources of precious metals and offer an environmentally superior alternative to batteries (DMFCs can be recycled). The first market will be chargers for batteries in two-way radios and cell phones. In addition, the printing process developed in this project could be used in making other products, such as batteries, supercapacitors, and sensors, and also in the electronics industry.

Cell phones used to be just phones, but now they are organizers, Web browsers, cameras, and music players, too. As the power-hungry functions pile up, running phones on batteries gets trickier. Cell-phone makers have been hoping micro fuel cells — tiny versions of the devices touted as a source of clean power for cars — would be the answer.

But problems with size and power have stalled early methanol-based versions in academic and industry labs. So Renew Power, a spin-off of the University of Illinois at Urbana-Champaign, is turning to formic acid, the chemical sprayed by black ants on the attack.[30]

In spring 2004, engineers demonstrated that they could power a cell phone with a fuel cell that actually fits in a phone.

Several venture capital firms hope to fund companies to begin pilot production of fuel cells for mobile handsets by early 2006.

Of all the markets for micro fuel cells, handsets are the big prize. The potential is huge. Nearly 500 million handsets were sold in 2003, and in 2005 predicted sales are expected to exceed 650 milliion handsets. But companies hoping to capture the huge market face huge challenges.

Only 3 years ago, industry watchers expected companies such as Samsung to sell methanol-powered cell phones by 2003. But problems with dynamic power demands, and with operating temperature and size, have stymied their development, and none have made it from lab to store.

Using formic acid as the fuel can solve all these problems. For starters, although formic acid yields less electricity per molecule than methanol, it can deliver energy more rapidly than a comparable methanol fuel cell, getting around the dynamic-power issue. Formic-acid fuel cells also operate just fine at room temperature; to achieve the same level of power, methanol fuel cells must work at a scalding 60°C and up: impractical for a device used near the face. And methanol must be used in a diluted form in fuel cells; handling it requires tiny pumps and pipes that increase the devices' size. Formic acid doesn't face that problem, so Renew's fuel cells require no moving parts, just a replaceable fuel cartridge.

A single cartridge should power a cell phone at least twice as long as the typical Li-ion battery used today, according to specialists. Some experts, however, are skeptical that formic acid will beat methanol into portable electronics. Two of the fuel's biggest problems are availability and toxicity. Methanol is a more plentiful fuel than formic acid. You can buy it on the drugstore shelf, and you can wash your hands in methanol; you can't in formic acid, because the concentrated acid would burn the skin.

Predictably, Renew Power says it is well on its way to solving such problems. The real competition is powerful, established Li-ion batteries. But as cell phones grow more complex, the need for more power in a small space should eventually push the industry toward fuel cells.

A Casio prototype laptop[31] has been run for more than 20 hours on one refueling of its fuel cell power supply.

DC Rotary Motors

The designers and manufacturers of fuel cells are continually striving to improve the efficiency of their products. A design challenge facing the engineer relates to reducing the power demands of the many electrical systems that support fuel cell operation.

Blowers, compressors, and pumps are necessary for fuel pump operation, but are considered parasitic electrical loads. Because they require power from the fuel cell itself, they impact overall system efficiency. To be truly effective in fuel cell applications, the motors used to drive the pumps must be lightweight, compact, and as efficient as possible in order to provide more energy for the intended application.

A patented electromotive coil technology eliminates the Fe core and wire windings of conventional motors. The result is an innovative motor with excellent power conversion efficiencies, higher power-to-weight ratios, and lighter weight. The technology is applicable to brush as well as brushless DC motors.

High inertia gives the motor the ability to operate over a larger dynamic range, which is always a desirable feature in fuel cell applications. The high inertia thus gives the motor the ability to push through the power stoke at lower speeds The motor also features smooth, quiet operation and low heat generation.

Office Buildings/Residential Homes/Hotels

A commercial fuel cell installed in a Japanese office building set a world record for continuous use. The fuel cell operated for 9500 hours before being shut down for scheduled inspections. The cell uses an electrochemical process to directly convert chemical energy into electricity and hot H_2O. The chemical energy normally comes from H_2 contained in natural gas and, for automotive applications, would come from regular gasoline. However, the cells do not burn the gas, so they operate virtually pollution-free. The units are part of more than 100 fuel cells in operation around the world.

A simulated blackout in a high-rise building showed the prowess of an uninterruptible power supply (UPS) powered by fuel cell modules. A 3 KVA unit provided power to a rack of computer networking equipment and a large plasma-screen display, while a portable fuel-cell generator kept a small refrigerator, lighting, and other small electronics operating during the hour-long demonstration.

The fuel cells will operate as long as they have fuel, so the duration of the backup depends on the size and quantity of fuel storage tanks. Typical range is 4 to 24 hours. H_2 is fed to the fuel cells via an adapter kit from standard 31-in.-high Q- or 51-in.-high K-style pressurized gas cylinders. The fuel cell modules produce a DC voltage that is then bused to the UPS.

Each 1 kW fuel cell module has a ±24-VDC output and produces 40 A of current. The UPS 1 uses 3.5 in. of rack space and weighs 80.4 lb. It includes connectors to accept DC input from the fuel cell modules and receptacles. Commercial production is slated for 2005.

Mitsubishi Heavy Industries, Ltd.[32] has announced plans to begin offering a commercial building cogeneration system comprising a 50 kW SOFC and a gas turbine by next year. The cogeneration system, which is expected to be priced at about several hundred million yen, is designed for use in such large commercial buildings as offices, hospitals, and hotels.

Residential fuel cells sound almost too good to be true. Take a hydrocarbon fuel such as natural gas, use a catalyst to extract H_2 from it, react the H_2 with air, and you have a home power plant!

As the H_2 and the O_2 in the air combine, they produce electricity. The primary "waste products" of the whole process are water and heat. But that's not all. The "waste" heat can be captured to provide space or H_2O heating for the home.

Residential fuel-cell systems can produce about 5 kW of power or 120 kW hours of energy a day: more than enough to operate the average household. But a lack of performance data on how well fuel cells work under different conditions is one of several factors slowing marketplace acceptance of the new technology.

Researchers at the National Institute of Standards and Technology[33] have launched an effort to supply the needed information. They are studying how changing electrical and heating demands, outside temperatures, humidity, and power systems affect the efficiency of fuel cells made by different manufacturers.

Ultimately, consumers will be able to use NIST-developed performance ratings to understand the financial costs and benefits of fuel cells operated in specific geographic and climate conditions, at different times of the year, and for different purposes such as heating or electricity generation.

Two 200 kW fuel cell systems have been installed in Sheraton hotels in northern New Jersey.

Under the New Jersey Clean Energy Program (NJCEP),[34] Merck and Company, Inc. has installed a UTC fuel cell at its research facility. The fuel cell was installed under a statewide initiative to promote the use of clean and renewable energy technologies. The 200 kW fuel cell was installed because the NJCEP rewards businesses in the state that install new energy-generating equipment utilizing alternative technologies, and the company receives rebates from federal and other state programs for using alternative technologies.

Communications Equipment

Micro fuel cell systems for portable military communications equipment have been developed by MTI Micro Fuel Cells, Inc. and Harris Corp.[35] These DMFC system power-pack prototypes are for use with tactical handheld radios.

Forklifts

A consortium of U.S. and Canadian corporations[36] have joined together to develop, demonstrate, and precommercialize fuel cell-powered forklifts. The project involves outfitting two Class-1 forklifts with 10 kW fuel cell propulsion systems and metal hydride H_2 storage systems; developing refueling capabilities; gathering market research; and demonstrating the forklifts to industrial end users.

Power Plant

Yale University[37] has installed a fuel-cell power plant that will provide heat and power for the university's Environmental Sciences Building. Electricity from the unit will provide approximately 25% of the building's electricity needs, with the heat being used primarily to maintain tight temperature and humidity controls at the storage facility, where rare bones and artifacts are kept and preserved.

Medical Implants

Microfluidic fuel cells developed by researchers at Brown University[38] may make it practical to devise long-running medical devices, such as implants that monitor glucose levels in diabetics. The Brown fuel cells do not require an ion-conducting membrane or selective catalysts at the electrodes to separate fuel-containing fluids. Instead, the cells exploit the fact that fluids do not mix under certain conditions. They take advantage of how fuels flow in small channels: They do not mix, which means one can keep fuels separated without a membrane.

These cells work in tandem to provide power under pulsating conditions that mimic blood flow in the body. Until now, fuel-cell makers have fallen short in their efforts to produce a membraneless device that didn't short-circuit under pulsed flow. One of the microfluidic cells features a branched channel, which encloses six electrodes. This cell is suitable for generating electrical power under conditions of pulsed flow, and the design of the device makes possible the delivery of power to a chip as a result of changes in the concentration of a fuel, such as glucose. This power feedback is a necessary component in an imbedded sensor for diabetes.

Power Supply Equipment

A smart fuel cell (SFC)[39] has been put into a product for a portable stand-alone power supply powered by an exchangeable methanol cartridge.

The system is already used by a large variety of business users in fields like traffic systems, environmental sensors, and camping and outdoor equipment, and it is now available commercially to other industrial customers.

This first commercial SFC is equipped with an exchangeable 2.5 L methanol cartridge, which yields 2500 W electrical energy. Power output is 25 W continuous and up to 80 W peak. The system operates in a wide temperature range and is thus well suited for a broad range of applications including outdoor situations in rough climates. Campers appreciate being independent from the grid. Sailing boats have been equipped with SFCs and were able to cover their complete electricity requirement from the fuel cell for 4 days using just one 2.5 L fuel cartridge. The SFC is dramatically lighter and more compact than Pb-acid batteries, more reliable than photovoltaic systems or wind generators, and silent and pollution-free compared to diesel generators.

Life-Cycle Cost Savings

In stationary applications like remote sensors or security systems, significant life-cycle cost advantages result from the use of an SFC. Operation and maintenance costs that are frequently associated with the use of batteries can be reduced. Furthermore, the SFC can be combined with traditional systems such as batteries or solar cells. An intelligent combination of several technologies often results in superior products.

Musical Band

A group of students in a band in Rhode Island performed at a concert with PEM fuel cell-powered electricity.[40] The band powers all its electric guitars, amplifiers, and PA entirely with H_2 fuel cell electricity.

Sailboats/Powerboats

A sailboat, X/V-1, has been outfitted with H_2-based technology systems with a fully self-contained, on-board, zero-, or ultralow-emission power system that may utilize fresh or saltwater and electricity from renewable technologies (e.g., wind and solar) to produce, store, and consume H_2 as fuel.[19]

Future

By this time next year, you may be able to buy laptops equipped with methanol cartridges that offer longer-lasting power than traditional batteries. In the future, be careful in getting your PC on a plane.

Japanese electronics giants Toshiba and NEC recently demonstrated working prototypes of portable devices that use methanol-based fuel cells to generate electricity. Motorola, Samsung, Sony, Hitachi, Casio, and NTT DoCoMo are pursuing similar efforts, and several plan to commercialize the technology in 2006. According to New York-based ABI Research., shipments of micro fuel cell-generating batteries could top 3 million units by 2008, producing $500 million in sales.

Early versions can power a laptop for 5 h — about as long as today's Li-ion batteries — but NEC claims to have developed a 40-h cull that will be ready for release in 2006.

The downside: Users have to replace the cartridges (or pour methanol directly into the cells) every time the laptops run out of gas. But the biggest barrier to adoption may come from regulators. The methanol reaction requires a potentially high concentration of volatile fuel, which makes the Federal Aviation Administration nervous The FAA hasn't certified micro fuel cells for use in aircraft, and the approval process could take a few years.[41]

AeroVironment[42] has flown a small drone powered by a fuel cell to demonstrate the potential for longer-endurance electric flight. Company officials believe it is the first time an H_2 fuel cell has solely propelled an aircraft.

The 6 oz, 15-in.-span "Hornet" micro air vehicle (MAV) in theory could fly for several hours: about twice the endurance of a similar "Wasp" MAV powered by high-performance Li-ion polymer batteries that the company previously flew for 107 min.

But the fuel-cell tests showed that development remains to be done. The Hornet could fly only for about 5 min at a time before the fuel cells became too dry to operate in the arid California climate.

There are 18 fuel cells along the upper surface of the wing, wired in series, and each producing 0.5–0.6 V. The bottom of the wing is standard structure. The upper surface is the thin fuel cell, which combines low-pressure H_2 on the inside of the wing with O_2 from the air on the outside to make electricity and H_2O. The outer layer is the air cathode that produces the H_2O.

The H_2 comes from a chemical hydride mixed with H_2O. The generator is hidden underneath the wing. Average power output was over 10 W, which directly fed the propeller motor without any intervening batteries, capacitors, or other storage devices. It also fed the radio control and other systems.

Under laboratory conditions, the Hornet fuel cell can produce 400 watt-hr/kg energy density, including the fuel cell and related systems in the weight, compared with 143 watt-hr/kg for the Wasp's rechargeable Li-ion polymer battery. But the battery can achieve that number in real-world conditions, whereas the fuel cell has yet to show its potential.

The Naval Research Laboratory is very interested in fuel cell-powered MAVs and will pursue development of this technology.

Passenger jets on the tarmac are noisy and smelly, kicking out 20% of all airport emissions. That's because jets burn fuel in APUs to generate the electricity that keeps air conditioners and other equipment humming. To take just one example, the APU in a Boeing 777 produces as much smog-causing nitrogen oxide as 155 Chevrolet Impalas.

But a new class of fuel cells, which use solid-oxide technology rather than the PEM technology favored for cars and small electronics, may eventually do the same job more cleanly — plus allow greater efficiencies in-flight. SOFCs can use jet fuel as an energy source, and while they are still at the laboratory prototype stage, recent advances have led Boeing and NASA to consider them serious contenders for planes. If all goes well,

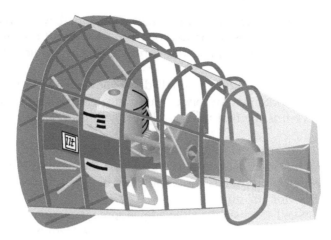

FIGURE 10.13
An envisioned fuel cell generator, shown here with blue housing, would replace a gas-powered auxiliary power unit in the tail of a passenger jet.[43]

Boeing will begin developing a tail-mounted fuel cell prototype in 2005 that could be commercialized in about 10 years. It would provide about a 2 times improvement in efficiency and will totally eliminate ground-based emissions coming out of the APU (see Figure 10.13).

The aircraft application for SOFCs is especially compelling because less fuel lightens the plane, meaning less power is needed to take off and fly. The cells would eventually generate power in-flight, too, replacing the pneumatic systems that suck energy from a jet's engines to power components like cabin-pressurization and anti-icing systems. What's more, aircraft can make use of the fuel cells' chief byproducts: H_2O and heat. But despite recent advances, to make it into an aircraft, completed fuel cell generators will need to become lighter, cheaper, and more powerful — by about 450 kW, enough to power about 20 houses. But with research being done by NASA and industrial partners, the improvements are expected in several years. Volume production and modular designs should make the fuel cells affordable. Much further out, they could power electric jet engines. But for now, it's enough to make airports quieter and cleaner.

According to researchers at Lawrence Livermore National Laboratory, Livermore, CA, miniature fuel cells may one day replace Li-ion batteries in consumer electronics.

The method for making these thin-film cells combines microcircuit processes, micro-fluidic components, and MEMS technology. Thin layers of electrolyte are sandwiched between electrodes with proportioned amounts of catalysts. Microfluidic-control elements distribute methyl-alcohol mixtures through an Si chip over one electrode surface, while air flows over the other electrode. Integrated resistive heaters allow heating of the electrolyte-electrode layers. This increases the conduction of cata-lytically generated protons from the fuel supply across the electrolyte to an air-breathing electrode, where they combine with O_2 to generate electrical current.

Researchers predict the MEMS-based fuel-cell power source will replace rechargeable batteries, such as Li-ion and Li-ion-polymer, in consumer electronics that include cell phones handheld computers, and laptops. The MEMS fuel cell is said to be half the cost and 30% of the weight, size, or volume of existing rechargeable power sources.

Auto designers and engineers must start from scratch in developing hybrid electric and fuel cell powertrains. Although automobile chassis can be adapted for similar-model cars within an automaker's product line, the powertrains for fuel cell and hybrid vehicles are completely different from gasoline-powered engines and require a somewhat steep learning curve on the part of the engineers.

Designers must begin with the idea that there are electrical machines in place, and they must be able to manage electric power as well as deal with the specific requirements of a different type of motor. Building electric vehicles is currently a costly process. Over the last decade, automakers have significantly reduced the weight and size of the power electronics from what used to be the size of a refrigerator.

Fortunately, auto designers have a wealth of simulation available to help them create new internal systems for electric and hybrid cars.

Fuel cell vehicles are the cars of the future, but how far off is that future? Daimler Chrysler has produced six test cars with fuel cell systems since 1994 and plans to introduce fuel cell cars to the market in 2005 with production line cars. The company's NECAR 5 (New Electric CAR), introduced in 2000, is a zero-emission vehicle with a driving range of up to 280 miles.

General Motors shipped its first fuel cell demonstration vehicle, the Hydro-Gen1, in 2003 and has a 5-year collaboration with Toyota to speed the development of advanced technology vehicles. The companies are working on vehicle and system projects such as a common set of electric traction and control components for future battery electric, hybrid electric, and fuel cell vehicles.[44]

GM's fuel cell costs are below the $500/kW it cites and moving toward its $50 production target.

GM must reach that benchmark in approximately 2006 in order to begin the 4-year process required to get plants built and cars rolling off assembly lines by the end of the decade.[45]

In the years to come, GM's squad of electrochemists and engineers will race to perfect the fuel cell. They will spend hours peering into $250,000 microscopes at atoms dancing on the surface of fuel cell membranes. The membranes coated in Pt react with H_2 to create electricity. One goal is to wring out the greatest yield of electrical current using the smallest amount of $629/oz Pt. They believe that they are still not using every particle of Pt effectively.

The race for fuel cell cars and their production and acceptance by the populations of the world is on[45] (see Figure 10.14).

Microtubular fuel cells that are said to operate at 10 times the power levels of conventional plate-and-frame fuel cells are under development by Physical Sciences Inc., Andover, MA, for the NASA Johnson Space Center in Houston, TX.[46] The main advantage of the microtubular construction is power density of 1 W/g, compared with 0.1 W/g for conventional fuel cells.

The fuel cell contains multiple tubular membrane/electrode assemblies that operate in an O_2/H_2 cross-flow. Each membrane/electrode assembly is composed of a tubular PEM, with the anode on the inner surface and the cathode on the outer surface. Targeted dimensions include an inner membrane diameter of 600 µm, membrane thickness of 50 µm, and anode thickness of 23 µm.

One end of each assembly is closed, while the other end is open and connected to a current-collection manifold. The H_2 manifold and the edge-tab electrical current collectors take up much less space than the bipolar plates of conventional plate-and-frame fuel cells.

An Ni-Sn catalyst may replace Pt in a low-temperature process for making H_2 fuel from plants. According to engineers at the University of Wisconsin–Madison,[47] the new catalyst, together with a second innovation that purifies H_2 for use in fuel cells, offers a new method for making the transition from fossil fuels to renewable H_2-based power.

Pt is very effective but it is also very expensive and it is also problematic for large-scale power production because Pt is already in demand for fuel-cell anode and cathode materials.

The single-step process uses temperature, pressure, and a catalyst to convert hydrocarbons such as glucose. Resulting products consist of 50% H_2 with the rest being CO_2 and gaseous alkanes. More refined molecules, such as ethylene glycol and methanol, almost completely convert to H_2 and CO_2. Because plants grown as fuel crops absorb the CO_2 released by the system, the process is greenhouse-gas neutral. The process takes place in a liquid phase at 227°C without vaporizing H_2O. This is said to represent a major energy savings compared to ethanol production or conventional fossil-fuel-based H_2-generation methods that boil away H_2O.

Researchers are trying to create a combined process where the Ni-Sn catalyst reforms oxygenated hydrocarbons to produce relatively clean H_2. The H_2 then passes to a second-stage ultrashift catalyst that further purifies it and removes CO.

An improved design concept for direct methanol fuel cells makes it possible to construct fuel cell stacks that can weigh as little as one third

Gentlemen, start your fuel cells
who's got pole position in the race to market a hydrogen car.

General motors
Hy-wire

Hy-wire has established GM as the company to beat.Given the progress it is quietly making on fuel cells, its next skateboard car should have longer range, higher top speed, and better handling-and be cheaper to make.

Toyota
FCHV

Heavy research spending for more than a decade makes Toyota GM's toughest rival. It already has eight fuel-cell-powered vehicles with Highlander SUV bodies on the road. Toyota has also unveiled a skateboard-based show car, though it's not yet drivable.

Ford/DaimlerChrysler
Ford: Focus FCV;
Daimler: F-cell, Citaro bus

Together, Ford and DaimlerChrysler own 36 percent of Ballard, a pioneering fuel-cell maker. Daimler fields many fuel-cell prototypes and Ford will continue testing 15 cars during the next year. But the companies' efforts have been hurt by Ballard's problems, as well as their own.

Nissan
Xterra

Nissan has collaborated on fuel-cell technology with Ballard and United Technologies, and is now developing engines with Renault, which owns a 44 percent stake in the company.

Honda
FCX

The city of Los Angeles is testing three FCX vehicles equipped with Ballard fuel cells; the cars, based on the Honda Civic, use conventional chassis and technology from existing Honda vehicles.

FIGURE 10.14
Gentlemen, start your fuel cells.[45]

as much as do conventional bipolar fuel cell stacks of equal power. The structural support components of the improved cells and stacks can be made of relatively inexpensive plastics. Moreover, in comparison with conventional bipolar fuel cell stacks, the improved fuel cell stacks can be assembled, disassembled, and diagnosed for malfunctions more easily. These improvements are expected to bring portable direct methanol fuel cells and stacks closer to commercialization.

In a conventional bipolar fuel cell stack, the cells are interspersed with bipolar plates (also called biplates), which are structural components that serve to interconnect the cells and distribute the reactants (methanol and air). The cells and biplates are sandwiched between metal end plates. Usually, the stack is held together under pressure by tie rods that clamp the end plates. The bipolar stack configuration offers the advantage of very low internal electrical resistance. However, when the power output of a stack is only a few watts, the very low internal resistance of a bipolar stack is not absolutely necessary for keeping the internal power loss acceptably low.

Typically, about 80% of the mass of a conventional bipolar fuel cell stack resides in the biplates, end plates, and tie rods. The biplates are usually made of graphite composites and must be molded or machined to contain flow channels, at a cost that is usually a major part of the total cost of the stack. In the event of a malfunction in one cell, it is necessary to disassemble the entire stack in order to be able to diagnose that cell. What is needed is a design that reduces the mass of the stack, does not require high pressure to ensure sealing, is more amenable to troubleshooting, and reduces the cost of manufacture.

The present improved design satisfies these needs and is especially suitable for applications in which the power demand is ±20 W. This design eliminates the biplates, end plates, and tie rods. In this design, the basic building block of a stack is a sealed unit that contains an anode plate, two cathode plates, and two back-to-back cells (see Figure 10.15). The structural-support and flow-channeling components of the units are made from inexpensive plastics. Each unit is assembled and tested separately, then the units are assembled into the stack. The units are joined by simple snap seals similar to the zipper-like seals on plastic bags commonly used to store food. The cathode and anode plates include current collectors, the inside ends of which are electrically connected to the electrodes and the outer ends of which can be used to form the desired series or parallel electrical connections among the cells. Because the stack need not be clamped or otherwise held together under pressure, the slack can easily be disassembled to replace a malfunctioning sealed unit.[48]

Finally, if gasoline and diesel engines are the current automotive power source of choice, then fuel cells are claimed to be the future based or their environmental friendliness. With H_2 widely considered in be the likely

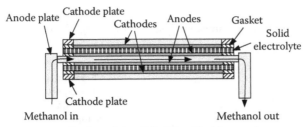

Partly sealed unit (building block)

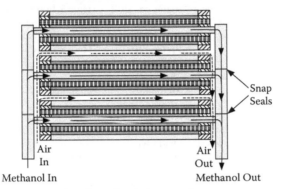

Stack without electrical connections

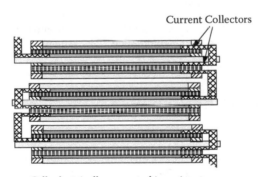

Cells electrically connected in series

FIGURE 10.15

Partly sealed units are joined by snap seals and their current collectors are connected together to form a fuel cell stack. Each sealed unit contains a back-to-back pair of fuel cells.[48]

long-term fuel choice for vehicles, there is probably a need for a number of new commercial catalysts in this system for H_2 generation, for H_2 purification, and of course within the fuel cell itself for catalyzing the H_2 plus O_2 reaction, the basis of fuel-cell energy generation. It is believed that there are significant opportunities for Au catalysts in this industry (see Table 10.3).

TABLE 10.3

Potential Areas within Automotive Systems for Using Nanoparticulate Gold Catalysts

Automotive Power Source	Potential Application for Gold-Based Catalysts	Reaction(s)	Main Characteristics of Supported Gold Nanoparticulate Catalyst	Remaining Technical Barriers
Fuel Cell	Fuel processing systems for clean H₂ production	Water gas shift reaction for H₂ production	High activity at low temperatures	Durability unclear
		Preferential oxidation of CO for H₂ cleanup	High activity at low temperatures	Durability unclear
		Methanol decomposition for H₂ production	High activity	Limited published studies to date
	Fuel cell catalyst	Oxidative removal of CO from H₂	Improvement in electrical conductivity	Limited published studies to date
Diesel Engine	Component of TWC for diesel engine emission control	CO and HC combustion and NOx reduction HC	Low-temperature activity (and high for NOx reduction)	Thermal durability and SO₂ poisoning
Petrol Engine	Low light off catalyst for petrol engine	CO and HC combustion	Low-temperature activity	Thermal durability and SO₂ poisoning

References

1. Valero, G., Wanted for Fuel Cells: Highly Conductive Compounds, *Modern Plastics*, Jan 2004, pp. 28–29.
2. Stone, B. Alternative Energy, *Newsweek*, Nov 17, 2003, pp. E22–E26.
3. McConnell, V.P., Composites and the Fuel Cell Revolution, *Peinforced Plastics*, Jan 2002, pp. 38–44.
4. Burns, L. D., McCormick, J. B., and Barroni, C. E., Bird, Vehicle of Change, *Scientific American*, Oct 2002, pp. 64–73.
5. Katz, R. N., Solid Oxide Fuel Cells, *Ceram Industry*, Oct 1999, pp. 25–26.
6. Yang, Z. G., Stevenson, J. W. and Singh, P., Solid Oxide Fuel Cells, AM&P, June 2003, pp. 34–37.
7. http://link.abpi.net/l.php?20030320A4, 3/23/03 (4p) and http://science.nasa.gov/headlines/y2003/18mar_fuelcell.htm?list911586(6p); 7/7/03.
8. Small, Ultra Efficient Fuel Cell Systems, 8/8/03, 2p, http://jazz.nist.gov/atpcf/prjbrief.cfm?ProjectNumber=00-00-4739.

9. Bansal, N. P. and Goldsby, J. C., Formulations for Stronger Solid Oxide Fuel-Cell Electrodes, NASA Tech Briefs, Feb 2004, p. 43.
10. http://link.abpi.net/l.php?20031009A4; 10/9/03, 6p.
11. Johnson, J., Fuel Cell Catalysts Made of Enzymes Found in Nature, Iiis Innovation Ltd, AM&P, Jan 2004, p. 19.
12. Narayanan, S., Haile, S., et al. Improved Polymer/Solid-Electrolyte Membranes for Fuel Cells, *NASA Tech Briefs*, Sept 2002, p. 52.
13. http://www.japantoday.com/e/?content=news&cat=4&Id=265139.
14. http://www.fuelcellpropulsion.org/army_loco_1aug2003.htm.
15. http://energyinfosource.com/dg/news.cfm?id=20341.
16. *CompositesWeek* 5(16), April 21, 2003.
17. http://marad.dot.gov/nmrec/energy%20&%20emissions/images/Newsletter%20Fall%202002.pdf.
18. http://www.anavu.com/trucknews.html.
19. http://haveblue.com/news/currentpr/020703.htm. 4/01/03, p. 2 of 8.
20. http://www.media.gm.com/news/releases/030109_mili.html.
21. DaimlerChrysler Pushes Fuel-Cell Cars toward the Commercial Marketplace, Scientific American, Jan 2004, p. 63; www.sciam.com.
22. Coalition of Industry Leaders Launches National Fuel Cell Bus Initiative, *Composites Week,* 5(46), 2, 2003.
23. http://www.utcfuelcells.com/news/archive/111402.shtml.
24. http://www.utcfuelcells.com/news/archive/111102.shtml.
25. http://www.ballard.com.
26. http://www.palcan.com/s/NewsReleases.asp?ReportID=44100.
27. http://biz.yahoo/com/prnews/02129/nyf025_1.html.
28. Mini Fuel Cells Slated to Power Cell Phones, Machine Design, Nov 6, 2003, pp. 43–45.
29. http://jazz.nist.gov/atpcf/prjbriefs/prjbrief.cfm?ProjectNumber=00-00-4822; 8/8/03, 2p.
30. Jonietz, Ant Power Packs, Technol. Review, Sept 2004, p. 89.
31. http://science.nasa.gov/headlines/y2003/18mar_fuelcell.htm?list911586; July 7, 2003, p. 1.
32. http://www.mhi.co.jp/power/e_power/techno/sofc/index.html, 08/02/04, p. 2 of 4.
33. Blair, J., Rating the Performance of Residential Fuel Cells, NIST Tech Beat, p. 1, 09/26/03; www.nist.gov;
34. http://www.utcfuelcells.com/news/archive/103002.shtml, 11/01/02, p. 2.
35. http://www.mechtech.com/newsandevents/article.cfm?A_ID=14105, 11/01/03, p. 2.
36. http://www.hydrogenics.com/ir/NewsReleaseDetail-1.asp?RELEASED=120530, 11/01/03, p. 1.
37. http://biz.yahoo.com/prnews/030403/nyth073_2.html.
38. Fuel Cells May Power Medical Implants, Machine Design, May 8, 2003, p. 30.
39. http://www.smartfuelcell.de/en/presse/c020911.html, 7/5/03, p. 2.
40. mailto:rkmccurdy@yahoo.com>rkmccurdy@yahoo.com;
41. Freund, J., Fuel-Cell Supercharge!, Business 2.0, Oct 2003, p. 36.
42. Dornheim, M. A., Fuel Cells Debut, *Av. Wk. Sp. Technol.*, June 2, 2003, p. 52.
43. Talbot, D., Flying the Efficient Skies, *Technol. Review*, p. 26, 2003.

44. Automakers See Fuel Cells as the Next Best Power, *NASA Tech Briefs*, July 2003, pp. 24–26.
45. King, R., Auto Fuel Cells, *Business 2.0*, Oct 2003, pp. 96–98.
46. Kimble, M., Microtubular Fuel Cells Have Ten Times the Power Density, *AM&P*, p. 35, 2004.
47. Fuel Cells May Say Goodbye to Platinum Catalysts, Machine Design, p. 50, 2004.
48. Narayanan, S. and Valdez, T., Lightweight Stacks of Direct Methanol Fuel Cells, *NASA Tech Briefs*, August 2004, pp. 38–39.

Bibliography

ASM International, Stuart Energy and Hamilton Sundstrand Complete Significant Milestone in Strategic Alliance — Unveil First Jointly Developed PEM-based H_2 Fueling Prototype, http://www.asminternational.org/Content/ NavigationMenu/News/HeadlineNews/. 12/11/02, 7 p.

Fuel Cells: Applications and Opportunities, http://www.escovale.com/fc/ fc_management_report.htm, p. 1, 7/5/03.

Jennifer@fuelcells.org, and fuelcell@listbox.com, Jennifer Gangi, Fuel Cells 2000s Fuel Cell Technology Update — Oct 2003, p. 1.

11

Liquid Crystal Polymers/Interpenetrating Network for Polymers/Interpenetrating Phase Ceramics

Liquid Crystal Polymers (LCP)

Introduction

Liquid crystal polymers (LCPs) are a class of advanced thermoplastics that are united not by chemical structure, but by a tendency to form ordered systems in both melt and solid states. The ordered structure of these liquid crystalline wholly aromatic copolyesters allows exceptional mechanical, chemical, and thermal properties. It has made them a high-growth segment in advanced polymers.

LCPs are distinguished from other plastics by their rod-like microstructure in the melt phase. Other resins have randomly oriented molecules in the melt phase, but when LCPs melt, their long, rigid molecules can align into a highly ordered configuration that produces a number of unique features. These include low heat of crystallization, extremely high flow, and significant melt strength. The molecular structure has such a profound effect on properties and processing characteristics that LCPs are best treated as a polymer category separate from amorphous and semicrystalline resins. In spite of their unique properties, however, LCPs can be processed by all conventional thermoplastic forming techniques.

They can, for instance, easily flow into molds for very thin or highly complex parts, reduce cycle time in molding, produce parts with little molded-in stress, eliminate deflashing and other secondary finishing steps, and reduce part breakage during assembly. These benefits can raise finished assembly or system yields 3% or more, which can more than offset the higher resin cost.[2]

Types (Classes) and Properties

The unique morphology is responsible for most of the properties that set LCPs apart from other polymers. Typically, the polymers of a plastic are pictured as limp, randomly oriented chains entangled in one another like a plate of cooked spaghetti. This is a fairly accurate image when the polymers are in solution or the melt phase. A special class of polymers whose chains are long and rigid, however, actually exhibits domains of parallel ordering in the liquid state. A good picture is that of tightly packed logs floating on the surface of water. A close-up view will show regions where many logs are packed side-by-side. In an adjoining region, logs will again be packed, but pointing in a different direction. An overview reveals many ordered regions randomly oriented to each other.

Polymer Classes

Polymers are classed as either thermosets or thermoplastics, based on their response to heat. Thermosets cannot be reheated or remolded because they undergo an irreversible two-stage polymerization reaction. In the first stage, the material supplier forms a linear-chain polymer with reactive end groups; in the second, the molder applies heat in the molding press, which cross-links the end groups. Shorter chains with many cross-links form rigid thermosets; longer chains with fewer cross-links form more flexible ones. Exposure to high temperatures for too long a time degrades the plastic by breaking its chains.

Thermoplastics, by contrast, can be remolded because they remelt when heated beyond a certain point. They have linear molecular chains with some branching, but little or no cross-linking. Those with a wide range of chain lengths are easier to process, while those with a narrow range have more uniform properties.

Thermoplastics generally have higher impact resistance, are easier to process, and are better suited to complex designs than thermoset polymers. Thermoplastics are divided into three groups based on molecular structure: amorphous, semicrystalline, and liquid crystalline (see Figure 11.1).

Amorphous polymers have randomly oriented chains in both melt and solid phases. Semicrystalline polymers have randomly oriented chains in the melt phase, but upon cooling can become partially ordered, resulting in a structure consisting of both crystalline and amorphous regions.

Semicrystalline polymers have lower melt viscosities than amorphous polymers. They also have a more compact molecular structure and have higher shrinkage and warpage in molded parts. LCPs are highly ordered in both liquid and solid states. As a group, LCPs have excellent processability that leads to short cycle times, high flowability in thin sections, and exceptional repeatability of dimensions. Molded parts exhibit very

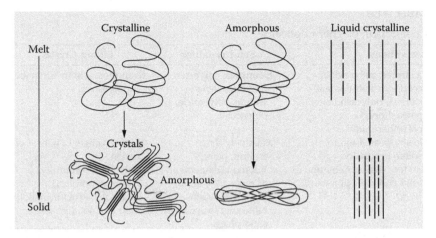

FIGURE 11.1
Two-dimensional representatives of the structures of crystalline, amorphous, and liquid crystalline thermoplastics.[2]

low warpage and shrinkage, along with high dimensional stability, even when heated to 200–250°C. The rigid-rod nature of LCP molecules results in a microstructure that resembles the physical orientation of the fibers in a reinforced thermoplastic (see Table 11.1).

Classes

The three most important classes of LCPs are aromatic polyamides, rigid rod polymers, and the aromatic copolyesters. They have in common the ease of orientability resulting in a high degree of orientation and high modulus properties. Lyotropic aromatic polyamides, poly(p-phenylene terephthalamide) (PTTA), were first commercialized under the Kevlar trademark.[3-5] The aromatic polyamides (aramids) are produced by a dry-jet wet spinning process where the nematic structure in solution is responsible for the well-known high performance. Variations in structure can be processed by annealing at elevated temperature, under slight tension, which is known to increase the fiber modulus due to a more perfect alignment of the molecules. Production of commercial aramid fibers has provided the impetus for many others to attempt to process LCPs into high modulus fibers.

The rod-like polymers are another class of lyotropic fibers, also produced by the dry-jet wet spinning process. Poly(p-phenylenebenzo-bisthiazole) (PPB), the most complex of the ordered polymers, is difficult to process, but exhibits the highest tensile modulus of all the LCP fibers produced to date.[6-11]

TABLE 11.1

Comparison of Thermoplastics[2]

Amorphous	Semicrystalline	Liquid Crystal
Examples: polyarylate, acrylonitrile-butadiene-styrene, polycarbonate, polysulfone, polyetherimide.	**Examples:** polyester, polyphenylene sulfide, polyamide, polyacetal	**Examples:** wholly aromatic copolyesters.
No sharp melting point, Soften gradually.	Relatively sharp melting point.	Melting range. Low heat of fusion.
Random chain orientation in solid and melt phases.	Ordered arrangement of molecules, and regular recurrence of molecular structure in solid phase.	High chain continuity. Extremely ordered molecular structure in both melt and solid phases.
Do not flow as easily as other two types.	Flow easily above melt temperature.	Flow well under shear or around melting range.
Reinforcement improves load bearing capability only below T_g.[a]	Reinforcement lowers anisotropy and improves load bearing capability between T_g and T_m.[a]	Reinforcement improves load bearing capability, esp. in highly crystalline forms between T_g and T_m.[a]
Can given transparent parts.	Part is usually opaque.	Part is always opaque.

[a] T_g is the glass transition temperature, above which the polymer is ductile and below which it is brittle. T_m is the melting point.

One of the major challenges for NASA's next-generation reusable-launch-vehicle (RLV) program is the design of a cryogenic lightweight composite fuel tank. Potential matrix resin systems need to exhibit a low CTE, good mechanical strength, and excellent barrier properties at cryogenic temperatures under load. In addition, the resin system needs to be processable by a variety of nonautoclavable techniques, such as vacuum-bag curing, RTM, vacuum-assisted resin-transfer molding (VARTM), resin-film infusion (RFI), pultrusion, and advanced tow placement (ATP).

To meet these requirements, NASA Langley Research Center has developed a new family of wholly aromatic liquid-crystal oligomers that can be processed and thermally cross-linked while maintaining their liquid-crystal order All the monomers were polymerized in the presence of a cross-linkable unit by use of an environmentally benign melt-condensation technique. This method does not require hazardous solvents, and the only side product is acetic acid. The final product can be obtained as a powder or granulate and has an infinite shelf life. The obtained oligomers melt into a nematic phase and do not exhibit isotropization temperatures greater than the temperatures of decomposition ($T_i > T_{dec}$). Three aromatic formulations were designed and tested and included esters, ester-amides, and ester-imides.

One of the major advantages of this development, named LaRC-LCR, or Langley Research Center-Liquid Crystal Resin,[12] is the ability to control a variety of resin characteristics, such as melting temperature, viscosity, and the cross-link density of the final part. Depending on the formulation, oligomers can be prepared with melt viscosities in the range of 10–10,000 poise (100 rad/s), which can easily be melt-processed using a variety of composite-processing techniques. This capability provides for custom-made matrix resins that meet the required processing conditions for the fabrication of textile composites. Once the resin is in place, the temperature is raised to 375°C, and the oligomers are cross-linked into a high-glass-transition-temperature (T_g) nematic network without releasing volatiles. The mechanical properties of the fully cross-linked, composite articles are comparable to typical composites based on commercial available epoxy resins.

LaRC-LCR can also be used in thermoforming techniques where short holding times are desired. The resin can be used to spin fibers, extrude thin films and sheets, or injection-mold complex parts. Although LaRC-LCR has been developed to meet NASA's needs towards the development of a next-generation launch vehicle, other applications can be envisioned as well. The thermal and mechanical behavior of this material are ideally suited for electronic applications and may find use in flexible circuits, chip housings, and flip-chip underfills. Another area where thermal stability and chemical resistance are highly desirable is the automotive industry. Distributor caps, fuel tanks, air-intake manifolds, rocker covers, and ignition systems are among the potential applications. The low viscosity of this resin makes this material ideal for coating applications, as well. Fine powders have been used in plasma-spray applications, and well-defined thin coatings were obtained. LaRC-LCR can also be used as an adhesive. Lap-shear values of 22.683 MPa were easily obtained. In contrast, these values are ≈20 times higher than those observed in commercially available LCP resins.

Another interesting class are unfilled (neat) LCPs.[13] Neat LCPs have excellent thermal and dimensional stability, as well as high rigidity, strength, and chemical resistance. LCP parts can be molded with great precision, are flame-retardant, absorb little moisture, are highly impermeable to gases, withstand cryogenic temperatures without growing brittle, and have a low CTE.

LCPs also offer many processing advantages. As an LCP melt is processed, its surface molecules line up in the flow direction to form a skin that usually amounts to 15 to 30% of part thickness. This provides for high melt strength and stability, high flow under shear, and a low and predictable shrinkage (see Table 11.2). The oriented structure reinforces the final material, giving it high flexural and tensile strength, elastic modulus, and dimensional stability.

TABLE 11.2

Advantages of LCPs[13]

End-Use Advantages	Processing Advantages
Excellent thermal stability. Relative thermal index of up to 240°C (464°F).	Extremely low shrinkage and warpage during molding.
High rigidity and strength. Typical modulus from 10 to 24 GPa, tensile strength from 125 to 255 MPa, and notched impact strength of 5 to 85 kJ/m^2.	Exceptional repeatability of dimensions.
Little dimensional change. Coefficient of thermal expansion is less than 10×10^6/°C (6×10^6/°F) in the flow direction and moisture adsorption is less than 0.03%.	High melt strength and stability.
Cryogenic strength. Retains 70% of its notched of its notched impact value down to –270°C (–450°F).	Low melt viscosity for high flow in thin sections and intricate parts.
Excellent chemical resistance. Inert to acids, bleaches, chlorinated organic solvents, alcohols, and fuels.	Low heat of fusion for fast melting and rapid cooling.
Low gas permeation rates compared to commercial packaging films.	High stiffness at high temperatures.
Low ionic extractables. These are well below requirements for corrosion-free environments in integrated-circuit chip applications.	Can be reprocessed many times without significant property change.
Inherently flame retardant, low smoke density and relatively nontoxic products of combustion.	Little or no flash.

Generally, the processed material is anisotropic so that strength and stiffness are higher in the flow direction, although some filled products are less anisotropic. Neat LCP applications range from injection-molded components to blown and cast film, either in monolayer or coextruded form, to extruded sheet and injection blow-molded parts. Neat LCPs also can be blended with other polymers to enhance flow, stiffness, and other properties.

Forms

Unfilled, standard LCPs, both wholly aromatic polyester and copolyester-amide grades, have better chemical resistance and mechanical properties than many glass-fiber-reinforced thermoplastics. In addition, their dimensional stability and barrier properties in film, sheet, and laminates are equal to or better than most other melt-processable polymers.

The high process temperature of standard LCPs (thermal transitions of 280 to 335°C) has limited the neat resins to some degree, especially in blends and alloys with lower-temperature polymers. This has been overcome by recent advances, such as wholly aromatic copolyester LCPs and ester-amide LCPs, which have thermal transitions as low as 220°C.

Except for their processing temperatures, both low-temperature and standard-temperature grades have similar mechanical performance, chemical resistance, stability, and other properties.

In general, wholly aromatic (closed ring), lower-temperature LCPs function better than LCPs containing aliphatic (open chain) components.

Films and Multilayer Structures. LCP films and sheets are well suited for many medical, chemical, electronic, beverage, and food packaging applications. They are more impermeable to H_2O vapor, O_2, CO_2, and other gases than typical barrier resins. When LCP film is biaxially oriented, it forms a high-strength material with relatively uniform properties and low fibrillation. Also, its high-temperature capability enables it to meet the needs of thermally demanding applications, such as films for printed wiring boards.

In barrier applications, even plastics considered to have a good combination of low H_2O vapor transmission and O_2 permeability, such as ethylene-vinyl alcohol copolymer (EVOH) and polyvinylidene dichloride (PVDC), do not match that of LCP resins. LCP films can be thinner than those of EVOH or PVDC at the same barrier level, making LCPs cost-effective components of multilayer laminates for containers and trays. LCPs also can serve as barrier films in blow-molded automotive gas tanks and as liners for metal, plastic, and composite storage tanks for chemicals and fuel.

Blends and Process Aids. Wholly aromatic copolyester, ester-amide, and ionomeric LCPs have been blended with many thermoplastics, including polyolefins, polycarbonates (PC), polyesters, polyamides, polyetherimides and even intractable, high-temperature resins such as polyimides. Such blends improve the base resin by increasing stiffness, impact resistance, thermal resistance, impermeability to H_2O vapor and O_2, and adhesion between components in laminates.

In one case, standard LCPs blended with PC at addition rates of 5 to 30% reduced viscosity versus PC alone by up to about 25%. For example, when a PC with a melt viscosity of about 3000 poise (1000 sec^{-1} at 290°C) was blended with a standard copolyester LCP (30% addition rate), viscosity dropped to about 2250 poise. The LCP also increased tensile modulus as wall thickness decreased. At a wall thickness of 0.8 mm, the 70/30 PC-LCP blend had a tensile modulus of about 5100 MPa vs. about 2300 MPa for PC alone.

In another case, a low-temperature LCP was blended into a polyolefin at a rate of 5 to 15%. Film blown from the blend had several hundred times more stiffness than film made with the polyolefin matrix alone. The orientation of the LCP molecules in such films can be controlled with a counter-rotating annular die.

The morphology and properties of LCP blends vary with the formulation, processing method and parameters, and other factors. They also are affected by the compatibility of the other polymers and compatibilizers. Because compatibilizers can change blend viscosity, dispersion, and interfacial and interpolymer adhesion, blends may need to be optimized with block copolymers and other additives that modify melt rheology and interfacial tension.

When LCPs are added at low levels as processing aids, they improve the melt strength and flow of low- and high-melt-temperature resins during extrusion. This can make film and fiber processing easier and enhance fiber strength and stiffness. LCPs as process aids also can enhance stiffness in polyolefins and reduce viscosity in polyetherimide, polyetheretherketone, and other viscous, high-temperature resins.

Films and Nonwovens. LCP fibers are melt-spun by conventional polyester extrusion methods. They have high strength and stiffness, so that drawing after spinning to build tensile strength is not needed. Spun fiber tensile strength can be increased significantly by heat treatment at 10 to 20°C below their melting points in an inert atmosphere for several minutes to several hours. LCP fibers and yarns typically have a heat-treated tenacity of 23 to 28 g/denier, an elongation of 3.3% or less, and an initial modulus of 585 g/denier (see Table 11.3).

LCP fibers are 5 times stronger per unit weight than steel and retain their strength over a broad temperature range. They offer all the traditional benefits of LCPs, such as low moisture pickup, high chemical resistance, a CTE close to that of glass, and little dimensional change as temperature varies from cryogenic levels to over 200°C.

TABLE 11.3

Properties of LCP Fibers[13]

Physical Forms	Mechanical Properties	Thermal Properties
Denier (continuous filament) 25–3750	Tenacity, 23–28 g/denier 2850–3470 MPa	Melting point, 330°C
Filament denier 5	Elongation, 3.3%	Boiling water shrinkage,
Filament diameter, 23 μm		≤0.5%
Density, g/cm³	Initial modulus,	Hot air shrinkage
1.4	585 g/denier	@ 177°C
	70 GPa	0.5%
Moisture regain, <0.1%		Limiting oxygen index, 30
Extractables (methyl chloride),		Coefficient of axial thermal
<0.1%		expansion, m/m-°C×10⁻⁶
		20–145°C, –4.8
		145–200°C, –14.8
		200–290°C, –26.7

Note: Test conditions 25 cm gage length, 10% strain rate, 98 tpm twist, five breaks/sample, ASTM D885 for I.M.

LCP fibers also have no measurable creep under load at up to 50% of threadline breaking point, and they exhibit good fiber-to-fiber abrasion resistance. Their fold endurance exceeds that of aramid: after 1000 flexing cycles; LCP yarns retained 93% of their initial tensile strength vs. 50% for aramid.

They have excellent cut resistance. Their Sintech cut resistance is 3.4 vs. 1.1 for aramid and 1.0 for ultrahigh molecular weight polyethylene (UHMW-PE). They are transparent to microwave energy, virtually unaffected by exposure to 500 megarads of cobalt 60 radiation, and able to withstand gamma-ray sterilization. However, they are affected by ultraviolet radiation and should be protected from long-term UV exposure by a jacket or protective film.

LCP fibers have many specialty applications. In safety garments and gloves, for example, they promote cut and stab resistance and withstand multiple wash/dry cycles in the presence of bleach. By comparison, aramid fibers have poor resistance to bleach, and UHMW-PE fibers are sensitive to the high temperatures encountered during machine drying.

In ropes and cables, LCP fibers are suitable for such applications as cargo tie-downs, tow lines, trawl ropes, helicopter rescue hoists, parachute cords, center core strength cables, and optical fiber tension members. Where space is critical, small LCP braided ropes can be terminated in eye-splices having pin-diameter/braid-diameter ratios as low as 1:1.

In industry, LCP fibers are applied in heat-resistant belts, tape reinforcement, pressure vessels, and chemically resistant gaskets and valves. They add excellent flex/fold characteristics and tear strength to composite fabrics for inflatable structures. This is the reason that LCP fiber was used for the airbags that successfully landed the Pathfinder craft on Mars and for similar but larger systems on the Mars Rover Mission.

In recreational equipment, LCP fibers improve vibrational damping and provide exceptional strength in bicycle components, hockey sticks, tennis rackets, golf clubs, skis, snowboards, fishing poles, bow strings, boat hulls, and sails (they were in sails for America's Cup sailboats). And, in nonimplant medical applications, such as catheters and surgical device control cables, they are inert, nontoxic, and withstand all common sterilization methods.

Properties Especially Affecting Design

The big difference between LCPs and other semicrystalline polymers is molecular structure. The ordered structure of LCP, present even in the melt phase, affects material properties and processing methods. By better understanding how the plastic reacts to molding operations, engineers can process LCPs using the same techniques as for other semicrystalline resins.

Designers generally specify LCPs because of dimensional stability, toughness, thin-walled requirements, and their ability to be processed using a broad range of parameters. The challenge comes in properly designing parts to take advantage of these characteristics.[14]

Besides using standard design practices that ensure smooth and uniform resin flow, engineers must consider the anisotropy of LCPs: the difference between properties in the flow and cross-flow directions. Such behavior is not new to plastic molders. Many experienced similar challenges when molding glass-fiber-reinforced resins. One key to successful designs uses wall thickness and gate location to ensure that flow does not degrade strength or promote unwanted shrinkage. A few other design criteria to observe include the following:

Shrinkage. One of LCP's advantages is an ability to maintain close tolerances. The degree of shrinkage after molding depends on molecular orientation in molded parts rather than processing conditions. Shrinkage in the flow direction is close to 0 as orientation increases. Cross-flow shrinkage, however, varies from 0.0762 to 0.1270 mm/mm for wall thickness between 0.7620 and 3.1750 mm — a typical level for conventional engineering plastics. To avoid warping, therefore, designers must adjust material flow in part molds to balance shrinkage in the flow and cross-flow directions.

Weldlines. LCP parts lose 50% of bulk strength at weld or knit lines. Therefore, high-stress areas should be designed away from weld lines. Another solution to minimize weld lines avoids multiple gates in part molds. With good flow properties, a single gate is often sufficient for LCP mold designs.

However, when multiple gates must be used, engineers should locate them so that flow fronts come together in low-stress areas. It is also important to give flow fronts room to flow together after they join. The process forms a seam weld. The opposite condition is a butt-weld line. These form when flow fronts approach each other from opposite directions and stop when they meet. Butt-weld lines reduce part strength to 20% of bulk properties. To encourage seam welds, position gates so that lines develop when the mold begins filling with resin and continue to flow through the weld.

Gates. Most conventional gates work well for LCPs. Designers must select a gate that matches part geometry and provides a balanced flow front. Controlling flow through gate position helps optimize mechanical properties, particularly in high-stress areas.

Gate dimensions also influence part performance. LCPs swell little when moving from the gate to the die, therefore gates should be 90 to 100% of part wall thickness. In terms of location, engineers should design gates opposite a core, rib, or cavity wall to change the direction of resin flow. Without an obstruction, resin shoots straight into molds, forming a stalactite structure. This rope-like formation, called jetting, weakens parts by creating weldlines.

Gate locations should also let the mold fill uniformly from thick to thin sections. When flatness is critical, gate position helps control warpage and counteract shrinkage differences across parts by creating complex flow patterns that change the orientation of flow direction across parts. With random orientation, shrinkage in the cross-flow direction has less effect on flatness.

Vents. Molds should have vents in several locations to avoid air entrapment during high-speed filling. With extremely low melt viscosity, LCPs will fill vent holes if they are deeper than 0.0127 to 0.0254 mm.

Runner system. Full-round or trapezoidal runners work best with LCPs. For multicavity molds, a carefully balanced runner system in the tool avoids overpacking and ejection problems. LCPs work well with hot-runner systems because of the resin's thermal stability. However, it is important not to use runners that are oversized.

Processors use large runners for most thermoplastics to boost flow and prevent crystallization before resins enter the mold. LCPs flow more easily, however, and therefore don't need large-diameter runners, which only create more waste and regrind material.

Molding. Besides flowing well in injection-molding machines, LCPs tolerate short cycle times and produce parts with low residual stress, little or no flash, high dimensional stability, and good toughness. Cycle times are usually 5 to 30 sec, depending on part design: less than half of most engineering plastics. Cycle times are short because the material has a low heat of fusion and keeps its ordered structure when melted and, therefore, requires little or no crystallization time. These qualities let parts cool quickly, thereby cutting the time needed to withdraw heat from parts and providing stiffness at high temperatures. Workers can therefore eject parts while they are still hot.

Low melt viscosity also requires less injection pressure, which helps reduce molded-in stress and cooling times needed to prevent warping. LCP viscosity is highly shear-sensitive. This means molders can improve melt flow into long, thin parts by increasing injection velocity rather than melt temperature. This technique pays off because it avoids unusually high melt temperatures that cause surface defects, such as blisters, that render parts useless.

However, it important not to use equipment that is oversized. Oversized molding-machine barrels increase resin residence time by holding several shots in the barrel, which overheats resins. Engineers avoid such overheating by using smaller machines or increasing the number of cavities in the mold.[15]

Properties and How They Affect Molding

LCPs are the easiest to process of all high-performance plastics, in part because they have the highest melt flow and shortest cycle time. They can be processed in conventional equipment by all conventional methods and can be compounded with most reinforcements and fillers. The resin's unique properties affect processing in several important ways.[2]

Melt viscosity is extremely low because the rigid-rod molecules move past each other easily. The high flow allows the resin to fill intricate parts and thin walls at relatively low injection pressures. Although pressure varies with part design, mold layout, and machine settings, the resin generally needs cavity pressures of only 7 to 35 MPa, well below those of other engineering resins. Low pressure reduces molded-in stress, so molders need not extend cooling time to counter warpage from such stress.

Mold cycle time for LCPs is generally half that of other high-performance resins: 10 to 30 sec for LCPs vs. 20 to 60 sec or more. Exact cycle time for an LCP part depends on parameters such as size, shape, thickness, mold layout, and machine settings.

One reason for the short cycle time is that LCPs can rapidly transition between melt and solid phases because both have a similar ordered structure. The rod-like microstructure in both phases shortens the time needed to melt the resin before injection and the time required for it to solidify in the mold.

Many other factors inherent to LCPs also contribute to the short cycle time. Low viscosity speeds mold fill, and good stiffness at elevated temperatures permits parts to be ejected while still hot. Low molded-in stresses eliminate the need for slow cooling to prevent stress build-up.

Shear thinning is a decrease in melt viscosity as shear increases. LCPs are more sensitive to shear thinning than other engineering plastics, which gives molders a powerful way to alter flow. Molders can improve the resin's ability to fill long, thin, or highly complex parts by increasing injection velocity. With other polymers, molders generally must increase temperature to reduce viscosity, which can degrade the polymer.

Use of regrind lowers material and disposal cost because sprees, runners, and other production scrap can be reground and reused. The reground LCP retains most of the properties of the virgin resin, except that the impact resistance of reground glass-filled grades may decrease because of fiber breakage. In practice, manufacturers mix 50% regrind with virgin resin. Parts should be tested to verify satisfactory performance.

Other Properties

LCPs have an exceptional property mix. Their rigid-rod chains act like fibers to reinforce one another. As a result, LCPs boast exceptional tensile strength and rigidity. Their stiffness at 200°C exceeds that of many engineering resins at room temperature. They stay tough even when bathed in liquid N_2.

LCPs resist chemical attack by most acids, bases, and solvents — even at high temperatures. Water absorption at room temperature runs about 0.2%, and 0.1% for some filled grades with virtually no dimensional change. LCPs showed some surface chalking after a year's outdoor exposure in a hot climate, but only a modest property reduction after 1000 hours in 121°C H_2O.

Dielectric constant, dielectric strength, volume resistivity, and arc resistance meet or exceed those of other advanced polymers over a wide range of operating conditions. Barrier properties exceed those of other melt-processable polymers.

Processing

Under the influence of shear (molding, extrusion, fiber spinning), polymer bundles slide on each other like dry spaghetti, with melt viscosity dropping unusually low. Domains align, giving rise to bundles of rigid molecules oriented in the flow direction, exhibiting extremely anisotropic properties. These rope-like bundles resemble and function like built-in reinforcing fibers, thus the nickname "self-reinforcing" (or rigid-rod) polymers.

There is substantial variation in how easily LCPs are processed, however. This is due to the combination of monomer units making up the backbone of the polymer. Although LCPs gain their strength stiffness, and thermal integrity, as well as their linearity, from rigid aromatic ring structures in the main chain, they must also have flexible links or spacers between these aromatic groups to be melt-processable. This bridge is typically an ester linkage in commercial LCPs (though a wide variety of links have been successfully synthesized), giving the chain more freedom of movement, and thus easier melt processability.

Compared to the highly viscous melts of some advanced resins, LCPs process easily. Their low enthalpy melting point transition creates little crystallinity, and they run like water when sheared. LCPs easily fill thin-walled molds (down to 10 mils) and cycle quickly with little shrinkage and high dimensional stability.

Fortunately, low viscosity makes LCP easier to load with up to 50% glass, talc, or other filler. The fillers help reduce LCP anisotropy. Although oriented molecules provide strength, too much of a good thing weakens transverse properties and weldlines. Fillers also cost significantly less than neat LCPs.

Processing LCPs in magnetic fields, for example, can more than double their tensile strength, according to LANL, which has also been working with industry to develop a method for using a magnetic field to produce three-dimensional composites. The field orients reinforcing fibers in the z-direction.[16]

Osaka Municipal Technical Research Institute[17] has developed and produced recyclable reinforced plastics that retain the initial strength even after recycling 5 times.

Fiber-reinforced plastics (FRPs) so far have contained glass fibers as reinforcement, and were incompatible with the recycling process in which fibers were broken The broken glass also damaged the extrusion and injection molding machine. In contrast, the new reinforced plastics used LCP reinforcement instead of glass fiber, although its LCP reinforcement forms during processing.

The production method used either polyethylene terephthalate (PET) or polybutylene terephthalate (PBT) resins of engineering plastics as a matrix polymer. The composition ratio of PET or PRT/LCP was fixed at 70/30 wt%. A large amount of the mixture was kneaded at 280°C to produce the pellets by using a twin-screw extruder. Small amounts (1/10) of the pellets were injection-molded to prepare specimens for the measurement of mechanical properties. All residual pellets were again extruded under the same conditions, and then a small amount of the pellets were injection-molded to prepare the second recycling specimens. Thus, specimens were prepared until the tenth recycling. At that time, the LCP was transformed into fiber reinforcement during the processings until the fifth recycling, but the LCP was transformed into droplets from the sixth recycling to the tenth recycling. Products until the fifth recycling are usually called "in situ composices." The mechanical properties and other properties of the specimens thus obtained were measured.

The tensile strength of the first recycling specimens, 93 MPa, is about 1.8 times greater than only PET resin. Further, the tensile strength was retained until the fifth recycling. Similar results were also obtained in the system PBT/LCP blend.[18] Promising applications include housings of TVs, personal computers, and bathtubs.

There has long been a desire to deposit conductive paths directly on molded plastic structures, combining the electrical and mechanical functions in one component to form an injection-molded circuit carrier. This technology has been accomplished by using the Laser Direct Structuring (LDS) process from LPKF on modified Vectra® LCP material.[19] The electronics housing can substitute for the conventional circuit board, encouraging miniaturization. This technology is suitable for mobile communication devices, hearing aids, and sensory technology for automobile electronics, just to name a few applications.

This laser-based process consists of a minimal number of manufacturing steps: injection molding, laser structuring, and metallization (see Figure 11.2).

FIGURE 11.2
Injection molded circuit carriers, produced using a laser direct structuring process on modified liquid crystal polymer, simplify the miniaturization of electronics by combining the electrical and mechanical functions in one component.

Basically, the structure is molded in a standard mold using polymer material that can be laser-activated, the desired interconnect pattern is directly written on the resulting molded part utilizing LPKF's scanner-based Microline 3D IR laser system, and the conductive paths are plated using industry-standard electroless plating technology from Enthone, Inc.[19] The plating adheres only where the plastic has been activated by the laser. Structures with a current resolution of > 150 μm can be produced. Due to the high-temperature resistance, the circuit structures on this LCP material are solderable.

The Microline 3D IR laser system contains an Nd:VO$_4$ laser source for activating the plastic surface and a galvanometer scanner that guides the laser beam along the three-dimensional component surface. The LCP material (plastic granulate shown supporting MIDs in Figure 11.2) consists of standard plastics, additives, and laser-fissionable metal-complexes.

Applications

The major penetration of LCPs is in the consumer market for dual-use cookware. Other appliance uses include knobs, handles, and microwave oven components.

LCPs were developed for use in microwave cookware with Xydar-resins for use in Tupperware dual oven/microwave cookware. However, LCP's high prices and competition from other resins have limited growth in this business.

Aircraft/aerospace applications are represented by rivets, aerodynamic fins, and radomes for smart bombs and missiles.

Transportation uses include under-the-hood encapsulants, fuel handling systems, and electrical components where increasing service temperatures demand higher performance.

LCPs act as reinforcement fibers in polyethylene film to extend the lives of He-filled balloons used by NASA for atmospheric experiments. The balloons are about 200 m in diameter, carry payloads of up to 3500 kg, and reach altitudes of 40 km.

These He-filled balloons stay up for only 2 or 3 days, and NASA would like to extend their lives to 2 or 3 months, or even years.

Increasing the internal pressure of the balloons, which are currently not pressurized at all, would prolong the time aloft. However, polyethylene by itself is not strong enough to withstand pressure. Therefore, the development was to fabricate polyethylene reinforced with LCP fibers, but the fibers tended to line up in the direction of flow, strengthening the film only in that direction.

One way to solve the problem was to laminate two sheets having fibers oriented at right angles. Another approach was to use a counter-rotating die, so that as the polymer is extruded, the fibers on the outside of the film become oriented in one direction, while those on the inside become oriented at right angles.

In the process, liquid polyethylene and LCP are mixed, and the LCP becomes suspended as tiny drops in the polyethylene. The drops are elongated, by passing them through a gradually narrowing path, and are then frozen in shape by cooling the mixture below the LCP freezing point (200°C), but above the freezing point of polyethylene (120°C). The mixture is then extruded on the counter-rotating die into films about 20 μm thick.

Electrical/electronic and telecommunication uses include interconnect insulators, high-density connectors, light-emitting diode cases, bobbins, coil forms, and other critical applications.

LCPs make excellent connectors, chip carriers, sockets, and relay and capacity housings because users can mold thin walls to separate closely spaced pins without sacrificing electrical insulation. More importantly, LCPs have the highest heat deflection temperature (HDT) of any thermoplastic. They also have high dimensional stability and withstand infrared and vapor-phase soldering. The introduction of platable grades opens the door to printed circuit boards and multichip modules. Designers have also considered LCPs for high-voltage air conditioners, small appliances, and business machines.

LCPs are among the many materials considered for under-the-hood sensors and controllers in automobiles. They not only have the required heat resistance, but the chemical resistance to stand up to mixed gasoline-alcohol fuels and corrosive fumes. They may also have the mechanical properties for small power train and fuel system parts.

Additionally, automotive applications for LCPs include fuel, ignition, and transmission system components; lamp sockets, reflectors, and bezels; coil forms; sensors; and motor components.

Industrial uses include mechanical and dimensionally critical components of floppy and hard disk drives, CD players, and other audiovisual devices. Chemical-resistant uses such as pumps, meters, and tubing are in development or commercialized. In addition, the chemical process industry is using distillation column packings of LCPs with impressive cost/performance benefits. Fiber-optic connectors, couplers, and stiffening members are in production, where the controllable CTE can be tailored to match the glass fiber or other components of the system.

Finally, in the medical equipment industry, the resin is replacing stainless steel in a growing number of surgical instruments and sterilization trays because it can be molded into long, complex shapes, and it can withstand repeated sterilization cycles.[20]

LCPs are finding their way into more medical applications due to their superior chemical resistance. A new dental syringe gun is one example of LCPs withstanding strong chemicals in sterilization. The switch from polyarylsulfone (PAS) to LCP eliminated a 2% return due to environmental stress cracking from repeated sterilization. PAS weakened after 500 to 1000 cycles; LCP is expected to achieve a minimum of 2000 cycles. LCP resins are suited for a variety of medical, dental, veterinary, and laboratory hardware and equipment applications, such as instrument handles and surgical trays, dental trays and cassettes, and hospital and operating room components.

Eight Vectra MT (LCP) grades are candidates for use in medical devices including drug packaging and delivery systems. They can be injection molded and extruded, and have various flow qualities and additives that yield parts with low friction and wear, improved surface appearances, and greater stiffness. They resist high temperature and chemicals and withstand numerous sterilization cycles. They can replace metal, provide for finely structured parts in drug-delivery systems, and be used in minimally invasive surgery and other areas.

The LCP grades comply with USP 23 Class VI for biocompatibility with skin, blood, and tissue. The grades also meet European Directive 2002/72/EC for food contact applications and BfR standards.[21]

Interpenetrating Network (IPN)

Introduction

Network polymers are formed by cross-linking one linear homopolymer with a second linear homopolymer using grafting techniques. In contrast, IPNs are generally composed of two or more cross-linked polymers with no covalent bonds between them. They are often incompatible polymers in a network form. However, at least one is polymerized or cross-linked in the immediate presence of the other with the result that the chains are completely entangled.[22-25]

IPNS are heterogeneous systems comprised of one rubbery phase and one glassy phase, where one usually is a hybrid resin. This combination produces a synergetic effect on mechanical properties, yielding either high impact strength or reinforcement that are both dependent on phase continuity. Depending on composition, thermal properties can also be improved.

Types

Four types of IPNs are

- Sequential IPNs. These start with one polymer already cross-linked. It is then swollen in another monomer, initiators and cross-linking agents are added, and the system polymerized in place. Example: butadiene cross-linked with divinylbenzene (DVB) forms the first polymer. To this, styrene, DVB, and appropriate initiators are added, and the system is polymerized to the IPN.

- Simultaneous IPNs. These are formed by a homogeneous mixture of monomers, prepolymers, linear polymers, initiators, and cross-linkers. Polymerization occurs by designed independent, noninterfering reactions. Example: urethane/methyl methacrylate copolymer and urethane/styrene copolymer are prepolymerized, dissolved together, and reacted to form the IPN.

- Semi-IPNs. These are made from one linear polymer and one cross-linked polymer. Examples: Urethane/silicone and silicone/polyamide.

- Homo-IPNs. These are prepared from polymers that are generically identical but retain specific characteristics when interpolymerized. Example: epoxy/epoxy IPNs.

Processes

A novel approach has been developed by production of IPNs or semi-IPNs with greater fracture toughness and resistance to microcracks by Pater, Smith, and Razon of Langley Research Center.[26]

Traditionally, IPNs have been made by using a common solvent for two different organic constituents and then evaporating the solvent. One or both of these constituents may have reactive end caps, producing semi-IPNs or full IPNs, respectively. During the evaporation process, the oligomer and polymer components separate because of differences in solubilities as the solution is concentrated. The thermodynamic dependence of the solubility of a constituent with removal of solvent strongly influences separation of phases and thus the microstructures of the IPNs.

The novel process involves freeze-drying. The solvent is removed by rapidly removing energy from the liquid at ambient pressure. The solution of oligomer and polymer constituents is thermally quenched to a solid in a very short time. The thermodynamic state of the two organic constituents in the solid is metastable.

A solvent suitable for use in the freeze-drying process can be either a solid, liquid, or gas at ambient temperature and pressure. As an example, 1,3,5-trioxane is an excellent solid solvent for the freeze-drying process. It can easily be sublimed at room temperature at a subambient pressure and can dissolve both an imide thermoset oligomer and a thermoplastic polyimide in the amic acid form. Suitable liquid solvents are numerous. CO_2 is a good example of a suitable gas solvent.

In a control experiment, IPNs and semi-IPNs were formed by dissolving the constituents in another common solvent in the traditional method. The neat resins and composite materials thus produced were tested for both the traditional and freeze-drying processes. In all cases, glass-transition temperatures and data from dynamic mechanical tests of the freeze-dried IPNs and semi-IPNs demonstrated that the freeze-drying process provided much improved IPNs and semi-IPNs.

Chen and Ma[27] reported on their development to manufacture poly(methyl methacrylate) (PMMA)/polyurethane (PU) IPNs pultruded composites. They demonstrated in their study the feasibility of pultruding fiber-reinforced PMMA/PU IPN composites, and from the viscosity study, the optimum temperature for the PMMA/PU IPN prepolymer was in the temperature range ~40–60°C. A high reactivity of the prepolymer was observed from gel time studies at elevated temperatures (>160°C). The morphological study based on SEM photographs showed that good fiber impregnation by the PMMA/PU IPN resin was obtained. From their studies, they concluded that PMMA/PU IPN prepolymer was suitable for the PMMA/PU IPN pultrusion process (see Figure 11.3).

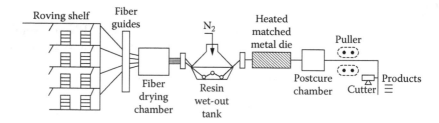

FIGURE 11.3
Flow chart of the Pultrusion machine.

The flexural strength, flexural modulus, and hardness of IPN composites increased with PMMA content. However, the impact strength and swelling ratio decreased with increasing PMMA content.[27]

Applications

IPNs are now used actively worldwide for structural adhesives and sealants, such as reactive hot melts and new insulated glass sealants.

A semi-IPN based on a thermoplastic polymer incorporated into a urethane network is one of the materials developed in the United States, Europe, and Asia. The urethane confers moisture-cure properties so that the final composition cures in place as a high-strength, heat-resistant hot melt with reduced cold flow. Appropriate application equipment is suggested to ensure that the composition remains stable and moisture-free during application and that sufficient water is present after application to produce the required moisture cure.

The resultant products from IPNs are composites that combine the high modulus and temperature properties of the engineering thermoplastic matrix with the improved lubricating and release properties of a silicone. But what may be the biggest plus commercially for these IPNs is their much lower mold shrinkage and warpage.

A high-performance digital tape drive IPN provides the warpage control and dimensional stability necessary for the component to replace the originally designed C fiber-reinforced polycarbonate carriage.

The IPN consisted of a statically conductive compound based on thermoplastic polyester (PBT) with C fiber reinforcement and PTFE lubrication. The material had a low coefficient of friction and low wear, and met all other property requirements but one: minimal warpage.

By modifying the original IPN with PBT and 15% silicone/PTFE and 30% C fiber, the postmold warpage in the part was reduced by 22% and a production carriage meeting the stringent quality standards of the manufacturer was produced. In fact, before commercial production, several of the carriage/head assemblies were subjected to rigorous testing up to 100,000 cycles with no failures.

Interpenetrating Phase Ceramics (IPCs)

IPCs developed at the Colorado School of Mines use one-step, in-situ simultaneous combustion to synthesize dense, affordable composites.

IPCs incorporate a ductile metal network into a porous but continuous ceramic matrix. The ceramic provides the advanced mechanical properties, whereas the metal reduces brittleness. What makes the production unique is the use of self-propagating high-temperature synthesis (SHS) to generate all composite components at the same time. Applying consolidation pressure just prior to or immediately after SHS produces dense systems.

Developers have used the technique to make $TiC-Al_2O_3$, $TiC-Al_2O_3-Al$, and $TiC-Al_2O_3-Ti$. Adding Al_2O_3 lowers the combustion temperature and controls particle size and density (which in turn improves stiffness and compressive strength). Even with Al_2O_3, the exothermic reaction generates enough heat for the Al to infiltrate the ceramic and form a ductile metal network. The result is a potentially low-cost composite.[28]

References

1. Savage, P., Market Forecast: Liquid Crystal Polymers (LCP), *High-Tech Mater. Alert, Technical Insights,* December 22, 1995, pp. 2–5.
2. Kaslusky, A., Liquid Crystal Polymers, A Primer, *AM&P,* December 1993, pp. 38–41.
3. Kerai, A., *Japan Chemical Week,* 1, November 11, 1985.
4. Brown, A. S. (Ed.), *High-Tech Matls. Alert,* 3(6), June 4, 1986.
5. Chung, T.-S., *Polym. Eng. Sci.,* 26, 901, 1986.
6. Baer, E. and Moet, A., *High Performance Polymers,* Munich: Hanser Publishers, December 1990, p. 329.
7. McChesney, C. E., Dole, J. R., Higher Performance in Liquid Crystal Polymers, *Mod. Plast.,* January 1988, pp. 112–118.
8. Stevens, T., The Unusual World of Liquid Crystal Polymers, *ME,* January 1991, pp. 29–32.
9. English, L. K., The New Look of LCPs, *ME,* June 1989, pp. 29–33.
10. Teague, P., Liquid-Crystal Polymers Produce Liquid Assets, *Des. News,* October 24, 1994, pp. 48–51.
11. Frost, D., Vectra Liquid Crystal Polymers, *AM&P,* July 1994, pp. 13, 16.

12. Dingemans, T., Weiser, E., Hou, T., Jensen, B., and St. Clair, T., Liquid-Crystal Thermosets, A New Generation of High-Performance Liquid-Crystal Polymers, *NASA Tech Briefs,* February 2004, pp. 42–43.

13. Sawyer, L., Shepherd, J., Kaslusky, A., and Knudsen, R., Unfilled Liquid Crystal Polymers, *AM&P,* June 2001, pp. 61–63.

14. McChesney, C., Bowers, D. A., and Kaslusky, A., Resin Selection for High-Performance Connectors, *SMT,* May 1996, pp. 70–74.

15. McChesney, C., Dole, J., and Brand, M., Flat Connectors, Right Out of the Mold, *Mach. Design,* October 9, 1997, pp. 108–116.

16. *AM&P,* February 1994, p. 49.

17. Recyclable Liquid Crystalline Polymer-Reinforced Plastics, *JETRO,* March 1996, p. 29.

18. *JPRS Rept.,* June 1994, p. 53.

19. Leggett, T., Laser Direct Structuring & Liquid Crystal Polymer Simplify Circuit Miniaturization, *Photonics Tech Brief,* January 2004, pp. 12a–13a.

20. Braeckel, M. J., Liquid-Crystal Polymers Cut the Cost of Medical Equipment, *Mach. Design,* December 10, 1998, pp. 60–66.

21. Medical-Grade Liquid Crystal Polymers, *Mach. Design,* September 18, 2003, p. 146.

22. Sorathia, U. A., Yeager, W. L., and Dapp, T. L., Interpenetrating Polymer Network Acoustic Damping Material, Dept. of the Navy, Washington, D.C.

23. Interpenetrating Polymer Networks, NERAC, Inc., Tolland, CT, April 1994.

24. Interpenetrating Polymer Networks, NERAC, Inc., Tolland, CT, June 1994.

25. Pater, R. H., Tough, Processable Simultaneous Semi-Interpenetrating Polyimides, NASA, Hampton, VA, Langley Research Center, p. 14.

26. Pater, R. H., Smith, R. E., Razon, R. T., et al., Freeze-Drying Makes Improved IPN and Semi-IPN Polymers, *NASA Tech Brief,* p. 29, May 1993.

27. Chen, C.-H. and Ma, C. M., Pultruded Fiber-Reinforced PMMA/PU IPN Composites — Processability and Mechanical Properties, *Composites Part A, Applied Sci. and Manufacturing,* 28(1), 65–72, 1997.

28. Interpenetrating Phases Form Low-Cost Ceramic Composites, *High-Tech Matls. Alert,* Tech Insights, Inc., November 17, 1995, pp. 10–11.

Bibliography

Baird, D. G., Processing Studies on In Situ Composites Based on Blends of Thermotropic Liquid Crystalline Polymers with Thermoplastics, Virginia Polytechnic Inst. and State University, Blacksburg. VA, *NTIS Alert,* 95(20), 5, 1995.

Chik, G. L., Li, R. K. Y., and Choy, C. L., Properties and Morphology of Injection-Molded Liquid-Crystalline Polymer/Polycarbonate Blends, *J. Mater. Proc. Technol.,* 63(1–3), 488–493, 1997.

Conductive Polymers Ready for Commercial Use, High-Tech Matls. Alert, *Tech Insights,* June 24, 1996, pp. 1–2.

Dole, J. R., LNP and IPN, *SAMPE Journal,* July/August 1986, pp. 134–136.

Downs, R., LCP "Micro-Parts" Cut Costs, *Mod. Plast.,* November 1995, p. 92.

English, L. K., The New Look of LCPs, *ME*, June 1989, pp. 29–33.

Gagne, J., Liquid Crystal Polymers, *Plast. World*, July 1992, p. 57.

Interpenetrating Polymer Networks, NERAC, Inc., Tolland, CT, *NTIS Alert*, p. 31, June 15, 1995.

Interpenetrating Polymer Networks, NERAC, Inc., Tolland, CT, *NTIS Alert*, p. 11, June 15, 1995.

Interpenetrating Polymer Networks, NERAC, Inc., Tolland, CT, *NTIS Alert*, p. 47, October 15, 1993.

Klempner, D. and Frisch, K. C. (Eds.), Advances in Interpenetrating Polymer Networks, Polymer Technologies, University of Detroit, Vol. 31991, p. 283, Vol. 4, 1994, p. 363.

Langerand, P., Dupont Joins Field of LCP Players, *Mod. Plast.*, July 1994, p. 48.

LCP, *Mod. Plast.*, January 1995, p. 54–55.

Li, J. X., Silverstein, M. S., Hiltner, A., and Baer, E., Morphology and Mechanical Properties of Fibers from Blends of a Liquid Crystalline Polymer and Poly(ethylene Terephthalate), Case Western Reserve University, Cleveland, OH, *Jn I. Appl. Polymer Sci.*, 44, 1531–1542, 1992; *NTIS Alert*, 96(3), 28, 1996.

Magnuson, B., LCP Fibers for High-Performance Needs, *ME*, June 1989, p. 10.

McChesney, C. E. and Dole, J. R., Higher Performance in Liquid Crystal Polymers, *Mod. Plast.*, January 1988, pp. 112–118.

Myers, J. and Moore, S., As Electronics Get Smaller, Resin Specs Get Tougher, *Mod. Plast.*, March 1994, pp. 60–61.

New, Higher-Performance Plastic Tubing, *Av. Week & Sp. Technol.*, May 13, 1996, p. 13.

O'Brian, J., LCPs: Heat Resistant, Tough, and Moldable, *Plast. World*, November 1985, p. 65.

Owens, J. O., Adoption of Recyclable Liquid Crystal Polymer FRP, June 27, 1994, p. 12.

Plastics Technology, July 1994, p. 21, 104.

Semi-Interpenetrating Polymer Network (IPN) Type Ultrafiltration (UF Membrane), p. 1, http://www.yet2.com/app/utility/external/indextechpak/28871, 7/7/03.

Singh, J. J., Pater, R. H., and Eftekhari, A., Microstructural Characterization of Semi-Interpenetrating Polymer Networks by Positron Lifetime Spectroscopy, NASA, Hampton, VA, Langley Research Center, *NTIS Alert*, 97(14), 1996.

Stevens, T., The Unusual World of Liquid Crystal Polymers, *ME*, January 1991, pp. 29–32.

Teague, P., Polymer Housing Surpasses Tight Connector Tolerances, *Des. News*, July 8, 1996, p. 25.

Teague, P., Liquid-Crystal Polymers Produce Liquid Assets, *Des. News*, October 24, 1994, pp. 48–51.

Thermoplastic Interpenetrating Polymer Networks, NERAC, Inc., Tolland, CT, *NTIS Alert*, p. 74, Vol. 96, No. 21, November 1, 1996.

Thermoplastic Interpenetrating Polymer Networks, NERAC, Inc., Tolland, CT, *NTIS Alert*, p. 31, September 15, 1994.

Xu, Q. W., Man, H. C., and Lau, W. S., Melt Flow Behavior of Liquid-Crystalline Polymer In-Situ Composites, *J. Mater. Proc. Technol.*, 63(1–3), 519–523, 1997.

12

Processes and Fabrication

Introduction

Most people understand that the well-worn Aesopian proverb, "Slow and steady wins the race," is flawed. Its laudable call to hard work ignores the fact that a flash of inspiration can propel one scientist's research ahead of another's. The trick, of course, is to back the flashy move with the proper amount of preparatory and follow-up effort.

Along with much hard work, most process/fabrication/manufacturing advances include some flashy moves — in the best sense. The resultant mix of invention and steady improvement is precisely what is needed to keep the various fields of processing and materials technological development moving along at a perennial brisk pace.

The processes and emerging technologies described in this chapter have been through the research, development, prototyping, evaluation and testing, and commercialization production phases.

Composite Processes

Dry Process for PI/C and B: Fiber Tape

A dry process has been invented as an improved means of manufacturing composite prepreg tapes that consist of high-temperature thermoplastic PI resin matrices reinforced with C and B fibers. Such tapes are used (especially in the aircraft industry) to fabricate strong, lightweight composite-material structural components. The inclusion of B fibers results in compression strengths greater than can be achieved by use of C fibers alone.[1]

Until now, PI/C- and B-fiber tapes have been made in a wet process: B fibers are calendared onto a wet prepreg tape comprising C fibers coated with a PI resin in solution. In the calendaring step, the B fibers, which typically have relatively large diameters, are pushed only part way into the wet prepreg. As a result, the B fibers are not fully encapsulated with resin. In addition, the presence of solvent in the prepreg contributes significantly to the cost of the finished product in two ways: (1) The tackiness and other handling qualities are such that the prepreg tape must be laid up in a labor-intensive process, and (2) the solvent must be removed and recovered before or during the final cure of the PI.

The present dry process is intended to enable the manufacture of prepreg tapes (1) that contain little or no solvent; (2) that have the desired dimensions, fiber areal weight, and resin content; and (3) in which all of the fibers are adequately wetted by resin and the B fibers are fully encapsulated and evenly dispersed. Prepreg tapes must have these properties to be usable in the manufacture of high-quality composites by automated tape placement. The elimination of solvent and the use of automated tape placement would reduce the overall costs of manufacturing.

In this process, a layer of parallel B fibers is formed and sandwiched between two layers of parallel C-fiber tows coated with a powdered PI resin. The layers are then heated and pressed together to form composite tape. As shown in Figure 12.1, the B fibers and the powder-coated C-fiber tows are pulled off reels and through combs that form the groups of fibers into the various layers with the lateral spacings consistent with the desired areal densities of C and B fibers. The three layers are pulled through a furnace and maintained parallel until they reach a position where each layer of coated C-fiber tows slides against a set of impregnation/spreader bars. The temperature zones in the furnace are set to provide enough heat to melt the PI before arrival at the bars. The bars are heated to promote

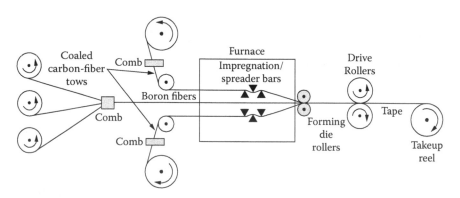

FIGURE 12.1
A layer of boron fibers is formed between two layers of resin-coated carbon-fiber tows. The fibers in each layer are spaced apart in the direction perpendicular to the page. The layers are heated and pressed together to form a composite tape.[1]

the flow of the resin system while facilitating the spreading of the tows as they slide over and under the bars.

After passing the bars, the layers are brought out of the furnace and pressed together between two forming die rollers. The speed of pulling of the tape and its fiber constituents is controlled by means of a pair of drive rollers downstream of the forming die rollers. The speed is chosen such that the time at temperature is adequate for the required melt flow. After passing through the drive rollers, the finished tape is wound on a take-up reel.[2]

Metal Matrix Composites (MMC)

The first production implementation of a Ti matrix composite (TMC) component in aircraft gas turbine engines was earmarked for use in the GE-built F110 engine for the F-16, a TMC nozzle compression link.

Modern aircraft performance is directly related to thrust-to-weight ratio of engines and the combined weight of the aircraft structure, systems, subsystems, and fuel. TMC is an advanced composite material of Ti reinforced with either SiC particulate or filament. Parts fabricated with TMC are significantly stronger, lighter, and considerably more resistant to the stress of extreme temperatures than conventional Ti or superalloys. They also provide increased performance (range, payload, and fuel efficiency).

This technology is key to improvements in propulsion systems for the next generation of commercial and military aircraft. Substantial cost, performance, and durability benefits are expected from the use of TMC components in transport and fighter aircraft engines. Other potential applications for TMCs include airframes, medical equipment, and chemical processing.

The technology to create items for military and commercial usage using TMCs is not new. Up to now, unfortunately, TMCs have been cost-prohibitive because a production base has not existed to affordably and routinely produce high-quality components. However, an Air Force and industry consortium was found, and they were to mature the TMC fabrication industry and deploy TMCs in advanced gas turbine engines in the form of fan blades, fan frames, actuators, rotors, vanes, cases, ducting, shafts, and liners.

A major step in that direction was reducing the cost of manufacturing the nozzle compression link by more than 66%. Although its competing metal part remains less expensive, continuing cost reduction actions, and improvements in manufacturing processes, will make TMC components more economically attractive for a broader range of military and commercial applications.

FIGURE 12.2
The U.S. Air Force F-16 jet fighter includes silicon carbide particulate-reinforced aluminum MMCs in the ventral fin and the fuel access door covers. The arrow indicates the aluminum composite ventral fin.[3]

The compression link deployment over the next 8 to 10 years could mean the manufacture of more than 20,000 parts.

The use of Al alloys in MMCs has been accepted more easily than Ti.

SiC particulate-reinforced Al MMC has been used in the U.S. Air Force F-16 aircraft. One application is in the ventral fins for the aircraft, which provide lateral stability during high angle of attack maneuvers (Figure 12.2). A 6092/SiC/17.5p MMC sheet material replaced the unreinforced Al skins in the honeycomb structure of the fin and, due to the increased specific stiffness and erosion resistance of the MMC material, increased service life significantly. The same material is used in the fuel access door covers for the F-16. Unreinforced Al access covers were experiencing cracking due to overload, and again the increased specific stiffness property of the MMC allowed this material to be successfully substituted.

Replacement of a C-fiber reinforced polymer tube by an SiC particulate reinforced Al extruded tube for a floor support strut for Airbus has also been reported. The driving force is improved damage tolerance and reduced cost.[3]

The Eurocopter rotor sleeve is made of a forged SiC particulate-reinforced Al 2009 alloy having good stiffness and damage tolerance. The Al MMC replaced Ti, reducing both weight and production cost.

In the automotive market, properties of interest to the automotive engineer include specific stiffness, wear resistance, and high-cycle fatigue resistance. Although weight savings is also important in automotive applications, the need for achieving performance improvements with much lower cost premiums than tolerated by aerospace applications drives attention toward low-cost materials and processes. MMCs have been successfully introduced where the combination of properties and cost satisfied a particular need.

- Engine: Replacement for steel and cast Fe in engine applications relies on the increased specific stiffness, on improved wear resistance, and in some cases, on the increased high cycle fatigue resistance provided by MMCs. A watershed application for Al MMC was the Toyota piston for diesel engines. The part consists of selective reinforcement of the Al alloy by a chopped fiber preform in the ring groove area that provides improved wear and thermal fatigue resistance. These pistons were placed into commercial production in Japan.

- Another component with selective reinforcement is in the Honda Prelude 2.3-l engine. In this case, hybrid preforms consisting of C and Al_2O_3 fibers were infiltrated by molten Al to form the cylinder liners during the medium-pressure squeeze casting process for the engine block.

- Brake system: Al-based MMCs offer a very useful combination of properties for brake system applications in replacement of cast Fe. High wear resistance and high thermal conductivity of Al MMCs enable substitution in disk brake rotors and brake drums with an attendant weight savings on the order of 50 to 60%. Because the weight reduction is in unsprung weight, it also reduces inertial forces and provides additional benefit.

- Driveshaft: Al MMCs in the driveshaft take advantage of the increased specific stiffness in these materials. Current driveshafts, whether Al or steel, are constrained by the speed at which the shaft becomes dynamically unstable. Driveshafts have been made of Al $6061/Al_2O_3$ materials produced by stir casting and subsequent extrusion into tube.

In the general sector, there are commercial and industrial products:

- Nuclear shielding: B_4C reinforced Al has promise in nuclear shielding applications because the isotope B10 present in B_4C naturally absorbs neutron radiation. As a result, this type of MMC is being considered for storage casks that will contain spent fuel rods from nuclear reactors.

- Precision parts: Al MMC in an industrial application requiring a lower weight material for improved precision has been reported.[3] Specifically, a 6-m-long MMC needle replaced a steel needle in a carpet weaving machine.

- Electrical conductors: An innovative yet specialized application of continuous fiber reinforced Al is for overhead electrical conductors. By infiltrating Nextel 610 continuous-fiber reinforcement with Al, the resulting composite wire can replace the conventional steel-reinforced wires. The performance of this product significantly improves the ampacity, or current-carrying ability, by 1.5 to 3 times compared with the steel-reinforced Al construction. In addition, because of the higher strength-to-weight ratio of the composite conductor, existing infrastructure such as transmission towers can remain, and longer spans between towers are possible. This advanced concept is being tested in field trials.

- A very important market area for Al matrix MMCs is in their application in electronic packaging and thermal management. The primary property capabilities of MMCs that are of interest in this market are the ability to tailor the CTE while retaining good thermal conductivity, and in some cases electrical conductivity, along with light weight. These components are typically high value-added.

- AlSiC: The key material of commercial importance has been coined "AlSiC" by the industry. Although not representing any specific formulation, in general AlSiC covers particle-reinforced Al MMCs in which the SiC volume fraction ranges from 20% to over 70% by volume, depending on the specific needs of the application.

- Microwave packaging: An early application of a 40 vol% reinforced Al MMC was in replacement of Kovar, a heavier Ni-Co-Fe alloy, in a microwave packaging application. The major drivers here were weight savings in this part, with a 65% reduction realized, along with improved thermal conductivity over the baseline material.

- Microprocessor lids: The application of AlSiC in microprocessor lids, specifically in flip chip packages, is driven by the need for a lightweight and potentially lower-cost replacement for Cu. The appropriate CTE in the MMC can be tailored to match the adjoining package by controlling the reinforcement volume fraction.

- Other applications: Printed wiring board cores are also being made of AlSiC materials, as replacements for conventional Cu or Al cores. In addition to the CTE matching property, the higher specific stiffness reduces thermal cycling and vibration-induced fatigue. Yet another application in which the thermophysical properties of AlSiC are important is in carriers for hybrid circuits for power amplifiers in cellular phone base stations.

Superior thermal conductivity and light weight in a CTE-matched material are also exploited in power module base plates, which are made of both AlSiC and Al-graphite MMCs.

RTM/CoRTM/VARTM/RARTM

Engineers have been looking for alternative processing methods that can reduce costs while maintaining the high performance of autoclave-cured components. Within the last few years, liquid composite molding (LCM) technologies have advanced to the point where they can provide that alternative. LCM processes are characterized by the injection of a liquid resin into a dry fiber preform and include resin transfer molding (RTM) and vacuum-assisted RTM (VARTM).

In conventional RTM, the preform is placed into a closed, matched tool and resin is injected under pressure on the order of 0.69 to 1.38 MPa. This low-pressure molding process contains the mixed resin and catalyst, which are injected into a closed mold containing a fiber pack or preform. When the resin has cured, the mold can be opened and the finished component removed.

A wide range of resin systems can be used including polyester, vinylester, epoxy, phenolic, and and methyl methacylates, combined with pigments and fillers including Al trihydrates and Ca carbonates if required.

The fiber pack can be glass, C, aramid, or a combination of these.

Early RTM processes, however, lacked die consistency needed for aerospace components, in both dimensional tolerances and mechanical properties. Fiber volume fractions were significantly lower than the 60 to 65% typical of prepregs. Problems with predicting flow fronts as well as flaws that were introduced into the preform when closing the matched metal molds often led to high void contents and dry spots.

Additionally, there has been a constant battle between open and closed mold processes for leading the composites market and with the new EPA and MACT standards. There is also a growing trend in switching to closed molding processes. Marine, wind, and some other industries, once dominated by open mold processes, are now moving to closed molding processes such as VARTM and RTM. The recent market study *Global Composites Market 2003–2008* breaks down the global composites industry as well as many of the market segments such as marine, automotive, and construction in terms of various open and closed molding processes and provides a clear picture on the size of open and closed molding processes in various market segments.[4]

Improvements in both materials and processes, though, have recently made RTM a viable option for aerospace manufacturing, where it normally takes 10 to 15 years for a new technology to become accepted. With RTM, the breakthrough began when Lockheed Martin selected RTM for many of the F/A-22 Raptor's structural components. Composites comprise approximately 27% of the F/A-22's structural weight (24% thermoset and 3% thermoplastic). RTM accounts for more than 400 parts, made with both BMI and epoxy resins. The wing's sine-wave spars were probably the first structural application of RTM composites in an aircraft. For a vertical tail on another Lockheed Martin aircraft, the RTM process reduced the part count from 13 to 1, eliminated almost 1000 fasteners, and reduced manufacturing costs by more than 60%.[5]

Because the RTM process is more complex than autoclave curing, it is more difficult to develop a general qualification methodology. With prepregs, the material manufacturer mixes the resin and impregnates the tape or fabric under highly controlled conditions. Once a material is qualified, for example, an end user just has to demonstrate site equivalency of its manufacturing process. With RTM, however, both the resin mix and the resin content are more variable. In particular, the final resin content depends on maintaining a good flow front.

The high cost of tooling also limits the adoption of RTM. A set of production tools can cost on the order of $500,000. Although the price is competitive with autoclave tools, autoclave programs can usually get by with a single set of tooling for both development and production. With RTM, the resin flow front (and hence the part quality) is highly dependent on the tooling geometry. Often it is necessary to build one or more sets of prototype tools, to develop and test the process, before the production tooling can be built. Although prototype tooling is less expensive than production tooling, it is not so inexpensive that it can be considered expendable.

Simulation of the molding process can predict flow fronts, allowing engineers to virtually test different mold designs without building expensive hardware. For example, you can model the resin flow and be able to identify preferred injection sites and sequences to achieve complete mold filling with no dry spots.

If one would ask 100 people to define RTM, you will get 100 different answers. Each manufacturer has its own version of RTM, and in many cases the differences are great enough to consider the process unique. The variations are even greater for VARTM, which can include a boat hull manufacturer using SCRIMP and an aerospace manufacturer using all-steel molds with vacuum only. For example, V System Composites (VSC)[5] has three versions — classical RTM, HyPer-VARTM, and HyPerRTM — each tailored to a different need.

In addition to classical RTM, VSC also has developed the HyPerVARTM process: a resin infusion system targeted to the aerospace industry. Most VARTM systems use a flow medium to get good resin coverage. This works well for relatively simple structures like boat hulls, but cannot handle the features amid complexities common in aerospace components, such as multiple buildups and ply drop-offs. HyPerVARTM is a single-point-of-injection system. Resin delivery is designed into the tool, eliminating the infusion medium and significantly reducing other consumables and touch labor, which translates into much lower costs. Selective control of permeabilities throughout the part provides excellent control over fiber volume fractions. The process is ready for production and has been proven on the CH-47 helicopter's forward pylon deck and on a Ka band deep space reflector built for NASA's Jet Propulsion Laboratory. Airbus is currently evaluating HyPerVARTM for production use, as well.

The third system in VSC's process development is the HyPerRTM process, which the company says combines the best of HyPerVARTM with classical RTM. HyPerRTM molds are built using the HyPerVARTM process and can incorporate some metallic details where necessary. Resin distribution details also are built into the HyPerRTM mold. HyPerRTM enables one to produce parts with the high quality of classical RTM but at the price of HyPerVARTM.

A new innovative process has been developed through the efforts of government and industry that reduces fabrication and assembly costs.

The process is called CoRTM (Co-curing of an uncured skin to a Resin Transfer Molded substructure). It was developed by Northrop Grumman and produces large, integrated, weight-efficient, precise, and repeatable structures.[6]

A vertical stabilizer from the F-35 was used to demonstrate the technology. These results, using CoRTM in the manufacturing of the part, revealed that nearly $14,000 in savings could be derived through reduced tooling, part count, fastener count, and the associated fit-up, liquid shimming, and surface mold line treatments for air vehicles.

Traditional aircraft structures consist of multiple piece assemblies that are prefit together; gaps between mating surfaces are filled with shim materials to create a snug fit and then mechanically fastened in place. This results in very lengthy manufacturing flow times and high acquisition costs.

Now, the CoRTM process has been proven to be a viable and promising alternative for affordable composite structures. CoRTM combines two cost-effective processes: fiber placement (the automated placement of bands of high-strength fibers combined with resin onto a tool) for skin structures, currently used on the F-35, F-18, V-22, F/A-22, and so on; and RTM (the injection of high-strength resin into a mold containing high-strength fibers formed to a specified shape) for substructures currently used on the F/A-22 Raptor as well as other aircraft.[7]

Instead of fastening the skin to the substructure, the CoRTM process enables the skin and the substructure to be designed and fabricated as a single component, eliminating the need to fasten them together. This creates structures with fewer parts and minimal fasteners, resulting in reduced assembly costs.

The savings, vs. the baseline construction costs, for the F-35 tail represent a 52% reduction in part count, a 38% reduction in tool count, a 7% reduction in weight, and a 17% overall cost reduction when compared to the typical F-35 construction process.

Vacuum-assisted resin transfer molding refers to a variety of related processes that represent the fastest-growing new molding technology. The salient difference between VARTM-type processes and standard RTM is that in VARTM, resin is drawn into a preform through use of a vacuum, rather than pumped in under pressure. VARTM does not require high heat or pressure. For that reason, VARTM operates with low-cost tooling, making it possible to inexpensively produce large, complex parts in one shot.

VARTM is best described as a complimentary process to RTM. VARTM mold costs are basically half the price of equivalent RTM molds but they produce, at best, at half the rate of RTM; however, the process provides molders an attractive introductory route into closed mold production.

In VARTM, resin flow rates cannot be speeded up above an optimum level in order to fill the mold more quickly, as the recommended VARTM mold construction and the atmospheric mold clamping pressures limit overall in-mold pressures to less than 0.5 bars.

As with any composite closed-mold production technique, VARTM is no exception to the rule in demanding high-quality accurate composite molds in order to provide good mold life and consistent production of good parts (see Figure 12.3).

In the VARTM process, fiber reinforcements are placed in a one-sided mold, and a cover (rigid or flexible) is placed over the top to form a vacuum-tight seal. The resin typically enters the structure through strategically placed ports. It is drawn by vacuum through the reinforcements by means of a series of designed-in channels that facilitate wet-out of the fibers. Fiber

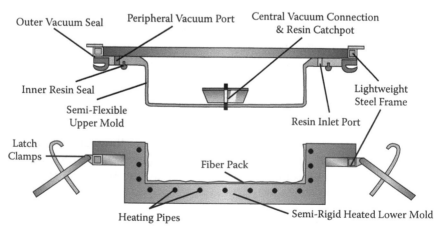

FIGURE 12.3
VARTM process.[8]

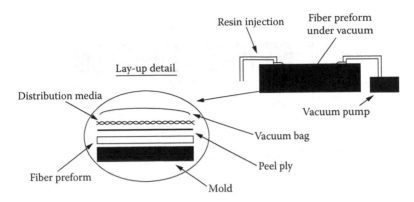

FIGURE 12.4
Schematic of VARTM process with lay-up detail.[8]

content in the finished part can run as high as 70%. Current applications include marine, ground transportation, and infrastructure parts.

Recently, an advanced VARTM system called Sequential Multi-Port Automated Resin Transfer Molding (SMART-Molding) was developed to fully automate the infusion process. The system incorporates sequential injection VARTM processing, actuators to control the flow, sensors to detect the flow behavior during resin injection, and on-line control to optimize the opening and closing of the actuators. The system has been successfully demonstrated during laboratory-scale process trials[8] (see Figure 12.4).

For the first time, the SMART-Molding approach completely automates this process. The injection gates are opened in sequence when mold-mounted flow sensors detect the tool-surface flow front at a particular gate. The process is completely autonomous and does not depend on resin permeability variations, changes in preform characteristics, or other processing factors. In addition, the SMART-Molding system does not require costly trial-and-error development of the conventional sequential scheme and therefore does not demand the engineering knowledge normally required during resin impregnation. However, the number of inlets and sensor locations must be specified.[8]

The RARTM process uses trapped rubber and a pressure relief valve with a closed mold. This provides

- Easy tool closing
- Improved resin flow and fiber wetting
- High fiber volume, low void content laminate

Figure 12.5 reflects the RARTM process, which is compared to conventional RTM (Figure 12.6), which is a closed-mold, low-pressure process.

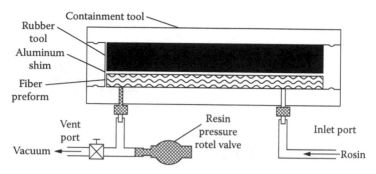

FIGURE 12.5
RATM process.

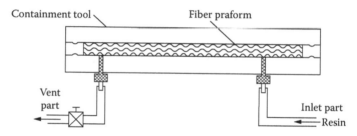

FIGURE 12.6
Conventional RTM.

A progressive injection method is used in the RARTM process with multiple injection/vent ports.

The part is positioned vertically during resin injection, and the resin is injected at the lowest geometric portion of the part and vented from the highest position.

When the resin reaches the port, the vacuum is removed and the resin is injected, and the procedure is repeated as the resin progresses up the part.

Infusion Processes: Resin Film Infusion (RFI)/DIAB Method

Although there are many variations of the infusion process, some proprietary and some generic, the basic principle is the same. Dry materials (reinforcements and core) are placed into a mold and then covered with a vacuum bag. A vacuum is then applied, resin is fed into the mold and the vacuum draws it through the pan until saturation occurs. The vacuum is then maintained until the part is cured.

Normally the process yields high-performance parts; that is, low weight due to a high fiber fraction. It is a repeatable process that is easy to control in a quality management system. Thus, the process is not reliant on the individual skill of the workers. It produces parts with a consistent and even quality. The working environment can be drastically improved and if done utilizing a mixing and dispensing machine, emissions can be decreased by more than 95% compared with open mold methods.

One crucial factor is to achieve a good flow of the resin in the process. This can be done utilizing different types of distribution media. Basically the resin flow can take place at three different locations in the material lay-up:

- Above the structural laminate. By using a separate distribution media with a low permeability above the structural laminate, the resin is distributed over the surface. This method yields high fiber fractions but produces a large amount of waste. Above the structural fibers, a peel-ply, sometimes a release film, and then the distribution media is put down before the vacuum bag is placed on top. All these materials are normally consumables and they are thrown away after each "shot." As they are not cheap either to buy or to lay-up, take away, and then dispose of, this increases the cost for the product. The method is known under many different trademark names.

- In the structural laminate. By using a distribution mat with a low permeability in the structural laminate, the resin is distributed over the surface. This method yields lower fiber fractions and thus a higher weight than surface infusion but does not produce large amount of waste. Also, this method makes it possible to get a gel coat on both surfaces if a light upper mold is used. In the structural fibers, a distribution media is put down before the vacuum bag is placed on top. Different types of these distribution materials are available on the market.

- Below the structural laminate. By using a grooved core, the resin is distributed over the surface of the core. This method yields high fiber fractions and thus a lower weight. It does not produce large amount of waste. Also, this method makes it possible to get a gel coat on both surfaces if a light upper mold is used.

RFI is a process whereby a resin film is impregnated in a preform by melting and infusing it into the preform.

The infusion process utilizing grooved core materials is an innovative, faster, and cheaper closed manufacturing method for producing sandwich constructions with fiber composite skins.[9]

FIGURE 12.7
Example of groove pattern in the core surface.[9]

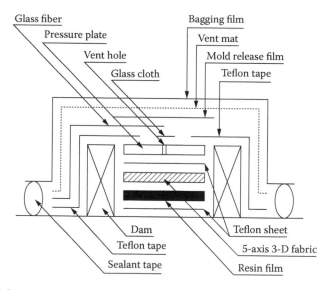

FIGURE 12.8
Bagging configuration for RFI processing.[10]

The grooved core method, however, utilizes an optimized fine groove pattern in the core surface to facilitate resin distribution (see Figure 12.7).

One RFI bagging configuration used is shown in Figure 12.8. A resin film was placed under the five-axis 3D fabric to effectively remove the air inside the five-axis 3D fabric. The resin film was prepared beforehand by melting bismaleimide at approximately 100°C and molding into the shape of the flat panel. A bismaleimide in the melt state has very low viscosity of 100 centipoise or less, and it is very easy to impregnate. Because it may leak from a clearance around dams, the clearance is scaled by the Teflon tape. The amount of resin was adjusted so that the fiber volume content of the fabricated composite might become 50%. The cure process was carried out at a temperature of 191°C and at a pressure of 0.62 MPa for 6 hr.[10]

The DIAB Infusion Method[11] is an innovative, faster, and cheaper closed manufacturing process for producing sandwich construction with composite skins.

The DIAB method utilizes an optimized fine groove pattern in the core surface to facilitate resin distribution. This has the following implications:

- Grooved core = > faster flow and no extra distribution media;
- Faster flow = > larger panels possible;
- Larger panels = > less extra materials and minimum waste;
- Less extra materials = > faster lay-up and a less costly product.

The DIAB method has the advantage of being extremely fast, making large structures feasible in one shot. It also does not need additional consumables like resin distribution materials, release films, and peel-plies, which makes it more environmentally friendly and less costly. Significant cost savings can also be made compared with traditional methods if molds and production equipment are adapted accordingly.

Examples of successful implementation in applications are boat hulls, wind turbine nacelles, shelters, containers, and storage tanks.[12]

C-fiber composite panels are an established technology in motor racing and luxury sports cars. This type of material allows lighter and stiffer components for low-volume production. New materials have been developed to offer low-cost alternatives to conventional prepreg processing and a route to higher production volumes.

Frost, Solanki, and Mills[13] conducted a study to evaluate three novel materials — Sprint CBS from SP systems; Zpreg from the Advanced Composites Group (ACG), Figure 12.9; and Carboform from Cytec, Figure 12.10 — for application in the ultralightweight demonstrator the Aerostable Carbon Car (ASCC). Manufacturing time, surface finish, and impact behavior were assessed and compared to prepreg technology. The results demonstrated that the novel resin film infusion systems were effective replacements for prepreg materials and their use should result in substantial manufacturing cost savings.

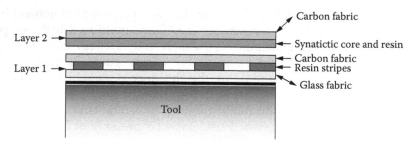

FIGURE 12.9
Schematic of Zpreg.[13]

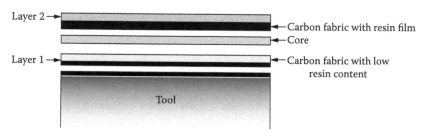

FIGURE 12.10
Schematic of carboform.[13]

Ceramic Matrix Composites (CMC)

Ceramic matrix composites (CMC) are materials that incorporate a reinforcing phase within a matrix phase during fabrication of the product. CMCs are fabricated using techniques such as sintering or hot-pressing. Sometimes a reinforcing preform is made of long, high-strength ceramic fiber, and then the ceramic matrix is infiltrated into the preform by methods such as chemical vapor deposition (CVD), sol gel, or polymer infiltration.

CMCs are under development at numerous government laboratories as well as industrial facilities to provide an alternative to monolithic ceramics with greater toughness that fractures in a noncatastrophic rather than a brittle mode. The CMC materials that presently have the best chance to nonbrittle failure modes are continuous fiber ceramic composites (CFCCs).

Currently, most of the precursor materials for advanced CFCCs are not in full-scale productions. Thus, prototypes of CFCC components are expensive in the near term. As the technology evolves, the baseline is continually being replaced by "new and improved" versions that generally are more expensive. R&D is needed to better the cost issue and to reduce the costs of precursors. Nonoxide fibers cost thousands of dollars per pound, and oxide fibers that have been in the marketplace for years sell for hundreds of dollars per pound because the production volumes remain small.

The optimization of the interphase (or interface) layer between the ceramic reinforcing fiber and matrix is critical to producing viable (commercial) CFCC materials. Currently, the interface layer is deposited by laboratory techniques that are an expensive step in the fabrication process. A technology challenge is the increased temperature stability to 1200–1500°C that is needed. Inadequate stability at high temperature is one of the major deficiencies of current CFCCs.

Significant improvements have been achieved in the past 5 years. Some CFCC materials have been demonstrated to survive in severe application environments such as an industrial gas turbine for about 2500 h at temperatures in the 1100–1200°C range. Environmental barrier coatings (EBCs) applied to the surface have increased the life to over 5000 h, with projections to about 10,000 h.

Target industrial applications require lifetimes of 20,000 h or greater, and they require higher temperature performance in the 1200–1500°C range. Nonoxide composite such as those containing SiC are not thermodynamically stable in air of high vapor atmospheres at high temperature.

However, progress is being made. Use of oxide EBCs has provided further improvement in CMC composite stability and an increase in life. Based on research in the past 3 years, EBCs are now required to obtain an acceptable component life for SiC-SiC CFCC material in a gas turbine atmosphere above 1100°C.

But, for the near term, the high price of finished CFCC components remains as a major barrier to commercial applications growth. Research to reduce the cost of precursor will increase the production volume, and the scale of manufacturing would also substantially reduce unit costs.

Trends in semiconductor electronics are challenging not only chip manufacturers but the makers of circuit boards as well. Circuit boards used for more than 95% of electronic products today are fiber-reinforced glass laminates. These boards are inexpensive and can be processed in large sizes (exceeding 60.96 cm), but they cannot support the high-speed and ultra-high-density integrated circuits (ICs) that are coming. Ceramic boards offer the needed thermal and mechanical stability but cannot be processed in large-area sizes, making ceramic too expensive for most applications. A government-sponsored development program involves a low-cost, reliable polymer-derived ceramic composite circuit board capable of supporting microminiaturized electronic systems of the future. The board material will have the required mechanical, thermal, and electrical properties — such as high stiffness, low moisture absorption, resistance to warpage, low thermal expansion, and dimensional stability — to support extremely high wiring density, four to eight layers of fine wiring, and multiple embedded functions.

Already developed is a low-cost SiC ceramic-forming polymer that can be chemically tailored to meet the demanding temperature and performance requirements of advanced electronic systems. The processing technology for fabricating thin CMC structures reinforced with microfibers will use this polymer, thus enabling low-cost manufacturing of smooth, flat boards for advanced electronic systems. By designing a board material that matches Si in expansion, an expensive underlining process can be eliminated. The program will utilize already developed electronic systems expertise to develop process and computational models to predict board behavior and properties, layer fine wiring and electronic assemblies on finished boards, and conduct tests to validate that this new technology fulfills the requirements of advanced electronic packaging applications. This work will support board technology that will play a key role in next-generation innovation of system-on-a-package technology, which will provide tenfold improvements in size, cost, reliability, and performance. System-on-a-package can deliver the advantages of system-level integration quicker and more cheaply than the problematical system-on-a-chip.

Metal Matrix Composites

Isotropic composites of Al-alloy matrices with particulate Al_2O_3 have been developed as lightweight, high-specific-strength, less-expensive alternatives to Ni-based and ferrous superalloys. These composites feature a specific gravity of about 3.45 g/cm^3 and specific strengths of about 200 $MPa/(g/cm^3)$. The room-temperature tensile strength is 689 MPa and stiffness is 206 GPa. At 260°C, these composites have shown 80% retention in strength and 95% retention in stiffness. These materials also have excellent fatigue tolerance and tribological properties. They can be fabricated in net (or nearly net) sizes and shapes to make housings. pistons, valves, and ducts in turbomachinery, and to make structural components of such diverse systems as diesel engines, automotive brake systems, and power-generation, mining, and oil-drilling equipment. Separately, incorporation of these MMCs within Al gravity castings has been demonstrated.

A composite part of this type can be fabricated in a pressure infiltration casting process. The process begins with the placement of a mold with Al_2O_3 particulate preform of net or nearly net size and shape in a crucible in a vacuum furnace. A charge of the alloy is placed in the crucible with the preform. The interior of the furnace is evacuated, then the furnace heaters are turned on to heat the alloy above its liquidus temperature. Next, the interior of the furnace is filled with Ar gas at a pressure about ($\approx$6.2 MPa) to force the molten alloy to infiltrate the preform. Once infiltrated, the entire contents of the crucible can be allowed to cool in place, and the composite part recovered from the mold.

Ceramic particulate fillers increase the specific strengths and burn resistances of metals. Researchers have theorized that the inclusion of ceramic particles in metal tools and other metal objects used in O_2-rich atmospheres (e.g., in hyperbaric chambers and spacecraft) could reduce the risk of fire and the consequent injury or death of personnel. In such atmospheres, metal objects act as ignition sources, creating fire hazards However, not all metals are equally hazardous: Some are more burn-resistant than others are. It was the researchers' purpose to identify a burn-resistant, high-specific-strength ceramic-particle/MMC that could be used in O_2-rich atmospheres.

The researchers studied several metals. Ni and Co alloys exhibit high burn resistances and are dense (ranging from 7 to 9 g/cm^3). For a spaceflight or industrial application in which weight is a primary concern, the increased weight that must be incurred to obtain flame resistance may be unacceptable. Al and Ti are sufficiently less dense that they can satisfy most weight requirements, but they are much more likely to combust in O_2-enriched atmospheres: In pure O_2, Al is flammable at a pressure of (absolute pressure $\approx$170 kPa) and Ti is flammable below (absolute pressure $\approx$14 kPa).

The researchers also examined ceramics, which they knew do not act as ignition sources. Unlike metals, ceramics are naturally burn resistant. Unfortunately, they also exhibit low fracture toughnesses. Because a typical ceramic lacks the malleability, durability, and strength of a metal, ceramics are seldom used in outerspace and industrial environments. The researchers theorized that a ceramic-particle/MMC might provide the best of both classes of materials: the burn resistance of the ceramic and the tensile strength of the metal. They demonstrated that when incorporated into such low-burn-resistance metals as Al and Ti, ceramic particles increase the burn resistances of the metals by absorbing heat of combustion. In the case of such high-burn-resistance metals as Ni and Cu, it was demonstrated that ceramic particulate fillers increase specific strengths while maintaining burn resistances.

Preliminary data from combustion tests indicate that an A339 Al alloy filled with 20 vol% of SiC is burn-resistant at pressures up to (absolute pressure $\approx$8.3 MPs) — that is, it has 48 times the threshold pressure of unfilled Al. The data show that of all the composites tested to date, this composite has the greatest burn resistance and greatest specific strength and is the best candidate for use in O_2-enriched atmospheres.

Advanced Fibers/Whiskers/Particulates

Siboramic Fibers

Next-generation combustion systems in power generation, aeronautics, and automotive applications are putting heavy demands on materials and components as the need for greater fuel efficiency and reduced emissions pushes combustion temperatures higher. Manufacturers have been forced to use expensive superalloys and ceramic fibers in high-temperature combustion applications.

Siboramic is a ceramic fiber that provides superior thermomechanical properties without the costly follow-up coating treatments necessary with conventional ceramic fibers. The low-density ceramic fiber does not require electron radiation curing and remains viable at up to 1500°C in an O_2 atmosphere.

These fibers are used to manufacture CMCs that have major applications in aerospace and power generation, where it is anticipated that significant operational cost reductions can be achieved through weight saving and higher thermal efficiencies.

Siboramic is corrosion-resistant to liquid metals such as Cu and Si. Applications include automotive engine components, valves, pistons, and brake discs, as well as waste incineration systems, turbine blades, and combustion liners for aircraft engines.

TiB Whiskers

TiB as reinforcement is attractive because of the absence of an intermediate phase between Ti and TiB, the requirement of a lower amount of B, and the relatively lower temperatures involved in composite processing. Ti-TiB composites may be processed at temperatures ranging from 900 to 1300°C.

Another major advantage is that TiB forms as long, pristine single crystal whiskers in the Ti matrix. This means that with a small amount of reinforcement, large increases in composite modulus and strength are possible, according to theories of whisker-induced stiffening and strengthening. On the other hand, compounds such as TiC, TiN, Ti_3Si_5, and TiB_2 do not have this advantage, because these compounds do not exist in thermodynamic equilibrium as whiskers in the Ti matrix.

TiB whiskers can be formed in Ti by solid-state reaction. Although the energy of formation G of TiB_2 is most negative, Ti and TiB_2 can react to form TiB because of the small negativity of the free energy of this reaction. This means that TiB whiskers can be formed in Ti by the reaction of Ti and TiB_2 particles, as long as the average amount of B is small. This reaction can be given as

$$Ti + TiB_2 \rightarrow 2TiB$$

The average amount of B required to form TiB is much smaller, because the compound requires one B atom for every Ti atom, while TiB_2 requires two B atoms for every Ti atom.

Methods of manufacture:*

- Wrought
- MA + HIP
- PM + HIP
- GA + HIP/Extrusion
- VAR + Hot swaging
- MA + CIP + Sintering + Hot swaging
- PM

Because of this reaction, solid-state fabrication techniques based on PM and sintering can be applied to manufacture Ti-TiB composites (see above). Both commercial powders and prealloyed gas atomized powder, as well as mechanically alloyed powders, are suitable. Following initial alloying or CIP, the composites are usually free-sintered or HIPed to enable the densification as well as the reaction between the Ti alloy and TiB_2 powders to form the TiB whiskers inside a continuous matrix.

Because of high stiffness, high hardness, and the low density of TiB whiskers, Ti-TiB composites can be effective contenders for applications requiring high strength and light weight. With improvements in composition, processing, and microstructure control, it is realistic to expect about 180 GPa in elastic modulus, about 1200 MPa in strength, and about 2 to 3% ductility with ~30% TiB whisker loading in the microstructure.

Therefore, one near-term possibility is the replacement of high-strength steels with these composites in some applications. This can reduce weight by about 45% in a structure, at the same time providing significant increases in resistance to creep, oxidation, wear, and corrosion. This is a promising outlook, considering the early stage of development.

Some notable current applications of Ti-TiB composites include golf-club heads as well as exhaust valves in SUVs made by Toyota. Some published research demonstrates that strength retention at high temperature is also superior, indicating potential for possible elevated temperature

---

* MA: mechanical alloying; PM: powder metallurgy processing; GA: argon gas atomization; VAR: vacuum arc melting; CIP: cold isostatic pressing; HIP: hot isostatic pressing.

applications. Although the cost of making these composites may be initially high, because of the powder metallurgical nature of processing, cost per part in large-scale commercial manufacturing will surely decline considerably.

Biocomposites

Investigations and research programs have been under way to develop biocomposites based on bioactive ceramic HA and bone cement (PMMA). The objective is to intensity the interaction and fixation between bone cement and tissue. A ceramic composite combining the attractive properties of hard Al_2O_3 and tough Y-TZP ZrO_2 may offer even better properties.

Other research is analyzing the potential of using zirconia-toughened alumina (ZTA) for total hip replacement. This new biocomposite, which consists of about 75% Al_2O_3, offers mechanical strength and fracture toughness similar to that of ZrO_2 and is nearly as hard as Al_2O_3. This advanced biocomposite is biocompatible: In vitro and animal tests have demonstrated that the wear couple offers better tribological properties for total hip arthroplasty than the standard Al_2O_3-on-Al_2O_3.

There is a great deal of potential for bioceramics in the future. Past experience has shown that all successful applications of bioceramics are based on interdisciplinary work starting with simple research and development. As we gain more of an understanding about designing with ceramics, an unlimited number of new applications will emerge.

One major application is ceramic hip replacement. Concerns over polyethylene (PE)-mediated osteolysis (the dissolution of natural bone) are driving the development of alternative bearing combinations for total hip replacements (THRs). In the past several years, an innovative new product has been developed: a CMC that can eliminate osteolysis caused by PE.

Test efforts focused on examining the wear behavior of the ceramic-on-ceramic articulating surfaces. Femoral heads and acetabular cups, fabricated from the ceramic composite, were tested on a hip simulator to 4.2 million cycles. The composite bearing couples and Al_2O_3 bearing couples were run at the same time to compare directly, under the same conditions, the wear resistance of the ceramic components. Additionally, the matrix wear resistance was compared to other articulating surface material combinations run under similar test conditions.

Test data for the CMC heads and cups indicate that very low wear rates of $0.016 \ mm^3/Mc$ were achieved. These data indicate the positive potential of this CMC to design without using ultrahigh molecular weight polyethylene (UHMWPE), therefore eliminating the disease related to the PE.

The low wear rates achieved also indicate the low level of ceramic particles produced, further reducing the possibility of bioactive reactions from this material.

Other test efforts were to determine the wear potential of the CMC while maintaining minimal influence of the fabrication procedures on the results.

Crystaloy™ or CxA

This initial work identified the potential to achieve 57% lower wear in 10 years and 75% lower wear in 22 years compared to Al_2O_3. During steady-state tests, the new composite demonstrated 38% lower volumetric wear compared to Al_2O_3, 29 times lower wear compared to metal-metal bearings, and 3529 times lower wear compared to metal-PE bearings. Lower protein buildup was also observed with the new composite, possibly due to lower frictional heat from inherent higher thermal conductivity (3 times) vs. Al_2O_3.

These results, combined with the new composite's higher strength (1.5 to 1.8 times), fracture toughness (2.4 times), and ability to be fabricated into complex shapes, indicate the high potential of the ceramic to solve limitations and concerns with many of the current ceramics used in THRs.

Laser Processes

Researchers at the Chiba Institute of Technology (Chiba, Japan)[14] are laser-forming thin metal foil into three-dimensional structures by first cutting them out and then causing them to fold by irradiating certain portions with lower-power light. The foil is held flat between two glass plates during the scanning; after release, the foil bends. Postrelease scanning is sometimes necessary to fold the foil further.

For 10-μm-thick stainless-steel foil, scanned with an Nd:YAG laser, the beam power, beam.diameter, and number of scans for a sharp bend are 0.3 W, 25 μm, and 5, respectively; for 20-μm foil, the values are 0.8 W, 50 μm, and 10. Broadly scanning an area on a ribbon curls the ribbon. When scanned, flat foil bends toward the scanned surface; however, prebent foil always bends in the direction of the bend. Cut and bent shapes such as coils, spirals, deformed squares, and microcubes can be fabricated. Metals such as Cu and Al can be bent, and even glass can be caused to bend by a 12° angle.

Accurate shaping of metal parts traditionally has relied on skilled craftsmanship and complex tooling, both costly. High-intensity laser beams can be used to heat and bend sheet metal, but the mechanisms of the laser forming process are not well understood or precisely controllable. A joint venture led by General Electric[15] plans to develop technologies for a controllable, repeatable laser forming process that shapes and reshapes a wide range of complex sheet-metal and tibular parts meeting specific material and mechanical requirements. The research team plans to develop and integrate modeling, metallurgical, and controllable technologies into a system that can accommodate variations in workpiece geometry, material properties, system control, and component complexity. The researchers will seek to model laser forming processes used with various types of parts, understand how material characteristics affect and are affected by the process, and develop the capability to sense and adaptively control the process.

If successful, the project will reduce the need for highly specialized metal-forming tools and provide for increased design flexibility and adaptive low-volume manufacturing of consistently high-quality products with little or no waste. The intelligent platform technology will benefit a broad range of applications, principally in the automotive and aircraft industries. The shipbuilding, heavy equipment, bridge construction, and sculpture/architecture industries also will benefit. With time to market reduced by up to 50% and production costs of some parts reduced by up to 80%, U.S. manufacturers in the metal-forming industry could save as much as $320 billion by 2010. In addition, new products could be produced that cannot be manufactured economically with existing methods.

Optoelectronic (OE) modules serve many existing and emerging applications including telecommunications, data storage, and the "lab-on-a-chip" biosensors used for biological threat detection or in point-of-care diagnostic systems. Many of these OE devices can now be made as monolithic glass modules because of new, pulsed-laser micromachining technologies that allow efficient all-laser fabrication as well as offering certain performance advantages. Several steps are involved, including singulation of the glass modules, marking them with unique identifiers, and the creation of subsurface features such as optical waveguides and Bragg gratings. Micromachining of external features — such as fiber socket holes and grooves — is also necessary so that light can be coupled into and out of these modules (see Figure 12.11).[16]

The process of cutting and separating large glass sheets into smaller individual glass substrates is called singulation and traditionally has been carried out using a mechanical scribe-and-break method. A diamond blade scribes the glass before it is mechanically broken along the scribed lines. However, in industries in which singulation is common — such as the fabrication of flat-panel displays — 250–500 W sealed CO_2 lasers are now used to cut glass at a rate of up to 300 mm/s.[16] Unlike the mechanical method, laser cutting does not create microcracks that can propagate over

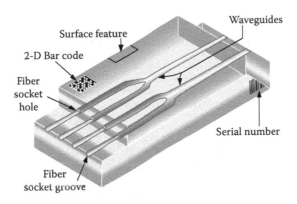

FIGURE 12.11

Fabrication of an optoelectronic device as a monolithic glass module involves several steps. In this example a fiber splitter is made by first using a sealed CO_2 laser to separate the glass substrate from a larger sheet of glass before a nanosecond-pulse UV solid-state laser marks the substrate with a unique serial number. Next, a Q-switched CO_2 laser (or a 266-nm UV solid state) laser machines in the fiber socket holes. Finally, a femtosecond-pulse Ti:sapphire laser is used to create a single layer (or multiple layers) or subsurface waveguides.[16]

time within the glass substrate. Furthermore, laser cutting has the potential to increase overall yield and productivity by eliminating most of the post-process polishing and cleaning necessary to smooth rough edges and by eliminating the glass dust that would otherwise be created during scribing.

Table 12.1 reflects a comparison of recent studies of material processing rates to determine the best material and laser combinations for machining surface features.

Laser Engineered Net Shaping (LENS)

A new low-volume production technology reduces the manufacturing-related limitations on designs. Laser Engineered Net Shaping, or LENS, makes near-net-shape metal parts directly from computer-aided design (CAD) data. Depositing metals in an additive process, it produces parts with material properties equal to or better than those of conventional wrought materials.

Unlike machining, which makes parts from the outside in, LENS parts are built from the inside out. This lets designers prototype and manufacture metal parts with shapes and features that would be difficult or impossible to machine. The technique, for example, can build ultrathin parts more than 2.54 cm tall with depth-to-diameter aspect ratios up to 70:1. This contrasts to machining aspect ratios that are limited to about 10:1.[17]

TABLE 12.1

Machining Parameters and Material Removal Rates in Various Materials for Several Types of Lasers[16]

Material	Laser	Rep. Rate (KHZ)	Power (W)	Pulse Energy (MJ)	Processing Rate (mm³/s)	N. Rate
UV fused silica	Avia 266	30	0.6	0.020	0.0007657	196
UV fused silica	Q-3000	1.5	2.65	1.767	0.018488	473
UV fused silica	GEM-100	5	9	1.800	0.089202	2284
UV fused silica	RegA	250	0.6	0.002	0.000039	1
IR fused silica	Avia 266	30	0.6	0.020	0.014379	345
IR fused silica	Q-3000	1.5	2.65	1.767	0.027489	660
IR fused silica	GEM-100	250	0.6	0.002	0.000042	1
BK7	Avia 266	60	0.55	0.009	0.001713	38
BK7	GEM-100	1	2	2.000	0.031612	694
BK7	RegA	25	0.09	0.004	0.000046	1
Fused Quartz	Avia 266	30	0.84	0.028	0.010580	226
Fused Quartz	Q-3000	1.5	2.65	1.767	0.019105	408
Fused Quartz	GEM-100	5	9	1.800	0.098319	2100
Fused Quartz	RegA	250	0.6	0.002	0.000047	1

In addition, it easily manufactures products from hard-to-machine materials such as Ti, and it deposits material combinations that were once impractical or prohibitively expensive. In a single step, the process produces multimaterial structures with gradual material transitions and negligible internal stress.

Process Basics

Originally developed at Sandia National Laboratories, Albuquerque, NM, LENS machines dispense metal powders in patterns dictated by 3-D CAD models. Guided by these computerized blueprints, the systems create metal structures by depositing them a layer at a time.

To start the process, a high-powered Nd:YAG laser beam strikes a tiny spot on a metal substrate, producing a molten pool. A nozzle blows a precise amount of metal powder into the pool to increase the material volume. A layer is built to the CAD geometry as the positioning system moves the substrate under the beam in the XY plane. The lasing and powder-deposition process repeats until the layer is complete. The system then refocuses the laser in the Z direction until the unit builds layer upon layer a metal version of the CAD model (see Figure 12.12).

The deposition process takes place inside a sealed chamber, where environmental variables are tightly controlled. For example, the chamber can maintain an Ar atmosphere with O_2 levels of less than 10 ppm. This is

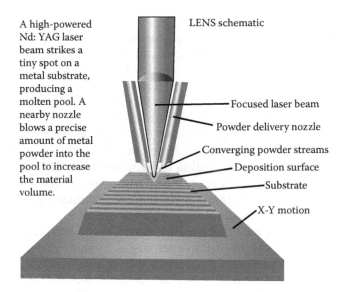

A high-powered Nd: YAG laser beam strikes a tiny spot on a metal substrate, producing a molten pool. A nearby nozzle blows a precise amount of metal powder into the pool to increase the material volume.

LENS schematic

— Focused laser beam

— Powder delivery nozzle

— Converging powder streams

— Deposition surface

— Substrate

— X-Y motion

FIGURE 12.12

A high-powered Nd:YAG laser beam strikes a tiny spot on a metal substrate, producing a molten pool. A nearby nozzle blows a precise amount of metal powder into the pool to increase the material volume.

essential when manufacturing parts made from Al, which are plagued by an oxide that prevents the material from properly wetting the deposited layer. Due to the almost O_2-free atmosphere inside the chamber, the system produces Al parts that can't be made by conventional manufacturing processes.

A variety of materials, including Ti, stainless steel, tool steel, Co, and Inconel, are candidates for the process. These metals cool quickly and solidify with fine-grained microstructures. They have greater strength than wrought materials, with no loss of ductility (see Table 12.2).

TABLE 12.2

Material Properties

Material Type	Ultimate Tensile Strength (kpsi)	Yield Strength (kpsi)	Elongation (% in 1 in.)
LENS 316 stainless steel	115	72	50
316 stainless steel wrought stock	85	35	50
LENS Inconel 625	135	84	38
Inconel 625 wrought stock	121	58	30
LENS Ti-6Al-4V	170	155	11
Ti-6Al-4V wrought stock	130	120	10

Process Pluses

Besides good material properties, the process offers other advantages over conventional manufacturing techniques. In contrast to machining, which removes material from a block of metal, LENS deposits materials only where needed, thus eliminating material waste. In addition, it makes parts directly from CAD files, without intermediate steps. This reduces manufacturing costs, product development time, and time to market.

The process also dispenses different metals in combination to create mixed-material parts. This lets designers specify different materials for different areas of a part, depending on requirements. Designers needn't build an entire part from an expensive, wear-resistant material. Instead, they can deposit wear-resistant materials on part surfaces, where needed, using a less-expensive material for the part interior.

Consider a rotating part that must be strong at the attachment point in the center and lightweight at the outer edges. In some cases, a designer could alter the part geometry to produce the required qualities. But manufacturability considerations may limit design freedom. Part geometry needs no alteration with LENS: The process can build a two-material part with strong centers and lightweight outer edges.

LENS can produce both abrupt and gradual transitions from one material to another. The latter is advisable when the two materials have large disparity in their CTEs. For example, consider a mold of Cu and steel. Cu expands twice as much as tool steel when heated. If the mold has an abrupt transition between the two materials, their interface will see a great deal of stress as the mold thermally cycles during the molding process.

LENS as a Tool

When developing most metal components, designers often make prototypes to evaluate product form, fit, and function. But most conventional rapid-prototyping (RP) systems make sample parts out of paper, polymers, ceramics, and porous metals. These prototypes serve well as a study for product form and fit. But in many cases, they fall short during functional tests because their materials aren't the same as those in the actual products.

In contrast to other prototyping processes that produce parts that ultimately must be made some other way, LENS can also handle low-volume production runs. Fully functional metal prototypes generated from CAD solid models let designers develop more innovative product designs as well as modify or rework existing prototypes to test design changes: Designers needn't fabricate an entirely new part with each minor design iteration.

Many low-volume end products are made of specialty materials that are difficult to fabricate. High-volume tooling is often prohibitively expensive to justify in these cases. Parts, therefore, are often the product of costly, labor-intensive manual processes. In a variety of industries, however, automated LENS machines cost-effectively manufacture small numbers of metal parts quickly.

Unlike conventional milling and machining operations, LENS machines can make parts with a wide range of complex internal geometries, including hollow or honeycomb interiors. These parts are lighter than their solid counterparts, but still strong enough for application needs.

There is a large and growing demand for medical implants that are made of lightweight, high-strength Ti alloys shaped in a forging process. Hip and knee prostheses, for example, are each expected to be implanted 650,000 times in the United States by 2006. But many of these procedures are actually to replace failed implants. To improve implant designs and properties, Optomec, Inc.[18] is developing an accurate, laser-based technology for depositing Ti alloy materials to make complex, three-dimensional parts. Laser deposition offers the option of adding material to a starting piece rather than removing it by, for example, machining, reducing waste (important in the case of Ti, which is expensive), and enabling new design concepts (see Figure 12.13). Already commercialized is a laser-engineered net shaping (LENSTm) technology to deposit 3D shapes from a variety of metal powders. This development will enable the company to develop a much higher-speed process for low-cost deposition of Ti alloys with exceptional mechanical properties.

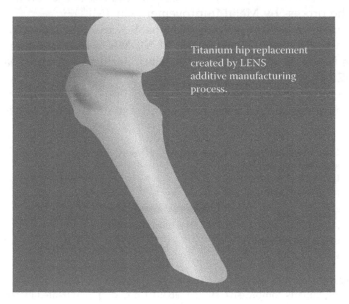

Titanium hip replacement created by LENS additive manufacturing process.

FIGURE 12.13
Titanium hip replacement created by LENS additive manufacturing process.

The technology development is to upgrade the existing LENS machine, develop the hardware and software for a four-/five-axis deposition head that can swivel and deposit material in any direction, develop an understanding of the relationships among processing conditions and material microstructure and mechanical properties for the LENS-deposited Ti alloy, and find ways to reduce the cost of raw materials. If successfully developed and commercialized, the proposed machine would enable low-cost manufacturing of high-quality Ti components and lead to medical benefits worth up to $2 billion over 10 years by reducing the number of surgeries for failed implants and improving implant designs and mechanical properties. The machine could also be used in rapid prototyping and to produce 3-D parts made of other materials, such as specialty steels and superalloys. The new process could also produce components for the petrochemical, defense, automotive, and other manufacturing industries.[19]

Peening

Laser peening is a process in which a laser beam is pulsed upon a metallic surface, producing a planar shockwave that travels through the workpiece and plastically deforms a layer of material. The depth of plastic deformation and resulting compressive residual stress are significantly deeper than possible with most other surface treatments.

The successful development of laser peening led to the commercialization of the process by Metal Improvement Co. (MIC).[20]

There are several laser peening systems; however, the technology is not significantly different. There is the LLNL-MIC[20] laser peening system, which is based on a novel Nd: glass slab, flashlamp-pumped laser (see Figure 12.14).

The peening process strengthens metal by applying compressive stress, which penetrates about 4 times deeper into a material when applied using

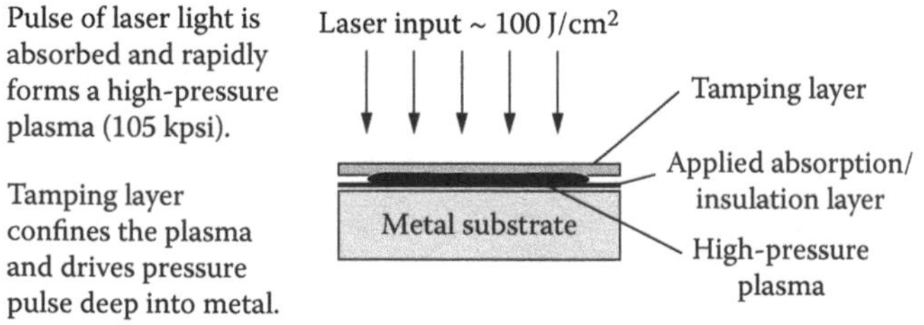

Pulse of laser light is absorbed and rapidly forms a high-pressure plasma (105 kpsi).

Tamping layer confines the plasma and drives pressure pulse deep into metal.

Laser input ~ 100 J/cm^2

Tamping layer

Applied absorption/ insulation layer

Metal substrate

High-pressure plasma

FIGURE 12.14
Laser peening imparts compressive stresses about 1 mm deep into treated metals — about 4 times as deep as traditional metal-shot peening — and can increase lifetimes of jet-aircraft turbine blades, for instance, by a factor of 10.

a laser than with metal shot. The laser also leaves a smoother surface, while allowing for more precise targeting of the compressive force and improved heat resistance in the finished material.

Square Beam

The actual laser-peening process imparts energy densities on the order of 50 to 250 J/cm^2 to a metal surface through laser-impact spot sizes ranging from 3 to 20 mm^2, with pulse durations on the order of 10 to 30 ns. Optics developed at MIC deliver a square beam with a flat profile (10% root-mean-square nonuniformity) across about a 1-m distance to the image plane at a spot size, duration, and energy density desired for a specific material and treatment objective — somewhat similar to a lithographic process.

Peening an Al alloy might only require a laser fluence of 60 J/cm^2, while high-strength steel might require 200 J/cm^2.

The square beam shape provides much more efficient surface coverage than a round beam, and the flat beam profile allows the same compressive stress to be delivered using about a third the power that would be required using a Gaussian beam.

Each laser pulse imparts a 10^6 psi pressure pulse at the component surface by generating a plasma in a thin layer of protective tape or paint on the metal surface. The plasma is spatially contained from spreading across the surface area by a 1- to 3-mm-thick layer of "tamping" water and thus transmits a shock wave directly forward into the metal (see Figure 12.14).

A complete laser-peening system, including robotics for positioning materials during treatment and computer-control systems, costs about $2 million: 4 to 8 times as much as metal-shot-peening systems.

The tighter process control available with laser peening is also extending potential applications into areas — such as metal shaping of Ti airfoils in high-performance aircraft — that would exceed the capabilities of metal-shot peening. Applications currently being developed other than military and aviation include treatment and forming of components used in health-care delivery, nuclear-power generation, and drilling for petroleum products.

A laser shock peening process (LSP) developed by the Air Force Laboratory Materials and Manufacturing Directorate (AFRL/ML) is being used for toughening damaged aircraft turbine engine blades greater than 5 times their normal fatigue strength. Tougher engine blades result in greater resistance to foreign object damage (FOD), resulting in less risk to an aircraft and its flight crew and cutting maintenance costs.

Prior to the advent of LSP, and in lieu of grounding the entire B-1 fleet, the Air Force required a manual inspection of fan blades before each flight.

These time-consuming preflight leading-edge inspections included rubbing the leading edge with cotton balls, cotton gloves, and even dental floss. If a single snag was detected, the blade was replaced prior to the next flight.

LSP uses a strong laser impulse to impart high-compressive residual stresses in the surface of metal leading-edge components. The laser pulse ignites a blast or shockwave from the specially coated surface of the component. The expansion of the blast wave then creates a traveling acoustic wave that is coupled into the component, thereby compressing (strengthening) the material's lattice structure. The results are a significant improvement in the high cycle fatigue properties of the component and greatly increase resistance to blade failure caused by FOD.

Results of high cycle fatigue tests on LSP-treated blades have been remarkable. A typical, untreated, undamaged fan blade has a fatigue strength of 690 MPa. Using LSP had demonstrated the process was capable of restoring the structural integrity of a damaged fan blade.

Testing was accomplished through a cooperative effort between the Air Force Manufacturing Technology (Man Tech) Division and General Electric Aircraft Engines (GEAE), which conducted engine tests using damaged LSP-treated blades and new, untreated blades. The damaged LSP-treated blades performed equal to the new untreated blades, retaining their fatigue strength of 690 MPa. This test also proved LSP could improve fleet performance by displaying its ability to virtually eliminate sensitivity for FOD defects up to one-quarter-inch in an F-101 fan blade, if not eliminate current FOD limitations altogether.

Although production costs for use of the new technology were relatively high, the benefits far outweighed the required manual inspection and blade replacement costs. Reduction of FOD sensitivity on the F-101 engine eliminated the need for the manual preflight inspections that had cost more than $10 million per year. It also avoided projected engine losses that conservatively would have cost in excess of $40 million over the life of the F-101 engine.

What benefits have been found through LSP operations are above and beyond the savings in preflight inspections and the avoidance of aircraft losses, and the technology has resulted in a cost avoidance of more than $100 million.

If the impact of LSP is calculated over all the engines in the Air Force fleet, the potential savings could easily approach $1 billion.

Welding

The housing, the cover, and the optical/electronic-device carrier of an automotive distance control system has been successfully joined by laser welding. The challenge was to join the parts without damaging the sensitive measuring equipment mounted on the carrier.

To be successful, the materials for the three components had to be compatible with laser welding technology. For the equipment carrier and the housing, the designers chose a PC/PBT blend that absorbs laser light and converts it to heat. For the cover, engineers selected a special grade of polycarbonate that appears black in the visible spectrum but allows infrared light to pass through.

The component is laser-welded in two steps. First, the laser-transparent cover is welded to the carrier, which absorbs the laser light and melts, joining the carrier and cover. This assembly is then placed in the housing, which absorbs the laser light and is then welded to the assembly.

The high specific weight, excellent corrosion resistance, high-temperature performance, and biocompatibility of TI make it attractive to many different industries (aerospace, defense, petrochemical, medical). Typically, Ti structures are either machined from cast, wrought, or forged materials, or are conventionally welded with high heat input/high distortion processes. GTAW welding produces acceptable welds but it is slow and uneconomical. GMAW may increase efficiency and decrease cost but cannot produce high-quality welds at higher processing speeds. The drawbacks of GMAW and GMAW-P processes are related to the instability of the arc and the cathode spot location in the weld pool at higher processing speeds.

The combination of laser beam welding (LBW) with either GMAW or GTAW is referred to as hybrid welding. Hybrid welding has been extensively investigated on a wide range of materials over the past few decades. Most hybrid welding research has utilized high laser powers for increased travel speed and penetration. While hybrid welds appear stable, there has not been any work to investigate the laser conditions that make the GMAW-P stable. If the GMAW-P process could be stabilized with a very low-cost laser system, then the utilization of Ti fabrications may increase.[21]

Experimental work demonstrated a dramatic change in the weld quality that was observed when laser power was added to GMAW-P welding of Ti. This work was accomplished with a Lincoln Electric PowerWave 455/STT GMAW-P power supply and a Trumpf/Haas 4006D Nd:YAG laser. The lamp-pumped 4006D can deliver 4 kW in CW mode through a 0.6-mm-diameter fiber-optic cable. For this effort, a four-axis CNC table produced the motion; the laser focus head and GMAW-P torch were mounted to the z-axis. The laser focus head was at 90° and the torch at 31° from vertical. Bead-on-plate welding trials were performed on 6.35-mm-thick Ti-6Al-4V by translating the plates in the x-axis. A Lincoln Electric PowerFeed 10-wire feeder was used to feed 0.889-mm-diameter ERTi-5 wire to a Binzel 401D torch. The gas cup and a 50 × 60 mm trail shroud supplied pure Ar at a flow rate of 65 CFH. A range of laser powers were investigated in order to stabilize the weld. The ability to stabilize the weld pool at such a low laser power was unexpected.

To determine the operational range of the stable process, a test matrix was performed. High-speed video (HSV) and data acquisition (DAQ) monitored the process and were used to help determine when the process was stable. The HSV and DAQ acquired data at a rate of 12,000 and 24,000 times per second, respectively. Various focal lengths (100, 150, and 200 mm) were used to produce a range (0.3, 0.45, 0.6, 3.18, and 5.58 mm) of focus spot sizes. Laser power was adjusted for these spot sizes until the arc was stabilized.

The results indicated that it was possible to use a low-power, low-cost laser with GMAW-P to make high-speed, low-heat input welds on Ti alloy structures. This could have a major future impact on economic decisions concerning machining vs. fabrication of components for a number of different applications. By using a very low-power, low-cost laser, it may be possible to economically produce very high-quality, low-heat input fillet and multipass welds in Ti alloys. This could drastically decrease the fabrication cost for Ti structures, making it more appealing to use Ti materials not only in aerospace and military applications but also automotive and consumer products.

Continued investigations and development has resulted in a number of important conclusions, including determining where additional efforts are needed. A better understanding of the relationship between the laser spot size, energy input, and cathode stability is needed to determine the limitations of the approach. Further, with this understanding, it may be possible to transfer this approach to other alloys where such stabilization has not been noted at low laser powers. Additionally, because the laser power is low and the laser spot size large, it may be possible to use lower-cost diode lasers or even focused "white light" to produce similar results.

Surface Treatment

The purpose of LSA technology[22] is to add elements to the surface of a material to enhance selected properties.

The metal to be alloyed with the substrate can be coated onto the substrate by many different means, including vacuum evaporation, plating, powder coating, thin foil application, or ion implantation. Also, the alloying metal can be applied in a powder form through a jet and sprayed into the melt. Powder techniques are very beneficial with a CO_2 laser, because the powder increases the absorption coefficient. With liquid state diffusion techniques, LSA can create a surface alloy by mixing the added elements with the substrate to depths of a few fractions of a micrometer to a few hundred micrometers.

Pulsed modes work best with Nd:YAG ruby and Nd:glass lasers, while continuous wave modes are more effective with the CO_2 lasers. For LSA, laser powers in the multiple kilowatt range are necessary to properly melt the material with a CO_2 laser.

One benefit of a continuous wave CO_2 laser is the control it provides over the amount of time that a point on the surface is in contact with the beam. Dwell times depend on the beam diameter and on the speed of the beam relative to the material. A larger beam site and a slower velocity produce a longer dwell time. As the dwell time lengthens, the penetration depth increases, thus diluting the concentration at the surface.

For example, LSA can alloy Cr, Mo, and Ni with a ferrous alloy substrate to improve its hardness, wear resistance, and corrosion resistance. Along with ferrous alloys, Ti alloys also have been studied for LSA technology.

LSA has many advantages over bulk alloying. If high corrosion resistance is needed, LSA would be a good route to examine. Instead of manufacturing the part from a highly alloyed material, it can be built from pure metal and the surface may be alloyed later, saving cost and effort. Also, LSA is more resource-efficient, because surface alloying requires much less of an expensive alloying element than bulk alloying.

However, LSA has disadvantages. For example, surface-alloyed parts provide a shorter fatigue life and lower strength due to the rough surface finish that is created when the molten alloy freezes. In an experiment on Ti-6Al-4V specimens, researchers found that after the molten material had resolidified, its fatigue strength was half that of a polished specimen, and its fatigue life was 100 times lower.

With current technology, LSP can treat surfaces to depths of micromillimeters to a couple of millimeters (see Figure 12.15). Unlike LSA, most LSP processes do not melt the metal, but instead heat-treat the surface.

Such treatment may be limited to only a specific area of the part while leaving the remaining areas in their original condition. In many circumstances, wear properties can be increased by LSP without degrading strength. Much less energy is needed to treat a part with a laser than a furnace. Also, if the part is large or has an irregular geometry, a new furnace may have to be purchased to treat it by conventional means. Quench tanks must also be large enough to hold the part, which along with the furnace can become quite large and cumbersome.

A practical application of LSP is hardening the bearings on a crankshaft. Crankshafts can become quite large, and the bearings must be hard to resist wear. It is possible to harden the entire part, but this would be very costly. However, LSP treatment could be focused on only the small percentage of the shaft that requires high hardness, thus consuming only a

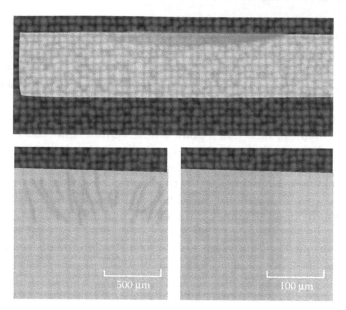

FIGURE 12.15
Laser surface processing heat treats the metal without melting it. Treatment depths range from micro-millimeters to a couple of millimeters, and can be focused on specific areas that need enhanced properties.

fraction of the energy. Furthermore, the capital investment made with the purchase of the laser could be offset because it could also weld, cut, and add alloys to surfaces.

LSP also enables immediate control over the heating and cooling rates of the materials. A common laser can irradiate a surface with 10^6 W/cm². To grasp the amount of energy being delivered, it would take a 60 W lightbulb held 10 cm from the area over 242 days to deliver the same amount of energy to the same area as the laser does in 1 sec. Because the rate of energy transfer to a material from a laser is so high, many metals do not melt as expected. Superheating is possible in many cases, up to 226.85°C in gallium arsenide, because the material is heated so rapidly that the average number of atomic displacements is small. Also, because the laser treats only a very small volume of the metal, thermal conduction brings about a very fast quench to the material when the laser beam is removed. Because the cooling rate for many LSP operations is very high, a fine crystalline or even amorphous structure is left on the surface.

LSP is a very useful process, but it does have disadvantages. The depth of treatment is limited to only a few millimeters at best, which may not provide a sufficient treatment for the given specifications. However, a great deal of research continues, and the processes are being improved.

Glazing and Cladding

Increasing enviromental regulations and rising waste disposal costs are driving the need for economic chrome plating replacement technologies. One of several chrome replacement research and development programs has focused on two laser-based coating processes — laser cladding and laser glazing — for the application of a thick (5–25 mils) uniform coating with similar if not superior mechanical performance to that of hard (wet) chrome plating. The work has been performed using a flat-top continuous wave CO_2 laser, operating at 10.6 μm.

Conventional chrome plating is used to apply a wide range of thicknesses, corresponding to different applications. Thin chrome coatings (1–10 mils) can be used for corrosion protection, while thick chrome coatings (10–40 mils) are often used to build up surfaces that have worn down through extended use. The primary challenge in developing a substitute process for chrome plating is to provide most, if not all, of the desirable properties of chrome — hardness, adhesion, corrosion protection, lubricity, surface finish, and the ability to build up worn surfaces — with one alternative process. Providing one or two of these characteristics is relatively easy; to simultaneously provide all of these characteristics is a formidable task.

This laser processing work at AFRL/ML (the Laser Hardened Materials Evaluation Laboratory, or LHMEL), both with laser cladding and with laser glazing of thermally sprayed coatings, has shown that laser processing is a viable and feasible process for producing well-consolidated, corrosion-, and wear-resistant coatings. The common feature of the two processes is that they each melt the coating layer, reducing its porosity and sealing the surface against air or water permeation, a critical requirement for preventing corrosion of the underlying metal. Results also showed that these laser-processed coatings could satisfy the majority of the performance requirements currently fulfilled by electrodeposited chrome plating, especially where large thickness buildup is necessary.

In summary, it was demonstrated that flame-sprayed coatings can be successfully applied to a substrate, then laser-glazed to produce excellent high-density, corrosion-resistant, wear-resistant coatings. Successful laser glazing was performed on rods coated with Versalloy-50 and on rods coated with a mixture of Versalloy/WC-Co. The reproducibility of the process was far greater for the mixture, which melted at a much higher temperature and introduced more heat into the substrate. It has also been shown that the laser-glazing process has very little effect on the underlying substrate microstructure and that the integrity of the flame-sprayed coating plays an important role in determining the integrity of the resultant laser-glazed coating. Microstructures produced by laser glazing of the WC-Co/Versalloy coating resulted in partial dissolution of the carbides into the Ni rich nalrix. This created a very strong and hermetically sealed coating that should withstand corrosion and thermal cycling environments very well.

Laser cladding was also demonstrated as a viable chrome-plating alternative. This method produced pore-free, well-bonded coatings of varying thicknesses on a variety of substrates. The cladding experiments gave better results with Versalloy-50 powder alone than with the WC-Co/Versalloy powder mixture. Corrosion and wear tests on the clad samples showed comparable performance to that of chrome.

This work, according to Hull and colleagues,[23] also demonstrated the successful repair of worn or damaged parts using either laser glazing or laser cladding. The ability to apply a second layer of coating without creating any apparent interface between it and the original suggests that cladding or glazing can be used to build up or repair worn parts without the added costs of prior coating removal and the subsequent disposal of the waste stream.

Coatings

Traditionally, coatings and substrates have developed independently. Coatings have also traditionally done an excellent job of doing what they were designed to do: prolong the life of turbine engines by protecting component parts from oxidation and corrosion, erosion by particulate debris, and other potential hazards. Engineers now face a challenge, however. With new technologies creating a broad range of heat-resistant materials, turbines now operate at temperatures that are significantly higher than a decade ago. The new demands on turbine coatings and substrates make it imperative that the two be designed interdependently; each must go hand-in-hand into the regime of ever-increasing temperatures. In this harsh environment, a failure in one quickly leads to a failure in the other. Indeed, in some proposed designs, the coating and substrate form a continuum, literally blurring the boundary between the surface deposit and the material it coats.

In future years, turbine engines will have the potential to reach new heights of efficiency and service life. But to keep pace, coating technologists will have to continue moving away from the traditional way of designing coatings. The bottom line is that coatings must be integrated into the total component design, taking into full consideration the alloy composition, casting process, and cooling scheme.

The efficiency of gas turbines, whether for industrial power generation, marine applications, or aircraft propulsion, has steadily improved for years. These advances have come about, in large part, because the means have been found to operate the gas-generator portion of the engine at increasingly higher temperatures. The need for greater performance from advanced turbine engines will continue, requiring even higher operating efficiencies, longer operating lifetimes, and reduced emissions. A large share of these improved operating efficiencies will result from still higher operating temperatures. Better engine durability would normally require lower operating temperatures, more cooling of the hot structure, or structural materials possessing inherently greater temperature performance. Because the first two options cause a penalty in operating efficiency, the last approach is preferred. Achieving greater temperature performance has made imperative the use of surface protection to extend component life and the concurrent development of the advanced structural materials and the coatings that protect the structure from environmental degradation.

Trends

High-Temperature Coatings Design

In the past, high-temperature coatings were selected predominantly after the component design was finalized. Current designs require that the substrate (typically a Ni-based superalloy) have sufficient inherent resistance to the degradation mechanisms to prevent catastrophic reduction in service lifetime in the event of coating failure. Because the materials considered for future substrates may possess less inherent environmental resistance at higher temperatures, the importance of coatings in achieving performance will continue to grow. In future turbine designs, coatings will be increasingly viewed as an integral portion of the design process to meet the high demands for system performance.

High-Temperature Coating Types

Although many types of high-temperature coatings are currently in use, they generally fall into one of three types: aluminide, chromide, and MCrAlY.* The family of coatings that insulate the substrate from the heat of the gas path (i.e., thermal barrier coatings, or TBCs) is increasing in importance as they begin to be used for performance benefits. TBCs are ceramic coatings (e.g., partially stabilized ZrO_2) that are applied to an oxidation-resistant bondcoat, typically an MCrAlY or aluminide.

---

* MCrAlY is a type of metallic coating in which M is a metal, usually Co, Ni, or a combination of the two; Cr is chromium; Al is aluminum; and Y is yttrium.

Processes for Applying Coatings

A wide variety of processes are used to apply coatings, although they rely on one of three general methods: physical vapor deposition (PVD), chemical vapor deposition (CVD), and thermal spray. These processes deposit a wide range of coatings between the extremes of diffusion coatings (i.e., the deposited elements are interdiffused with the substrate during the coating process) and overlay coatings (i.e., the deposited elements have limited interdiffusion with the substrate). Diffusion coatings are well bonded to the substrate but have limited compositional flexibility; their usefulness is strongly dependent on substrate chemistry. Overlay coatings are typically well bonded and have broad compositional flexibility; however, they are more expensive and thicker than diffusion coatings. TBCs are overlay coatings and as such can be deposited on a variety of substrates. The main difficulty with TBCs is that the abrupt change in composition and properties at the interface tends to promote ceramic layer spallation.

Electron-beam PVD is often favored over plasma deposition for TBCs on turbine airfoils because it applies a smooth surface of better aerodynamic quality with less interference to cooling holes. However, the widely used plasma-spray process has benefits, including a lower application cost, an ability to coat a greater diversity of components with a wider composition range, and a large installed equipment base.

Degradation Modes

A primary consideration in selecting a coating system is determining if it provides adequate protection against the active, in-service, environmentally induced degradation mechanism(s) experienced by the component. These degradation modes are a function of the operating conditions and the component base materials. The degradation modes common to superalloy hot-section components include — to varying degrees — low-cycle thermomechanical fatigue, FOD, high-cycle fatigue (HCF), high-temperature oxidation, hot corrosion, and creep.

Because of the use of thin walls and compositional design for highest strength, aircraft turbine blades with internal cooling passages have historically had insufficient high-temperature oxidation resistance to meet required lifetimes without the use of a coating. Coatings have been used in these circumstances to extend overhaul limits and useful life of the component. Although the latest generation of single-crystal blades has excellent oxidation resistance compared with conventionally cast industrial engine blades and aircraft gas-turbine blades with moderate to high Cr contents, the blades have less tolerance for hot corrosion once the coating has been breached. Industrial gas-turbine blades, which use thick walls and lower-strength alloys with higher corrosion resistance, generally have significant service life after the coating is breached.

During service, coatings degrade at two fronts: the coating/gas-path interface and the coating/substrate interface. Deterioration of the coating surface at the coating/gas-path interface is a consequence of environmental degradation mechanisms. Solid-state diffusion at the coating/substrate interface occurs at high temperatures, causing compositional changes at this internal interface that can compromise substrate properties and deplete the coating of critical species. In the worst case, interdiffusion leading to the precipitation of brittle phases can cause a severe loss of fatigue resistance.

Engineering Considerations

Given that a coating system is required and that one has been identified that provides environmental protection, six significant engineering factors must be evaluated:

1. Chemical (metallurgical) compatibility. The coating must be relatively stable with respect to the substrate material to avoid excessive interdiffusion and chemical reactions during the service lifetime. An unstable coating can lead to premature degradation of both the coating and the substrate through lower melting temperatures, lower creep resistance, embrittlement, etc.

2. Coating process compatibility. The coating material may be completely compatible with the component, but the coating process may not be compatible. This would usually occur when process conditions require high temperatures or special precoating surface treatments.

3. Mechanical compatibility. Coatings resistant to oxidation and corrosion maintain their protectiveness only if they remain adherent and free from through-thickness cracks. Important considerations include close match of the CTE of the coating with the substrate, strain accommodation mechanisms within the coating, coating cohesion, and coating adhesion. CTE match is the most important factor, closely followed by the need for strain tolerance in the coating.

4. Component coatability. The ability to deposit a coating on the required surface is a function of the geometry and size of the component, as well as the capability of the coating process. Accessibility of the surface is a consideration. For example, some processes are line-of-sight and thus cannot coat internal passages. Size of the component is important because some processes must

be done inside an enclosed tank or reactor. The ability to apply a uniform coating must be evaluated, particularly at edges, inside corners, and for irregular part contours. The change in part dimensions and surface characteristics because of the coating must also be taken into account.

5. Contaminants in air and fuel (and water and steam for industrial turbines). Contaminants can combine in the hot section to produce corrosion, erosion, and deposition under certain temperature and pressure conditions; they contribute to accelerated degradation of high-temperature components. Limits on allowable concentrations must be established in order to assure the effectiveness of a coating system.

6. Turbine emission levels. Gas turbines can produce harmful emissions as part of the combustion process. As combustion technology has improved, emission levels have been reduced. These emissions include nitrogen oxides (NO and NO_2, commonly called NOx), CO, unburned hydrocarbons, sulfur oxides (mainly SO_2 and SO_3), and particulate matter. Coatings affect emissions primarily by reducing the need for cooling air.

In addition to the factors pertaining to the selection of an appropriate coating system, the following general engineering considerations are also important:

- Available databases of coating and coated structure properties. Traditionally, engineering property data for high-temperature coatings are generated after the mechanical properties of the uncoated substrate have been well characterized. These data are generally specific to the application domain and process conditions and are usually proprietary. Long-term data (i.e., performance of coated structures for durations greater than 50,000 hours) is sparse and related to old technologies.

- Coating standardization. Generally, each component in the hot section of the engine has a particular coating system optimized for the prevailing conditions. Greater consideration should be given to optimizing a coating system for many components because of the wide variety of alloys and component systems.

Life-Cycle Factors

Hot-section structures are designed to operate at the highest possible temperatures and stresses in order to maximize performance. As a consequence, these structures continuously degrade during service. The rate at which this degradation occurs is crucial to the function of the component and, ultimately, to the performance and longevity of the gas turbine.

The role of the hot-section coating is to protect the substrate from the gas-path environment in order to meet performance objectives, as manifested in the time between overhauls or the designated service. Component refurbishment involves the economical and timely restoration of part integrity.

The types of repairs allowed to coated structures are dictated first by safety and reliability and second by economic benefit. Repairs therefore vary greatly depending on the type of component to be refurbished. Although the replacement of the coating is generally a small portion of the overall repair, it can be critical to meeting the intended life of the component after it is returned to service. The wide variety of coating systems and the lack of standard designations adds complexity to the logistical task of maintaining an engine's coated structure complex. This task will only become more difficult as advanced coatings find their way into service.

In the past, coatings had to be capable of being removed and reapplied. This may not continue to be a requirement for industrial turbines. If a new coating could allow higher-temperature operation (for increased efficiency), the savings in fuel costs could possibly outweigh the extra expense of purchasing new parts vs. repairing old parts. The future trends for aircraft engine repairs will tend to parallel those of the industrial turbines with the further complication of thinner walls and more sophisticated cooling passages. Thinner walls in advanced components may preclude any stripping of the prior coating, potentially leading to a non-repairable part, as is the case with many of the current turboprop and turboshaft high-pressure turbine blades.

Chemical Vapor Deposition

CVD coating technology has proven effective at forming coatings on both internal and external surfaces of turbine airfoils. Furthermore, thanks to computer control of all process variables, the quality of CVD diffusion coatings is often beyond that possible with other coating technologies — even when the most reactive materials are deposited. This is especially important for the design of surface-film cooling airflow patterns.[24]

The CVD low-activity aluminizing process yields much lower Al m concentrations in the coating (<22 wt %), and so, CVD aluminides exhibit much higher levels of ductility. This superior ductility makes possible what's called "cast-coat" production routing: Parts are coated as cast, then machined and assembled.

In general, a coating better resists fatigue cracking as it gets thinner and its ductility increases. CVD low-activity coatings are more ductile than high-activity diffusion aluminides made by either pack cementation or above-the-pack processes. Also, because CVD coatings have better oxidation and corrosion resistance, a thinner coating can give the same component life. Hence, the fatigue-crack resistance of CVD low-activity coatings far exceeds that of high-activity products. But the extreme performance demands of the most advanced engine designs are rapidly approaching the fatigue-crack resistance limits of CVD low-activity aluininide coatings.[24]

Rapid prototyping is essential to fast-paced, results-oriented development programs.

A new ultrasonic direct-injected, pulsed-CVD system gives designers a robust, versatile, and cost-effective tool for producing prototype devices requiring thin films. The system can be used to make films from metalorganic compounds, polymers, ceramics, and nanophase suspensions. The system called ThinSonic uses in situ process controls, which lets designers create a variety of structures including MEMS, organic light-emitting diodes (OLEDs), conformal coatings, thermal barriers, surface acoustic-wave (SAW) chemical sensors, touch screens, and many other thin-film circuits and devices.[25]

In contrast to other CVD systems, this process uses a precursor (metal, polymer, nanophase, or macromolecule) that is delivered to the ultrasonic nozzle through a series of automatically controlled solenoid valves. An adjustable pump with a ceramic piston determines the shot size and delivers a predetermined amount of precursor to a collector tube. Films as thick as several microns can be produced in minutes. Here are a few examples of current applications using ultrasonic direct-injected, pulsed-CVD: devices such as tissue scaffolds, biosensors, biochips, and photonic devices, which all require thin organic films that cannot be exposed to heat. These films and membranes can range in thickness from several angstroms to many microns. Such devices are extremely important in contributing to biomedicine and electronics.

After the attack on the World Trade Center, interest in airport security focused on sensors that could detect explosive agents. One candidate is a so-called surface acoustic wave chemical agent detector (SAWCAD). These piezoelectric-crystal devices resonate in the megahertz range, and when coated with an appropriate polymer, they capture molecules from the environment.

The ultrasonic pulsed CVD has been tested using many different oxide-producing precursors. The ultrasonic nozzle atomizes the <15 µm precursor droplets, which vaporize completely in the vacuum before arriving at the heated substrate. Molecule-by-molecule, crystallographic growth produces high-quality single-crystal thin films in the low-pressure reactor. Metal-organic materials tested include titanium oxide (TiO_2), lithium tantalate ($LiTaO_3$), indium tin oxide (ITO), and yttria-stabilized zirconia (YSZ). Total film thicknesses up to 45 µm are possible.

A new process called hot filament chemical vapor deposition (HFCVD) has been used to deposit nanolayers of polytetrafluoroethylene (PTFE, a.k.a Teflon) from the vapor phase, offering the potential to coat materials that cannot be immersed in a solution.

The process involves building up the PTFE coating one molecule at a time, much like beads on a necklace. This nanoscale control may allow researchers to tailor the head at the outer surface to provide additional properties beyond water-repellency.

The technique is also capable of coating unusual geometries, like fine wires, on which traditional PTFE deposition processes (involving baking a thick layer of powder) do not work. HFCVD process can go beyond the exterior surface of a material to coat interior cavities, such as those in a porous substance.

The technique has been shown to deposit a very thin layer of a water-repellent coating that will make it possible to waterproof new kinds of materials and offer ways to combine waterproofing with antibacterial and other active coatings. Potential applications include fabric coatings for soldier uniforms, coatings for fine wire neural probes, and insulation for integrated circuits.

Ion Beam Enhanced Deposition (IBED)

Ion beam enhanced deposition (IBED), also termed ion beam assisted deposition (IBAD), is a thinfilm technique that allows deposition of a variety of optical, dielectric, semiconductor, and tribological coatings. The nature of the process enables deposition of coatings having a wide range of metallurgical compositions, and with a high degree of control over nanostructure. Film-substrate adhesion is achieved without the external application of heat, and processing temperatures can be held below 93°C.

These process features are important for optical and semiconductor thin-film coatings and can have significant benefits when utilized for the deposition of tribological coatings. In contrast to semiconductor and optical coatings, which are always thin (submicron) and most often deposited on flat substrates, tribological matings must be thicker (5 to 10 μm) and must be deposited on three-dimensional mechanical components. Therefore, the processing requirements and equipment are quite different from those needed for semiconductor or optical element coating.

IBED is a physical, nonequilibrium coating process implemented by the simultaneous bombardment of a growing film with an independently controllable beam of energetic atomic particles. The growing film is generated either by vacuum evaporation or ion beam sputtering. The independent beam of particles consists primarily of charged atoms (ions) extracted

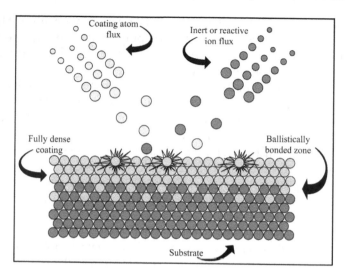

FIGURE 12.16
Schematic diagram of the implementation of the IBED process.

at high energy from a broad beam ion source. Beams of either inert spades (Ne⁺, Ar⁺, or Kr⁺) or reactive species (N⁺ or O⁺) can be utilized.

The IBED process is carried out in a high vacuum environment, at pressures of 0.00013 Pa or below. With proper choice of deposition parameters, the temperature rise in components being processed can be held below 93°C. A general diagram of the implementation of the IBED process is seen in Figure 12.16.

The types of coatings that have been deposited include metals (Cr, Ni, Ti, Al, Cu, Ag, Pt, Au), metallic nitrides (TiN, Cr_2N, Si_3N_4, AlN, ZrN), oxides (SiO_2, Al_2O_3), and carbides (diamond-like coatings, or DLCs). Coating thicknesses in the range of 1 to 10 µm are technically and economically feasible. Because coating nanostructure is highly uniform, the mechanical properties of IBED-deposited coatings are better than the equivalent coatings deposited by plating or PVD. The IBED coatings exhibit improved adhesion and cohesion, are significantly smoother both macroscopically and microscopically, and are free of voids and pinholes. Importantly, the coating properties are highly repeatable — a must for high-volume production.

Applications include precision engineered components and manufacturing tooling found in industries including aircraft/aerospace, automotive, medical, pharmaceutical, chemical, oil/gas, industrial equipment, refrigeration, and electronics. IBED coatings are ideal when high-performance surfaces are required for as-finished, precision components when conventional coatings and coating technologies cannot deliver an acceptable quality level.

Tribological coatings such as TiN and Cr_2N are usually deposited by glow-discharge, physical vapor deposition processes such as activated reactive evaporation, cathodic arc sputtering, or magnetron sputtering. In these processes, coating reactants are transported and combined in a glow-discharge carrier gas plasma. The reactions that occur when the reactants are combined in the glow-discharge carrier gas are driven primarily by the energetics of the glow-discharge. Therefore, the metallurgical, morphological, and mechanical properties are determined by the thermodynamic environment in the glow-discharge. Once submerged in the glow-discharge, the PVD reactants cannot be controlled individually, and one coating feature (such as grain size) may have to be optimized at the expense of another feature (such as density or adhesion).

The IBED process differs from the PVD process in that the reactants are not first reacted in a glow-discharge, but instead are delivered individually, directly to the surface. In addition, the energy of reaction is supplied by kinetic energy provided to one of the reactants. This provides more control over the reaction process and therefore more flexibility in the final morphology of the coating.

This process allows deposition of tribological coatings that perform better than the equivalent PVD-deposited coating. The major differences between the IBED and PVD processes are summarized in Table 12.3.

In addition, no toxic materials are required as feedstocks, and no waste products are produced by the process. Therefore, no special design criteria are needed to comply with environmental regulations. Electrical safety considerations include the need for high-voltage power supplies (up to 100,000 VDC) required for the high-energy ion source, and high-current power supplies (15 kW) required for the electron gun evaporator.

Plasma Vapor Deposition

Harder and tougher nanocrystalline Cr_2N coatings said to have superior adhesion have been developed by Phygen Inc.[27] This ST.3 (Super Tough, Variation 3), the thin-film Cr_2N coating, is applied by a proprietary PVD process.

Because the coatings are so thin, they do not affect critical dimensions of properly heat-treated tools or precise machine components. They provide much higher abrasive wear resistance and can withstand higher mechanical loads than possible with many conventional PVD coatings.

The coatings consist of a single-phase, stoichiometric, Cr_2N with excellent adhesion to virtually any substrate material. They have a noncolumnar equiaxed grain structure, nanocrystalline microstructure, and high cohesive strength.

The nanoindentation hardness reaches as high as 40–45 GPa, and elastic modulus is typically around 450 GPa. The coatings are chemically and

TABLE 12.3

IBED and PVD Process Comparison

PVD	IBED
Reactant Delivery	
Into plasma atmosphere surrounding parts	Simultaneous, directly to surface of parts
Reaction Chemistry	
Thermally driven by plasma temperature	Kinetically driven by kinetic energy of ions
Reaction Atmosphere (Pressure)	
10^{-3} Torr with high partial pressures of (H_2, H_2O, O_2, CO_2)	10^{-6} Torr with low partial pressures of (H_2, H_2O, O_2, CO_2)
Reaction Vessel Temperature	
>400°C	<93°C
Coating Morphology	
Crystalline, grains 1–50 micron dimensions	Semi-amorphous, grains sub-micron dimensions
Coating Adhesion	
Thermal diffusion driven, interlayer needed	Ballistic-alloyed, no interlayer needed

thermally stable in air up to at least 840°C. The Cr_2N coatings are chemically inert and provide excellent barrier protection against corrosion because of their dense and void-free noncolumnar structure. The coatings also possess low friction properties (a coefficient of friction less than 0.1) under proper lubrication and oxidation conditions.

Applications include a variety of machined components and tools for Al die casting, plastic injection molding, glass manufacturing, deep drawing, and similar components.

The primary tooling concerns when machining Al are minimizing the tendency of Al to stick to the tool cutting edges; ensuring there is good chip evacuation from the cutting edge; and ensuring the core strength of the tool is sufficient to withstand the cutting forces without breaking.

Technological developments have concentrated on minimizing the built-up edge due to the tool coating. Tool coating choices include TiN, TiCN, TiAlN, AlTiN, Cr_2N, ZrN, diamond, and diamond-like coatings. With so many choices, aerospace milling shops need to know which one works best in an Al high-speed machining application.

The PVD coating application process on TiN, TiCN, TiAlN, and AlTiN tools makes them unsuitable for an Al application. The PVD coating process creates two modes for Al to bond to the tool: the surface roughness and the chemical reactivity between the Al and the tool coating.

The PVD process results in a surface that is rougher material to which it is applied. The surface "peaks and valleys" created by this process causes Al to rapidly collect in the valleys on the tool. In addition, the PVD coating is chemically reactive to the Al due to its metallic crystal and ionic crystal features. A TiAlN coating actually contains Al, which easily bonds with a cutting surface of the same material. The surface roughness and chemical reactivity attributes will cause the tool and workpiece to stick together, thus creating the built-up edge.

In testing, it was discovered that when machining Al at very high speeds, the performance of an uncoated coarse-grained carbide tool was superior to that of one coated with TiN, TiCr, TiAlN, or AlTiN. This testing does not mean that all tool coatings will reduce the tool performance. The diamond and DLC coatings result in a very smooth chemically inert surface. These coatings have been found to significantly improve tool life when cutting Al materials.

The diamond coatings were found to be the best performing coatings, but there is a considerable cost related to this type of coating. The DLC coatings provide the best cost for performance value, adding about 20–25% to the total tool cost. But, this coating extends the tool life significantly as compared to an uncoated coarse-grained carbide tool.[28]

Koval[29] has shown that temperature-sensitive parts that cannot stand a lot of heat are now fair game for high-performance coatings.

Vapor-deposition processes let designers apply finishes that are both functional and aesthetic. The process deposits metals and refractory compounds such as ZrN, TiN, CrN, TiCN, and TiAlN that cannot be applied easily by other means. The low-friction coatings have good wear and corrosion resistance. They are uniform and provide fine metallic finishes that can resemble Au, Ni, and stainless steel. Vapor deposition is also eco-friendly. It has none of the environmental limitations or hydrogen embrittlement associated with platings. Faucet and door hardware, interior automotive parts, cutting tools, bakeware, and surgical implements look and perform better with vapor-deposition coatings.

Vapor deposition generally divides into two broad, sometimes competing, categories. CVD is typically associated with an application process employing extreme-temperatures where hot erosion is a problem. High temperatures (750°C) relegate most commercial CVD processes to coating high-temperature materials such as cemented carbides. In contrast, PVD processes can apply decorative and functional coatings on low-temperature materials, including polymers and Al alloys. PVD is more commonly used for aesthetics and mechanical components.

PVD can deposit almost any metal or refractory-metal compound. Refractories are the usual choice where a combination of properties such as extreme hardness, corrosion resistance, and aesthetics are important. Coatings are ultrathin, typically ranging from 50 nm to 5 µm. Many PVD films are biologically compatible with the human body and find use on implants (see Table 12.4).

Recent developments in PVD now let vapor-deposited coatings go on at low temperatures. The technique, known as low-temperature arc-vapor deposition (LTAVD), can now apply both refractory metals and conventional metal coatings at near-ambient temperatures. Parts to be coated go in a chamber and revolve around a cathode that is the metallic source of the coating (often Zr). A vacuum is drawn on the chamber, and a low-voltage arc is established on the metallic source. The arc evaporates the metal from the source temperatures, which are rarely above 100°C.

The chamber gets charged with a mixture of common inert and reactive gasses, such as Ar and N_2, and an arc-generated plasma surrounds the source. Arc-evaporated metal atoms and reactive-gas molecules ionize in the plasma and accelerate away from the source. Arc-generated plasmas are unique in that they generate a flux of atoms and molecules that have high energies and are mostly (>95%) ionized. The high energy causes hard and adherent coatings to form on parts mounted to fixtures that rotate around the source. A bias power supply can be used to apply a negative charge to the parts, which further boosts the energy of the condensing atoms.

A range of forged and tempered parts can now be coated without warping or compromising their hardness or grain orientation. Surface properties can be tailored for appearance, wear resistance, release, corrosion resistance, biomedical compatibility, or friction coefficient. Electroplated plastics, Al, Ti, or steel parts can now have the look of brushed Ni, Au, Ag, and stainless steel. In addition, coated Al, Ti, or high-Ni steel parts can work together without galling. All in all, the process makes it possible for multipart assemblies with components made from different materials to all have the same finish.

Vapor Plasma Spraying (VPS)

Vacuum plasma spraying (VPS) has been demonstrated to be an effective technique for the fabrication of refractory-metal components of solar-thermal engines. Heretofore, such components have been fabricated by specialized techniques that include electrical-discharge machining (EDM), shear spinning, sintering under pressure, CVD, and ECM.

Though effective, these specialized techniques are time-consuming and costly. On the other hand, VPS makes it possible to fabricate components with complex shapes, simply and a relatively low cost.

VPS is a thermal-spray process conducted in a low-pressure, inert-gas atmosphere. Hot plasma is generated by making an inert gas flow through

TABLE 12.4

Comparison of Metal-Coating Processes[29]

Process	Application Temp. (°c)	Suitable Substrates	Common Film Materials	Characteristics
PVD sputtering	50–500	Stainless steel, glass, plated zinc, plated brass, high-carbon (tool) steel, structured plastics, chrome, ceramics	TiN, CrN	Nonthermal vaporization process in which surface atoms are physically ejected by momentum transfer. Requires finely tuned control of all variables such as gasses used. Deposition rates are low. Deposition is directional.
PVD LTAVD	40–180	Aluminum, forged steel, titanium, plated plastics, unplated plastics, zinc, brass, stainless steel, tool steel	TiN, CrN, ZrN, TiCN, TiAlN, pure metals or alloys	Lower processing temperatures enable it to be used with wider array of substrates, particularly those that would be degraded by high temperatures. More easily controlled. Fast. Coating 3D parts.
CVD	350–1500	Superalloys, tool steel	TiN, WC, platinum-aluminide	Penetrates blind holes and channels with uniform coating. Materials used to create coating are expensive and hazardous.
PACVD	Room temp. to 500+	Thermoplastics, structural plastics, metals, glass, ceramics	DLC, a-C:H:Si, TiN, TiC Ti(C,N), Zr(C,N)	Compatible with thermally sensitive substrates. Film composition and morphology can be tuned easily. Coating of 3D parts. Corrosion-resistant coatings, hard coats, optical coatings, diffusion-barrier coatings, and high/low surface-energy coatings.
Hardchrome plating	Elevated room temperature	Wrought and forged steel	Cr	Electrochemical in nature. Can expose substrates to hydrogen embrittlement.

a DC arc. The plasma flows through a nozzle to a downstream vacuum chamber, where the pressure is maintained at about 13 kPa and the deposition substrate is located. A powder of the material to be deposited is injected into the plasma, causing the material to be heated and accelerated toward the substrate. The process is continued until the deposit reaches the desired thickness.

In the present application of VPS, the substrates used to form solar-thermal-engine components are graphic mandrels. Each mandrel acts as a male mold to form the inside of the deposit to the required size, shape, and surface finish. Inasmuch as graphite is soft and easy to machine, complex shapes and precise surface textures can be achieved without difficulty.

After the refractory metal has been deposited to the desired thickness, the mandrel is removed by blasting with acrylic or other polymeric beads, which are hard enough for eroding the graphite but soft enough not to damage the refractory-metal deposit. Removal of the graphite in this manner is rapid and simple and complies with environmental laws.

Because the interior of a component formed by VPS is inherently of net size, shape, and surface finish imparted by the mandrel, no additional machining or surface conditioning of the interior is necessary. The exterior dimensions can be controlled by controlling the deposition rate and time, so that the exterior of the VPS-formed component is near net size and shape. Thus, little or no subsequent machining is necessary; this is especially advantageous for fabricating components of W, Rh, and Mo, which are difficult to machine by conventional techniques. Even in cases in which tolerances are tighter than are achievable by VPS, the amount of subsequent machining and grinding needed is less than in older methods and is confined to exterior surfaces, which are more accessible than interior surfaces are. Although VPS shares these advantages with other vacuum-vapor-deposition processes (CVD and PVD), VPS deposits materials at rates orders of magnitude greater than those of the other processes.

Thermal Spray Processes

In the flame-spraying process, O_2 and a fuel gas, such as acetylene, propane, or propylene, are fed into a torch and ignited to create a flame. Either powder or wire is injected into the flame where it is melted and sprayed onto the workpiece.

Flame spraying requires very little equipment and can be readily performed in the factory or on-site. The process is fairly inexpensive and is generally used for the application of metal alloys. With relatively low particle velocities, the flame spray process will provide the largest build-ups for a given material of any of the thermal spray processes. Low particle velocities also result in coatings that are more porous and oxidized as compared to other thermal spray coatings. Porosity can be advantageous in areas where oil is used as a lubricant. A certain amount of oil is always retained within the coating and thus increases the life of the coating. The oxides increase hardness and enhance wear resistance. With regard to hardfacing, self-fluxing alloys are typically applied by flame spraying and then fused onto the component. The fusing process ensures metallurgical bonding to the substrate, high interparticle adhesive strength, and very low porosity levels.[30, 31]

Thermal spraying processes deposit metallic or nonmetallic materials in a molten or semimolten state onto a prepared substrate to form a coating. The coating is excellent for repairing worn parts and improving new or used equipment performance.

There are five commonly used processes for thermal spraying. The arc wire process (Figure 12.17) utilizes an electric arc to melt continuously fed metallic wires. Compressed air or an inert gas atomizes the molten metal and propels it to the substrate.

In the flame wire process (Figure 12.18), a continuously fed wire is brought to a molten state by an oxygen/fuel flame. The molten particles are atomized and propelled onto the substrate by the force of burning gases and compressed air.

The plasma powder process (Figure 12.19) uses a W electrode to generate an arc. A gas or gas mixture passes through a Cu nozzle at the tip of the electrode. The arc excites the gas in this channel into a plasma state,

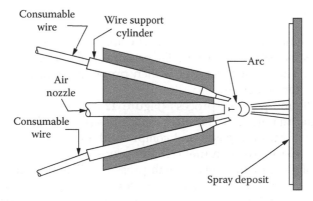

FIGURE 12.17
Arc wire process.

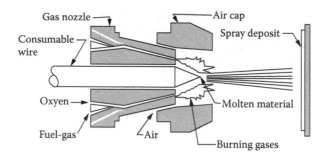

FIGURE 12.18
Flame wire process.

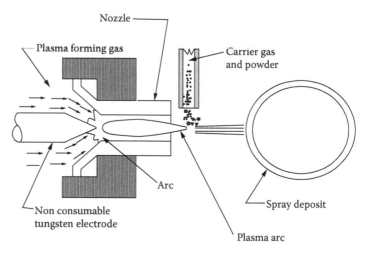

FIGURE 12.19
Plasma powder process.

creating extremely high temperatures. The arc is forced outside the nozzle, where a mechanism feeds a metallic or nonmetallic material in powder form into it. The molten powder is then propelled onto the substrate at sonic or greater velocities.

With the flame powder process (Figure 12.20), an oxygen/fuel flame melts powder material. This molten material is propelled to the substrate by the force of burning gases and compressed air.

The high-velocity oxygen fuel (HVOF) process (Figure 12.21) combusts a mixture of O_2 and fuel gas at high pressure. This burning mixture is accelerated to supersonic speeds by specially shaped constricting nozzles. Metallic or nonmetallic powders are fed into this high-velocity flame, where they are propelled onto the substrate, impacting to form a dense coating with high bond strength.

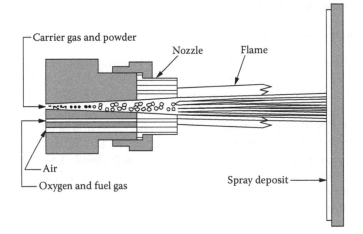

FIGURE 12.20
Flame powder process.

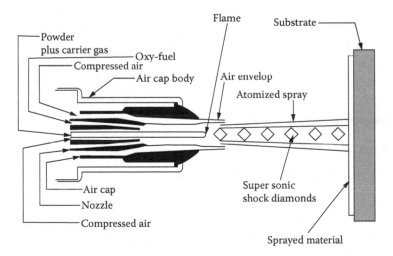

FIGURE 12.21
High-velocity oxygen fuel process.

The HVOF process is the preferred technique for spraying wear-resistant carbides and is also suitable for applying wear- or corrosion-resistant alloys like Hastelloy, Tri-balloy, and Inconel®. Due to the high kinetic energy and low thermal energy the HVOF process imparts on the spray materials, HVOF coatings are very dense, with less than 1% porosity, and have very high bond strengths, fine as-sprayed surface finishes, and low oxide levels.

These properties have enabled HVOF sprayed coatings to become an attractive alternative to cladding and chrome plating.

Metal Spray Processes

Inovati[32] has successfully developed a low-temperature metal deposition technique, Kinetic Metallization (KM). KM is capable of depositing fully dense, adherent coatings of a variety of metals on standard metal surfaces without costly surface preparation. Coatings of pure Cu, stainless steel, Ni, Cr, Al, Co, Ti, Nb, and other metals, as well as alloys based on these metals, are possible on such surfaces as steel, Al, Ti, Cu, and brass. Additionally, braze powders (e.g., Ag-, Cu-, Al-, or Ni-based) can be sprayed out onto parts to be joined, and coatings have also been demonstrated on ceramic substrates. The feedstock material for KM is powder. The cost of KM is comparable to competitive processes. Applications include the preparing of corrosion or wear-resistant surfaces for parts, machinery, and equipment. Decorative coatings are also deliverable.[33]

Because the powders are deposited at well below their respective melting points, the coatings exhibit very fine grain size and one can avoid heat distortion of the workpiece being coated and interdiffusion of multilayer coatings. Spray forming of such metals as pure Al and Al-SiC composite has also been successfully carried out with fine microstructure in the final material.

Friction Stir Processes (FSP)

Friction Stir Processing for Superplasticity

Friction stir processing (Figure 12.22) for superplasticity has the potential to expand the domain of superplastic forming to several new enabling concepts, such as selective superplasticity, thick-sheet superplasticity, and superplastic forging, according to R. S. Mishra.[34]

Two microstructural features have been identified as critical for superplasticity in commercial Al alloys: grain size and constituent particle size. Friction stirring produces very fine-grained microstructure in the stirred region. The grain refinement results from intense plastic deformation associated with the movement of material from the front to the back of the rotating pin. It follows that a very fine-grained microstructure is developed along the FSP path. Table 12.5 summarizes the grain size data for various commercial Al alloys after friction stir welding/processing. It should be noted that grain size of <10 μm is readily achieved.

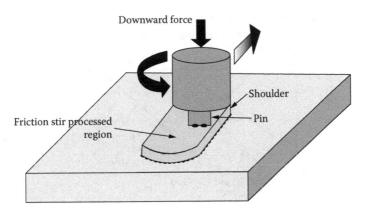

FIGURE 12.22
Schematic of friction stir process shows the rotating tool with pin and shoulder. The heating is localized and generated by friction between the tool and the work piece, resulting in very fine grain size.

TABLE 12.5

A Summary of Grain Size in Nugget Zone of FSW/FSP Commercial Aluminum Alloys[34]

Material	Plate Thickness, mm	Tool Geometry	Rotation Rate, rpm	Traverse Speed, mm/min	Grain Size, μm
1100A1	6.0	Cylindrical	400	60	4
2017A1-T6	3	Threaded, cylindrical	1250	60	9–10
2024A1	6.35	Threaded cylindrical	200–300	25.4	2.0–3.9
2095A1	1.6	—	1000	126–252	1.6
2519Al-T87	25.4	—	275	101.6	2–12
5054Al	6.0	—	—	—	6
5083A1	6.35	Threaded cylindrical	400	25.4	6.0
6061A1-T6	6.3	Cylindrical	300–1000	90–150	10
7010A1-T7651	6.35	—	180, 450	95	1.7, 6
7050A1-T651	6.35	—	350	15	1–4
7075A1-T651	6.35	Threaded cylindrical	350, 400	102, 152	3.8, 7.5
7475A1	6.35	—	—	—	2.2

In commercial Al alloys, the onset of cavitation is linked to primary constituent particles. Figure 12.23 shows a comparison of cavitation in FSP and TMP 7XXX alloys. FSP significantly increases the critical strain value and greatly reduces the extent of cavitation.

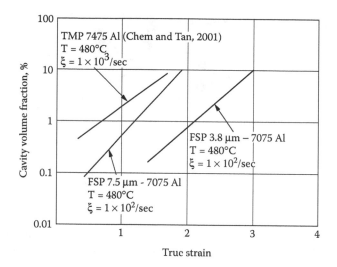

FIGURE 12.23

A comparison of cavitation during superplastic deformation in commercial 7XXX alloys. Note that the FSP 7075 alloy shows lower cavitation and higher critical strain for onset of cavitation.[34]

Several innovative concepts have been developed to take advantage of the superplasticity induced by FSP.[35] Among these are selective superplasticity, thick-sheet or plate forming, and superplastic forging.

- Selective superplasticity: The idea of selective superplasticity is based on its unique capability to locally modify the microstructure. Many components have design features that require high formability in a few limited areas. With this selective superplasticity approach, designers can adopt superplasticity for thick sheets and plates. Conceptually, a continuous-cast sheet or plate can be friction stir-processed in local regions and superplastically formed for tremendous design flexibility and cost advantages.

- Thick sheet or plate forming: The current practice of superplastic forming is limited to sheet metal forming of <3 mm thicknesses. However, FSP of plates up to 25 mm thick has been demonstrated, and over 65-mm-thick Al alloy plates have been friction stir-welded (FSW) with special tooling or by multiple passes from two sides. Conventional TMP does not introduce enough processing strain to impart fine grain size in 25-mm-thick plate, or break down the constituent particles significantly from the cast stage. Conversely, FSP can induce fine grain size and break up the particles. This opens up the possibility of thick-plate forming via FSP of selective regions.

FSP can produce a microstructure conducive to high strain-rate super-plasticity in commercial Al alloys. With local microstructural modifications, a three-step manufacturing process can be envisaged: Cast (Friction Stir Process) superplastic forge/form. In addition, FSP combined with several innovative new concepts can give designers even more flexibility and cost advantages.

Friction Stir Welding (FSW)

In the friction stir welding (FSW) process, heat is generated by friction between the tool and the surface of the workpiece. The amount of heat conducted into the workpiece determines the quality of the weld, and the amount of heat generated in the tool determines the life of the tool.

Chao and Liu[36] have described an experiment in which a butt joint was made by FSW on two long strips of Al alloy 6061-T6, a heat-treatable Al alloy. The experiments were performed on an FSW machine equipped with sensors and controllers for achieving vertical force control or displacement control and measurement of the vertical and horizontal forces, torques, and total power applied by the machine

The two workpieces of Al plates were clamped tightly by anvils along the far edges to thick low-C steel backplates, and then to the table of the FSW machine. A specially designed cylindrical tool with a pin protruding from the shoulder rotates rapidly and then slowly plunges into the weld line to yield the joining process. There are two different runs during the welding: One welding process uses the rotation tool with full-length center pin, and the other welding process uses a very short center pin.

A large testing program has shown that FSW offers tremendous potential for low-cost joining of lightweight Al airframe structures for large civil aircraft such as the Airbus A380, reports D. McKeown of TWI.[37] Researchers at Airbus Deutschland have also presented data showing that the mechanical and technological properties of FSW welds for skin-to-skin fuselage connections approach the properties of the parent material.

The Phantom Works of Boeing is pursuing FSW of curvilinear joint configurations and has demonstrated its feasibility on a complex aircraft landing gear door. The company has also flight-tested FSW sandwich assemblies for a fighter aircraft fairing.

To achieve consistency of operation, the FSW process is highly automated, with setup being critical. It is therefore essential that sufficient data are known about the process so that the parameters can be accurately set. The trend is now towards on-line force and torque measurement for data monitoring and closed-loop control. Especially on robots and transportable machines, which may not be rigid enough to withstand deflection, load cells or more complex systems can be installed.

Magnetic Pulse Welding (MPW)

The lowering of manufacturing costs through the development, implementation, optimization, and advancement of leading-edge welding technologies, such as magnetic pulse welding (MPW) has occurred.

MPW, a solid-state welding process, uses magnetic pressure to drive the primary metal against the target metal, sweeping away surface contaminants while forcing intimate metal-to metal contact, thereby producing a weld. Because of its numerous advantages, MPW is a viable option for a variety of applications to reduce manufacturing cost, achieve higher weld quality, and enhance productivity.

For several years, weld engineers have used MPW to weld a variety of similar and dissimilar material combinations. Equipment for MPW includes 20 KJ, 90 KJ, and also 100 KJ machines.

MPW has been used to produce tubular components, such as fuel lines. MPW can be used in applications ranging from small-diameter lines used in jet engines to the larger diameter Al tubes typical of launch vehicles. More potential application areas also include ducting, fittings to tubes, and weapon systems.

Al, Ti, stainless steels, and Ni alloys are frequently used materials in aerospace applications. Using MPW, engineers have successfully welded Al-to-Al alloys, Al-to-W, 321 stainless steel-to-321 stainless steel, and Ti-to-Ni alloys (see Figure 12.24).

MPW is a promising technology that has the potential to significantly reduce manufacturing costs, especially for joining tubular components in the aerospace sector. Although this technology is flexible and simple to use, the complex electromagnetic reactions behind it makes the design process a challenge for researchers and engineers. Moreover, the process has several flexible variables that need to be carefully designed and calibrated to ensure consistent quality and optimal robustness for each application.

An outgrowth of MPW is electromagnetic (EM) pulse technology, which is an efficient, elegant, and low-cost solution for joining dissimilar materials and for applications where metallurgical bonding is not required. Compared to welding, equipment for crimping is more robust, lower cost, and smaller in size.

The structural crimping of joints uses grooves, nuts, knurls, and threads to provide strength values that are comparable to those of the parent material.

Crimping is a joining technique in which the plastic deformation of the components produces mechanical interlocking. EM process uses high-frequency currents to produce high-intensity ($\sim$290 MPa) and short-duration (10–20 μs) magnetic pressure to cause plastic deformation of the workpiece.

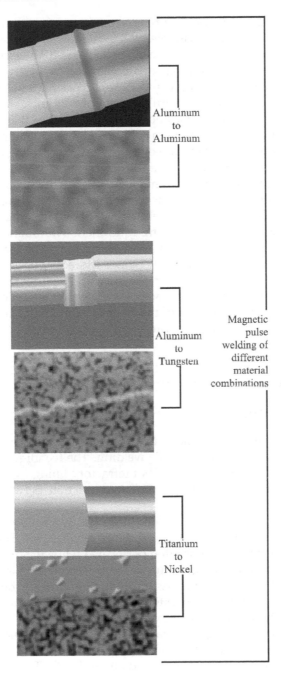

Aluminum
to
Aluminum

Aluminum
to
Tungsten

Titanium
to
Nickel

Magnetic
pulse
welding of
different
material
combinations

FIGURE 12.24

Most applications of EM crimping are tube joining to mandrels with surface features. The tube is accelerated to high velocities prior to contact with the mandrel. There are three major benefits for high-velocity crimping. First, in high-velocity crimping, two pieces can be easily conformed to each other to cause macroscopic interlock. Second, high-velocity crimping almost always produces a natural interference fit, or a state of residual stress that holds the two components together. Lastly, there is an opportunity to use surface indentation to hold components together, in a manner similar to the operation of threaded fasteners. EM crimping normally uses a 16 KJ capacitor bank with single turn coils.

EM crimping is a promising technology that has the potential to replace riveted assemblies, welding applications, and traditional mechanical joining processes. This process can help reduce manufacturing cost, increase joint reliability, and simplify joining equipment. Although this process is flexible and simple to use, the complex electromagnetic reactions behind it makes the process design challenging. Moreover, there is very little engineering science information in the literature on how to design an EM crimp joint and predict joint characteristics. Having modeling capabilities (including electromagnetic and mechanical simulation) are critical for extending this technology to a variety of applications.

Ceramic Welding

The refractory linings of glass furnaces are often repaired using an advanced technique called ceramic welding. The technique feeds O_2-rich crushed refractory particles to the hot refractory lining, where the metals oxidize and fuse in a highly exothermic reaction between 1600° and 2200°C.

The high-temperature combustion melts the surface areas of both the particles and the lining, creating a durable repair bond. The key benefit of ceramic welding is that it is often performed while the furnace or vessel is at or near operating temperatures. There is no need for shutdown or cooldown, and minimal or no loss of production time. By detecting problems early, the technique can extend the life of a furnace or vessel and save the sometimes-staggering cost of a rebuild or reline.

FuseTech, Inc.[38] is using an advanced approach to ceramic welding that includes inspecting the refractory lining before, during, and after the welding with an advanced water- and air-cooled CCTV monitoring system. The solid-state system can withstand temperatures as high as 1925°C while penetrating the flames inside operating furnaces and other vessels.

This CCTV system has been successfully used as part of a ceramic welding process for 3 years.

The ceramic welding process was originally designed for in-situ repair of glass furnaces. The technology was introduced in the United States in 1979 as a method of repairing the walls of coke ovens in the steel industry. It is now employed primarily in the glass, Al, Cu, cement, and coke oven industries.

Designed to withstand, without failure, all hostile conditions inside furnaces or boilers, the new-generation CCTV system uses a quartz lens that can operate up to 1925°C. This lens, at the tip of a periscope, extends out into the furnace interior for optimal viewing. It captures clear color images of the refractory liner, combustion at burner nozzle, and other conditions. A fail-safe air filtration system removes aerosols, vapor, oil, and particles as small as 0.03 μm from the compressed air, which cleans and cools the lens.[38–39]

Ceramic Joining

A robust, affordable ceramic joining technology has been developed at NASA Lewis Research Center for joining SiC-based ceramics and fiber-reinforced composites.[40] Called ARCJoinT, the technique produces joints that maintain their mechanical strength up to 1350°C in air. Large and complex-shaped components may be joined, and thickness and composition of joints may be tailored. In addition, the technique may be adapted for repairing ceramic and composite components in the field.

Within the aeronautical, energy, electronics, nuclear, and transportation industries, SiC-based ceramics and fiber-reinforced composites are either in use or being considered for a wide variety of high-temperature applications. These include engine components, radiant heater tubes, heat exchangers, heat recuperators, and land-based power generation turbines. Other products are the first-wall and blanket components of fusion reactors, furnace linings and bricks, and parts for diffusion furniture (boats, tubes) in the microelectronics industry.

In these cases, the engineering designs often require expensive fabrication and manufacturing operations. It is often more cost-effective to build complex shapes by joining simple, geometrical shapes. Therefore, joining technology is a key to successful utilization of SiC-based ceramics and fiber-reinforced composites in high-temperature components. The joints must retain their structural integrity at high temperatures and have good mechanical strength and environmental stability, comparable to the bulk materials. In addition, the joining technique should be robust, practical, and reliable. For electronics applications, joint composition can also be critical.

Most current techniques involve the joining of monolithic ceramics and fiber-reinforced composites (with metals and ceramics) by diffusion bonding, metal brazing, brazing with oxides and oxynitrides, or diffusion welding. Some of these techniques require high temperatures for processing or hot pressing. In other instances, the joint operating temperatures are lower than the temperature capability of the base ceramics or composites. However, joints produced by brazing techniques can have thermal expansion coefficients different from the parent materials. This differential contributes to stress concentration in the joint area. Therefore, the service temperature for brazed joints is typically limited to ~700°C.

ARCJoinT is a reaction forming approach. It is unique because it produces joints with tailorable microstructures. Specifically, the thermomechanical properties of the joint interlayer are tailorable to those of the SiC-based materials. An additional benefit is that ARCJoinT requires no high-temperature fixturing to hold parts at the infiltration temperature.

The ARCJoinT process begins with placing a carbonaceous mixture into the joint areas, as shown in Figure 12.25. These are held in a fixture and cured for 10 to 20 min at 110 to 120°C: This step fastens the pieces together. Si or an Si alloy in the form of tape, paste, or slurry is then applied around

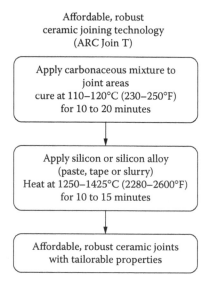

FIGURE 12.25
Schematic of ARCJoinT process for joining SiC-based ceramics and fiber-reinforced composites.[40]

the joint region. It is then heated for 10 to 15 min at 1250 to 1425°C, depending on the type of infiltrant. The molten Si or Si-refractory metal alloy reacts with C to form Si with controllable amounts of Si and other phases as determined by the alloy composition. Joint thickness is readily controlled by regulating the properties of the carbonaceous paste and the applied fixturing force.

NASA Lewis researchers have joined a wide variety of SiC-based ceramics and fiber-reinforced composites, of different sizes and shapes, including tubular components. They include Cerastar reaction-bonded (RB) SiC,[41] Hexoloy sintered alpha (SA) SiC, and a variety of C and SiC fiber-reinforced SiC matrix composites (C/SiC and SiC/SiC). After joining, microstructure and mechanical properties of joints were characterized at several different temperatures.

The ARCJoinT process is able to fabricate reaction-formed joints with different thicknesses. Joint thickness and composition strongly influence both the low- and high-temperature properties of the joined materials.

The flexural strength of as-received and joined bars increases at high temperatures. Healing of machining flaws is one possible explanation. The flexural strengths of joined bars are comparable to those of as-received materials. The fracture origins appeared to be inhomogeneities inside the parent material.

The joined Hexoloy-SA specimens with 45–50 μm joint thickness had strengths of 275 ± 13, 302 ± 17, and 297 ± 15 MPa. In the joined Hexoloy-SA materials, fracture does initiate in the joint region. Inhomogeneous Si distribution was observed in certain areas of the joint.[42–45]

Adhesives

Which adhesive, where? Knowing the eight basic kinds of adhesives helps in quickly sizing up potential uses.

It is possible to find hundreds of different assembly adhesives in manufacturing today. But most fall into eight basic categories: anaerobics, cyanoacrylates, light-cured acrylics and cyanoacrylates, hot melts, reactive urethanes, epoxies, polyurethanes, and two-part acrylics.

Anaerobics are one-part adhesives that remain liquid in air but cure into tough thermoset plastics when confined between metal substrates. Anaerobic adhesives typically are used to lock and seal threaded assemblies, retain bearings and bushings on shafts or in housings, and seal metal flanges in place of cut gaskets. Anaerobics can enhance, and in some cases replace, mechanical joining methods, lowering manufacturing costs and prolonging equipment life.

Cyanoacrylates are high-strength, one-part adhesives for bonding plastic, metal, and rubber. Cyanoacrylates cure rapidly at room temperature to form thermoplastic resins when confined between two substrates in the presence of microscopic surface moisture. Cure initiates at the substrate surface, limiting cure-through gaps to about 0.0254 cm.

Cyanoacrylates build fixture strength in seconds and full strength within 24 h, making them well suited for high-volume, automated production. Polyolefin primers applied to substrates before assembly let cyanoacrylates better adhere to difficult-to-bond plastics such as polypropylene and polyethylene, while accelerators help them cure rapidly in low humidity. So-called "surface-insensitive" cyanoacrylates also hasten cure in low humidity as well as on acidic surfaces. And special low-blooming types minimize white haze (blooming or frosting) around bond lines.

Early cyanoacrylate adhesives were not noted for having good bond strength and chemical resistance, and they did not work at temperatures above about 83°C. However, newer formulations such as rubber-toughened cyanoacrylates have better peel and impact strength. And recent thermal-resistant types withstand continuous 121°C temperatures.

Light-curing acrylics are one-part, solvent-free liquids that turn to thermoset plastics in about 2 to 60 sec when exposed to light of the proper wavelength and irradiance. Cure depths can exceed 1.27 cm. This cure-on-demand approach lets parts be positioned and repositioned before cure. Some hybrid varieties use a secondary cure mechanism such as heat or chemical activators to completely cure adhesive in shadowed areas. Like cyanoacrylates, light-curing acrylics come in thin liquids (about 50 cP) to thixotropic gels. The cured adhesives, depending on formulation, form soft elastomers to glassy plastics that remain clear for improved aesthetics. Light-curing acrylics bond well to a variety of substrates and withstand chemical attack, elevated temperatures, and other environmental assaults.

Light-curing cyanoacrylates are a recent development that combines the benefits of cyanoacrylates and light-curing acrylics while overcoming many of the limitations of each. Light-curing cyanoacrylates fixture almost instantly when exposed to the proper light source and reach about 60% of final strength after just 5 sec of exposure. The remaining adhesive in shadowed areas cures by the normal residual-moisture mechanism, eliminating second-step accelerators or activators. Light-curing cyanoacrylates can also bond overlapping, nontransparent parts.

These adhesives produce minimal vapors and are insensitive to substrate surface condition. They lower stress cracking on crack-sensitive substrates such as polycarbonate and acrylic. And they bond polyolefin plastics when used with special adhesion promoters compounded into the molded parts or applied to part surfaces before bonding. The use of light-curing cyanoacrylates is growing rapidly in high-volume production of medical devices, cosmetic packaging, speakers, electronic assemblies, and small plastic parts.

Hot-melt adhesives have been used for decades to assemble industrial and consumer products. Conventional hot melts (basically thermoplastic resins) are heated to reflow onto a bonding surface. Hot-melt adhesives fill large gaps and build high bond strengths upon cooling. Some of the higher performance varieties include ethyl-vinyl acetate (EVA), poly-amide, polyolefin, and reactive urethane. EVA hot melts are typically used for low-cost potting applications. Polyamides replace EVAs for more-demanding temperature and environmental conditions. Polyolefins adhere well to polypropylene substrates and resist moisture, polar solvents, acids, bases, and alcohols.

Reactive-urethane adhesives represent the latest in hot-melt technology. Unlike traditional thermoplastic hot-melt resins that can be repeatedly reheated, reactive urethanes form thermoset plastics when fully processed at about 121°C, or roughly 93°C cooler than other hot-melt chemistries. Initial strength develops a bit slower than traditional thermoplastic hot melts, though for structural bonding and difficult-to-bond plastics, reactive urethanes generally get the nod.

Epoxies are one- or two-part structural adhesives that bond well to a wide variety of substrates, emit no byproducts, fill large volumes and gaps, and shrink minimally when cured. Cured epoxies typically have excellent cohesive strength, exceptional chemical resistance, and good heat resistance. A major disadvantage of epoxies is a longer curing time than other adhesive types. Typical fixture times are between 15 min and 2 hr. Heat accelerates the curing process, though certain substrates such as plastic limit cure temperatures.

Polyurethane adhesives have greater flexibility, better peel strength, and lower modulus than epoxies. These one- or two-part systems consist of a soft core for joint flexibility and a rigid skin for cohesive strength, temperature, and chemical resistance. Varying the ratio of hard to soft regions makes possible a range of physical properties. Like epoxies, polyurethanes work on several different substrates, though a primer is sometimes required to prepare surfaces. However, polyurethanes must be applied to dry substrates because moisture can compromise bond strength and final appearance. Fixture times are on par with epoxies (15 min to 2 h), so racks are often needed to hold parts while the adhesive sets. Polyurethanes resist chemicals and elevated temperatures, though long-term exposure to high temperatures degrades them more rapidly than epoxies.

Two-part acrylics easily fill gaps and withstand elevated temperatures and environmental factors similar to epoxies and polyurethanes. However, two-part acrylics can be formulated to fixture faster and adhere better to difficult-to-bond substrates than these other adhesives. And they remain highly flexible for long-term fatigue resistance.

In conclusion, most adhesive-joint failures are not caused by lack of adhesive strength. More likely culprits include poor design, inadequate surface preparation, inappropriate operating conditions, or simply using the wrong adhesive for the job. Always test designs thoroughly before production begins.

A bonding technology in which polymers are joined through the inter-diffusion of adhesive monomers and thermoplastic polymer chains, and their subsequent entanglement upon low-temperature cure, has been developed. Called diffusion-enhanced adhesion (DEA), the low-pressure, low-temperature process also enables reduced tooling and assembly costs. It eliminates the need for pretreatnent and offers virtually unlimited shelf life.

Researchers determined that adhesion of thermosets to thermoplastics could be greatly improved by selecting process parameters and adhesives that enhance interdiffusion across the interface. In the case of thermosetting composites such as epoxies, which do not melt, an amorphous thermoplastic polymer is cocured to the matrix during fabrication, yielding a surface that offers the potential for future fusion bonding or adhesive bonding, with minimal surface preparation.

DEA is useful for joining a variety of materials, such as thermoplastics to each other, thermosets to thermoplastics, and either material to metals. For example, the technique was successfully applied to the U.S. Army Composite Armored Vehicle program.[45]

Wire Bonding

Most semiconductor packages use very fine Au wires to make electrical connections between the semiconductor chip and the leadframe within the package.[46] These wires are typically 2 mm long and 25 µm thick, so only about 20 ng of Au is used per wire. However, some integrated circuits (ICs) can have over 200 wire bond connections, and it is estimated that about 180,000 million ICs were manufactured in 2003,[47] which in total accounts for a large quantity of Au. The quantity of Au used for wire bonding is continually increasing as more and more ICs are manufactured. Gold Fields Mineral Services estimate that demand for Au bonding wire has doubled since 1994 to about 125 tons in 2003 and the continuing long-term demand for new consumer electronics will ensure that this growth continues.[48]

Wire bonding is carried out thermosonically, which uses a combination of heat and ultrasonic energy, or ultrasonically without heat. The use of heat alone is rarely used and is relatively slow. Au is used for several reasons. Bonding with Au can be carried out thermosonically with very high reliability. This is essential where possibly millions of devices having over 100 leads are manufactured. Manufacturers do not want to carry out complex and costly tests on every device and so rely on being able to achieve a very high yield of good ICs. Another advantage with Au is that ball bonds can be made at a rate of 20 bonds per second. Al is also used for wire bonding (ultrasonically) but reliable ball bonds cannot be made and so wedge bonds have to be used, which occupy more space than ball bonds and can be bonded at a rate of only 8 per second. Au does not oxidize or corrode so packages do not need to be hermetically sealed.

There is a rapid growth in the use of MEMS devices for a wide range of applications.[49] These include crash sensors in automobiles, optical switches for the Internet, and pressure sensors for automobiles and medical applications. Their use in mobile phones is predicted to grow rapidly as MEMS switches and other devices are introduced. Most optical switches are coated with a thin layer of Au, using a vacuum technique, to make a reflecting mirror surface. Most MEMS switches operate with a very low contact force, and so only Au can be used to obtain a low contact resistance.

Nitrocarburizing

Salt bath nitrocarburizing is a thermochemical process for improving the properties of ferrous metals. However, some tool and other high-alloy steels are susceptible to reductions in core hardness after standard nitrocarburizing. To prevent such losses, a low-temperature salt bath nitrocarburizing process has been developed. With treatment temperatures as low as 480°C, this process not only maintains core hardness, but also can sometimes increase core hardness.

In addition, recent advances have made salt bath nitrocarburizing environmentally friendly. Modern salt bath nitrocarburizing involves non-cyanide salts, a completely regenerable bath (no waste salts produced), and typical process times of 60 to 90 min. Fully automated equipment, designed with zero water discharge, can control and document the process as well as eliminate waste water. By comparison, the old process operated with high-cyanide baths, required very long treatment times, and generated large amounts of waste salts.

Process

During salt bath nitrocarburizing, the part is immersed in a vessel of molten salt. N_2 and C in the salt react with the Fe on the surface, forming a compound layer with an underlying diffusion zone. The compound layer consists of iron nitrides, chromium nitrides, or other such compounds, depending on the alloying elements in the steel, and small amounts of carbides.

Ranging in depth from 2.5 to 20 μm, the compound layer provides improvements in wear and corrosion resistance, as well as in service behavior and hot strength. Hardness of the compound layer, measured on a cross-section, ranges from 700 HV on unalloyed steels, up to 1600 HV on high-Cr steels. Note that this layer is formed from the base metal and is an integral part of it and is therefore not a coating. The diffusion zone can extend as deep as 1 mm, depending on the steel. This diffusion zone causes an increase in rotating-bending strength and rolling fatigue strength as well as pressure loadability.

Salt bath nitrocarburizing may be applied to a wide range of ferrous metals, from low-C to tool steels, cast Fe to stainless steels. Specifically, the process

- Improves wear and corrosion resistance
- Reduces or eliminates galling and seizing
- Increases fatigue strength
- Raises surface hardness
- Provides highly predictable, repeatable results
- Performs consistently, even with varying contours and thicknesses within the same part or load
- Maintains dimensional integrity
- Shortens cycle times
- Offers flexibility and ease of operation

Conventional treatment temperatures are in the range of 580°C, but for highly alloyed steels as well as stainless and tool steels, this temperature can cause a reduction in core hardness. The above benefits, derived both from the N_2 and C diffused into the metal surface, as well as the processing in a liquid bath, are often necessary for applications in which a reduction in core hardness is not acceptable. For this reason, a new low-temperature process was developed, known by the trade name Low Temperature Melonite.[50]

The Low Temperature Melonite process normally takes place at 480°C, although it can operate at 480 to 520°C. Specific advantages of this process are

- Core hardness and tensile strength are maintained in the tempered condition.
- Very thin compound layers can be formed.

- Distortion is extremely low.
- Formation of a compound layer on high-speed steels can be suppressed.
- Hardness of surface and diffusion layers can be customized.

This low-temperature process is beneficial for high-alloy steels such as stainless, tool, die, and high-speed steels.

By treating tools at a lower temperature, such as 520°C for periods of 30 to 60 min, the necessary hard nitride layer is developed, but without any brittleness.

Vacuum Carburizing

It should make no difference if a carburized case is developed by atmosphere or vacuum carburizing, because the end result is the same, right? Wrong! Almost all products today are being engineered to achieve increased performance at lower cost in smaller and smaller footprints. For example, transmission specifications call for higher horsepower ratings produced from smaller packages. Extending life and improving performance are necessities for today's manufacturer. These demands can only be met by using low-pressure vacuum carburizing.

Vacuum-carburized and oil-quenched components have been shown to last longer by approximately 20,000 more cycles and exhibit higher bending fatigue values than their atmosphere-carburized counterparts. In addition, the best part of the case, the region of high hardness (>58 HRC) is much deeper on a vacuum-carburized gear (Figure 12.26). There is also a tendency to reduce heat-treat distortion, not only due to gas quenching but with oil quenching as well.[51, 52]

VringCarb is a new low-pressure carburizing technology that uses cyclohexane (C_6H_{12}), which is a very stable, high-purity, high-density liquid. A saturated hydrocarbon, cyclohexane is readily available at low cost. This liquid is introduced into the heating chamber using a proven, precisely controlled liquid injection delivery system at pressures of approximately 13.3 mbar. No C dropout (sooting) is observed.[51]

Vacuum carburizing with this process technology has proven to deliver the following advantages:

- Extremely clean parts
- Elimination of intergranular oxidation (IGO)
- Greater than 90% root to pitch hardness profiles

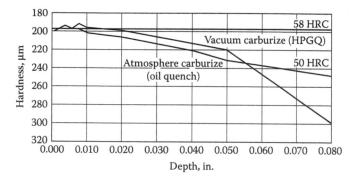

FIGURE 12.26
Improved depth of high hardness with vacuum carburizing. Source: Twin Disc, Inc.[51]

- Very low consumption of cyclohexane
- Uniform case depths
- Lower distortion
- Higher carburizing temperatures

Another development is the Infracarb process to precisely control its low-pressure vacuum carburizing. Carburizing takes place in the 1 to 8 mbar range and alternates between the addition of a hydrocarbon gas during the boost stage and a neutral gas (N_2) during the diffusion stage. Approximately 95% of installations worldwide use propane as the C source; the other 5% use acetylene. This exclusive control system allows the selection of heating, carburizing, and quenching parameters, which enables the equipment to run like a CNC machine tool. Extensive field-work has been done with high-pressure gas quenching using N_2 to understand and predict the transformation and distortion of parts.

Designs of furnaces are available in dual- and multichamber configurations. Oil and high-pressure gas quenching cells are available. In fact, 45 installations have 20 bar (N_2) high-pressure gas quench capability, and 11 are vertical designs. The modularity of the design allows for easy expansion of additional carburizing or quench cells in the future.

Advancements in computer process control, the development of new steels, continuous improvement in high-pressure gas quenching technology, and vacuum equipment design innovations are the driving forces behind the resurgence of vacuum carburizing as a process capable of carburizing production workloads. Using a unique combination of ethylene (C_2H_4), acetylene (C_2H_2), and H_2 at pressures of 1.3 to 40 mbar allows this technology at low cost to the heat-treat industry.

Electrospinning

Electrospinning is a process for making extremely fine submicron fiber by a process of charging polymer solutions to thousands of volts. This method of manufacturing artificial fibers has been known since 1934, when the first patent on electrospinning was filed by Formhals.[53] Since that time, many patents and publications have been reported on electrospinning.

Process

Nanofibers — about 100 times smaller in diameter than typical textile fibers — have a very high surface-area-to-mass ratio, potentially offering performance advantages in fields ranging from tissue engineering to protective clothing to catalyst supports and membranes for chemical processing. Because of high costs and other factors such as health and safety issues, there are currently few efficient and economic processes for making high-quality nanofibers at a commercial scale. A development is under way to design, build, and demonstrate a high-speed machine for low-cost electrospinning of mats made of polymeric nanofibers with diameters 10 to 100 times smaller than today's conventional fibers for industrial, military, consumer, health care, and environmental applications. Electrospinning is a complicated process that merges aspects of polymer science, electrostatics, fluid mechanics, and several subfields of engineering. In electrospinning, a high voltage is applied to a thin tube (capillary) filled with polymer solution, a continuous stream of liquid is ejected, the stream is split into fine jets, and after the solvent evaporates, the resulting fibers are solidified and collected.[54] To transform the existing low-volume process into a method for making sufficiently high volumes of nanofiber materials for commercial development, the key technical challenge is to get a higher mass flow rate of polymer through each capillary jet. Other challenges include managing the electrical charge, component design, solvent evaporation, uniformity of fiber collection, and postprocessing techniques.

Applications

Despite the long history of electrospinning technology, it has never been applied to fabrics as a protective membrane layer. This new application has been under development at the U.S. Army Natick Soldier Center for the purpose of providing protection from extreme weather conditions, enhancing fabric breathability, increasing wind resistance, and improving the chemical resistance of clothing to toxic chemical exposure.

In summary, electrospun fibrous membranes are highly porous structures that can be produced from a number of polymer/solvent combinations. Porosities ranging from 0.1 to 2.6 μm in diameter can be produced from solvent electrospinning. Air and moisture transport measurements on experimental electrospun fiber mats compare favorably with properties of textiles and membranes currently used in protective clothing systems. The electrospun layers present minimal impedance to moisture vapor diffusion required for evaporative cooling. Experimental measurements show that electrospun fiber mats are extremely efficient at trapping airborne particles. The high filtration efficiency is a direct result of the submicron-sized fibers generated by the electrospinning process. Electrospun nanofiber coatings have been applied directly to a spunbonded fabric and an open cell foam. The airflow resistance and aerosol filtration properties correlate with the electrospun coating add-on weight. Particle penetration through the foam layer, which is normally very high, is eliminated by extremely thin layers of electrospun nanofibers sprayed on to the surface of the foam. Electrospun media possess filtration efficiencies as high as electrostatically charged olefin fibers.

In addition to providing outstanding aerosol protection, electrospun membranes also appear to enhance the activity of reactive additives for chemical protection. New elastomeric membranes prepared from thermoplastic polyurethanes are about half the strength and elongation of a continuous film of the same material, so improvement in the tensile properties of these nanofiber membranes is required. Emerging manufacturing techniques are currently being developed that may someday enable us to produce these novel multifunctional liner materials of nano- and microfibers.[55–58]

Potential future applications of electrospun layers include direct application of membranes to garment systems, eliminating such costly manufacturing steps as laminating and curing. It may be possible to electrospin fibers directly onto three-dimensional (3D) screen forms obtained by 3D body scanning. In the future, clothing manufacturers will be able to use a laser-based optical digitizing system to record the surface coordinates of a wearer's body.[59] This information could be integrated with computer-aided design and manufacturing processes to allow electrospun garments to be sprayed onto the digitized form, resulting in custom-fit, seamless clothing. This concept has already been in practice for new racing swimwear produced in solid elastic film form, rather than utilizing fibrous structures. The MACH1 swimwear design by Dianasport in Italy uses an elastic silicone molded into just two sizes on a computer-designed steel mold to obtain desired surface shapes and symmetry for reduced drag in water.

Clearly, the manufacturing capability to produce head-to-toe body suits for sport clothing and protective clothing utilizing computer-generated molds is making strides and may soon be ready for elastic, microfiber spinning using a method like electrospinning.

Spinning Spider Silk

Spider silk is among the strongest materials known, but an artificial version whose strength matches the original's has proven difficult to produce commercially. The U.S. Defense Department has been encouraging research into developing web-slinging weapons from spider silk.

Scientists have long been trying to unravel the molecular secrets of spider silk itself. The hope is that the silk could some day be woven into objects ranging from impregnable body armor to wear-resistant ropes, parachutes, and uniforms.

Thinner than human hair and lighter than cotton, the strongest silk is 3 times tougher than Kevlar, the human-made material used in bulletproof vests, and 5 times stronger than steel cable.

Unlike Kevlar, which requires intense pressure and poisonous H_2SO_4 to produce, spiders can unspool silk at room temperature, under normal pressure, using little more than proteins and water.

But efforts to produce large quantities in the laboratory have progressed slowly. In fact, spiders are only ahead of us by a few million years.

Scientists long ago ruled out the most obvious solution to harvesting silk: "spider farms." Unlike mulberry-munching silkworms, whose shimmering fiber is prized in the fashion world, spiders are territorial cannibals. Cage them in a pen and they will soon consume each other. So scientists have turned to genetics. A pair of spider silk genes was decoded in 1990.

The genes produce :dragline" silk. The strongest of the spider's seven silks, it is used to frame webs and as a lifeline when the arachnid dangles.

The next trick is to produce the proteins made by these genes. A flock of genetically engineered goats has been created and designed to churn out silk proteins in their milk. But the silk is not cheap, at $1500/gm.

Churning out cheap silk protein is only part of the challenge; scientists must then spin it. In reality, spiders don't spin, they squeeze. Liquid silk protein is forced through fingerlike glands on a spider's abdomen called spinnerets. Spiders then stretch these filaments with their legs to align the silk protein molecules.

In 2002, researchers mimicked this squeezing process mechanically, and they created a filament that was a little stretchier than the real thing and not quite as strong.

Some engineers are side-stepping the challenges of natural silk and devising applications that are only loosely spider-inspired. As a result, spider experts in Oxford, England, have developed a device they believe holds promise for breaking that barrier by mimicking arachnoid spinning. The researchers, who plan to commercialize the technology, start with proteins from silkworms, and the device produces highly lustrous fibers only about 15 to 20 μm in diameter. The device uses a special membrane to mimic the conditions in a spider's silk-making organ; under these conditions, the proteins self-assemble into nanofibrils. Making larger fibers out of the nanofibrils is simple because all you do is just pull it out of the device. The molecules stick together in the actual device and come out as a beautiful thread. Possible uses include medical implants, safety belts, composite materials for car body parts, protective clothing, durable sneakers — just about anything that could benefit from an ultrastrong, supertight material. The researchers are working to refine the spinning device and believe that artificially spun fibers as strong and flexible as spider silk could be on the market in 2007 to 2010.

Press Forming Thermoplastics

Fiber-reinforced thermoplastics have been recognized as having a number of advantages over thermosets in material properties. Furthermore, although the materials are relatively expensive, manufacturing methods specific to thermoplastic-matrix composites offer potential for creating lower-cost structures. A number of simply contoured parts, such as landing gear doors, floor panels, ribs, and skin panels, have been produced.[60,61] Among the processes that have been used to create aerospace structural components from thermoplastic-matrix composites is press forming, which combines elements that reduce labor and processing time. Beginning with either a preconsolidated sheet or a stack of individual plies, the material is heated, transferred to the press, formed under pressure, and a semifinished product is ready for the next operation. However, successful fabrication of larger, more complex structures is challenging. Although many materials can drape and conform to complex contours under slower forming processes such as diaphragm forming, stamping a flat sheet of continuous-fiber composite into a complex contour can result in wrinkles, bridging, or fiber breakage. A trial-and-error process that involves tooling redesign and remanufacture is time-consuming and expensive. Therefore, the use of process simulation in part and tooling design may prove beneficial in determining whether a part of a given material, lay-up, and shape can be successfully manufactured.

Continuous fiber-reinforced thermoplastics are much more difficult to form than nonreinforced or short fiber-reinforced materials due to the

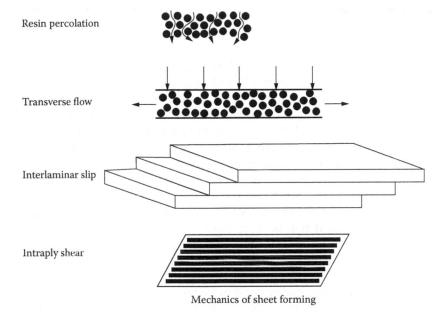

Resin percolation

Transverse flow

Interlaminar slip

Intraply shear

Mechanics of sheet forming

FIGURE 12.27
Mechanics of sheet forming.[62]

presence of the stiff reinforcement. For a continuous fiber-reinforced thermoplastic sheet, the complexity of the parts that can be formed is limited mainly by the movement capability of the fiber reinforcement, although mold compliance may also have a bearing. Four different processes must take place within the sheet, three of which involve moving the fibers. The following processes, as described by Cogswell,[62] are illustrated in Figure 12.27:

- Resin percolation
- Transverse flow
- Interlaminar slip
- Intralaminar shear

Resin percolation refers to flow of the viscous polymer matrix through or along the network of reinforcing fibers. Occurring to some extent in all types of forming operations, it is the process in consolidation of a laminate that enables bonding of plies and removal of voids. Transverse flow is the process by which a prepreg layer spreads locally, usually due to an applied pressure. Interlaminar slip is the sliding of individual plies relative to one another, and, finally, intralaminar shear is the in-plane shearing distortion within a ply.[63–70]

Ion Implantation

Ion implanters are essential to modern integrated-circuit manufacturing. Doping or otherwise modifying Si and other semiconductor wafers relies on the technology, which involves generating an ion beam and steering it into the substrate so that the ions come to rest beneath the surface. Ions may be allowed to travel through a beam line at the energy at which they were extracted from a source material, or they can be accelerated or decelerated by DC or radio-frequency (RF) electric fields.

Semiconductor processors today use ion implantation for almost all doping in Si ICs. The most commonly implanted species are As, P, B, boron difluoride, In, Sb, Ge, Si, N_2, H_2, and He. Implanting goes back to the 19th century and has been continually refined ever since. Physicist Robert Van de Graaff of MIT and Princeton University helped pioneer accelerator construction and the high-voltage technology that emerged from this effort.[71]

Ion-implantation equipment and applications gradually came together in the 1960s. Experience gained in building research accelerators improved hardware reliability and generated new techniques for purifying and transporting ion beams. Theorists refined the hypothesis of ion stopping, which enabled the precise placement of ions based on the energy and angle of implantation, and experimenters determined that high-temperature postimplant annealing could repair implantation-induced crystal damage. Initially, these anneals were done at a temperature of 500 to 700°C, but after several years, semiconductor processors found that the optimum annealing temperature ranged from 900 to 1100°C. After the resolution of process integration issues, ion implantation rapidly displaced thermal diffusion of deposited dopants as the dominant method of semiconductor doping because it was more precise, reliable, and repeatable.

Among semiconductor-processing techniques, ion implantation is nearly unique in that process parameters, such as concentration and depth of the desired dopant, are specified directly in the equipment settings for implant dose and energy, respectively. This differs from CVD, in which desired parameters such as film thickness and density are complex functions of the tunable-equipment settings, which include temperature and gas-flow rate. The number of implants needed to complete an IC has increased as the complexity of the chips has grown. Whereas processing a simple n-type metal oxide semiconductor during the 1970s may have required 6 to 8 implants, a modern complementary metal-oxide semiconductor (CMOS) IC with embedded memory may contain up to 35 implants.

The technique's applications require doses and energies spanning several orders of magnitude. Most implants fall within one of the boxes in

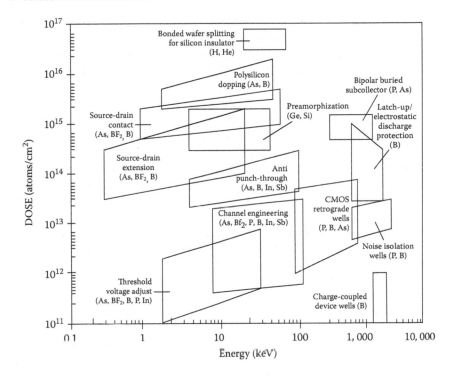

FIGURE 12.28
Dose and energy requirements of major implantation applications (species shown roughly in order of decreasing usage).[71]

Figure 12.28. The boundaries of each box are approximate; individual processes vary because of differences in design trade-offs. Energy requirements for many applications have fallen with increased device scaling. A shallower dopant profile helps keep aspect ratios roughly constant as lateral device dimensions shrink. As energies drop, ion doses usually, but not always, decline as well. The width of the statistical distribution of the implanted ions decreases with energy, and this reduces the dose required to produce a given peak dopant concentration. The result is the sloping lines in Figure 12.28: Implantation is actually extremely inefficient at modifying material composition. The highest ion dose implanted with an economical throughput is about $10^{16}/cm^2$, yet this corresponds to but 20 atomic layers. Only the extreme sensitivity of semiconductor conductivity to dopant concentration makes ion implantation practical.

Ion energy requirements vary from less than 1 keV to more than 3000 keV. Accelerating ions to higher energies requires a longer beam line, yet low-energy beams are difficult to transport intact over longer distances because the beam cross-section expands to a point where it can no longer travel down the beam tube. This fundamental physics makes it nearly impossible to construct a beam line capable of all required ion energies. Figure 12.28 indicates that the largest magnitude in required doses occurs in the middle of the energy range. Because dose is essentially the beam's charge multiplied by the implantation time, available beam currents in the 5–200 keV range must vary by at least four orders of magnitude to perform all required implants efficiently.

There are three applications, respectively, for ion implanters: high-current, medium-current, and high-energy.

As the name suggests, high-current implanters produce the highest beam currents, up to 25 mA. For high-dose applications, the greater the beam current, the faster the implantation, which means the output of more wafers per hour.

Medium-current implanters are designed for maximum dose uniformity and repeatability. Their beam currents are in the range of 1 μA to 5 mA, at energies of 5 to ~600 keV. The wafer-processing end stations can implant ions at angles up to 60° from the perpendicular to the wafer surface. This is essential for certain applications, such as anti-punch-through implants, for example, in which dopants must be implanted partially underneath a previously formed gate structure.

Last, only high-energy implanters can generate megaelectron volt ion beams. Commercial high-energy implanters produce beam currents for singly charged ions up to ~1 mA. Energies for multiple-charged ions can be up to ~4000 keV, with beam currents of ~50 μA. High-energy implanters can produce beams down to 10 keV, making them suitable for many medium-current applications as well. This additional functionality justifies the capital cost of these machines. High-energy implanters using both RF linear acceleration (Figure 12.29) and DC acceleration are used widely today in semiconductor manufacturing.

A modern ion implanter costs about $2–$5 million, depending on the model and the wafer size it processes. Of the three classes of implanters, the high-current machines have traditionally been the biggest market in terms of revenue and unit volume.[72-77]

Gelcasting

Gelcasting is a relatively simple near-net-shape process that is similar to traditional forming methods (see Table 12.6).[78] A slurry of ceramic powder in a monomer solution is poured into a mold and then gelled in situ to the shape of the mold at temperatures between 40 and 80°C. After the part is removed from the mold, it is dried and sintered. Binder burnout

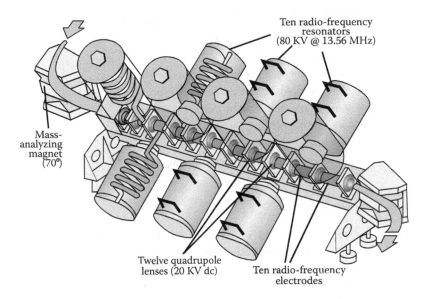

FIGURE 12.29

Schematic of a radio-frequency linear accelerator used in a high-energy ion implanter.[71]

TABLE 12.6

Comparison of Gelcasting with Other Forming Processes[78]

Property	Gelcasting	Slipcasting	Injection Molding	Pressure Casting
Molding time	5–60 min	1–10 hrs	1–2 min	30 min–5 hrs
Strength (as formed)	Moderate	Low	High	Low
Strength (dried)	Very high	Low	N/A	Low
Mold material	Metal, glass, plastic, wax	Plaster	Metal	Porous plastic
Binder burnout	2–3 hrs	2–3 hrs	Up to 7 days	2–3 hrs
Molding defects	Minimal	Minimal	Significant	Minimal
Maximum part dimension	>1 m	>1 m	~30 cm, 1 dimension must be </=1 cm	~0.5 m
Warpage (during drying or binder removal)	Minimal	Minimal	Severe at times	Minimal
Thick/thin sections	No problem	Thick section increases casting time	Problems with binder removal	Thick section increases casting time
Particle size	A decrease increases viscosity	A decrease increases casting time	A decrease increases viscosity	A decrease increases casting time

Source: Oak Ridge National Laboratory.

- Mix ceramic powder with water, disperant and monomer.
- Deair under partial vacuum.
- Add catalyst (gel initiator).
- Cast into metal, glass, plastic or wax mold.
- Heat the molds to create a gel (48–80°C).
- Remove the part from the mold.
- Dry at 90% humidity.
- Machine part, if required.
- Burn out binder (550°C) and sinter (up to 1800°C).

FIGURE 12.30
Schematic of the gelcast process.[78]

is similar to pressed parts (0.5 to 1°C per minute up to 650°C for Si_3N_4 and Al_2O_3). Either a water-based or organic solvent can be used, though an aqueous system is preferred for environmental reasons. Only a relatively small amount of monomer, a gel initiator, and in some cases, a crosslinker and catalyst are required (typically 3 to 4 wt%). Figure 12.30 shows a schematic of the process.

This process is applicable to a wide range of ceramic powders, including Al_2O_3, Si_3N_4, SiC, B_4C, ferrites, aluminum titanate, ZrO_2, sialon, and sodium zirconium phosphate. Gelcasting can also be used to produce complex-shaped ceramic composites. Because gel-cast green parts have high strengths (3 to 4 MPa), parts are easily machined, avoiding expensive postsintering grinding. For turbine engine components under development at Honeywell Ceramic Components (HCC), gelation times range from 10 to about 60 min, drying times range from 6 to 60 h, and sintering cycles are similar to slipcast $Si_3 N_4$.

Compared to injection molding, which requires as much as 20% binder, gelcasting only uses up to 3 to 4%. In injection molding, burning out the binder can take up to a week, compared to less than a day for gelcast ceramic. Because binder burnout is also much more difficult in injection-molded parts, defects can be a major problem. Such defects can be avoided in gelcast ceramics if they are properly dried.

Gelcasting also has a number of advantages over slipcasting. It is 50 to 80% faster, produces much more uniform powder packing, results in a much higher green part strength, and allows simpler molds (no dewatering of surfaces is required). These benefits also make gelcasting potentially easier to automate.

In addition, recent success with the performance of gelcast structural Si_3N_4 parts in aircraft auxiliary power unit (APU) turbine engine applications has resulted in the demand for larger components. Gelcasting has

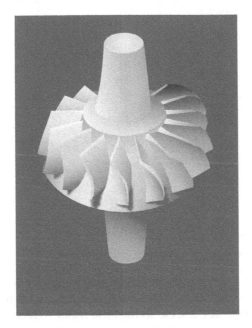

FIGURE 12.31
A gelcast AS800 silicon nitride turbojet turbine wheel.[78]

demonstrated a unique ability to form large size, complex, and thick cross-section components. For example, an approximate 40.64-cm-diameter (green) Si_3N_4 turbine blisk was gelcast.

The next step is automation. The automated process was developed for a turbine wheel representative of a component for potential use in small turbojets. The example part is shown in Figure 12.31 and the automated gelcasting system (Figure 12.32).

All steps except for removing the part from the forming machine and transferring it into the drying chamber are automated. The system incorporates a modular design with a conveyor system to transport the molds between individual specialized stations. The modular design allows stations to be easily added for increased capacity. Another advantage is that each station can be programmed to function as an individual unit independent of the other stations, so that if one station experiences difficulties, production levels can be maintained at the other stations.

The closed-loop automated system is composed of eight process stations and an upper and lower conveyor system connected by two lift stations. The mold is attached to a pallet and is transferred via the conveyor to each process station. There are eight stations: mold release, mold assembly, mold fill, mold heating, mold cooling, mold disassembly, part removal,

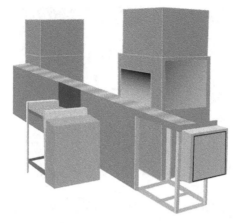

FIGURE 12.32
Honeywell ceramic components' automated gelcasting system.[78]

and mold washing/drying. Each station is equipped with integrated programmable logic controllers. Machine functions are performed by either pneumatic or electrical devices. To heat the mold (which reduces gelation time), an infrared heater was chosen because it provides excellent temperature control, can be scaled to each mold size, has a long life, and provides high output.

Applications

Gelcasting is being applied[78] to the development of powders or parts for automotive applications. For instance, gelcasting is used to make automotive parts from sodium zirconium phosphate.

Gelcasting is also being investigated for electronic materials. Researchers at ORNL[78] have shown that gelcast lead zirconate titanate (PZT) components made using water-based slurries have comparable properties to those made by die pressing and are superior to tapecast parts. Binder burnout of PZT gelcast parts also occurs at the same rate as die pressed parts. Chinese researchers have used gelcasting to make porous perovskite ceramics and complex Al_2O_3 parts.

Other nonautomotive applications are also being developed. The ORNL researchers are developing gelcasting to fabricate circular magnets — more than 50 cm in diameter from ferrite — for a high-energy physics accelerator.

Others[78] are collaborating with the ORNL group to produce complex-shaped SiC parts for semiconductor water processing. Yet another application under investigation is the use of gelcasting to manufacture artificial bone; ORNL is also investigating rapid prototyping using gelcasting.

The potential applications are virtually endless. Over the past decade, gelcasting has emerged as a viable production technology for high-performance ceramics. Future improvements are certain to provide the momentum required to bring gelcasting into the forefront of forming technology.[79]

Robocasting

Joseph Ceserano of Sandia National Laboratories (SNL)[80] has developed a concept that relies on robotics for computer-controlled deposits, through a syringe, of ceramic slurries, mixtures of ceramic powder, H_2O, and trace amounts of chemical modifiers. The material is deposited in thin sequential layers onto a heated base.

The robocasting technique uses a robot to squeeze the slurry out of the syringe and follows a pattern prescribed by computer software. The method allows a dense ceramic part to be free-formed, dried, and fired in less than 24 h, making it suitable for rapid prototyping.

It has proven difficult to develop ceramic slurries that actually contain more solid than liquid but that exhibit a fluidlike consistency and flow.

"By understanding the colloid science and manipulating the powder surface chemistry, ceramic slurries that are up to 85% ceramic powder and 15% H_2O are made," Ceserano says. "The high solid content minimizes drying and shrinkage."

Within 10 to 15 sec of being deposited, the slurry must dry from a fluidlike state into a solid state to permit adding the next layer. If the slurry is too fluid, the deposits will come out as liquid beads that spread uncontrollably. With the proper consistency, each deposited bead is a rectangular cross-section with relatively straight walls and flattened tops, a sound foundation upon which more layers can be deposited. After the part is formed and dried, firing in a furnace at high temperatures densifies the particles.

Along with the benefits of limited machining and no molds, the technique can apply more than one material at a time. Materials can be graded, as when going from a ceramic to a metal within one part, without causing structural damage.

More traditional ceramic fabrication may take weeks to go from a design to an actual part. For instance, the standard dry-pressing method requires ceramic powder to be compacted into a billet. To make a complicated part, the billet must be sculpted into the final shape with costly machines. Prototyping a ceramic part for use in fabricating fiber optics for advanced x-ray diffraction applications is being developed.[80]

Rapid Prototyping

Producing prototypes or finished manufactured parts directly from computer inputs (rapid prototyping or rapid manufacturing) has become an increasingly important technique for fabricating precision parts. Rather than making molds or dies — a process that often takes weeks or months — rapid prototyping fabricates parts by creating point-by-point patterns in layers of powder, solidifying them with heat, and building up layer upon layer into an object. Materials such as polymers, ceramics, steel, and Al are used.

For Al, a polymer-Al powder is fabricated using a rapid prototyping technique. Then the polymer is burned out and the remaining metal lightly melted or sintered together to form a solid part. However, as the metal skeleton melts together, its shape changes slightly, making it impossible to maintain high precision and to form complex parts.

In recent years, however, a new liquid infiltration method has allowed high-precision rapid prototyping of metal parts. In infiltration, a molten metal lightly coats a sintered skeleton and then solidifies to form the final part.

The rapid-prototyping process starts with a powdered mixture of Al, Mg, and nylon. A computer-guided laser polymerizes a pattern in a layer of the powder, and another layer is laid down until the part is completed. At this point, the part is about 50% metal by volume.

The challenge to forming Al parts by infiltration is that Al in air rapidly acquires an oxide layer that prevents the individual metal particles from binding together. To prevent this oxide layer, the part was heated at 540°C in a flowing N_2 atmosphere, using the Mg in the powder to scavenge O_2. This process resulted in an AlN layer that allowed the Al particles to join together into a skeleton so that the part could then be infiltrated without distortion.

After its initial bearing, the part was heated at 570°C for another 6 h, during which the liquid metal infiltrated it. For this procedure to work, the liquid melt must have a melting temperature above that at which the skeleton forms but below the skeleton's melting point. An alloy of Al, Si, and Mg has the right properties.

This process can yield complex parts with delicate features as small as 500 μm across and, theoretically, parts of any size. Tensile strength and ductility of the parts are about 80% of those of an Al casting.[81]

Builders of rapid-prototyping equipment have done a good job publicizing the value of parts made on their machines. After holding an RP part you have designed, it takes only seconds to tell whether or not it is big enough or shaped right. Handling the part tells right away what should be done next.

But RP parts can also be assembled into nearly complete products. To some extent, that unsung capability comes from RP equipment that builds several connected parts and materials that closely approximate the strength of production plastics.[82,83]

Parts with complex three-dimensional shapes and with dimensions up to 20.3 by 20.3 by 25.4 cm can be made as unitary pieces of a room-temperature-curing polymer, with relatively little investment in time and money.

The main advantages of this RP process over other rapid-prototyping processes are greater speed and lower cost. There is no need to make paper drawings and take them to a shop for fabrication, and thus no need for the attendant paperwork and organizational delays. Instead, molds for desired parts are made automatically on a machine that is guided by data from a CAD system and can reside in an engineering office.

The process centers around an office-compatible rapid-prototyping machine. This machine is essentially a three-dimensional printer that builds a part directly from CAD data that specify a solid mathematical model, in the same manner as that of a rapid-prototyping machine of the stereolithographic or fused-deposition-modeling type. A CAD operator merely builds a plot file and submits it to the machine (this submission takes approximately 1 min per part), then the machine builds the part. The time that it takes to build the part could be a few hours or as much as 30 h, depending on the size of the part.

The machine builds parts with extremely fine detail but with two severe drawbacks. One of the drawbacks is that it makes parts of a wax that lacks toughness and strength. The other drawback is that any surfaces that are facing down with respect to the machine are covered with supports. These supports can easily be cleaned off by a light manual brushing, but the resulting surfaces are not smooth.

The present rapid-prototyping process overcomes these drawbacks. The steps of the process are the following:

1. The CAD system is used to design the desired part.
2. Taking advantage of a solid-modeling subtraction capability, the CAD system is used to design a mold that contains a cavity of the size and shape of the desired part.
3. The CAD model of the mold is sliced into appropriate pieces to eliminate any downward-facing surfaces (to prevent the production of supports on surfaces of the molded part).
4. Filling ports and vents are added to the CAD model to complete the mold design.

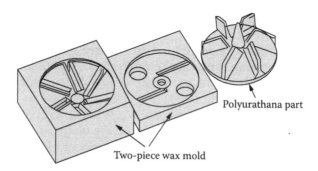

Polyurathana part

Two-piece wax mold

FIGURE 12.33
A two-piece mold is used to make a complex three-dimensional part.[84]

5. The data from the CAD model of the mold pieces are submitted as a print job to the rapid-prototyping machine, then the machine builds the mold pieces.

6. The mold pieces are taped together and filled with a room-temperature-curing polymer. The polymer used by the developer of this method is a durable polyurethane that becomes cured sufficiently for removal from the mold in about 0.5 h.

7. The mold is removed, and after removal of any minor flashing, the part is ready for use.

One advantage of using a wax as the mold material is that mold can be removed from the part by melting, if necessary (the melting temperature of the wax is less than that of the polyurethane). Of course, if the mold is melted and it is desired to produce more copies of the part, then more copies of the mold must be built from the CAD files (see Figure 12.33).[84]

It is clear that research into heterogeneous, functionally graded, composite, and multiple material RP has a long road to travel. The interest generated from high-tech industries is already apparent, and indications are that this could develop into a whole new branch of manufacturing technology. No doubt this technology will develop its own chasms along the way and be guided across them by engineers of similar caliber to those working tirelessly in the RP community.[85–91]

Plasma-Assisted Boriding

Boriding is a thermochemical heat treatment with a long history. The application of boriding in industry occurred around 1955. From Russian publications, it is known that the service life of parts from pumps used in the oil exploration industry (which were salt bath borided) was increased by 4 times compared to case hardening and induction hardening.[92] However, only when powder pack boriding was developed in the late sixties did it became more widespread.[93, 94] Although many efforts have been made since to develop a more efficient boriding process from the gaseous phase, until now no industrial application was known, which can be attributed to the choice of precursors used, namely highly toxic and explosive diborane (B_2H_6)[95] and boron halides (BCl_3 and BF_3). The latter is also highly toxic and promotes corrosion by the formation of HCl and HF. Recently, trimethyl borate, $B(OCH_3)_3$ was introduced as an uncritical alternative to the aforementioned precursors.[96] It was shown that using trimethyl borate, Fe_2B layers can be generated in a plasma-assisted process. The boride layers, however, are not completely dense but show a porous morphology. The amount of O_2 contained in trimethyl borate is responsible for this.

It can be concluded that by boriding a Ni-coated steel, a multiphase layer is formed. When the initial Ni coating is too thick, the resulting layer system can contain a comparably soft Ni-Fe layer in between two hard boride layers, which, in general, is to be avoided for wear-resistant purposes. More favorable is the multiphase layer that forms when the initial Ni coating is sufficiently thin. Then the superficial layer (consisting of Ni and Fe) is soft and the boride layer underneath is hard.

The mechanism of boride formation when boriding a Ni-coated steel is most likely the same as the three steps schematically described in Figure 12.34, namely:

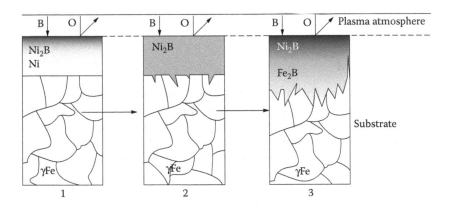

FIGURE 12.34
Most likely mechanism of boride formation during boronizing of a nickel-coated steel. The gray tone denotes the gradient of boron within the layer.

1. At the surface, Ni_2B is formed growing in direction of the core.
2. At the interface between Ni_2B and Fe Ni boride, (Ni_2B) transforms into thermodynamically more stable Fe boride (Fe_2B) according to:

$$Ni_2B(s) + 2\ Fe(s) \rightarrow Fe_2B(s) + 2\ Ni(s) \quad \Delta G°_{1200k} = -30,77 \text{ kJ/mol}$$

3. With increasing time, B continues to diffuse into the surface, and subsequently the boride layer grows.

The Ni coating is of great importance when boriding steel substrates with trimethylborate as it acts as a diffusion barrier for O_2. The Ni coating actually separates the O_2 containing plasma atmosphere from the location where Fe_2B formation takes place. O_2 cannot interfere with Fe_2B, thus. In contrast to the results presented in Reference 96, the boride layer is dense and free of pores.

Therefore, for the first time the generation of dense boride layers on steel substrates in a plasma-assisted boriding process using trimethylborate was successful by depositing the steel substrates with a thin coating of pure Ni prior to boriding. The Ni coating can be either deposited by PVD or by electroplating. To generate a thick boride layer (e.g., for wear-resistant applications), the thickness of the initial Ni coating is of great importance. Best results were received with an approximately 2-μm-thick Ni coating. The maximal boride layer thickness was then 160 μm, depending on boriding time. The boride layers are of type $(Ni, Fe)_2B$ and Fe_2B. The hardness of the boride layer was determined to be up to 2150 HV 0.025.[97]

Net-Shape P/M

Net-shape P/M has intrigued aerospace researchers and the industrial community for more than 40 years because it eliminates machining operations from component manufacturing processes, resulting in a significant reduction in parts cost. For example, the Air Force Research Laboratory wants to develop rotating-pump component materials and processes for turbine engines that trim current parts weight by 35%, reduce current production costs by 45%, and also lower the projected costs of fabricating a net-shape P/M manufactured bladed disk (blisk) by more than 40%.

Net-shape P/M is a manufacturing method for casting with solid metal powder and consolidating to a net shape to produce parts with structure and properties comparable to forgings. The method retains the inherent design and processing benefits of the P/M casting process and provides the performance benefits of forgings. Engine manufacturers currently fabricate turbine rotors, used to propel jet aircraft and rockets, from forged disks with blades either mechanically attached or machined from the forgings. The state-of-the-art method for manufacturing the disks is to either fabricate them from conventional high-strength, Ni-based superalloys and coat them for environmental protection or fabricate them from moderate-strength alloys fully compatible with the hostile environment. Coatings introduce reliability and cost issues, while the moderate-strength alloys limit turbine performance.

Figure 12.35 shows a high, specific-strength fabricated blisk using the selective net shape (SNS) P/M process and an environmentally compatible HIP-bonded surface layer (HBSL).

In early development, researchers detected both surface contamination and surface microroughness on parts fabricated with the net-shape P/M-HIP method. These defects were related to powder indentation and diffusion bonding with the soft sacrificial tooling during HIP. The addition of an HBSL not only solved the microroughness problem of the net-shape P/M-HIP surfaces but also offered a revolutionary opportunity to protect

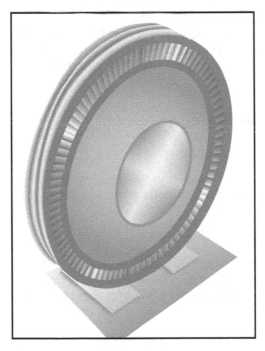

FIGURE 12.35
Prototype PM-HIP blisk.

very-high-performance alloys in turbine environments normally hostile to these alloys.

The environment in which a turbine disk must operate is most commonly H_2-rich or O_2-rich gas. H_2-rich gas embrittles high-strength superalloys, and O_2-rich gas burns high-strength superalloys. In the HBSL process, researchers deposit a layer of an alloy (selected for its environmental resistance and compatibility with the core alloy) using vacuum plasma spray on the machined surface of the low C steel mold. After tool assembly, powder filling, degassing, sealing, and HIP, the core superalloy powder consolidates and HIP bonds to the outer surface of the environmental alloy on the sacrificial tooling. This durable and damage-tolerant HBSL provides environmental compatibility and improves the as-processed surface finish. The improved surface finish enhances the ability to deter structural defects with nondestructive testing equipment.

The development of P/M blisk fabrication methods, incorporating high-strength, environmentally compatible superalloys, will enable revolutionary advances in turbopump rotating element materials and designs. The proposed blisk fabrication methods will enable the aerospace industry to build turbine blisks using the highest-performing, Ni-based superalloy turbine materials, achieving the reliability and maintainability of current low-strength superalloys and attaining compatibility with cryogenic rocket engine turbine environments.

Engine designers will incorporate this high-payoff P/M blisk technology into current and proposed O_2-rich turbine engines and then extend this high-payoff technology to the H_2-rich elevated temperature turbines.

Rheocasting Processes

Rheocasting processes create nondendritic, equiaxed microstructure suitable for semisolid forming directly from liquid Al alloy.

Semisolid Casting

Semisolid casting refers to any casting process that utilizes (a) partially solidified and (b) nondendritic material. Numerous process advantages are derived from the casting of alloy with these two characteristics, a few of which include

- Capability to produce high-quality, heat-treatable, complex parts with minimal entrapped air or shrinkage porosity because of planar front filling of the metal at relatively higher injection velocities compared with other high-integrity casting processes and enhanced feeding of the casting during solidification.

- Reduction in die dwell time because of reduced metal heat content.
- Increased die life because of reduced thermal shock and fatigue due to decreased casting temperature.

The most widely cited benefits of semisolid processing usually refer to the final part quality, complexity, and properties.[98] Although these are of extreme importance, the aforementioned advantages of reduction in solidification time and increased die life are of even greater significance to the majority of casters.

However, cost and processing issues of previous and existing semisolid processes have prevented casters from achieving the dramatic operational cost advantages associated with semisolid processes.

Semisolid processing is typically classified into two major categories: thixocasting and rheocasting. Thixocasting refers to any process that starts with a specially prepared alloy that is reheated from ambient to the desired semisolid forming temperature prior to casting (Figure 12.36). Rheocasting refers to any process that modifies liquid alloy into semisolid slurry that is then directly formed into a part (Figure 12.37). Thixocasting is more commercially practiced than rheocasting, but its growth has stagnated because of high raw material costs and the inability to recycle scrap from the process on site. Rheocasting addresses these problems by using ordinary foundry alloy, and scrap from the casting can be recycled and used again with the process.

A new rheocasting technology that efficiently creates nondendritic material was developed at MIT[99] and led to the realization that the critical factor for creating the nondendritic, semisolid slurry is the combination of rapid cooling and convection as the alloy cools through the liquidus,

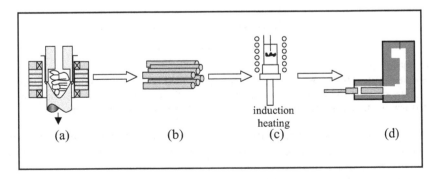

FIGURE 12.36
Thixocasting process. Specially prepared billet is cut to length at the foundry, reheated into the semi-solid state, and finally into a part.[99]

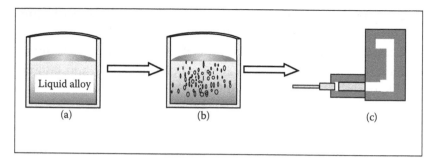

FIGURE 12.37
Rheocasting process. Molten aluminum is modified in the foundry to form suitable semisolid material and then immediately cast.[99]

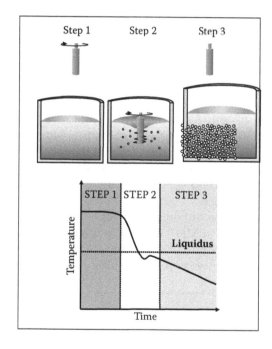

FIGURE 12.38
Schematic of the SSR™ process. (1) Molten alloy is held above the liquidus temperature. (2) A rotating, cool rod is descended into the melt, initiating solidification. (3) The rotating rod is removed, and the quiescent melt is cooled to the desired forming temperature.[104]

with agitation unnecessary after formation of only a very small fraction of solid. For greater process flexibility, a separate cooling and stirring device was used to remove heat and create convection in the molten alloy.[100–103] A schematic of the process can be seen in Figure 12.38.[104] The most remarkable finding from the experiments was that stirring after the metal temperature dropped below the liquidus temperature had no effect

TABLE 12.7

Chemical Composition of Four Alloys Tested with SSR™

Element (wt%)	Si	Mg	Zn	Cu	Fe	Mn	Sr	Ti	Other
380	7.80	0.04	2.73	3.18	0.85	0.25		0.10	0.25
365	11.19	0.17			0.11	0.62	0.0167	0.05	
Magsimal™	2.27	5.60	0.006	0.005	0.134	0.688		0.13	
356	6.96	0.10	0.01	0.65	0.32	0.01	0.06	0.0139	

on final microstructure. The SSR™ process has significant advantages compared with other rheocasting processes.

- Heat removal and convection are controlled with a separate cooling/stirring device, rather than relying upon temperature loss from pouring into a cold vessel. This allows a wider range of incoming melt temperatures and ensures a consistent slurry temperature after the cooling process.

- During the rheocasting process, heat is removed from the metal before it is transferred to the die casting machine, thus decreasing the cycle time within the die casting machine.

- Cooling occurs within the melt via the spinning rod, thereby ensuring a uniform cooling of the material. Other processes that rely upon heat removal through the outer surface of a container are more susceptible to formation of dendritic skin because of the localized rapid cooling on the surface.

- The rapid cooling and stirring of the rod creates a very fine microstructure that does not require lengthy coarsening to achieve a globular state and can be immediately cast.

A number of alloys and castings have been rheocast with SSR and produced on 800- and 1000-ton die casting machines. SSR has been used to produce slurry from 356, 380, 365, and Magsimal™-59 alloys. Castings have been produced with both liquid and slurry[105] (see Table 12.7).

Investment Casting

Single Crystals

In developing the engine that will power future 500-seat jumbo jets, technology has produced single crystal blades for the high-pressure turbines and thus the most fuel-efficient and most quiet class of engine power plant.

With other innovations, the engine will consume less than 1.2 gallons of fuel per passenger over 95 miles, a fuel consumption on par with that of the best turbo-diesel cars on the market (see Figure 12.39).

The initial solution to creating relatively few grains and have them grow or form along the length of the blade in a process called directional solidification (DS).[106]

The key to DS is cooling the molten metal in the mold unidirectionally, maintaining a uniform temperature gradient ahead of the solidifying interface and preventing new grain creation. Solidification begins at once and proceeds until the casting is solid. DS molds have openings at the bottom and a space called the starter, where solidification starts. The mold sits on a water-cooled Cu chill plate while heated to the alloy melting point inside a furnace. Metal hits the chill plate and solidifies, with grain growth in one direction. Meanwhile, metal in the rest of the mold remains molten. The mold and chill plate are slowly withdrawn from the furnace through a radiation baffle. This generates a thermal gradient along the vertical axis. The gradient causes grains growing vertically to outgrow those with any other orientation.

DS blades lack grain boundaries transverse to the direction of operational stresses, which increases strength and ductility. They also have lower elastic moduli in that direction. This quality improves thermal fatigue properties. DS castings can also have thinner walls than random-grain castings. With random-grain orientation, it is increasingly likely that a grain boundary completely spans cross-sections as those cross-sections gets progressively smaller.

For even stronger blades, manufacturers have modified DS casting to create single-crystal blades. The entire airfoil is made of one crystal with grain boundaries. To get only one crystal to grow in the molten metal, a helical selector or "pigtail" is inserted between the starter and the blade cavity. The selector takes crystals growing vertically and filters them until only a single crystal emerges. The crystal goes on to grow throughout the rest of the casting (see Figure 12.40).

Because there are no grain boundaries in single-crystal blades, there is no need for alloying material to strengthen grain boundaries. Most other superalloys, including those for DS blades, contain C, Zr, B, and Hf to make grain boundaries stronger, but this also lowers the melting point. Alloys developed for single-crystal blades, therefore, have melting points approximately 18°C higher than DS alloys. This makes them better suited to the high turbine inlet temperatures of modern jet engines.[106, 107]

Single Crystal Ceramics

Single crystal ceramics, including oxides, flourides, sulfides and carbides, are generally grown from seed crystals in a molten or gaseous environ-

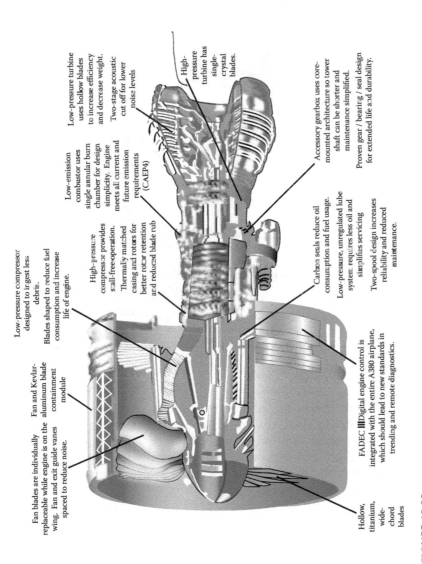

Low-pressure turbine uses hollow blades to increase efficiency and decrease weight. Two-stage acoustic cut off for lower noise levels

High-pressure turbine has single-crystal blades.

Low-pressure compressor designed to ingest less debris. Blades shaped to reduce fuel consumption and increase life of engine.

Low-emission combustor uses single annular burn chamber for design simplicity. Engine meets all current and future emission requirements (CAEP4)

High-pressure compressor provides stall-free operation. Thermally matched casing and rotors for better rotor retention and reduced blade rub

Accessory gearbox uses core-mounted architecture so tower shaft can be shorter and maintenance simplified. Proven gear / bearing / seal design for extended life and durability.

Carbon seals reduce oil consumption and fuel usage. Low-pressure, unregulated lube system requires less oil and simplifies servicing. Two-spool design increases reliability and reduced maintenance.

Fan and Kevlar-aluminum blade containment module

Fan blades are individually replaceable while engine is on the wing. Fan and exit guide vanes spaced to reduce noise.

FADEC III Digital engine control is integrated with the entire A380 airplane, which should lead to new standards in trending and remote diagnostics.

Hollow, titanium, wide-chord blades

FIGURE 12.39
Clean and most efficient jet engine.

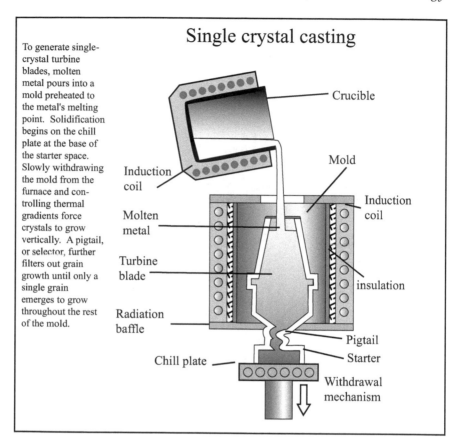

To generate single-crystal turbine blades, molten metal pours into a mold preheated to the metal's melting point. Solidification begins on the chill plate at the base of the starter space. Slowly withdrawing the mold from the furnace and controlling thermal gradients force crystals to grow vertically. A pigtail, or selector, further filters out grain growth until only a single grain emerges to grow throughout the rest of the mold.

Single crystal casting

Crucible

Mold

Induction coil

Induction coil

Molten metal

Turbine blade

insulation

Radiation baffle

Pigtail

Chill plate

Starter

Withdrawal mechanism

FIGURE 12.40
Single crystal casting.

ment The most prevalent techniques for seeded growth from the liquid phase are flame fusion, the Czochralski method, the heat exchanger methods, and edge-defined film-fed growth (EFG).[108] These materials have extraordinarily high levels of perfection and purity, and are critical for many optical applications, solid-state lasers, medical imaging components, and acoustic resonators.

Applications

Optical Window Applications

In this area, the two dominant single crystals are sapphire (Al_2O_3) and CaF_2. Verneuil grown sapphire has long been used as scratch-resistant watch crystals. Sapphire is widely used as high-temperature visible and infrared (IR) windows. Sapphire has been used as an IR missile dome.

Sapphire is also gaining an expanding niche as a window for super-market bar code scanners. Several supermarket chains have documented increased productivity and reduced maintenance by using scratch-resistant sapphire windows. Plates produced by EFG are uniquely suited to this application.

CaF_2 single crystals, long used in spectrascope optics, are finding a major new market in the semiconductor fabrication industry. The out-standing optical properties of CaF_2 at 193 and 157 nm have made it the preferred material for deep UV microlithography systems for the next generation of microelectronics.

Lasers

Solid-state lasers use single crystal rods or slabs of transparent materials doped with elements that possess the appropriate electron energy level transitions to produce laser light at desired frequencies and efficiencies. The uses for these ceramic laser materials vary widely. For example, Cr^{3+} doped sapphire (ruby), which lases at 0.694 μm, is used for welding and removing tattoos. Ruby's radiation does not adversely affect the normal skin next to the features to be removed.

Other materials include Ti^{4+} doped sapphire, which yields lasers that can have ultrashort, femto-second pulses and are tunable between 0.65 and 1.1 μm. These are useful for a wide variety of specialized uses, such as ultrafast spectroscopy, micromachining, and nuclear fusion R&D. Nd^{3+} doped $Y_3Al_5O_{12}$ (YAG), with a 1.06 μm wavelength, is used in a variety of applications ranging from welding, range finding, and cataract removal. And alexandrite (synthetic chrysoberyl), which lases at 0.755 μm, is used for removing unwanted body hair.

Medical Imaging

Single crystal ceramics are used as scintillators in medical imaging equip-ment such as positron emission tomography (PET) and x-ray computed tomography (CT). Scintillators are crystals that convert incident radiation (gamma rays, x-rays, etc.) into photons, which in turn excite a photo-detector array producing an image that can be stored and manipulated by a computer. Examples include bismuth germanate, used in PET, and cadmium tungstate for CT use.

Substrates

The use of sapphire and the newly developed SiC single crystals as sub-strates for GaN deposition is an exciting development. The close match in lattice parameters on appropriate planes of sapphire, SiC, and GaN facilitates epitaxial deposition of the GaN (or doped GaN). This epitaxially deposited GaN is the key to the newly commercialized blue LED.

Blue LEDs with an appropriate fluorescent coating — or alternatively, coupled with red and green LEDs — produce white light. These solid-state white light sources provide low power usage and long life. As costs fall and efficiencies increase, LEDs will become the preferred lighting source over the next 20 years, creating a very large market for single crystal ceramic substrates.

Other Applications and the Market

Another large market is synthetic gemstones. Cubic ZrO_2 has become the major substitute for diamond, and quartz crystals are used as resonators that control the timing of data flow in virtually every microcircuit, including electronic watches.

Surface acoustic wave devices are also used in many circuits. Lithium niobate is commonly used as a substrate for these SAW devices; in 2001 one company alone grew 60 tons of lithium niobate single crystals.

The market for ceramic single crystals used in the above applications can be estimated at between $125 and $200 million — and that may well be an underestimate. As microelectronics, medical imaging, laser surgery, and solid-state lighting markets grow, so will this fascinating area of ceramic technology.

Microwave Processing

Drilling

A microwave drill can bore through materials such as concrete and glass, silently and without creating dust. By heating a target to nearly 2000°C, the microwaves soften it up enough for a small rod to be pushed through.

This should provide a low-cost solution for a variety of needs.[109] It can make holes between a millimeter and a centimeter wide, and could find use, for example, in the production of ceramic components for cars and planes, in building construction, and in geological engineering.

Using heat to cut and drill materials is nothing new. Lasers are already widely used to make incisions or holes as narrow as a thousandth of a millimeter. But laser drilling is too expensive for many routine engineering jobs, whereas the microwave drill costs little more than a mechanical one.

The drill bit is a needle-like antenna that emits intense microwave radiation. The microwaves create a hot spot around the bit, melting or softening the material so that the bit can be pushed in.

But the drill can't bore through everything effectively. Sapphire's melting point, for instance, is too high. And steel conducts heat too well for a hot spot to develop. But the device works fine on rocks and concrete. In fact, the heat may even strengthen holes' walls in ceramics by welding together the fine grains in the material.

To use the drill, workers would have to be shielded from the intense microwave radiation it produces. A simple shielding plate put in front of the drill bit is enough to meet common safety standards.[110]

Casting and Sintering

Agrawal[111] and colleagues have shown that a blast of microwaves will cook a powdered metal into a solid metal object. By filling molds with the powder and treating it with a dose of microwaves, they were able to produce cast metal parts without any need for conventional melting and pouring.

The use of microwaves to process materials is nothing new, but it has not previously been possible to use it on metals. Ceramic powders are commonly formed into solid bodies using microwave energy to heat the grains, causing them to become welded together, a process called "sintering." Despite the intuitive association of ceramics with brittle, fragile artefacts, they include some of the hardest materials known. These high-performance ceramics, such as SiC, have been used for making components of jet engines, turbines, automobile turbochargers, and cutting tools.

Microwave sintering is really little different from heating up a pie. The powder is placed in a microwave field, and the grains heat up as they absorb the radiation. This can raise the temperature to over $1000°$: hot enough that the atoms in the material become mobile, so that adjacent grains can "flow" together. One advantage over conventional methods of heating is that the microwaves penetrate the entire sample and are absorbed throughout, rather than the heat being transferred gradually from outside inwards by conduction. This means that the energy is absorbed more efficiently, and the processing time can be short. Just a minute or so can be sufficient to sinter a ceramic powder into a solid mass of fused grains.

Microwave processing is now used for a huge variety of materials: It can be used to treat wood and textiles, to process food, to cure plastics and even to instigate chemical reactions. But for metalworking, there is a problem: Metals are reflective. Microwaves tend to bounce off them like light from a mirror. This has discouraged researchers from trying to use microwaves for forming metal objects.

But Agrawal and colleagues have shown that despite the fact that a sheet of metal will reflect microwaves, in powdered form it seems that metals are no longer so reflective. The researchers were able to make dense sintered bodies from powders of just about every metal they tried, including Fe, steel, Cu, Al, and Ni. Just a 10- to 30-min blast of microwaves was enough, and in several cases the microwave-sintered metals were denser and harder than those sintered by conventional heating.

Flowforming

Flowforming is an advanced, cold metalworking process that produces seamless, dimensionally precise, thin-walled tubing and cylinders. With the elimination of welding, the resulting components have increased mechanical properties, superior surface finishes, and excellent dimensional control. Features difficult to achieve by other forming methods, such as increasing or decreasing wall thickness, tapers, radii, and steps, are easily produced by this process.

The following materials have been successfully flowformed:

- All Ti including Ti-6Al-4V
- All Ni alloys
- All stainless steels, including PH stainless
- Carbon steels
- Nb, Zr, Ta
- All malleable metals

In summary, the flowforming process should be considered if any of the following characteristics are required for a component:

- Highly precise, seamless construction to net shape
- Increased mechanical properties
- Uniform and directional grain structure
- Varying wall thickness on the outside diameter
- Fine internal surface finishes
- Desire to eliminate welds

X-Ray Inspection

Today's trend of increased miniaturization of components and packages is continuing in the electronics and micromechanics industries. Design and manufacturing technologies such as optoelectronics, MEMS, and micro-optoelectromechanical systems (MOEMS) are creating new application requirements and inspection needs. In certain cases, inspection can be achieved only by using x-ray tubes with focal spot sizes of 1 µm and below. This has led to the development of nanofocus and multifocus x-ray sources. Multifocus tubes allow an operator to switch from either high-power, microfocus or nanofocus imaging, depending on the requirements of particular applications. Although the selection of microfocus or nanofocus depends on the required spot size, the high-power mode in a multifocus tube is for imaging and inspecting dense parts. The tube is designed to offer a particular advantage by delivering the highest intensity possible in a 160 kV transmission tube.

A visual image of a material's internal structure is produced when x-rays pass through the material and strike a photographic plate or fluorescent screen. Shadows that appear on the plate or screen depend on the relative opacity of different parts of the sample. A crack in a solder ball is easily visible, for example, because the ball itself is more opaque than the void created by the crack.

Microfocus X-Ray Inspection Systems

X-ray systems basically consist of a sealed or open x-ray source, a fixture for holding and manipulating the sample to be inspected, and a radiation detector (see Figure 12.41).

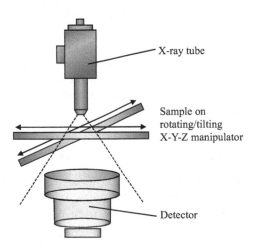

FIGURE 12.41
Diagram of a typical x-ray system used for industrial inspection.

- High-voltage generator. A high-voltage generator provides the required power for the electronics emitting from an electron gun within the x-ray tube, controls, computers, and image-processing software.

- Manipulator. An x-ray system's manipulator is used for high-precision x-y-z positioning and rotating/tilting of samples. The manipulator should be capable of directional and rotational speeds for requirements ranging from quick overview searches at low magnification to very low speeds at high magnification. For example, x-y speeds may range from a few micrometers per second to several hundred millimeters per second.

- Detector. The function of a detector is to process data of x-ray waves in real-time into an image of light visible to the human eye or to electronic vision systems. Although the most common detector is an image intensifier/video camera that converts x-rays into visible light, other recently developed detectors include high-dynamic cameras and flat-panel direct digital detectors.

Nanofocus X-Ray Inspection

The electronics industry's trend toward smaller and more densely populated components, and the emergence of MEMS and MOEMS, led to the development of nanofocus x-ray technology. Nanofocus technology is defined as having a focal spot of less than 1 μm in diameter, which enables the level of detail and resolution needed for the inspection of low-density structures and ultrasmall features common in today's electronics.

Nanofocus x-ray inspection systems are particularly suited for applications consisting of submicron components, circuitry, and assemblies such as with MEMS and MOEMS, but also with wafer-level packaging. In such instances, the resolution and sharpness required to detect defects in solder bumps and interconnects can be met only with nanofocus tube design and system technology.

Future of X-Ray Inspection

A need exists for both nanofocus and nanofocus tubes for real-time radiographic imaging. For contract manufacturers, where inspection requirements can vary from microfocus applications that demand high x-ray output to nanofocus applications that demand high resolution, multifocus x-ray tubes are suitable. The tubes incorporate a high-power mode for dense structures that require high intensities for inspection, such as castings, weldings, and machined parts. Switching modes is a matter of a keystroke or a mouse click at any time during the inspection process.

For many applications, microfocus tubes provide sufficient resolution, contrast, and magnification for the respective inspection task. For smaller components and denser circuitry where feature recognition requires a focal spot smaller than that of a microfocus tube, then nanofocus tubes are a preferred solution. The growing applications of MEMS and MOEMS indicate a continuing trend toward miniaturization, which can only mean a bright future for nanofocus radiography.

References

1. U.S. Patent # 6.500,370, July 2003, P 25, NASA Tech Briefs-LAR-15470-1.
2. Belvin, H. J., Cano, R. J., Johnston, N. J. et al., Dry Process for Making Polyimide/Carbon & Boron-Fiber Tape, *NASA Tech Briefs*, LAR-15470-1, 2003.
3. Hunt, W. H., Jr. & Herling, D. R., Aluminum Metal Matrix Composites, *AM&P*, vol. 162, #2, Feb 2004, pp. 39–42.
4. Patrickk@e-composites.com; p. 2 of 5 and *Composites Week*, vol. 5, #48, Dec 2, 2003.
5. Berenberg, B., Liquid Composite Molding Achieves Aerospace Quality, *High-Performance Composites*, pp. 18–22, Nov 2003.
6. AF Man Tech Highlights, CoRTM Process Reduces Fabrication & Assembly Costs, p. 10, Summer 2003.
7. Black, S., New Approaches to Cost-Effective Tooling, *High-Performance Composites*, July 2003, pp. 30–35.
8. Heider, D. and Gillespie, J. W., Jr., Automated VARTM Processing of Large-Scale Composite Structures, *Journal of Advanced Materials*, Oct 2004, 6 p.
9. Reuterlöv, S., Grooved Core Materials Aid Resin Infusion: Influence on Mechanical Properties, *SAMPE Jrl.*, vol. 39, #6, Nov/Dec 2003, pp. 57–64.
10. Uchida, H., Yamamoto, T., & Takashima, H., Development of Low-Cost, Damage-Resistant Composite Using RFI Processing, SAMPE Jrl., vol. 37, #6, Nov/Dec 2001, pp. 16–20.
11. Reuterlöv, S., Infusion: The DIAB Method, http://www.rpasia.com/rpasia/conf_sesle.html; p. 1 of 1, 07/28/02.
12. Black, S., An Elegant Solution for a Big Component Part, *High-Performance Composites*, May 2003, pp. 45–48.
13. Frost, M., Solanki, D., & Mills, A., Resin Film Infusion Process of Carbon Fibre Composite Automotive Body Panels, *SAMPE Jrl.*, vol. 39, #4, July/August 2003, pp. 44-49.
14. Yoshioka, S., Laser Cuts & Bends Foil, *Laser Focus World*, Sept 2003, p. 13.
15. http://jazz.nist.gov/atpcf/prjbriefs/prjbrief.cfm?ProjectNumber=00-00-5269; p. 1 of 2, 08/08/03. GE Global Research, One Research Circle, Bldg KW, Rm C293, Niskayuna, NY 12309.
16. Ozkan-Anderson, A. M., Pulsed Lasers Micromachine Photonic Integrated Circuits, *Laser Focus World*, Nov 2003, pp. S3–S7.

17. Grylls, R., Intricate Parts from the Inside Out, *Machine Design*, Sept 4, 2003, pp. 56–62.
18. http://jazz.nist.gov/atpcf/prjbriefs/prjbrief.cfm?ProjectNumber=00-00-5222; p. 1 of 2; Optomec Design Co., 3911 Singer, N.E., Albuquerque, NM 87109.
19. Grylls, R., Laser Engineered Net Shapes, *AM&P*, vol. 161, #1, Jan 2003, pp 45–46–86.
20. Hill, M. R., DeWald, A. T., Demma, A. G., Hackel, L. A. et al., Laser Peening Technology, *AM&P*, Aug 2003, pp. 65–67.
21. Jones-Bey, H. A., Laser Peening Enters Commercial Mainstream, *Laser Focus World*, Jan 2004, pp. 42–46.
22. Kopel, A. & Reitz, W., Laser Surface Treatment, *AM&P*, vol. 156, #1, Sept 1999, pp. 39–41.
23. Hull, R., Firisch, D., Woods, C., & Eric, J., Laser Cladding/Glazing as a Surface Coating Alternative, AF Pollution Prevention Grp., AFRL/ML & LHMEL, 8 p, Jan 2001.
24. Warnes, B. H., Cool Coatings Let Engines Run Hotter, *Machine Design*, Dec 12, 2002, pp. 58–61.
25. Leiby, M. W., A Better Way to Produce Thin-Film Prototypes, *Machine Design*, Feb 20, 2003, pp. 114–118.
26. Deutchman, A. H. & Partyka, R. J., Ion Beam Enhanced Deposition, *AM&P*, July 2003, pp. 33–35.
27. Bell, D., CrN Nanocrystalline PVD Coatings Are Harder, Tougher, *AM&P*, July 2003, p. 14.
28. New Tools Maximize New Machine Designs, Makino, Inc., *Modern Applications News* (MAN), Jan 2004, pp. 38–40.
29. Koval, M., High-Dazzle Coatings at Low Temperatures, *Machine Design*, Nov 20, 2003, pp. 66–72.
30. Degitz, T. & Dobler. K., Thermal Spray Basics, *WJ*, Nov 02, pp. 50–61.
31. Thorpe, M. L., Thermal Spray Industry in Transition, *AM&P*, vol. 143, #5, May 1993, pp. 50–61.
32. Inovati, Santa Barbara, CA, Aug 5, 2002.
33. Wichmanowski, S., Pseudo-Alloys for Spray Metal Tooling, *AM&P*, vol. 161, #4, April 2003, pp. 33–34.
34. Mishra, R. S., Friction Stir Processing for Superplasticity, *AM&P*, vol. 162, #2, Feb 2004, pp. 45–47.
35. Mishra, R. S., Friction Stir Processing Technologies, *AM&P*, Oct 2003, pp. 43–46.
36. Chao, Y. J. & Liu, S., Temperature, Force, & Power in Friction Stir Welding, *AM&P*, vol. 161, #5, May 2003, pp. 44–45.
37. McKeown, D., FSW Aluminum Sheet for Airliners Has High Strength, *AM&P*, Feb 2004, p. 19.
38. Lang, W. J., Ceramic Welding Used with High Temperature CCTV in Advanced Repair Technique for Glass Furnaces, *IH*, May 1998, pp. 51–52.
39. Zvosec, C., Using Ceramic Welding to Extend the Life of Gas/Oxy Furnaces, *Ceramic Industry*, June 1998, p. GMT-4-5.
40. Singh, M., Ceramic Joining Technology, *AM&P*, Oct 1998, pp. 89–90.
41. National Industrial Research Institute, Superplastic Compressive Deformation Process for Strengthening and Toughening Technology of Silicon Nitride, *JETRO*, May 1998, pp. 16–17.

42. Chiou, B.-S., Young, C.-D., & Duh, J.-G., Liquid Phase Bonding of Yttria Stabilized Zirconia with CaO-TiO$_2$-SiO$_2$ Glass, *J. of Matls. Sci.*, vol. 30n5, Mar.1, 1995, pp. 1295–1301.

43. Krajewski, A., Joining of Si$_3$N$_4$ to Wear-Resistant Steel by Direct Diffusion Bonding, *J. of Matls. Processing Technology*, vol. 54n1-4, Oct 1995, pp 103–108.

44. Nishi, H. & Araki, T., Low Cycle Fatigue Strength of Diffusion Bonded Joints of Alumina Dispersion Strengthened Copper to 316 Stainless Steel, *NTIS Alert*, June 1, 1996, vol. 96, n 11, p. 20.

45. Ohashi, O., Diffusion Welding of SUS304L Stainless Steel to Titanium, *Welding International*, vol. 10, n3, 1996, p. 188.

46. Simons, Ch., Schrapler, L.and Herklotz, G., *Gold Bull.*, 2000, 33(3), pp. 89–96 and 102.

47. Chipscale Review Website at http://chipscalereview.com/issues/0501/trends.html.

48. Gold Survey 2001, Gold Fields Mineral Services Ltd., April 2001.

49. Market Analysis for Microsystems 1996–2002, The Network of Excellence in Multifunctional Microsystems; (www.nexus-emsto.com).

50. Alwart, S., Baudis, U., Low-Temperature Nitrocarburizing, *AM&P*, Sept 1998, pp. 41–43.

51. Herring, D. H., Low-Pressure Vacuum Carburizing, *Heat-Treating Progress*, June/July 2003, pp. 56–60.

52. Houghton, R. L., Vacuum Carburizing: The View From a Commercial Shop, *Heat Treating Progress*, June/July 2003, pp. 32–35.

53. Formhals, A., Process and Apparatus for Preparing Artificial Threads, U.S. Patent 1,975,504 (1934).

54. Reneker, D. H., Yarin, A. L., Fong, H., & Koombhongse, S., Bending Instability of Electrically Charged Liquid Sets of Polymer Solutions in Electrospinning, J. Appl. Phys., 9, Part I, 2000, p. 87.

55. Gibson, P. W., Schreuder-Gibson, H. L., & Pentheny, C., Electrospinning Technology: Direct Application of Tailorable Ultrathin Membranes, *J. Coated Fabrics*, 28, 1998, p. 63.

56. Gibson, P. W., Rivin, D., Kendrick, C., & Schreuder-Gibson, H. L., Humidity-Dependent Air Permeability of Textile Materials, *Text. Res. J.*, 69 (5), 1999, p. 31.

57. Gibson, P. W., Rivin, D., & Kendrick, C., Convection/Diffusion Test Method for Porous Textiles, *International Journal of Clothing Science and Technology*, 12 (2), 2000, p. 96.

58. Tsai, P. P., Schreuder-Gibson, H. L., & Gibson, P. W., Fibrous Electret Filters Made by Different Electroprocesses, Proceedings of the 8th World Filtration Congress, Brighton, England, 2000.

59. Paquette, S., 3-D Scanning in Apparel Design and Human Engineering, *IEEE Computer Graphics Applications*, 16 (5), 1996, p. 11.

60. Offringa, A. & Davies, C. R., Gulfstream V Floors — Primary Aircraft Structure in Advanced Thermoplastics, *Jrl. of Advanced Materials*, 27(2), 2 (1996).

61. Offringa, A. R., Thermoplastic Composites — Rapid Processing Applications, Composites: Part A, 27A 329 (1996).

62. Cogswell, F, N., The Processing Science of Thermoplastic Structural Composites, *International Polymer Processing*, 1(4), 1987, p. 157.

63. PAM-FORM™, Engineering Systems International, 20 rue Saarinen, 94578 Rungis Cedex, France.

64. de Luca, P., Lefébure, P., & Pickett, A. K., Numerical and Experimental Investigation of Some Press Forming Parameters of Two Fibre Reinforced Thermo-plastics, Proc. Fourth International Conference on Flow Processes in Composite Materials (FPCM '96), Aberyswyth, U.K., 7–9 September 1996.

65. Cogswell, F. N. & Leach, D. C., Processing Science of Continuous Fibre Reinforced Thermoplastic Composites, *SAMPE Jrl*, 24 (3),11(1988).

66. Mantell, S. C. & Springer, G. S., Manufacturing Process Models for Thermoplastic Composites, *Journal of Composite Materials*, 26 (16), 1992, p. 2348.

67. Tam, A. S. & Gutowski, T. G., Ply-Slip During the Forming of Thermoplastic Composite Parts, *Journal of Composite Materials*, 23(6), 1989, p. 587.

68. van West, B. P., Pipes, R. B., & Keefe, M., A Simulation of the Draping of Bidirectional Fabric over Three-Dimensional Surfaces, *Journal of the Textile Institute*, 81 (4), 1990, p. 448.

69. Bergsma, O. K., Computer Simulation of the 3D Forming Processes of Fabric Reinforced Plastics, Proc. Ninth International Conference on Composite Materials, Madrid, Vol. IV, 12–16 July 1993, p. 560.

70. FiberSim, Composite Design Technologies, Wellesley Hills, MA.

71. Rubin, L. & Ponte, J., Ion Implantation in Silicon Technology, http://www.tipmagazine.com/tip/INPHFA/vol-9/iss-3/p12.html, pp. 1–7, May 28, 2003.

72. Rimini, E., *Ion Implantation: Basics to Device Fabrication*, Kluwer Academic Publishers: Boston, 1995; 393 p.

73. Rubin, L., Morris, W. High-Energy Ion Implanters and Applications Take Off, *Semiconductor International*, 1997, 20, p. 77–85.

74. Ryssel, H., Ruge, I., *Ion Implantation*, John Wiley and Sons, NY, 1986, 350 p.

75. Ziegler, J. F., Ed. *Ion Implantation: Science and Technology*, Ion Implant Technology Co., Edgewater, MD, 2000, 687 p.

76. Chason, E., et al., Ion Beams in Silicon Processing and Characterization, *J. App. Phys*, 1997, 81 (10), pp. 6513–6561.

77. Current, M. I., Ion Implantation for Silicon Device Manufacturing: A Vacuum Perspective, *J. Vac. Sci. Tech. A: Vacuum, Surfaces, and Films*, 1996, 14 (3), pp. 1115–1123.

78. Sheppard, L. M., Gelcasting Enters the Fast Lane, *Ceramics Industry*, April 2000, pp. 26–34.

79. Krause, C., ORNL's Gelcasting: Molding the Future of Ceramic Forming?, http://www.ornl.gov/ORNLReview/rev28-4/text/gelcast.htm, pp. 1 of 13, August 02, 2000.

80. Ceserano, J., Robocasting: A New Way to Fabricate Ceramics, *Machine Design*, April 8, 1999, p. 41; *Design News*, 04/05/99, p. 23.

81. Gibson, I., Materials for Rapid Prototyping: Looking to the Future, *Time-Compression Technologies*, vol. 4, #7, 2001, pp. 30–52.

82. Dvorak, P., The Next Step in Rapid Prototyping, *Machine Design*, August 21, 2003, pp. 59–62.

83. EWI Prompts Creative Commercialization, *EWI Insights*, Spring 2003, vol. 16, #2, p. 3.

84. Swan, S. A., Relatively Inexpensive Rapid Prototyping of Small Parts, *NASA Tech Briefs*, July 2003, pp. 48–49.

85. Ho, C. H., Gibson, I., Cheung, W. L., Effect of Energy Density on Property and Morphological Development of Selective Laser Sintering Polycarbonate, *J. Materials Processing Technology*, Elsevier, 89-90, April 1999, pp. 204–210.

86. Quian, X. & Dutta, D., Features in Layered Manufacturing of Heterogeneous Objects, Proc. Solid Freeform Fabrication Symposium, 1998, pp. 689–696.

87. Gervasi, V. R. & Crockett, R. S., Composites with Gradient Properties Form Solid Freeform Fabrication, Proc. Solid Freeform Fabrication Symposium, 1998, pp. 729–735.

88. Cooper, A. G., et al., Automated Fabrication of Complex Molded Parts using Mold SDM, Proc. Solid Freeform Fabrication Symposium, 1998, pp. 721–738.

89. Klosterman, D. A., et al., Development of a Curved LOM Process for Fiber Reinforced Composite Materials, Proc. 8th European Rapid Prototyping and Manufacturing Conf., 1999, pp. 353–365.

90. Lous, G. M., et al., Fabrication of Curved Ceramic/Polymer Composite Trans-ducers for Ultrasonic Imaging Applications by Fused Deposition of Ceramics, Proc. Solid Freeform Fabrication Symposium, 1998, pp. 713–719.

91. Ling, W. M. & Gibson, I., Possibility of Colouring SLS Prototype Using the Ink-Jet Method, *Jrl of Rapid Prototyping*, vol. 5(4), 1999.

92. von Matuschka, A. G., Borieren (in German), Carl Hanser Verlag Wien (1977).

93. Kunst, H. & Schaaber, O., Beobachtungen beim Oberflächenborieren von Stahl II — ber Wachstumsmechanismen und Aufbau der bei Eindiffusionvon Bor in Eisen bei Gegenwart von Kohlenstoff entstehenden Verbindungs — und Diffusionsschichten (in German), HTM 22, 1, 1967.

94. Kunst, H. & Schaaber, O., Beobachtungen beim Oberlachenborieren von Stahl III -Borierverfahren (in German), HTM 22, 275, 1967.

95. Casadesus, P., Frantz, C., & Gantois, M., Boriding with a Thermally Unstable Gas, Metallurgical Transactions 10 A 1739, 1979.

96. Kper, A., Stock, H.-R., & Mayr, P., Plasma-Assisted Boronizing Using Trimethyl Borate, Proc. 12th IFHTSE Congress, Melbourne, Australia (Eds. IME Australasia Ltd) 3, 177, 2000.

97. Kper, A., Plasma-Assisted Boronizing, *AM&P*, vol. 161, #3, March 2003, pp 20–22.

98. Brown, S. B. & Flemings, M. C., Net-Shape Forming Via Semi-Solid Processing, *AM&P*, vol. 143, #1, Jan 1993, pp. 36–40.

99. Yurko, J. A., Martinez, R. A., & Flemings, M. C., Commercial Development of the Semi-Solid Rheocasting (SSR™) Process, Metallurgical Science & Technology, Teksid Aluminum, vol. 21, no.1, pp. 10–15, June 2003.

100. Martinez, R. A., de Figueredo, A. M., Yurko, J. A., & Flemings, M. C., *NADCA Transactions*, Cincinnati, OH, pp. 47–54, 2001.

101. Adachi, M. & Sato, S., *NADCA Transactions*, Cleveland, OH, T99-022, 1999.

102. Shibata, R., Kaneuchi, T., Souda, T., Yamane, H., & Umeda,T., Proceedings of the 5th International Conference on the Processing of Semi-Solid Alloys and Composites, Golden, CO, pp. 465–470, 1998.

103. Flemings, M. C., *Met Trans. B*, vol. 22B, pp. 269–293, 1991.

104. Yurko, J. A., Martinez, R. A., & Flemings, M. C., Proceedings of the 7th International Conference on the Processing of Semi-Solid Alloys and Composites,Tsukuba, Japan, pp. 659–664, 2002.

105. Rice, C. S. & Mendez, P. F., Slurry-Based Semi-Solid Die Casting, *AM&P*, vol. 159, #10, Oct 2001, pp. 49–52.
106. Mraz, S. J., Birth of an Engine Blade, *Machine Design*, July 24, 1997, pp. 39–44.
107. Parker, I., Turnaround at Turbomeca, Helicopter World, May 1997, pp. 11–13.
108. Katz, R. N., Single Crystal Ceramics, *Ceramic Industry*, Feb 2000, p. 16.
109. Jerby, E., Dikhtyar, V., Aktushev, O., & Grosglick, U., The Microwave Drill, *Science*, 298, 2002, pp. 587–589.
110. Ball, P., Microwaves Drill Ceramics, *Nature Science Update*, pp. 1–2; http://www.nature.com/nsu, 18 Oct 2002.
111. Agrawal, D., *Nature*, Macmillan Magazines Ltd, 17 June 2002.

Bibliography

An Elegant solution for a Big Composite Part, *High-Performance Composites*, May 2003, pp. 45–48.
Anderson, T., New Developments in Aluminum Shipbuilding, *Welding Journal*, Feb 2004, pp. 28–30.
Balmforth, M.C., Hicken, G.K., and Puskar, J.D., Inertia Welding for High-Pressure Hydrogen Storage, SNL, ibid.
Campbell, F.C., *Manufacturing Processes for Advanced Composites*, N.Y., Elsevier 2004, 513 p, ISBN 1-85617-415-8.
Carmody, S., Parts Live Longer with Lasers, *Machine Design*, April 1, 2004, p. 24.
Carter, D.W. and Meltzer, Jr., P., Vacuum-Mold Repair System, *AFRL Technology Horizons*, vol. 5, #4, August 2004, p. 50.
Ceramics Take the Heat, *Machine Design*, Nov 20, 2003, p. 31.
Complex Composites Lighten NATO Copter, *High-Performance Composites*, March 2003, pp. 44–46.
Composite Nacelles: Flying Toward New Horizons, *High-Performance Composites*, May 2004, pp. 49–53,
Decker, R.F., Walukas, D.M., LeBeau, S.E., Vining, R.E., and Prewitt, N.D., Advances in Semi-Solid Modeling, *AM&P*, vol. 162, #4, April 2004, pp. 41–42.
Dupont Materials Help Keep Mars Rovers Moving, *NASA Tech Briefs*, May 2004, p. 18.
Europe's Infusion Pioneer Simplifies Process with Bottom Up Approach, *Composites Technology*, Oct 2003, pp. 34–38.
Feng, Z., Wang, X.L., and David, S.A., Prediction of Residual Stresses and Property Distributions in Friction Stir Welds of Aluminum Alloy 6061-T6, ORNL, ibid,
Hancock, R., Friction Welding of Aluminum Cuts Energy Costs by 99%, *Welding Journal*, Feb 2004, p. 40.
Hazen, J. ed., Sourcebook 2004, vol. 9, Dec 2003, 178 p, www.compositesworld.com.
High-Performance Composites: An Overview, *High-Performance Composites*, 2003 Sourcebook, pp. 7-19, www.hpcomposites.com; International Edition, Nov 2002.
http://link.abpi.net/l.php?20021024A6.

http://link.abpi.net/l.php?20021024A6; p. 3 of 4, 10/24/02.

http://lists.newsforindustry.com/cgi-bin3/DM/y/egG10EpoT40CKg0B1210AD.

Hy-Bor Provides Performance and Feel to World's Lightest Production Bike Frame, p. 1 of 4, *Composites Week*, no 36, vol. 5, Sept 8, 2003

Kimapong, K. and Watanabe, T., Friction Stir Welding of Aluminum Alloy to Steel, *Welding Journal*, pp. 277-s to 282-s, Oct 2004.

Kinney, P., Hunt, R.W., Optimization of a Laser + GMAW Hybrid Welding Process, Ohio State Univ., ibid.

Kren, L., New "Window" of Opportunity For E-Beam Welding, *Machine Design*, July 8, 2004, pp. 94–98.

Lerner, E.J., Biomimetric Nanotechnology, *The Industrial Physicist*, vol. 10, #4, Aug/Sept 2004, pp. 16–19.

Lienert, T. and Carpenter, R.W., Beam Profiling of a 3.3 kW CW Cd:YAG Laser, LANL, 2004 AWS Welding Conf., Chicago, IL, April 6–8, 2004.

MacCallum, D.O., Nowak-Neely, B.M., Knorovsky, G.A., and Hooper, F.M., Laser and E-Beam Joining Unusual Fe-Base, Ni-Base, and Refractory Alloys, SNL, ibid.

Manufacturers Welcome New Reinforcement Forms, *Composites Technology*, April 2003, pp. 12–21.

Marz, S., Using Microtechnology to Get to Nanotechnology, *Machine Design*, Sept 2, 2004, pp. 65–68.

Meador, M.A., Meador, M.A.B., Williams, L.L., and Jones, J.R., Room-Temperature, Ultraviolet Curing of Polyimides, *NASA Tech Briefs*, March 1999, pp. 52–53.

Messler, Jr., R.W., Bell, J., and Craigue, O., Laser Beam Weld Bonding of AA5754 for Automobile Structures, *Welding Journal*, June 2003, pp. 151-s to 159-s.

Messler, Jr., R.W., Johnson, E.A., and Levene, J.Y., A Novel Approach to Laser Weld-Bonding Al Autos, RPI, Univ. of Rochester, and Montana Tech Insti., ibid.

Microwave Steel, *Machine Design*, April 1, 2004, p. 40.

Milling Diamond Coatings, *Laser Focus World*, Nov 2003, p. 23.

Missile Composite Body Program Saves Millions, *AFRL Technology Horizons*, Sept 2002, p. 6.

Mohamed, M.H., Bogdanovich, A.E., Habil, I., Dickinson, L.C., et al., A New Generation of 3D Woven Fabric Preforms and Composites, *SAMPE Journal*, vol. 37, #3, May/June 2001, pp. 8–17.

New RTM/VM/RTM Light Injection Valve, p. 1, 11/27/03, http://www.netcomposites.com/news.asp?1880.

Nguyen, J.P., Posada, M., DeLoach, J.J., Fielder, R.S., and Palko, W.A., Friction Stir Processing of Nickel Aluminum Bronze, NSWC, ibid.

Non-Destructive Testing, *AM&P*, vol. 162, #1, Jan 2004, pp. 39–44.

Nordwall, B.D., Tiny Ceramic Spheres Used to Absorb Energy, *Av. Week & Sp.Technology*, Sept 6, 1999, pp. 68–69.

Nowak-Neely, B.M., MacCallum, D.O., and Knorovsky, G.A., Controlled Scanning Electron Beam Micro-to-Nano Welding: Micro-to-Nano Electron Beam Welding Via Adapted Scanning Electron Microscopy, SNL, 2004 AWS Welding Conf., Chicago, IL, April 6–8, 2004.

Occhialini, C., Robotic Welding of Aluminum Space Frames Speeds Introduction of Sports Car, *Welding Journal*, Feb 2004, pp. 24–27.

Opportunities in Continuous Fiber Reinforced Thermoplastic Composites 2003–2008, p. 2 of 6, *Composites Week*, no 25, vol. 5, June 24, 2003.

Rathod, M.J. and Kutsuna, M., Joining of Aluminum Alloy 5052 and Low-Carbon Steel by Laser Roll Welding, *Welding Journal*, Jan 2004, pp. 16-s to 26-s.

RTM Process Turns Out High-Quality Jet Engine Blades, *High-Performance Composites*, July 2004, pp. 38–41.

Schackleford, J.F., *Bioceramics: Applications of Ceramics and Glass Materials in Medicine*, Technomic Publishing Co., Inc., 120 p, 1999.

Schaeffer, R.D. and Stepanova, M., A Radiant Solution, *Ceramic Industry*, June 1999, pp. 34–47.

Schulze, K.-R. and Powers, D.E., EBW of Aluminum Breaks Out of the Vacuum, *Welding Journal*, Feb 2004, pp. 32–38.

Singh, M., Reaction-Forming Method for Joining SiC-Based Ceramic Parts, *NASA Tech Briefs*, March 1999, pp. 50–52.

Smoke Spun into Fiber, *Industrial Physicist*, vol. 10, #4, Aug/Sept 2004, pp. 13–14.

Thermal Diffusion Process Boosts Die Life by 10X, pp. 34–37, www.modernapplicationsnews.com.

Ultra-High Temperature Ceramics, http://link.abpi.net/l.php?20031104A2; pp 1–2 of 5, 11/04/03.

Weinberg, M. and Wyzykowski, J., Powering Unmanned Aircraft, *Aerospace Engineering*, Nov 2001, pp. 23–26.

Index

9 780367 391812